Practical methods in
electron microscopy
Volume 10

Practical Methods in
ELECTRON MICROSCOPY

Volume 10

Edited by
AUDREY M. GLAUERT
Strangeways Research Laboratory
Cambridge

ELSEVIER
AMSTERDAM · NEW YORK · OXFORD

Low Temperature Methods in
BIOLOGICAL
ELECTRON MICROSCOPY

A. W. ROBARDS
Department of Biology
University of York
York, England

U. B. SLEYTR
Zentrum für Ultrastrukturforschung
der Universität für Bodenkultur
Vienna, Austria

1985
ELSEVIER
AMSTERDAM · NEW YORK · OXFORD

© 1985, Elsevier Science Publishers B.V. (Biomedical Division)

ISBN Volume: 0–444–80684–9 (Paperback)
            0–444–80685–7 (Hardback)
ISBN Series: 0–7204–4250–8

Published by:
ELSEVIER SCIENCE PUBLISHERS B.V. (BIOMEDICAL DIVISION)
P.O. BOX 211
1000 AE AMSTERDAM
THE NETHERLANDS

Sole distributors for the U.S.A. and Canada:
ELSEVIER SCIENCE PUBLISHING COMPANY, INC.
52 VANDERBILT AVENUE
NEW YORK, NY 10017
U.S.A.

Library of Congress Cataloging in Publication Data

Robards, A. W. (Anthony William)
    Low temperature methods in biological electron microscopy.

    (Practical methods in electron microscopy ; v. 10)
    Bibliography: p.
    Includes index.
    1. Electron microscopy--Technique.   2. Cryobiology--Technique.   I. Sleytr, U. B. (Uwe Bernd)   II. Title.
III. Series.
QH212.E4P7   vol. 10 [QH201]   502′.8′25 s   85-15872
ISBN 0-444-80685-7   [578′.45]
ISBN 0-444-80684-9 (pbk.)

Printed in The Netherlands

Previous volumes in the series
Practical Methods in Electron Microscopy

v

General Editor's preface
to the series

Electron microscopy is a fundamental technique with wide applications in all branches of Science and Technology, and every year a large number of students and research workers start to use the electron microscope and require to be introduced to the instrument and to the many and varied methods for the preparation of specimens.

Many books are available describing the techniques of electron microscopy in general terms, but when the authors of Practical Methods in Electron Microscopy first met in 1970 they agreed that there was an urgent need for a comprehensive series of laboratory handbooks in which all the techniques of electron microscopy would be described in sufficient detail to enable the isolated worker to carry them out successfully. Now, some fifteen years later and with the fourteenth book in the series in production, Practical Methods in Electron Microscopy has an international reputation as a unique source of practical information for all electron microscopists. This achievement largely results from the great care that has been taken in the selection of the authors, since it is well known that it is not possible to describe a technique with sufficient detail for it to be followed accurately unless one is familiar with the technique oneself. This fact is only too obvious in certain 'one author' texts in which the information provided quickly ceases to be of any practical value once the author moves outside the field of his own practical experience. Perhaps the best compliment that Practical Methods in Electron Microscopy has received was a statement in a review of the book by Alan Agar, Ron Alderson and Dawn Chescoe: "The book leaves the reader with the impression of not having read a textbook, but having talked to a practising microscopist".

Our aim continues to be to help the worker in the laboratory (and particularly the isolated worker) to be a better electron microscopist. Each book of the series starts from first principles, assumuing no specialist knowledge, and is complete in itself. The books are planned to be guides through the often bewildering choice of techniques that are available, and every attempt is made to simplify this choice. Consequently only well-established techniques which have been used successfully outside their laboratory of origin are described in detail. We do not provide descriptions of the latest exciting 'break-through', although each book hopes to be able to indicate the most promising of future developments.

This series of books will eventually cover the whole range of techniques for electron microscopy, including the instrument itself, methods of specimen preparation in biology and materials science, and techniques for the analysis of the image. In addition, as envisaged by the authors in their original plan, each book will be revised, independently of the others, at such time as the authors and editor consider necessary, thus keeping the series of books up-to-date. Consequently a number of the earlier books are now in the process of being revised, and for most of them this revision will be so extensive that they will emerge effectively as new books and will therefore be given new titles.

This whole enterprise would not have been possible without the fullest support and collaboration of our publisher. North Holland/Elsevier have continuously helped us to maintain a high standard in all aspects of the books, from the accuracy of the text and figures to the high quality of the final volumes. I look forward to a future in which authors, editor and publisher continue to contribute to the Science and Art of Electron Microscopy.

Strangeways Research Laboratory   AUDREY M. GLAUERT, Sc. D.
Cambridge, England   *General Editor, March 1985*

Authors' preface

This book grew out of a long friendship and scientific collaboration with a special mutual interest in the use of low temperatures in ultrastructure research. Looking back some years, to when preliminary conversations were made in the appropriate setting of the ice caves under the Great Dachstein, we now realise that we were totally unaware of the vast amount of work that would be involved in surveying the techniques in such a rapidly developing field. In common with many of our fellow biologists, we have often found ourselves in the position that, while realising the immense potential importance of low temperature methods, satisfactory practical means for achieving acceptable results have not always been available. Indeed, despite much progress over the past few years, this is a circumstance that, to a considerable extent, still prevails. The present trend towards the collaboration of biologists, physicists, engineers and, not least, commercial people means that low temperature methods will become more effectively, more widely and more easily available.

This book does not and cannot exclusively present well tried and tested methods. Rather, it attempts to pursue a middle course whereby the reader is not only made aware of what *can* be done but is also informed of background details so that, as the subject develops further, he or she will have the scientific basis upon which to build. The purpose is to paint the background against which future developments will take place. The book tells the reader both *how* to chose and implement different methods as well as *why*. It also attempts to provide an adequate gateway into the literature so that those who wish to extend their knowledge further will be able to do so.

Low temperature methods are already used in very many electron microscope laboratories and it seems highly probable that, sooner or later, *all* such laboratories will call upon these methods as they become better established. They are often ignored for a variety of reasons, including that they are complex and expensive. This book will serve both to help the absolute beginner to progress into this fascinating field as well as to provide additional information to those who are already familar with some, if not all, cryotechniques.

York and Vienna, February 1985   A.W. ROBARDS and U.B. SLEYTR

Acknowledgements

We are very much indebted to our many friends and colleagues who have helped in the production of this book by reading through various Chapters, in whole or in part, correcting errors and suggesting changes and improvements. It is impossible to express our thanks to each and every one of you and we hope that it will suffice to know that the willing help that we have uniformly received when it has been requested has been greatly appreciated. We should, however, specifically like to thank Dr. John Baker, Prof. Stan Bullivant, Dr. Hugh Elder, Dr. Diana Harvey, and Dr. Walter Umrath for their invaluable help with the manuscript. Despite all the help that we have received we remain, of course, fully responsible for any errors that have not been eliminated.

In addition to those who have contributed illustrations, and are acknowledged in the accompanying text, we would like to thank Dr. P. Messner, the late Hilary Quine and Meg Stark for help with some of the diagrams as well as E. Pohoralek and Ch. Robien, who prepared the photographs for Chapters 5, 6 and 7. A special word of thanks goes to Sue Sparrow who has skilfully prepared the final version of the many diagrams presented here: her expertise and good humour, even when diagrams appeared yet again for still further changes, has been greatly appreciated. For additional general assistance with typing and word-processing we are most grateful to Joan Chambers and Wendy Crosby.

We should also thank the members of our laboratories for their unceasing support and understanding when we have, perhaps, been more occupied with this major project than they would have wished!

Finally, it is not inappropriate to say a word of special thanks to our Editor who, with expert guidance and encouragement, has undoubtedly caused us to produce a much better book than would have resulted if we had been left entirely to our own devices.

Nomenclature

A	=	area
α	=	thermal diffusivity
Bi	=	Biot modulus (dimensionless)
C	=	capacitance
c_p	=	specific heat capacity at constant pressure
D	=	length
d	=	thickness
ΔT	=	temperature difference
E	=	total emissivity
ε	=	dielectric constant
g	=	acceleration due to gravity
h	=	surface heat transfer coefficient
η	=	coefficient of evaporation
I	=	ice
I_c	=	cubic ice
I_h	=	hexagonal ice
I_v	=	vitreous ice
I	=	intensity
i	=	current
J	=	flux
J_a	=	actual sublimation rate
J_d	=	departing flux
J_i	=	impinging flux
J_s	=	absolute rate of sublimation of ice
j	=	$\sqrt{-1}$
k	=	thermal conductivity
λ	=	mean free path

λ_f = latent heat of fusion
λ_v = latent heat of vaporisation
M = molecular weight
m = mass
μ = viscosity
N_{Nu} = Nusselt's number
N_{Re} = Reynolds' number
N_{Pr} = Prandtl's number
n = number
P = pressure
P_c = partial pressure
P_s = saturation vapour pressure
π = pi (3.14)
$\dot{Q}$ = heat flux
R = outer radius
r = inner radius
ρ = density
S = concentration
σ = surface tension
T = temperature
T_c = temperature cold body
T_f = freezing temperature
T_h = temperature of homogeneous nucleuation
T_i = initial temperature
T_m = melting temperature
T_o = equilibrium temperature between solid surface and specimen
T_w = temperature of hot body
T_x = T at distance x after time t
T_α = temperature bulk fluid
τ = time
φ = gas constant (Stefan Boltzmann constant)
V = potential
v = velocity
v_c = critical cooling velocity
ω = frequency
x = distance

Contents

Chapter 3.   *Direct viewing and analysis of cold specimens* *147*

Chapter 1

Introduction

Life on earth began in water, evolved from water, and remains critically dependent on water. Most cells of living organisms contain more than 70% water. The interior of an electron microscope is evacuated to a very low pressure of less than 1×10^{-5} Torr (1.33×10^{-3} Pa). This is approximately 76 million times less than normal atmospheric pressure. It follows from this that, if a living cell is put into the vacuum of an electron microscope, the water usually evaporates extremely quickly with disastrous consequences for the preservation of structure. If we want to observe or analyse biological cells and tissues in electron-beam instruments it is therefore necessary to take steps so that such cataclysmic artefacts are avoided. In principle, there is a number of possibilities: the specimen could be contained at approximately atmospheric pressure within an electron-transparent envelope (the so-called 'environmental cell'); the material could be preserved (fixed) so that subsequent dehydration produces minimal interference with the *in vivo* structure; or the water could be frozen so that it would not evaporate in the electron-beam apparatus. Environmental cells (Butler and Hale 1981) pose many different problems and, while they have made valuable contributions in some areas of electron microscopy (mainly non-biological), they do not provide a routine means for the observation and analysis of biological specimens at high resolution. Fixation, dehydration and embedding do, of course, together constitute the major pathway by which the vast majority of biological specimen processing is carried out. Numerous books and papers have been devoted to this subject (e.g. Glauert 1974; Hayat 1981) and more will not be said here beyond noting that such techniques almost always involve the *removal* of water. Using 'wet' processing methods, the withdrawal of water inevitably leads to changes in the native biological structure; many soluble materials will be extracted, and, further, the *in vivo* distribution of ions and other mobile components will be radically changed. Such considerations have led to much greater attention being paid to an alternative possibility for preparing biological specimens: that of *immobilising* the water by freezing it. In principle this may sound straightforward but in practice it is not! This book is devoted to the practical assessment of low temperature techniques for the preparation of biological specimens for electron microscopy.

In learning how to freeze cells so that they are preserved in as life-like a state as possible, we could do much worse than to look at the responses of cells to natural chilling in Nature. Reduction of the ambient temperature below the norm for a particular cell type is accompanied by a reduction in the rate of energy-requiring processes – 'life' slows down. More severe (but above 0°C) cooling will, in many cells, result in a metabolically inactive state in which life can appear to be suspended. Ultimately, some cells in Nature are subjected to such low temperatures that water would normally freeze. Under these circumstances there can be a number of different responses: many cells do not possess cold tolerance at all and will die; others may contain relatively small amounts of water, thus minimising the problems caused by ice formation; and still others can actually produce chemicals that will act as antifreezes and will thus protect the cell and, in some situations, allow it to function normally at sub-zero temperatures (e.g. Zachariassen and Husby 1982). Only by appreciating the structure of liquid water, the structure and formation of ice and the response of cells to different temperature regimes can we hope to develop the optimum methods of artificially imposing freezing techniques for the preservation of cells.

Water is a curious and special substance. Many of its properties are quite dissimilar to any other molecule (Eisenberg and Kauzmann 1969; Franks 1972–1982). In order to understand the interrelationships between the processes of life and water we should thus attempt to gain the best possible insight into the structure and behaviour of water itself (Eisenberg and Kauzmann 1969; Fletcher 1970; Franks 1972–1982, 1983; Hobbs 1974). Only then can we logically attempt to adjust the normal relationship between a cell and its (liquid) water without creating excessive artefacts.

The pressure/temperature phase diagram for water (Fig. 1.1) illustrates the relationships between the three physical phases of the molecule. Water within cells is usually liquid but changing the ambient conditions of temperature and/or pressure can bring about a change to either the vapour or solid phases. Not only does absolute change of pressure or temperature have an

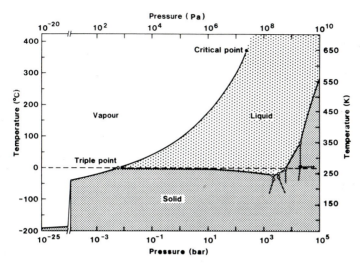

Fig. 1.1. Pressure-temperature phase diagram for water. There are many different units used for the citation of pressure. We generally use the pascal (Pa) which is the appropriate SI unit. However, because pressures are frequently cited in torr or bar, these are sometimes also quoted in this book. $1.0\,Pa = 7.5 \times 10^{-3}\,Torr = 1.0\,N\,m^{-2} = 1.0 \times 10^{-5}\,bar = 9.87 \times 10^{-6}\,atm$.

effect; the *rate* of change can also be important. Indeed, as we shall see later (§2.3.1), the rate of cooling is absolutely critical in determining the nature of the final, frozen cell. Although a sound understanding of the structure and behaviour of water is important, it should also be appreciated that the reactions of the pure liquid are not necessarily reflected within living cells. This is a point that is frequently overlooked and, as yet, inadequately studied. For example, while parameters such as the recrystallisation temperature (§2.1.1) or cooling rate to achieve vitrification can be relatively easily and precisely determined for the pure substance, they will certainly be quite different for water as a solvent within cells. Moreover, such parameters will vary greatly from cell type to cell type or even between adjacent cell compartments.

As this book progresses through the description of low temperature techniques, the relevance of these points and the need for further work will become evident. Nevertheless, there is now an adequate platform of sound theoretical and experimental knowledge so that many techniques can be described with complete confidence. In other instances (and cryoultramicrotomy is a good example, see Chapter 4), there remains much to be done before we can express, with conviction, an adequate understanding of the physical principles of the technique which would lead to its use as a more routine method than is the case today. Although low temperature techniques create their own artefacts, these are often more easily understandable than those arising during chemical treatments because only physical processes are involved.

Low temperatures can carry some degree of risk in their use which may not be encountered in other environments. For this reason a complete Appendix (Appendix 1) has been devoted to the topic of safety in the use of low temperature techniques. It is recommended that this is read by newcomers to the field before any practical work is attempted.

Finally, in this brief introduction, it is worthwhile pointing out that low temperature methods neither supersede nor replace conventional preparation techniques for electron microscopy: they should rather be thought of as augmenting them. There is now a vast range of microscopical methods. The most effective worker will be the one that has the knowledge, ability and (it is to be hoped) equipment to select the most appropriate course of action to achieve his particular end.

References

Butler, E. P. and K. F. Hale (1981), Dynamic experiments in the electron microscope, in: Practical methods in electron microscopy, Vol. 9, A. M. Glauert, ed. (North-Holland, Amsterdam).

Eisenberg, D. and W. Kauzmann (1969), The structure and properties of water (Oxford University Press).

Fletcher, N. F. (1970), The chemical physics of ice (Cambridge University Press).

Franks, F. (1972–1982), Water – a comprehensive treatise (Plenum Publishing Corporation, New York).

Franks, F. (1983), Water (Royal Society of Chemistry).

Glauert, A. M. (1974), Fixation, dehydration and embedding of biological specimens, in: Practical methods in electron microscopy, Vol. 3, A. M. Glauert, ed. (North-Holland, Amsterdam).

Hayat, M. A. (1981), Fixation for electron microscopy (Academic Press, New York).

Hobbs, P. J. (1974), Ice physics (Clarendon Press, Oxford).

Zachariassen, K. E. and J. A. Husby (1982), Antifreeze effects of thermal hysteresis agents protects highly supercooled insects, Nature, Lond. *298*, 865.

Chapter 2

Freezing

The process fundamental to the whole subject of low temperature micro-scopy is freezing: the conversion of water from a liquid to a solid state. This apparently simple phase transition is, in fact, a complex series of events which is modified by such factors as cooling rate, presence of solutes, and pressure. In biological systems, the physicochemical changes during freezing remain poorly understood and many of our present practical methods are based more on empirically judged success than on a true understanding of the changes taking place in the cooling tissue. The questions that need to be asked in the context of cryobiological research are primarily: what cool-ing *rate* is optimal and over what temperature range; can *chemical additives* be used to improve the quality of the frozen specimen without producing unacceptable side-effects; and what is the *highest temperature* to which we can safely subject the frozen specimen? To answer these questions it is necessary to understand something of the principles of freezing and thawing and we shall therefore consider these processes before going on to deal with more practical methods.

2.1 Principles of freezing

2.1.1 Ice nucleation

When pure water is cooled below its *equilibrium temperature*, T_f, the temper-ature at which two phases are in equilibrium (for water and ice $T_f = 0°C$, 273 K), it does not freeze immediately but remains in a metastable super-cooled state. Super-cooling cannot continue indefinitely and eventually spontaneous crystallisation occurs, accompanied by release of the latent heat of fusion which warms the water back towards its melting point (Fig. 2.1; see Stephenson 1956, 1960). The degree of super-cooling depends on the purity of the water but cannot fall below approximately $0.8\ T_f$ (measured in Kelvin; Fletcher 1970). The reason for this is that, as the water cools, small ice crystals form but, close to the equilibrium temperature, they immediately dissociate (melt) due to their energetically unfavourable surface/volume ratio. However, as cooling continues, these crystals become more stable until they eventually grow and act as the seeds for crystallisation of the whole volume. Such freezing is referred to as *homogeneous nucleation*. The subsequent rate of growth of ice crystals is temperature-dependent (Fig.

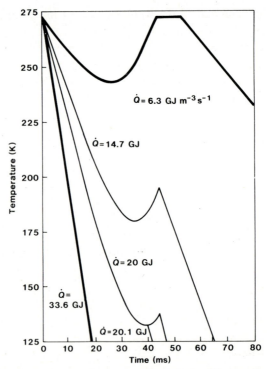

Fig. 2.1. Theoretically computed cooling curves for water at different rates of heat removal ($\dot{Q}$) (GJ = gigajoules = 1×10^9 Joules). The lower two curves illustrate partial vitrification of the water; the upper three curves illustrate complete transformation of water to crystalline ice. Thus, only for the two highest rates of cooling is the heat of fusion removed more quickly than it is produced. In the other three cases some degree of rewarming is evident. (Redrawn from Stephenson 1956.)

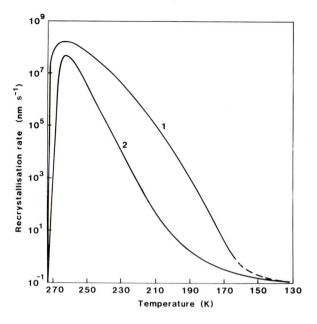

Fig. 2.2. Rate of recrystallisation of ice in relation to temperature. The rate of ice crystal growth is temperature-dependent with a maximum rate at about 260 K ($-13°C$). (Curve 1 redrawn from Riehle 1968a; curve 2 adapted and redrawn from Moor 1973.)

2.2), with a maximum at about 260 K, and falls as the temperature decreases and the liquid viscosity increases until, at about 140 K, ice crystals no longer grow. Franks (1980) has emphasised the contrast between the behaviour of water in the normal physiological temperature range (0 to 100°C) and super-cooled water. For example, the heat capacity of liquid water in the normal temperature range is essentially constant at 75 J mol^{-1} K^{-1}; at $-35°C$ super-cooled water has a heat capacity of 105 J mol^{-1} K^{-1}, whereas the heat capacity for ice at the same temperature is only 33 J mol^{-1} K^{-1} (Fig. 2.3). Such factors affect the rate of cooling and, therefore, the morphology and size distribution of ice crystals that appear at the super-cooled limit. Yannas (1968) gives the vitrification temperature of water (the temperature above which vitrified water would transform into the crystalline state) as 127 ± 4 K. The significance of this is that ice crystals will form and grow at all temperatures between the *homogeneous nucleation point* (T_h) and the *recrystallisation point* (T_r): i.e. between about 230 and 130 K. Only if this range is traversed sufficiently rapidly (i.e. faster than the *critical cooling velocity, v_c*), so that the heat of fusion is removed more quickly than it is produced, can

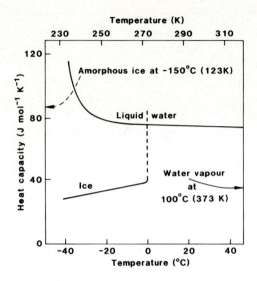

Fig. 2.3.   Heat capacity of liquid water and ice as a function of temperature. (Redrawn from Franks 1980.)

crystallisation be totally avoided. The virtue of maximum super-cooling prior to freezing becomes more apparent from the knowledge that, whereas the latent heat of fusion at 0°C is 334 J g^{-1}, it is only about half this value at the homogeneous nucleation point (approximately -40°C; Franks 1980). With the possible exceptions of the work of Brüggeller and Mayer (1980), Mayer and Brüggeller (1982) and Dubochet and McDowall (1981), such rapid heat extraction has never been achieved for pure water where the critical range is at least from 230 K to 130 K (Fig. 2.4). Indeed, Rasmussen (1982) concludes, from a re-evaluation of the energetics of homogeneous nucleation, that "the nucleation of ice cannot be avoided simply by rapid quenching". However, the addition of solutes and the consequent reduction in the total proportion of available water for freezing both lowers the freezing point and raises the recrystallisation point, hence reducing the critical temperature interval through which very rapid cooling must take place. Work such as that of the Dubochet group (Dubochet and McDowall 1981; Dubochet 1984) gives real hope that, at least for some very small specimens, true vitrification may be achieved.

    In 'typical' biological cells containing more than 80% water, the freezing point is only depressed by about 2 K below that of pure water, although the recrystallisation temperature is raised by about 50 K; nevertheless, the critical interval from approximately 270 to 180 K is probably still too large to be bridged in the super-cooled state (Fig. 2.4). In homogeneous nucleation, ice crystals themselves form the seeds for crystallisation but the stability of a 'crystal embryo' is considerably enhanced if it grows in association with the surface of some insoluble foreign particle $\geqslant 20$ nm in diameter (Fletcher 1970). In such a system super-cooling is much less and the freezing process is known as *heterogeneous nucleation*. It will now be evident that maximum super-cooling can only take place in extremely pure water and that small droplets can be super-cooled further than equivalent large ones because, although the total number of 'seeding' particles per total volume

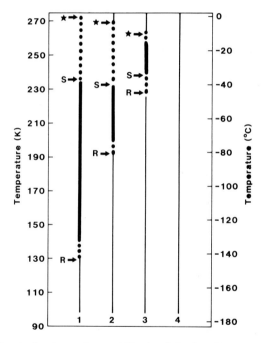

Fig. 2.4. Variation in freezing and recrystallisation behaviour for water and different cell types. (1) Pure water melts at 273 K (★) but may be super-cooled to about 235 K (**S**). Recrystallisation of ice can take place down to temperatures as low as about 130 K (**R**). Crystal growth is therefore possible anywhere within the range 273 to 173 K. (2) In active, living cells the melting point is depressed and the recrystallisation point is raised, thus reducing the critical temperature interval in which crystallisation can occur. (3) The situation is further improved in frost-hardy or cryoprotected cells, where the melting point can be lowered by about 10 K and the recrystallisation point raised to approximately 230 K. Under some circumstances, the super-cooling may be so great that the cells can be solidified without crystallisation. (4) In dry cells, where no water is present, ice cannot form and, therefore, there can be no crystallisation damage.

of liquid will be the same, the chance of having a seeding particle in any one droplet becomes less as the droplet size is reduced. For example, Heverly (1949) found that water droplets of about 0.4 mm diameter could be super-cooled to 257 K, whereas droplets of 50 μm diameter could be super-cooled by a further 14–19 K. It might be expected that biological specimens would freeze primarily through heterogeneous nucleation. This appears not to be so as most cells super-cool well below the predicted level (Mazur 1966; Rebhun 1972). The explanation for this probably lies in the fact that up to 15% of cellular water is bound to macromolecules and does not freeze at low temperatures, thus removing a large component of the potential heterogeneous nucleating sites. The number of water molecules arranged in a crystal-like form to provide a seed for homogeneous nucleation is inversely related to temperature so that the probability of nucleation increases as super-cooling continues until, at about 235 K ($-38°$C), spontaneous freezing occurs even in absolutely pure water. In the presence of heterogeneous nuclei, freezing occurs at higher temperatures during cooling at moderate (i.e. not ultra-rapid) rates at atmospheric pressure although the super-cooling temperature may be lowered by the presence of additives (Fig.

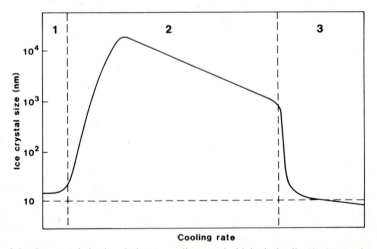

Fig. 2.5.   Ice crystal size in relation to cooling rate in biological cells. (1) At very low rates of cooling (K min⁻¹), water is withdrawn from cells; there is thus less water to crystallise and the ice crystals formed are small. (2) At intermediate cooling rates (K s⁻¹ to >1000 K s⁻¹), large intracellular ice crystals are formed. (3) Above a critical cooling velocity (which will vary from cell type to cell type, depending on, among other things, intracellular concentrations), ice crystals have inadequate time to grow and, therefore, are very small. (Ice crystal dimensions and cooling rates shown here are relative and not indicative of precise measurements for any one cell type.)

2.4). Under these circumstances the ice crystal size is inversely correlated with cooling velocity; that is, the slower the cooling rate, the larger the size of the ice crystals formed (Fig. 2.5). This poses one of the basic problems of low temperature biological microscopy.

2.1.2 Phase separation and eutectics

The process of ice crystallisation automatically involves phase separation: as ice crystals grow, so solutes become more concentrated. *This is an inescapable fact, no matter what the ice crystal size.* Phase separations can have serious repercussions on structural interpretation and can render microanalytical methods effectively impotent. Ideally, the solution to this problem would be to *vitrify* the specimen. Vitrification is the formation of a glassy or amorphous state without the production of ice crystals. (Strictly, we should not refer to 'vitrified ice' as 'ice' signifies a crystalline state. However, the term is used relatively indiscriminately in the literature and its meaning is clear enough. Furthermore, the distinction has sometimes been made between 'vitreous ice', only obtained by ultra-rapid freezing, and 'amorphous ice', which is formed by condensation at low temperature in a vacuum. Some properties of these two forms, for example the density, do differ but need not concern us here. A review of the properties of vitreous water is provided by Pryde and Jones (1964).) If such 'amorphous' solid water could routinely be produced in specimens, then many of the major limitations to low temperature ultrastructural research would immediately be overcome. However, critical cooling rates (v_c) to achieve true vitrification have rarely been reported (Brüggeller and Mayer 1980; Dubochet and McDowall 1981; Mayer and Brüggeller 1982) for pure water nor, under normal circumstances, have they been adequate to vitrify untreated hydrated biological specimens at normal atmospheric pressure (Table 2.1; Fig. 2.5). Menold et al. (1976) have reported that cooling rates of approximately 100 K s^{-1} give ice crystals of 4 to 5 μm while rates of several thousand K s^{-1} give crystals of 1 to 2 μm in aqueous suspensions. Biologists have frequently, and erroneously, used the term 'vitrification' to describe freezing where the ice crystals are smaller than the resolution of the measuring technique employed. For example, Riehle and Hoechli (1973) consider the vitreous state to be "a solid modification of the water with crystals less than 10 nm in size". Such ice crystals, while of small dimensions (5 to 10 nm in diameter), nevertheless indicate that phase separation has occurred and their presence is not to be confused with the real attainment of a vitreous

TABLE 2.1

Cooling rates required for vitrification

Rate (K s^{-1})	Specimen	Reference
1×10^{10}	Micrometre-sized droplets of pure water	Fletcher 1970
1×10^6	Pure water	Riehle 1986a
2×10^4	Pure water at 2.05×10^8 Pa (2045 bar)	Riehle 1968a
5×10^3	Hydrated specimens	Stephenson 1956
1×10^4	Hydrated specimens	Riehle 1968a
1×10^2	Hydrated specimens at 2.05×10^8 Pa	Riehle 1968a

These data provide a very approximate indication of the cooling rates required to vitrify water or hydrated specimens. However, it is important to appreciate (as stressed by a number of authors) that the rates cited depend heavily on the size of the specimen being cooled and may vary by a few orders of magnitude.

state. It might be less confusing if ice with crystals of < 10 nm was referred to as *microcrystalline*. Whenever phase separation occurs, other physiological effects within the cell, such as increase in solute concentration and change in pH, must necessarily follow. Ice crystals of about 10 nm or less do not usually interfere greatly with ultrastructural observations and it may well be that such small crystals still allow preservation of viability (Nei 1973). Amorphous (vitreous) ice (I_v) has so far only been produced by vapour deposition at very low temperatures (Burton and Oliver 1935; Lonsdale 1958; Dowell and Rinfret 1960; McMillan and Los 1965); by using special conditions for extremely small droplets (Brüggeller and Mayer 1980; Dubochet and McDowall 1981; Mayer and Brüggeller 1982; Dubochet et al. 1982; Dubochet 1984; see §3.4.4); or by melting ice under pressure below its glass transition temperature (Mishima et al. 1984). Whether true vitrification has been, or can be, achieved within normally hydrated cells remains an open question. For the moment we must usually content ourselves with the goal of producing extremely small, intracellular ice crystals. If these crystals are those of normal, hexagonal ice (I_h), then they are likely to be relatively large (of micrometre-sized dimensions as opposed to crystals of cubic ice with dimensions in the 0.1 μm range; Dubochet et al. 1982; Dubochet 1984) (Fig. 2.6). Even if vitrification of water is achieved within cells, some structural alterations during freezing will still occur because phase transitions can take place in liquids other than water; for example, in the fluid lipid components of the cytoplasm and membranes.

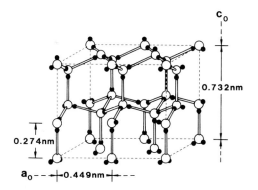

Fig. 2.6. Lattice structure of hexagonal ice (I_h). Oxygen atoms are shown white; hydrogen atoms are shown black. The lattice parameters are temperature-dependent. Those given apply at 110 K ($-163°C$). At this temperature, the density of ice is 933 kg m^{-3} and the linear coefficient of thermal expansion is approximately 2×10^{-5} (Dubochet et al. 1982). The oxygen atoms lie in crinkled sheets which are orientated normal to the c-axis and each position is surrounded tetrahedrally by four other oxygen positions, thus keeping four molecules per unit cell. (Redrawn from Fletcher 1970.)

As the freezing of biological cells and tissues is usually a process of ice crystallisation accompanied by phase separation, it is apparent that this physical method of 'fixation' can produce a spectrum of artefacts of its own. Extracellular crystallisation may lead to dehydration and shrinkage, so disturbing the osmotic balance across cell membranes; proteins may be denatured, and the conformation of polymers may be changed. In particular, mechanical rupture of cell membranes caused by ice crystallisation may lead to an irreversible loss of compartmentalisation (§2.3.1).

When pure water is frozen then, as already explained, ice crystals form by homogeneous nucleation. Adjacent crystals abut against each other with the crystal edges (or 'grain boundaries', to give them their correct nomenclature) in close contact. However, in biological samples, the likelihood of having a single component system is remote: we are usually dealing with complex mixtures of solutions and suspensions. In such specimens, the crystallisation of ice produces a phase separation such that the dissolved salts (and macromolecules) become concentrated into the liquid phase surrounding the growing crystal. This may be exemplified by considering a simple two-component (binary) system consisting of salt and water. If a solution of, say, common salt in water is cooled, then ice crystals eventually form. The ice crystals grow, excluding solutes from the crystal lattice and so concentrating the remaining, unfrozen salt solution until it reaches a critical concentration

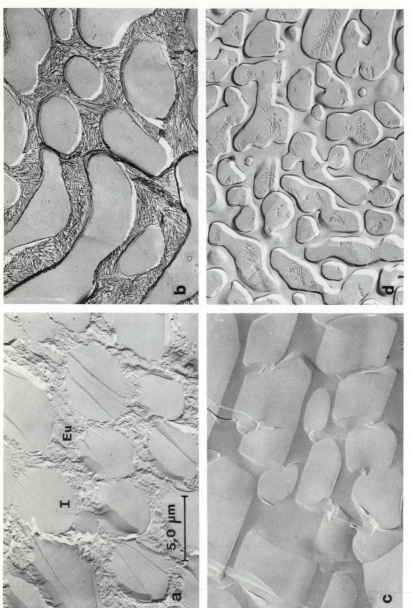

Fig. 2.7. Freezing patterns in simple solutions. (a) and (b) illustrate 20% solutions of sodium chloride that have been frozen and then freeze-fractured (a) and freeze-etched (b). The ice crystals (**I**) with the surrounding eutectic phase (**Eu**) can be distinguished. (c) and (d) illustrate a similar situation when a 20% solution of glycerol is frozen.

TABLE 2.2
Eutectic temperatures of some solutions

Substance	Eutectic temperature		Eutectic concentration
	(°C)	(K)	(%)
KNO$_3$	− 2.9	270.3	10.9
KCl	− 11.1	262.1	19.7
KBr	− 13.0	260.2	–
NH$_4$Cl	− 15.8	257.4	18.6
NaCl	− 21.8	251.4	23.6
MgCl$_2$	− 33.6	239.6	21.6
CaCl$_2$	− 55	218.2	29.8
MgSO$_4$	− 3.9	269.3	19.0
Glucose	− 5.0	268.2	32.0
Sucrose	− 13.5	259.7	62.5

Taken from Meryman (1966) and Lange (1961).

– the *eutectic concentration* – at which point the salt and remaining water freeze as a single component (Fig. 2.7). The phase around the ice crystals is referred to as the *eutectic* (a mixture of two or more substances with the lowest melting point); a specific eutectic will have its own constant freezing or melting temperature – the *eutectic temperature* (Fig. 2.8) (Rey 1960). If cooling rates are higher than those which allow equilibrium conditions to prevail, then supersaturation of the system can occur so that the eutectic point is moved to the so-called '*hypoeutectic point*' (Menold et al. 1976; Fig. 2.9). As, during cooling, the eutectic freezes last so, during warming, it will melt before the ice crystals. Eutectic temperatures of biologically important salts are usually relatively high; for example, NaCl, 252 K; KCl, 262 K; CaCl$_2$, 218 K (see also Table 2.2). Many solutes do not form true eutectics (Meryman 1966; Luyet and Rasmussen 1968). For example, glycerol solutions show progressive ice formation with lowering temperature until a concentration of 60–70% has been reached, beyond which point further cooling results in the formation of a glass; i.e., it vitrifies (Fig. 2.10). In fact, Rasmussen and Luyet (1969) determined the borderline between freezeable and non-freezeable glycerol to be at a concentration of 73%, the remaining solution then solidifying at 216 K. When a binary system is considered, such as H$_2$O/NaCl, the eutectic temperature may be lowered (from 252 to 193 K) by the addition of glycerol to produce a three-component (ternary) sys-

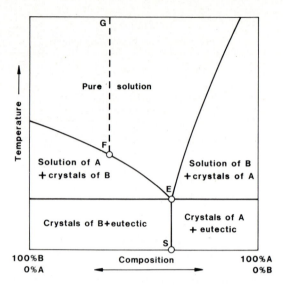

Fig. 2.8.   A simple phase diagram for a binary system (e.g. water and sodium chloride) frozen very slowly. The two substances (e.g. **A**, sodium chloride; **B**, water) form a homogeneous liquid, are intersoluble in all proportions but have negligible solid-solid solubility. When cooling such a system containing a relatively small proportion of substance **A**, the line from **G** to **F** is followed until, at **F**, the system starts to deposit crystals of **B**. As further cooling continues, more and more crystals of **B** are deposited so that the remaining solution becomes more concentrated in **A**. Thus, the line from **F** to **E** is followed until, eventually, **E** (the eutectic point) is reached. At this point, there is a eutectic mixture which, if cooled still further, freezes like a pure element or compound. Freezing along the line **F–E** is termed primary crystallisation; that along the line **E–S** (the eutectic mixture) is called secondary crystallisation. If the proportion of component **A** is relatively high, then crystals of **A** (sodium chloride) will be deposited first. The **A** crystals will push a solution in front of them in which the concentration of **B** becomes higher and higher. When the solution consists of a eutectic mixture from the very beginning, it solidifies as a whole as soon as the eutectic temperature is reached. The situation described here only prevails if the system is allowed to pass through states of equilibrium where the cooling rate is slow enough, and there are sufficient nuclei for crystallisation. If these conditions are not satisfied, the solidus line (saturation curve) will be crossed without solidification of either component **A** or **B** (see Fig. 2.9). (Redrawn from Menold et al. 1976.)

tem (Shepard et al. 1976; Boutron and Kaufmann 1979). Clearly the complexity of such multi-component systems is considerable and little information relevant to biological studies is available. Nevertheless, the magnitude of the changes that can be induced by adding further components is clearly evident.

The addition of a solute to water serves to lower the temperature not only of heterogeneous nucleation but of homogeneous nucleation as well

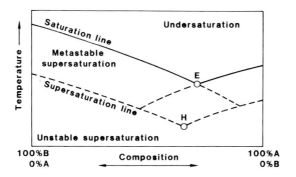

Fig. 2.9. At higher rates than those that allow equilibrium conditions to prevail (Fig. 2.8), the solidus line (saturation curve) is crossed without solidification of either component **A** or **B**. After crossing the solidus line and passing the zone of metastable supersaturation, nucleation occurs which leads to crystallisation. This means that homogeneous nucleation is a retarded process, the magnitude depending on both the composition of the system and on the cooling rate. Consequently, in such a phase diagram of a binary system, there is a line below the solidus line which is less sharp and which connects all points of spontaneous nucleation. This line is termed the supersaturation line or curve. When the system is in a state corresponding to a point within the zone between the saturation and the supersaturation lines, crystals may grow provided the system contains crystallites that are sufficiently large. If the original binary system contains no nuclei, there will be no spontaneous nucleation and consequently no crystal growth. The system is then in some sort of metastable state. Because the supersaturation curve runs parallel to the saturation line, the eutectic point will be found at another concentration than that corresponding to equilibrium conditions. This change in the eutectic point can be seen here. The modified point (**H**) is termed the *hypoeutectic point*. (Redrawn from Menold et al. 1976.)

(Rasmussen and Luyet 1970; Fig. 2.10); the relationship between the lowering of $T_f(T_m)$ and T_h is linear (Rasmussen and MacKenzie 1972). Thus, whereas the homogeneous nucleation temperature (T_h) of pure water is about 235 K, the addition of glycerol to give a 40% solution lowers T_h to 203 K. Absence of nucleating sites in both the solute and solvent phases leads both to super-cooling and to supersaturation of the solution. If the maximum degree of supersaturation corresponding to homogeneous nucleation runs through the lowest value of the eutectic point in the glycerol/water phase diagram (Lusena 1960; Fig. 2.10), then the most favourable situation for achieving the amorphous state would be by freezing a 60% glycerol/water solution. However, as we shall see later (§2.3.2a), biological considerations do not usually allow the use of such high concentrations of additives to facilitate such enhanced super-cooling.

When water is placed under pressure, its melting point is depressed (Fig. 2.11). This is an unusual phenomenon shared by water, bismuth and gal-

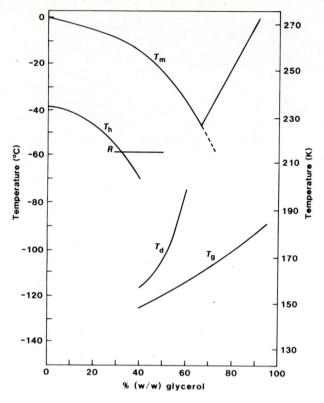

Fig. 2.10.   'Phase diagram' for the glycerol/water system. The temperatures at which different phenomena occur for different concentrations of glycerol are indicated. T_g, glass transition temperature; T_d, devitrification temperature; T_h, homogeneous nucleation temperature; T_m, melting temperature; R, recrystallisation. (Redrawn from data in Luyet 1970 and Rasmussen and Luyet 1970. See Lane 1925 for thermophysical data on glycerol solutions.)

lium. At sufficiently high pressures, different forms of ice are produced but these are not of general biological interest (Fig. 1.1). However, the fact that the critical cooling rate (v_c) required to achieve vitrification of water may be reduced by the application of 0.21 GPa (2,100 bar) pressure, which lowers the melting point to about 251 K, has been used by Riehle (1968a) as the basis for a high pressure freezing device (§2.4.1d).

   The conclusion to be drawn from these facts is that frozen biological specimens normally contain ice crystals of a size related to the composition of the cell fluids and to the local cooling rate. Various artefactual changes will have occurred, not the least of which is an increase in volume of about 9%

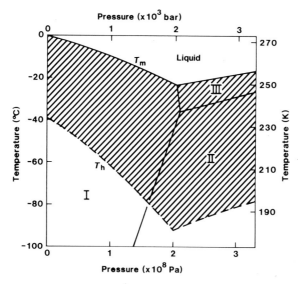

Fig. 2.11.   Part of the pressure/temperature phase diagram for water (see Fig. 1.1) showing melting (T_m) and nucleation (T_h) temperatures of ice as a function of pressure. Roman numerals indicate the different stable ice polymorphs. The cross-hatched area shows the region within which super-cooled water can exist.(Redrawn from Franks 1980.)

during the change of water from a liquid to a solid state (Fig. 2.12; Ushiyama et al. 1979).

It will be apparent from the above statements that *untreated* biological specimens normally freeze somewhere within the range 273–223 K. However, irrespective of the original cooling rate (and hence ice crystal size), ice crystals continue to grow if the specimen is maintained within this relatively warm temperature range. The temperature below which pure ice does not recrystallise is as low as 130 K but this is temperature-dependent and recrystallisation probably does not take place at significant rates until the temperature is above 180–190 K (Moor 1973; Nei 1973). However, few detailed studies have been made, and this is an area where much more work is required. If it is accepted that the frozen specimen must be retained below 180 K if ice crystal growth is to be avoided, then it is usual and convenient to use liquid nitrogen as the storage cryogen. Having frozen the specimen and maintained it at low temperature (77 K), the effect of once again bringing the temperature above the equilibrium freezing point for water can be considered.

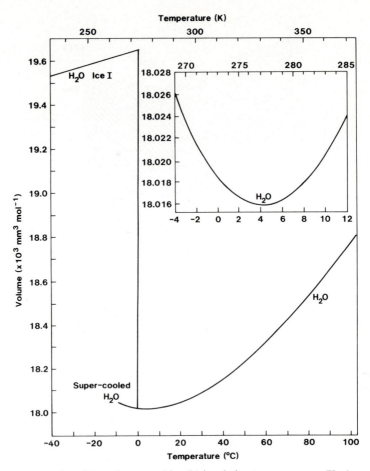

Fig. 2.12.   Specific volume of water and ice (I_h) in relation to temperature. The inset shows specific volume in relation to temperature close to 0°C (273K). (Redrawn from De Quervain 1975.)

2.1.3   Warming frozen specimens

Were the specimen to be vitrified, warming would produce a devitrification transition as the glass (I_v) transformed into cubic ice (for pure water this would be at about 135 K; Fig. 2.13). Subsequently, the cubic form (I_c) would change to hexagonal ice (Fig. 2.13; Dowell and Rinfret 1960; Dubochet et al. 1982). In normal, crystalline specimens continued warming takes the specimen through the recrystallisation temperature, after which ice crystals

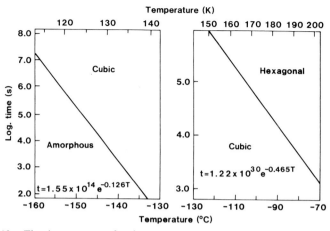

Fig. 2.13. The time necessary for the conversion of vitreous (I_v) to cubic (I_c) ice and cubic ice to hexagonal ice (I_h) as a function of temperature as determined by X-ray diffraction. (Redrawn from Dowell and Rinfret 1960; see also Dubochet et al. 1982.)

continue to grow at increasingly high rates. Finally, the crystals undergo a phase transition, with concomitant absorption of latent heat, and revert to the original liquid state. The condition of the thawed cell, however, greatly depends on the processes that have occurred during both the cooling and warming (thawing) sequences. In general, low temperature microscopy has not been greatly concerned with cells after they have been thawed (but see §2.6) and the following sections only consider the processes and practicalities of freezing biological specimens prior to microscopical examination and analysis.

2.2 Cooling mechanisms

As will have been seen from the previous section, the rate of cooling of a biological specimen has an enormous influence on the nature of the final frozen specimen. It is therefore important to know what rate of cooling should be aimed at, over what temperature interval, and at what position in the specimen. In a later section (§2.3.1) we will consider cooling rates in relation to both viability and structure but, for the moment, we shall assume that we usually require *rapid* cooling rates so that the critical range is traversed as rapidly as possible.

In considering cooling rates (i.e. the extraction of heat from a specimen) we enter the subject of *heat transfer* (§2.2.3). This is a field that has been intensively studied but remains inadequately understood, particularly in relation to heat transfer through and from complex biological specimens. Nevertheless, it is instructive to consider some of the phenomena involved and to establish how well our understanding will allow improvements in the practical treatment of real biological specimens. Before discussing further the theoretical background to and practical attainment of rapid cooling methods, it will be useful to consider the rates of heat extraction that are required and how they might be measured.

2.2.1   What is the actual cooling rate?

The range of cooling rates that is relevant and practicable for biological cells and tissues is from extremely slow (<0.1 K s^{-1}) to ultra-rapid ($>10^4$ K s^{-1}). The literature is full of citations of cooling rates expressed in K s^{-1} that, without further information, are almost worthless in either absolute or comparative terms. Cooling rates are modified by, among other factors, the non-linearity of many cool-down curves (Figs. 2.1 and 2.14); the nature of

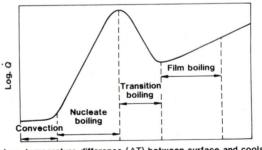

Log. temperature difference (ΔT) between surface and coolant

Fig. 2.14.   Pool boiling heat transfer regimes. The heat transfer rate ($\dot{Q}$) is expressed in relation to the temperature difference (ΔT) between the surface of an object being cooled and the temperature of the coolant. When the surface temperature of the solid object in contact with the liquid is at a relatively high temperature in relation to the temperature of the liquid, a stable vapour film forms at the interface. This film insulates the object (specimen) and reduces the rate of heat withdrawal (cooling rate); as ΔT becomes smaller, a transition region is encountered in which the film becomes unstable and a maximum transition boiling region is reached. Subsequently, bubbles of gas form at specific sites on the specimen surface, grow and detach themselves: a phenomenon known as nucleate boiling. (See Westwater 1959 for a general discussion of boiling heat transfer and Marto et al. 1968 for a discussion of nucleate pool boiling of nitrogen with different surface conditions.)

the specimen; the size and shape of the specimen; the nature and size of the monitoring element (usually a thermocouple); the rate of entry into the cryogen; and the position of the monitoring element within the specimen. Thus, to say that "sample x was cooled at n K s^{-1}" is really not in itself a meaningful statement. Suggestions have been made that some standard for citation of cooling rates should be adopted (Bald 1978b) but the enormous variability of biological samples and freezing methods makes the selection of a universally acceptable standard virtually impossible. The best that can be hoped for is that authors will make the methods for recording cooling rates as sensitive and reliable as possible and will then cite the fullest possible details of how these rates were attained.

It will be helpful to take an example of the difficulties encountered in attempting to use a single figure to cite a cooling rate over a substantially large (> 100 K) range when cooling in a liquid *cryogen*. Most biological specimens to be frozen are small (< 1.0 mm^3) and the monitoring element must therefore also be so small that it does not itself modify the cooling rate. In practice this means that cooling rates are almost always determined using thermocouples (but see the electrical capacitance techniques used by Heuser et al. (1979) and Van Harreveld and Trubatch (1979) in §2.2.2b, as well as the fluorescence label method of Aurich and Förster (1984)). When two dissimilar metals are joined together, an electromotive force (voltage) that is proportional to the temperature is produced across the junction (the *Seebeck* effect, which is the reverse of the *Peltier* effect). By measuring the open circuit voltage it is possible to determine the temperature with considerable accuracy (§2.2.2a). However, the thermocouple itself modifies the apparent cooling rate by increasing the heat capacity (§2.2.4) if it is large in relation to the size of specimen within which it is positioned. This phenomenon is more clearly demonstrated if bare thermocouples of different sizes are cooled in the same manner (Table 2.3). The greater heat capacity of the larger thermocouples reduces the perceived cooling rate. This indicates that, so far as possible, small thermocouples should be used when dealing with biological samples.

A further difficulty in citing cooling rates arises from a simple thermodynamic consideration of what happens when a warm object is placed into a liquid cryogen such as liquid nitrogen at its boiling point (Fig. 2.14). When there is a large temperature difference (ΔT) between the object and the coolant, the heat exchange is so great that a stable film of insulating gas forms around the specimen and serves to retard the cooling rate: this is known as the '*Leidenfrost*' phenomenon and arises from *film boiling*. As the tempera-

TABLE 2.3
Some cooling rates in relation to thermocouple size

Coolant	Diameter	Cooling range	Cooling rate	Reference
	(μm)	(K)	(K s^{-1})	
Liquid nitrogen	25	248–198	39,000	Luyet and Kroener 1960
	70	273–173	16,000	Costello and Corless 1978
	240	273–173	1,880	Glover and Garvitch 1974
	250	273–173	437	MacKenzie 1969
	500	273–173	80	Robards and Crosby 1979
Isopentane	25	279–176	167,000	Luyet and Gonzales 1951
	51	296–196	111,000	
	76	296–196	83,000	

These data, culled from various sources in the literature, serve to show the general dependency
of measured cooling rate on thermocouple size.

ture of the cooling object moves towards the temperature of the coolant
pool, it passes through a phase of *transition boiling* before moving into the
phase of rapid *nucleate boiling* where individual gas bubbles are formed at
the surface of the coolant but do not have time to form a stable layer. The
temperature of nucleate boiling depends upon the cryogen being used and
the temperature difference (ΔT) between object and coolant. For liquid nitrogen this temperature difference extends only 11 K above the temperature
of the coolant and is hence of little significance in improving cooling rates
over the range relevant to biological specimens. (Incidentally, the onset of
nucleate boiling is the explanation of the violent, terminal boiling of liquid
nitrogen as an immersed object finally achieves the coolant temperature –
a phenomenon commonly observed when cooling anticontaminator plates
in electron microscopes). It should now be clear that, if the temperature of
a specimen as it cooled were to follow a changing profile such as depicted
in Fig. 2.14, then any arbitrary cooling rate between temperature 1 and temperature 2 would merely give a misleading integration of what happens. For
specimens of significant size and relatively low thermal conductivity, the
cooling rate will vary greatly in relation to the position within the object:
at the surface the rate might be quite high but within the specimen much
lower rates will prevail. If extremely high cooling rates are achieved, then
it is possible to obtain curves of much greater linearity but many cooling

specimens undergo thermal histories (records of temperature against time) similar to that shown in Fig. 2.14 and the only way to be aware of this is to follow the temperature accurately and rapidly as the specimen cools.

2.2.2 Measurement of cooling rates

Although the importance of high cooling rates for the preservation of biological specimens has been well established for a long time (Luyet and Gonzales 1951; Stephenson 1956; Luyet 1960) and the absence of appropriate standards for measuring or reporting rates has not gone unnoticed (Meryman 1966; Bald 1978b), it remains a fact that relatively few authors have attempted to consider the whole topic of biological freezing as opposed to looking at one or two limited, specific aspects.

As explained earlier, the small possible size of thermocouples makes them the best means for monitoring rapid cooling rates and some techniques for making and using thermocouples will now be considered. It is also possible to make an indirect measurement of the rate of *freezing* of a hydrated specimen by measuring the change in its electrical capacitance during ice formation (Heuser et al. 1979; Van Harreveld and Trubatch 1979; §2.2.2b). Aurich and Förster (1984) have used a novel method for determining temperature during the rapid cooling of microlitre-sized droplets. The fluorescent dye, umbelliferone, fluoresces at two wavelengths, the intensity ratio of which is proportional to temperature. Thus the system allows sensitive and accurate monitoring of cooling rates without mechanical interference with the specimen.

2.2.2a Thermocouples and temperature measurement

Junctions between dissimilar metals produce voltages related to the temperature of the junction. This is the principle of temperature measurement using thermocouples. A range of different junctions can be used in practice (Table 2.4) but copper/copper-nickel (copper/Constantan; Type V) junctions have been much the most widely used for temperatures between 77 and 273 K, although Chromel/Alumel (nickel-chromium/nickel-aluminium; Type K) is also often used. Thermocouples can be bought ready-made in a range of metals, wire diameters and sheathing forms (see List of Suppliers in Appendix 2) but it is often more convenient (and less expensive) to make up thermocouples as required. For the moment we shall confine our attention to copper/Constantan (Cu/Con) thermocouples.

TABLE 2.4
Thermocouple materials for low temperature applications

Type code	Conductor combinations (alternative names)		Approx. working temperature range of junction (°C)		General standards for conductors	Uses
	Positive leg	Negative leg	Continuous	Short-term		
T	Copper	Copper-nickel (Nickel; constantan)	−185 to +300	−250 to +400	BS 4937 pt 5 ANSI type T DIN 43710	Ideal for most cryogenic uses
V	Copper	Copper-nickel (Nickel; Constantan)	−	−	BS 4937 pt 4	Compensating alloy for interconnecting type K thermocouples and instrumentation
K	Nickel-chromium (Chromel)	Nickel-aluminium (Alumel)	0 to +1100	−180 to +1350	BS 4937 pt 4 ANSI type K DIN 43710	Wide range, most commonly used
J	Iron	Copper-nickel (Constantan)	+20 to +700	−180 to +750	BS 4937 pt 3 ANSI type J DIN 43710	Less common in low temperature applications

The accuracy and reproducibility of thermocouple measurements depends primarily upon the purity and homogeneity of the wire and the absence of any further interfering junctions between the thermocouple junction itself and the terminals on the monitoring instrument. If a series of measurements is to be made, it is best to obtain enough wire from the same batch. In the U.K., thermocouple wire is manufactured to British Standards specifications (see Table 2.4; BS 4937, output tables; BS 1041, output tolerances; BS 1834, colour codings). A number of methods for making junctions

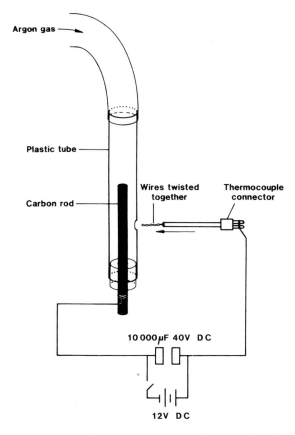

Fig. 2.15. A simple apparatus for preparing thermocouples from thin wires. The two wires (e.g. 0.1 mm diameter copper and 0.1 mm diameter Constantan are twisted together at one end and connected to a capacitor. This capacitor is then discharged across a connection made between the twisted thermocouple wires and a carbon rod. The discharge takes place in an atmosphere of inert (e.g. argon) gas to avoid oxidation of the thermocouple wires. A simple arrangement for achieving this is illustrated here.

has been described: some involve 'butt-welded' junctions using solder; others do not use any additional metal component. The use of solder is not to be recommended if it can be avoided. The simplest form of junction is the 'bead-welded' thermocouple that can be produced very simply by discharging a capacitor across the pair of wound wires as they are held in an atmosphere of argon gas. A simple device for doing this is illustrated in Fig. 2.15. Such a system can be used for thermocouples down to about 100 μm diameter but smaller junctions pose more of a problem and, in addition, often cannot easily be obtained from commercial suppliers.

Some methods of producing very small thin-film junctions have been described by Clark et al. (1976), while Escaig et al. (1977) have also used very thin film thermocouples, produced by the deposition of about 200 nm of copper onto 500 nm of nickel and supported by a 12.5 μm-thick polyimide layer (Bullis 1963). Escaig has subsequently used the same method to produce very small Chromel/Constantan thermocouples (Fig. 2.16). The response time ('thermal inertia') of such thermocouples is less than 1.0 ms and allows the accurate monitoring of cooling rates of the order of 10^5 K s^{-1}. This is important as the size of the thermocouple determines its heat capacity and, therefore, the response time (Balko and Berger 1968; Fig. 2.17).

Having once produced a thermocouple, the question of how accurate it is has to be considered. It is necessary to calibrate the junction against some known standards (such as the melting point of ice, the sublimation point of dry ice (solid CO_2) or the boiling point of liquid nitrogen) and to determine the precise voltage produced (with a calibrated DC voltmeter) (Fig. 2.18). The results can then be compared with the relevant voltage/temperature tables published for thermocouples (e.g. Weast 1971). If the thermo-

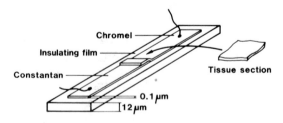

Fig. 2.16.   Preparation of thin film thermocouples (after Escaig et al. 1977). The thermocouple is made by sputtering a layer each of Chromel and Constantan (Table 2.4), each 100 nm thick, onto a 6.0 or 12.0 μm thick polyimide (§3.4.1) film. The termal conductivity is 3.12×10^{-4} cal s^{-1} K^{-1} (1.3 mJ s^{-1} K^{-1}) and the thermocouple is suitable for operation between 4 and 673 K. The junction surface area is approximately 2.0 mm^2. Between 293 and 173 K, such a thermocouple has an average sensitivity of 20 μV K^{-1}. (Diagram redrawn from an original kindly provided by J. Escaig.)

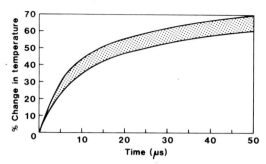

Fig. 2.17. Response of thermocouples. The stippled area indicates the range of computed temperature changes at the centre of 5.0 μm thick slabs of different materials after subjecting them to step-function temperature changes in the surrounding water. The stippled area embraces the responses of copper/Constantan and Chromel/Alumel thermocouples. (Redrawn from data provided in Balko and Berger 1968.)

couple is connected to a direct-reading thermometer, then the absolute accuracy should be checked both with the internal and external reference junctions if available. In the former system the thermometer contains an internal electronic circuit generating a constant voltage as a reference whereas, in the latter, a second, separate external thermocouple is maintained in a

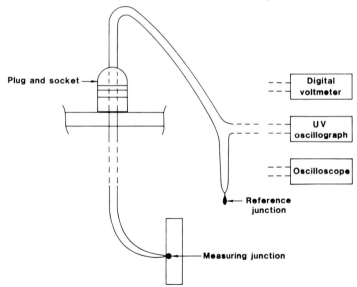

Fig. 2.18. A typical arrangement for measuring temperatures with an accuracy of ±1.0 K. Connections should be made using special plugs and sockets of compatible materials. A reference junction should be incorporated as shown. Examples of different monitoring devices are illustrated.

constant temperature medium (again, such as liquid nitrogen at its boiling point or melting ice) so that a precise reference temperature is available. In matters of calibration everything depends on the accuracy required. Many direct-reading thermocouple thermometers now give good accuracy with little attention but it is important that the need for calibration is understood, particularly when thermocouples are connected to 'non-standard' instrumentation or are used in rapid temperature change experiments. It is also important that the input impedance of the measuring device is very high. Useful publications describing the practicalities of temperature measurement are Hall (1966) and White (1979).

One aspect of thermocouple use in cool-down experiments that is of particular importance is the possibility of cooling rate modification by heat withdrawal along the thermocouple wire (Fig. 2.19). This effect is likely to be greater at low cooling rates than at high ones (Costello and Corless 1978). For this reason it is essential that the thermocouple is insulated as well as possible right up to the monitoring junction.

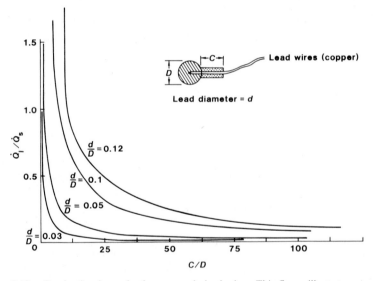

Fig. 2.19.   Conduction losses in thermocouple lead wires. This figure illustrates graphically an equation (Bald 1979) which shows that, if thermocouple lead losses are to be kept below 5% of the heat flow through the sample itself (i.e. $\dot{Q}_l/\dot{Q}_s < 0.05$), then the length of well-insulated wire must be quite large. Although these curves have been derived for a spherically shaped specimen of diameter D, similar curves would be expected for biological samples of cylindrical shape. $\dot{Q}_l$, heat flow through the leads; $\dot{Q}_s$, heat flow through the sample; D, spherical specimen diameter (m); d, lead wire diameter (m); C, length of insulated lead wire (m). (Redrawn from Bald 1979.)

The direct measurement of temperature now poses little problem down to liquid nitrogen temperature (77 K) and there are many inexpensive and efficient electronic thermometers commercially available (see Appendix 2) to carry out such a function. These are also able to monitor relatively slow cooling rates and some can be connected to a chart recorder so that a continuous thermal history can be obtained. However, once the range of fast/ultra-fast cooling rates is entered, an entirely different approach is required. The problem is basically one of finding a system that allows a permanent record of temperature fluctuations in the millisecond and even microsecond range. This means that recording instruments must have an adequately large *bandwidth*. Chart recorders with mechanically moving pens are generally incapable of responding at the rates required. Indeed, the thermocouple itself may not be able to respond adequately if its thermal capacity is too large. Furthermore, it may be necessary to use some sort of 'trigger' for the recording apparatus so that it starts operation in synchrony with the process to be monitored.

A well-tried and -tested system for recording cooling rates is the sequence: thermocouple; high impedance, high sensitivity (<1.0 mV) DC amplifier; oscilloscope (e.g. Luyet and Gonzales 1951; Rebhun 1972; Glover and Garvitch 1974; Menold et al. 1976; Escaig et al. 1977). An alternative arrangement, which avoids the problem of triggering the oscilloscope sweep at the appropriate moment, is to feed the amplified thermocouple output into an ultra-violet (UV) recording oscillograph in which very high paper speeds can be used (>400 mm s^{-1}) and to record the trace by a beam of UV light which is deflected onto UV-sensitive paper by a miniature mirror galvanometer (e.g. Bald and Robards 1978). Further advantages of this system are that a number of thermocouple outputs can be monitored simultaneously (UV recorders can usually accept 12 or more channels) and, by using a differentiator in association with the amplifiers, it is possible to feed to the recorder simultaneous temperature (T) and cooling rate (dT/dt) signals in relation to time (t). However, for very high rates ($> 1,000$ K s^{-1}) the response time of galvanometric mirrors may be limiting.

With the advent of modern semiconductor devices it is now possible to record rapid cooling rates even more easily. This can be done by using a transient recorder or with a purpose-designed cooling rate meter (Robards 1980; Appendix 2). This instrument can be programmed with the temperature range over which it is desired to monitor the mean cooling rate. Once the thermocouple has passed through this temperature interval, the mean rate is *immediately* displayed. Moreover, the whole time/temperature history during the cooling process is stored in a Random Access Memory so that it can be displayed or used for subsequent analysis. Such a device enables cooling experiments to be carried out both rapidly and accurately and is capable of monitoring rates up to about 10^5 K s^{-1}. It is not necessary to describe in detail the various amplifier/recording arrangements: commercial systems are available (see Appendix 2) or suitable combinations of amplifiers and recorders can be put together in the laboratory. However, it is necessary to make some comments on the possible pitfalls that may be encountered while attempting to make accurate determinations of cooling rates with thermocouples.

Assuming that suitably standardised thermocouples are available, these should, ideally, be connected to the amplifiers without any further junctions. (Additional junctions between dissimilar metals also produce voltages, thus giving rise to inaccuracies.) Often, however, it is necessary to make joints (as, for example, when taking a thermocouple lead out of a vacuum apparatus) and then particular care must be taken not to introduce errors in monitoring. It is possible to make vacuum-tight continuous thermocouple lead-throughs (Fig. 2.20) but these often cause inconvenience in assembly and thermocouple replacement. Special thermocouple connectors are commercially available so that the copper wire has a copper connecting pin/socket and the Constantan wire has Constantan connectors (see Appendix 2). This minimises the introduction of 'extraneous' voltages. If thermocouples must be soldered to connectors, then it is best to use low melting point solder and high quality connectors (gold-plated connectors are recommended). Even here, an additional electromotive force will be created but the effect can be minimised by ensuring that all leads passing through the

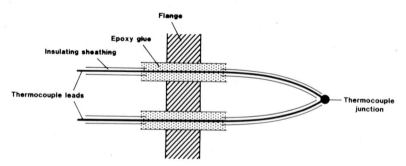

Fig. 2.20.   Simple method of making a jointless lead-through for thermocouple wires. This method is suitable for using thermocouples to monitor temperature in apparatus at low or high vacuum.

connector are maintained at a constant temperature so that no fluctuating, temperature-induced variations arise. This can be particularly important if junctions are positioned close to liquid nitrogen containers where cool gas will drift across the wires and keep the spurious voltages in a constant state of fluctuation. *All junctions other than the recording junction should be well protected from such changes in temperature.* It can be seen that, as with standardisation of thermocouples, ideal theory and actual practice can be some distance apart. Once again, the degree of accuracy required must be considered and, more importantly in this context, the *reproducibility* of measurements should be taken into account. If the ideal situation cannot be attained, then the system should be constructed with careful attention to the elimination of random or continuous fluctuations so that, once assembled, it can be standardised in relation to known temperatures. Quite frequently it is necessary to take thermocouple leads over considerable distances to the monitoring instruments: in such circumstances the appropriate *compensating wire* should be used.

In a later section (§2.4) the practical methods of freezing will be considered but it is necessary here to consider how the method of freezing can affect the accuracy of the determination of cooling rates, particularly when using liquid cryogens (i.e. when a specimen is immersed or 'quenched' in a coolant). It is well established that the nature and rate of movement of a specimen into a cryogenic liquid will modify the *Reynolds number* (N_{Re}) and hence the turbulence and thus may significantly affect its cooling profile. For example, Costello and Corless (1978) noted a 21% increase in cooling rate (from 36,800 to 44,400 K s^{-1} over the range 273 to 173 K) when a bare 50 μm thermocouple was plunged into propane at 0.74 m s^{-1} as compared with 0.39 m s^{-1}. Robards and Crosby (1983) found a similar linear relationship between cooling rate and entry velocity over the range 1.0 to 6.0 m s^{-1}.

The depth to which a specimen is plunged into a cryogen can also have an important effect on cooling velocity since it influences the amount of exposure to 'new' cold liquid. Costello and Corless (1978) cite an improvement in cooling rate from 4,400 to 4,900 K s^{-1} when a copper sandwich was plunged at a velocity of 0.49 m s^{-1} into liquid propane to depths of 5 or 15 mm, respectively. These authors have suggested that the optimal depth of plunge can be simply calculated as the product of the time that it takes for the specimen to cool to a prescribed temperature (e.g. 173 K) and the entry velocity. If the entry velocity were 0.49 m s^{-1} and the cooling rate 4,900 K s^{-1} then the time to cool from 293 K (ambient) to 173 K would be 24.5 ms and consequently the depth of plunge needed would be 12 mm. To this might be added a 25% increase in depth to allow for deceleration and other effects not directly taken into account. It should be noted that this calculation is only likely to be of practical use if high cooling velocities are applicable and relatively slow entry velocities are used. The combination of a slow cooling rate with a high entry velocity will produce an impracticable depth of plunge. In fact, the relevant cooling rate is probably the rate at the specimen surface.

If *reproducible* results are to be obtained, then it is essential that a standardised method of specimen insertion is used. The shape of the specimen as well as the angle, velocity and depth of plunge can all have a pronounced influence on the cooling rate and hence render determinations from specimens cooled in hand-held forceps totally meaningless, although such a

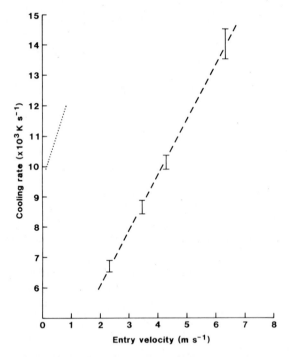

Fig. 2.21.   Cooling rates of model specimens plunged into liquid propane at various velocities. The copper/Constantan thermocouple (60 μm diameter with 25 μm leads) was positioned between a pair of 50 μm thick copper plates 2.0 mm × 5.0 mm. The means ($\pm$standard error) of 25 determinations at each entry velocity are indicated. The thin, dotted line shows the relationship between entry velocity and cooling rate obtained by Costello (1980) for a smaller specimen assembly than the one used here. (Redrawn from Robards and Crosby 1983.)

method may be satisfactory for non-quantitative work. A number of different reproducible systems based on plunging the specimens into liquid coolants or apposition onto cold blocks (see §2.4.2) have been used and are described in detail in §2.4. There is some indication that rates in excess of 5.0 m s^{-1} may be needed to take the maximum advantage of plunging into liquid cryogens (Fig. 2.21), in which case compressed air-actuated devices may be more convenient.

Using arrangements similar to those described above, precise information relating to cooling rates has been obtained, making it possible to select the preferred method of freezing for different specimen/freezing combinations.

2.2.2b   Electrical capacitance method for determining freezing rates

Although thermocouple measurements of temperature and cooling rates are straightforward, for very high rates and where it may not be desirable or possible to have a thermocouple within the specimen, some alternative means of monitoring the rate of freezing must be explored. Heuser et al. (1979) and Van Harreveld and Trubatch (1979) both described methods for measuring the change in electrical capacitance of a hydrated specimen as liquid water changes to ice (Fig. 2.22). The hydrated specimen is treated as a dielectric between two conducting plates of a capacitor: one of these is an electrically conducting specimen support, the other is a metal block against which the specimen is to be thrust. Both Heuser et al. and Van Harreveld and Trubatch describe the method for use in conjunction with cold block freezing, but it is possible that it could be adapted for other cooling methods, although this has not been reported.

The capacitance of the specimen can be calculated from:

$$C = \frac{\varepsilon A}{4\pi d} \qquad (2.1)$$

where
C	= capacitance
A	= area (cm^2)
ε	= dielectric constant
d	= thickness

Before freezing, the liquid water of the specimen has a very high dielectric constant (approximately 80) but this cannot be measured directly because the conductivity of the dilute electrolyte solution of the specimen provides for the flow of a large resistive current between the two plates. However, as soon as freezing commences, this flow is terminated and any subsequent current flowing must be capacitative and will decline as the ice thickness grows because the dielectric constant of ice is only about 2. Thus, as the ratio of these two layers (water and ice) changes, so will the total capacitance. Taking an example from Heuser et al. (1979): if the total sample thickness is 0.05 cm and its area is 1.0 cm² then, when the water is entirely in the liquid form, its capacitance is:

$$\frac{80 \times 1.0^2}{4 \times 3.14 \times 0.05} = 127 \quad \text{(in CGS units)}$$

As 1.0 Farad $(F) = 9 \times 10^{11}$ CGS units, this becomes 141 pF.

However, when a 10 μm layer of ice has formed, the capacitance will be:

$$\frac{1}{C_{total}} = \frac{1}{177_{ice}} + \frac{1}{144_{liquid}} = 79 \text{ pF}$$

and when a 20 μm layer of ice has formed it is:

$$\frac{1}{C_{total}} = \frac{1}{88_{ice}} + \frac{1}{147_{liquid}} = 55 \text{ pF}.$$

The conclusion is that the change in capacitance (C) will be dominated by the change in thickness of the growing ice layer.

Since the thickness of the layer of frozen material is a function of the reciprocal of the current passing across the specimen and:

$$C = \frac{i}{V j \omega} \tag{2.2}$$

where
  C = capacitance
  i = current
  j = $\sqrt{-1}$
  ω = frequency
  V = oscillating potential

it is possible to obtain a direct indication of the course of freezing by applying a high frequency oscillating potential (V) and measuring the voltage transmitted. Heuser et al. used a frequency of 100 kHz while Van Harreveld and Trubatch used 20 kHz. The change in voltage passing through the capacitor (specimen) during freezing is monitored using an oscilloscope. The moment of impact (first formation of an ice layer) is determined from the abrupt cessation in the flow of resistive current (Fig. 2.22). From these data it is possible to construct a detailed record of the passage of the ice front into a freezing specimen. Heuser et al. calibrated their results by using varying thicknesses of a plastic film having approximately the same dielectric constant as ice, while Van Harreveld and Trubatch used a specimen support that ensured that the *total* thickness was quite accurately controlled (50–60 μm) and obtained a direct record of the progress of freezing from their results.

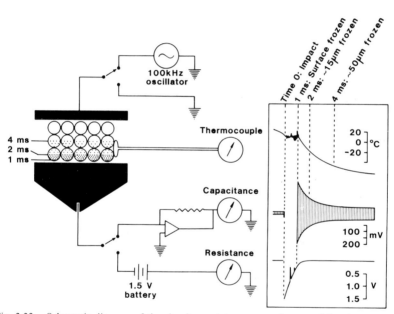

Fig. 2.22. Schematic diagram of the circuits used to measure the rate of freezing on a cold block as described by Heuser et al. (1979). The biological tissue (muscle) is shown as circles between the wedge-shaped copper freezing block (below) and the flat aluminium specimen mount (above). The simplest circuit is a thermocouple placed in contact with the surface of the muscle but it is difficult, if not impossible, to use a thermocouple to record what happens in the first 10–20 μm thickness of tissue where freezing is best. Alternatively, a simple battery circuit is used to pass current through the muscle after contact with the copper block and thus to measure its resistance. As soon as a thin layer of ice is formed on the muscle surface, the resistance becomes so high that the circuit is essentially broken. A better indication of freezing is given by following the changes in the *capacity* of the specimen by applying a 100 kHz oscillating signal to the muscle mount and recording transmission through the muscle with a virtual-earth connected to the copper block. Once the surface has frozen and the relatively huge resistive current is over, the decline in capacitance is a measure of the subsequent growth of ice into the muscle. (Redrawn from Heuser et al. 1979.)

2.2.2c   *Evaluation of cryofixation in frozen samples*

While the experimental determination of actual cooling rates, or rates of fusion, can be very helpful in optimising cooling procedures, ultimately it is the sample itself that will show whether or not the cooling method was satisfactory. For most frozen, uncryoprotected specimens, the ice crystal size is itself a sufficient indicator. This can be observed visually or by low temperature X-ray diffraction (Gulik Krzwicki and Costello 1978). Costello et al. (1982) have described a method of estimating the cooling rate by observing the morphology of freeze-fracture faces of a 30% dilauryl lecithin/ water system. At high cooling rates the lamellar phase of the lipid is preserved, leading to smooth fracture faces; at lower rates, an easily identifiable worm-like or rippled texture is seen. Costello et al. found that cold block freezing, rapid plunging into propane and propane jet-cooling all gave rates, measured using a thermocouple, of more than 10,000 K s^{-1} and all were able to maintain the lipid in its lamellar phase.

2.2.3   *Heat transfer in cryogenic fluids*

As previously mentioned, consideration of how a specimen cools (i.e. how heat is removed from it), involves the subject of *heat transfer*: this is usually expressed in units of W m^{-2} s^{-1}. As a first approximation, the heat flux ($\dot{Q}$) can be determined from Newton's rate equation:

$$\dot{Q} = h A (T_w - T_c) \tag{2.3}$$

where
$\dot{Q}$ = heat flux
h = surface heat transfer coefficient
A = area
T_w = temperature of hot body
T_c = temperature of cold body

The complexities of fluid flow, thermal conductivity and other parameters are combined together in the single factor h, the heat transfer coefficient. The value of h is significantly changed depending upon whether heat transfer is by *natural convection* or by *forced convection*, and both of these can be subdivided according to the flow characteristics (whether laminar, transition or turbulent). In natural convection, motion is primarily due to gravitational forces acting on different densities of fluids at different temperatures. Forced convection arises from motion that is primarily due to a superimposed flow. The flow characteristics depend on the *Reynolds' number* (N_{Re}):

$$N_{Re} = \frac{vD\rho}{\mu} \tag{2.4}$$

where

N_{Re} = Reynolds' number
v = relative velocity of specimen and fluid
D = length (characteristic dimension of specimen)
ρ = density of fluid
μ = viscosity of fluid

Take, for example, a 100 μm diameter ($D = 100$ μm) thermocouple plunging at 2.0 m s^{-1} (v) into liquid propane having a density (ρ) of 0.5 kg m^{-3} and viscosity (μ) of 5×10^{-3} N s^{-1} m^{-2}. Then $N_{Re} = 20$, which is very much below the value (approximately 2,300) above which turbulent flow exists. Thus, for most practical purposes, smooth-sided specimens plunging through liquid cryogens experience laminar flow.

While the thermodynamics of natural convection have been relatively well investigated, it is forced convection heat transfer that is applicable when a specimen is thrust into a coolant (or when a coolant is projected against a specimen). A theoretical appraisal of forced convection in relation to rapid cooling of biological specimens has hardly been made at all (see Silvester et al. 1982). When a fluid in laminar flow impinges on the surface of a specimen, a temperature gradient begins to become established, as does a velocity gradient as the fluid moves downstream. The rate at which heat is transferred from (or to) the surface varies with the distance from the starting point at the surface since the Fourier rate equation states that the rate of heat transfer by conduction is proportional to the temperature gradient:

$$\frac{\dot{Q}}{A} = k\left(\frac{dT}{dx}\right) \tag{2.5}$$

where

$\dot{Q}$ = heat flux
A = area
k = thermal conductivity
dT/dx = temperature gradient.

Laminar forced convection systems are among the most difficult to analyse, partly because the boundary layers may change in form from point to point and also because any disturbance in the impinging fluid may cause eddy formation closer to the leading edge than theory would predict. Further theoretical treatments can thus only provide data of limited qualitative significance. However, for a circular cross-sectional shape (Schenk 1959):

$$N_{Nu} = 0.023 \, N_{Re}^{0.8} \, N_{Pr}^{0.3} \tag{2.6}$$

where
$$
\begin{aligned}
N_{Nu} \ &= \text{Nusselt's number } (=hD/k) \\
N_{Re} \ &= \text{Reynolds' number (Eq. 2.4)} \\
N_{Pr} \ &= \text{Prandtl's number } (=\mu c_p/k) \\
h \ \ &= \text{surface heat transfer coefficient} \\
D \ \ &= \text{length (characteristic dimension of specimen: diameter if the} \\
&\quad \ \text{specimen is of circular cross-sectional shape)} \\
k \ \ &= \text{thermal conductivity} \\
\mu \ \ &= \text{viscosity} \\
c_p \ &= \text{specific heat capacity at constant pressure}
\end{aligned}
$$

$$N_{Nu} = \frac{hD}{k} = 0.023 \left(\frac{v D \rho}{\mu} \right)^{0.8} \times \left(\frac{\mu c_p}{k} \right)^{0.3}$$

$$h = 0.023 \, (v\rho)^{0.8} \times \mu^{-0.5} \times D^{-0.2} \times c_p^{0.3} \times k^{0.7} \tag{2.7}$$

with terms as in Eqs. 2.4 and 2.6.

From this it is evident that, to obtain a high surface heat transfer coefficient, v, ρ, c_p and k must all also be high. The last three terms are easily understood as they all derive from the previous consideration of conduction and it also seems obvious that increased fluid flow will improve the heat transfer coefficient. However, it is also seen that h depends on both μ and D which, ideally, would both be small. The viscosity, μ, is a measure of the 'fluid inertia' and a high fluid inertia would mean that less of the cooler liquid would be able to reach the specimen. High μ increases the tendency of a fluid to laminar flow. Thus, h increases for turbulent flow and this could be an important consideration when designing specimen holders (e.g. by including ridges or a rough surface to encourage a turbulent boundary layer). However, it is also important (as shown by Costello 1980) to have an overall streamlined shape to discourage trapping of vapour or restricting the free flow of liquid around the specimen. Turbulence is useful for reducing the incidence of film boiling and experiments have been carried out where the surface has been suitably modified to encourage this (e.g. Rebhun, 1972; see §2.2.3b). This is of less significance in cryogens such as propane where the tendency towards film boiling is much less marked.

So far, the discussion has centred on the *surface* heat transfer characteristics. However, the quantity that determines the *overall* ability of an object to transfer heat is the *Biot modulus* (*Bi*). The Biot modulus is a dimensionless number determined from the equation:

$$Bi = \frac{hR}{k} \tag{2.8}$$

where

h	= instantaneous surface heat transfer coefficient
R	= characteristic dimension of the specimen (outer radius of a sphere or cylinder; semi-thickness of a slab)
k	= thermal conductivity.

If $Bi < 0.1$, the specimen is classified as microscopic whereas if it is > 0.1 it is considered macroscopic (Bald 1978b). Microscopic samples, having a low characteristic dimension (R) are generally cooled rapidly and hence have a relatively high surface heat transfer coefficient (h); Bi can be > 0.1 and hence different parts of the specimen will cool at different rates. This is an important conclusion, because it implies that any specimen of large Bi will cool at different rates at different points within the specimen; hence even the citation of a precise cooling rate from a single thermocouple will only define the thermal history of that particular point *and no other*. This problem has been well reviewed by Bald (1975) who has suggested a method for defining the thermal history of *any* point within a specimen of cylindrical, spherical or flat geometry. Bald (1979) and Bald and Crowley (1979, 1981) have carried this type of theoretical analysis still further and demonstrated that the cooling rate profile into a homogeneous specimen is not linear and, moreover, may be higher at the centre than at points intermediate between the specimen surface and the centre. This theoretical prediction receives some support from the work of Venrooij et al. (1975) and Elder et al. (1982). It is clearly important that h, as a function of t, can be precisely reproduced and this is, in effect, the rationale for using devices that allow reproducible cooling rates to be achieved for this sort of work. A further theoretical analysis of ice crystal size distribution in frozen specimens has been made by Kopstad and Elgsaeter (1982), while Jones (1984) has made a detailed estimation of the time taken to freeze at different depths within a tissue block in relation to the method of cooling.

Bald and Robards (1978) investigated some characteristics of specimens of high and low *Bi* when subjected to a range of pool boiling heat transfer experiments. Some results from these and later experiments are instructive. As model systems, 6.0 mm-diameter, high *Bi* cylinders (e.g. polytetra-fluoro-ethylene, PTFE $\simeq 2.5$) and low *Bi* (e.g. high conductivity copper $\simeq 13 \times 10^{-4}$) were plunged into:

(*i*)   liquid nitrogen at its equilibrium boiling point (77 K at atmospheric pressure – 191.325 kPa), giving a typical surface heat transfer coefficient (*h*) of 180 W m^{-2} K^{-1};

(*ii*)   nitrogen sub-cooled by heat conduction into a surrounding nitrogen container pumped to the triple point of liquid nitrogen (about 13.5 kPa) to 63 K;

(*iii*)   pressurised (hyperbaric) nitrogen, where both specimen and cryogen were simultaneously subjected to pressures of up to 1×10^6 Pa; and

(*iv*)   Arcton 12 (CCl$_2$F$_2$; Freon 12) at its freezing point (approximately 118 K).

As has already been explained (§2.2.1), liquid nitrogen at its equilibrium boiling point is not a good coolant. However, by pumping liquid nitrogen down to its triple point it can be sub-cooled by about 14 K to 63 K. If a warm object is then inserted into the cryogen while this is at atmospheric pressure in a well-insulated container (so that it does not gain heat rapidly) the heat removed from the specimen raises the temperature of the pool back towards the equilibrium point rather than boiling the liquid pool (which produces the disadvantages of film boiling). The same thermodynamic trick can be played in the opposite direction by subjecting both cryogen and specimen to pressure (isentropic compression) in a well-insulated container prior to plunging the specimen into the pressurised coolant. If, for example, liquid nitrogen at its atmospheric equilibrium boiling point is subjected to a pressure of 1×10^6 Pa (10 bar) then a new equilibrium point approximately 30 K warmer will result. This cannot be immediately achieved by inflow of heat from outside the pool if the pool is well insulated and, consequently, if the specimen is inserted into the effectively sub-cooled liquid, the heat removed will again raise the pool temperature towards the equilibrium boiling point. This method of hyperbaric freezing has not been used to any great extent in the freezing of biological specimens because, from a cooling rate point of view, it offers no significant advantage over the use of sub-cooled nitrogen at atmospheric pressure and it has the disadvantage of requiring pressurisation of the specimen (although only to relatively low, non-damaging, levels). Nevertheless, it is of interest in comparison with other methods and its use derived from a suggestion by R. Thaine that freezing cells of high turgor pressure (e.g. plant phloem elements) under pressure might alleviate some of the difficulties associated with preserving such specimens: a hypothesis as yet untested. It should be noted that many authors refer to the use of liquid nitrogen 'slush' as a coolant: i.e. a mixture of solid nitrogen and the sub-cooled liquid. It would be impossible to achieve reproducible cooling rates in such a cryogen because of the uncertain locations of solid and fluid phases within the mixture; even sub-cooled liquids have substantial temperature gradients throughout the liquid depth.

The measured cooling rate can be related to the thermocouple size (Table 2.3 in §2.2.1). If rates are to be monitored *within* specimens then very small thermocouples (in fact, of low *heat capacity*) in relation to the size of the specimen must be used (see §2.2.4). The difference in cooling rates between that of a high thermal conductivity specimen (low *Bi*) plunged into liquid nitrogen at its equilibrium boiling point at atmospheric pressure and the same specimen plunged into hyperbaric nitrogen at 650 kPa is illustrated in Fig. 2.23. A number of important features are evident:

(*i*)   the *total* cool-down time is very much less in the 'sub-cooled' cryogen;
(*ii*) the cool-down curve is non-linear; and
(*iii*) the peak cooling rate is a function of either the 'entry phenomenon' (mechanical action of the specimen moving into the cryogen) or the transition to nucleate boiling (which is at far too low a temperature to have any biological relevance).

The cooling rates in liquid nitrogen sub-cooled by evacuation ('nitrogen slush') are very similar. Some representative cooling rates for low and high *Bi* specimens are listed in Table 2.5. Perhaps the most significant feature here is the small difference in rate of cooling of the PTFE specimen whether boiling liquid nitrogen or sub-cooled cryogens are used. This brings us to one of the main problems concerning cooling of biological specimens: if the thermal resistance of the specimen is large compared with the solid/fluid thermal resistance as heat is conducted away from the surface of the specimen, then any increase in surface heat transfer will not be manifested in the specimen centre (or detected by a thermocouple in that position; see Bald

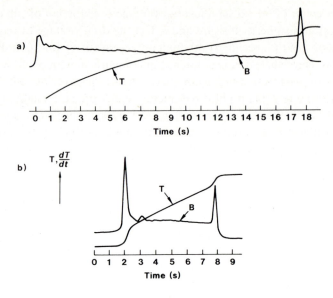

Fig. 2.23.   A comparison between a copper specimen cooled in liquid nitrogen at 100 kPa (1.0 atm) (a) and the same specimen cooled in sub-cooled liquid nitrogen at a hyperbaric pressure of 650 kPa (b). In each case the temperature vs. time curve is shown (**T**) as well as the cooling *rate* vs. time (**B**). (Redrawn from Bald and Robards 1978.)

and Robards 1978). The only ways of achieving an improvement in cooling rate are either to improve the thermal conductivity of the specimen (not practically possible for most biological specimens) or to reduce the total thermal resistance (heat capacity) of the specimen by making at least one of its dimensions smaller (the main method available for biologists). The significance of the results with the PTFE specimens becomes the more apparent when it is appreciated that, size for size and under the same conditions of cooling, the Biot modulus of a 6.0 mm diameter cylinder of PTFE (2.5) is very close to that of ice. Ice is an extremely poor conductor of heat (Bald 1975; Fig. 2.24) and, therefore, if a specimen of any appreciable bulk is considered (e.g. $>1.0 \, \mu$l), the cooling rate (except for the surface layer of about $10.0 \, \mu$m) will primarily be determined by the thermal conductivity of ice and not by the surface heat transfer characteristic of the specimen. In spite of a literature full of cooling rate citations, this point has not been fully appreciated. In so far as the production of very small (<10 nm) ice crystals and still more a vitrified state, throughout untreated, hydrated biological specimens of significant size (all dimensions >0.1 mm) is concerned, an enor-

TABLE 2.5
Some representative cooling rates (K s^{-1}) from model specimens
of high and low Biot number

Sample	Coolant			
	Liquid nitrogen	Sub-cooled liquid nitrogen	Hyperbaric liquid nitrogen	Halocarbon 12
Bare 0.5 mm thermocouple	80	317	319	714
2.4 mm diameter phosphor-bronze bead	34	79	116	63
2.4 mm diameter PTFE bead	43	79	46	37

This Table lists some 273–223 K cooling rates for a bare 0.5 mm diameter copper-Constantan thermocouple; a similar thermocouple embedded in a 2.4 mm diameter high conductivity speci men (phosphor-bronze; Biot number approximately 0.002); and another in a 2.4 mm diameter low conductivity specimen (PTFE; Biot number approximately 2.5). All cooling rates were determined as described in Bald and Robards (1978). It can be seen that the cooling rates for these large samples are very low although the higher conductivity phosphor-bronze bead cooled more rapidly in the more efficient cryogens.

mous amount of time has been spent in the practical pursuit of the theoretically unobtainable!

Assuming that we *do* have a specimen of significant size (and, unfortunately, most biological tissue specimens will need to be at least 0.1 mm in each dimension), then it is possible to calculate the relation between the cooling rate and the time (thermal history) for any point within the specimen if a few basic assumptions are made (Arpaci 1966; Bald 1975).

$$t = \frac{\rho \, \lambda_f}{k(T_f - T_\alpha)} \left[\frac{k}{2h} \left(R - \frac{r^2}{R} \right) \frac{R^2}{4} - \frac{r^2}{2} \left\{ \frac{1}{2} + \ln \left(\frac{R}{r} \right) \right\} \right] \qquad (2.9)$$

where

t = time
ρ = density of specimen
λ = latent heat of fusion
k = thermal conductivity

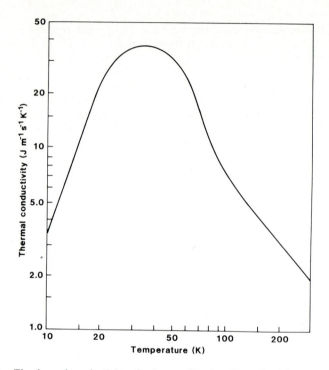

Fig. 2.24.   The thermal conductivity of polycrystalline ice. (Data simplified and diagram re-drawn from Fletcher 1970.)

T_f = freezing temperature
T_α = bulk temperature of cooling fluid
h = surface heat transfer coefficient
R = radius of specimen
r = distance of cell from central axis of specimen ('inner' radius).

This formula has been used effectively by Johnson (1978) to calculate values of t for various values of h, R, r, and T and we have used it to construct a series of curves showing the time taken to freeze, at different depths within a specimen, using different surface heat transfer coefficients (Fig. 2.25).

The fact that the low thermal conductivity of ice (Fig. 2.24) so greatly modifies the cooling rate within any biological specimen of significant size does not mean that we should totally neglect further efforts to optimise the heat transfer characteristic at the specimen surface. It is still often useful

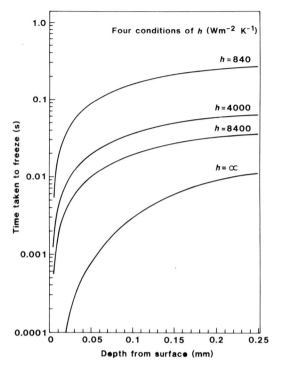

Fig. 2.25. The time taken to freeze (i.e. to pass 0°C) at different depths within a biological specimen in relation to four different values for surface heat transfer coefficient (*h*) calculated using Eq. (2.9). The lowest value of *h* approximates to that obtained by cooling in liquid nitrogen; the intermediate ones represent cooling in halocarbons or propane. Even at an infinite rate of heat withdrawal, points deeper than about 0.1 mm from the specimen surface take a significant time to freeze.

even if only the outer few micrometres of material are frozen well and for very small specimens (droplets or thin films) the correct selection and use of cooling procedures is critical. Some further attention should therefore be given to the question of maximising surface heat transfer. Let us first consider the situation for liquid cryogens.

2.2.3a Characteristics of liquid cryogens

While sub-cooled nitrogen is a convenient cryogen (§2.4.1a), other liquid coolants have also been used. In the main these have been selected on the basis that they have a low melting point and high boiling point so that the phenomenon of film boiling is minimised. These cryogens will be considered in detail later (§2.4.1) but suffice it to say for the moment that dichlorodifluoromethane (CCl_2F_2 – halocarbon 12, Arcton 12, Freon 12, Genetron 12, etc.) is typical of this class of compound: it has a melting point of approximately 118 K and a boiling point of approximately 243 K. (The generic name 'halocarbon' is used here for the halogenated methane compounds, such as dichlorodifluoromethane, which are sold under various trade names.)

The liquids suitable for rapid cooling must have a number of essential characteristics: they must be liquid at a suitably low temperature (at least 140 K) and must thus have a low melting point at atmospheric pressure; they should ideally have good heat conduction, heat capacity, fluidity, and high density; and their boiling point should be well removed from their melting point. It is also preferable that they should be safe to use, relatively inexpensive and should not contaminate specimens to be used for microanalysis with extrinsic elements. A limited range of potential cryogens is available at present but some cryogenic liquids are as yet untested. Liquid helium II, with its very low temperature (<2.19 K), very high heat conduction capacity (8.03 mW m^{-1} K^{-1} at 4.0 K) and superfluidity has been suggested as a potentially excellent cryogen and tested by Fernández-Morán (1960). However, experiments on heat transfer in superfluid helium (Rinderer and Haenseler 1959), as well as actual cooling rate determinations (Bullivant 1965) show that a stable film of gaseous helium insulates the specimen from the coolant and thus retards the cooling rate. This, together with the difficulty and expense of obtaining liquid helium II, as well as the fact that the liquid helium (and the specimen) must be subjected to a vacuum of 70 μm Hg (9.33 Pa), has meant that little further use has been made of liquid helium II as a direct liquid coolant (but it has a use in cold block devices; §2.4.2).

Rebhun (1972) has given a useful assessment of some different cryogens using a simple 'ballasted' thermocouple (a thermocouple inside a piece of thin hypodermic tubing) as a test probe. He stresses the very great difficulties involved in using biological materials in comparative experiments: the size and geometry of tissue and its surface characteristics, as well as the influence of the thermocouple itself, can all seriously affect measured rates. Table 2.6 gives some physical characteristics of a range of quenching agents. For many fluids the temperature of the object to be cooled exceeds the temperature to which the cryogenic fluid may be super-heated and this results in bubble formation at the object surface as the liquid boils. Rebhun (1972)

TABLE 2.6
Some characteristics of liquid cryogens

Liquid	Melting point °C (K)	Boiling point °C (K)	Specific heat (J g^{-1}K^{-1})	Thermal conductivity (mJ m^{-1}s^{-1}K^{-1})
Ethanol C_2H_5OH	− 117.3 (156)	78.5 (352)	1.89	206
Isobutane $CH_3CH(CH_3)_2$	− 159.2 (114)	− 11.7 (261)	1.68	180
Isopentane $(CH_3)_2C_3H_6$	− 159.9 (113)	27.8 (301)	1.72	182
Propane C_3H_8	− 189.6 (84)	− 42.1 (231)	1.92	219
But-1-ene $C_3H_6 = CH_2$	− 185.4 (88)	− 6.3 (267)	1.94	222
n-Pentane $CH_3(CH_2)_3CH_3$	− 129.7 (143)	36.1 (109)	1.89	177
Ethane CCH_3CH_3	− 183.5 (90)	− 88.8 (184)	2.27	240
Halocarbon 12 CCl_2F_2	− 158.0 (115)	− 29.8 (243)	0.85	138
Halocarbon 22 $CHClF_2$	− 160.0 (113)	− 40.8 (232)	1.08	152
Liquid nitrogen N_2	− 210 (63)	− 195.8 (77)	2.0	153
Liquid oxygen O_2	− 218.9 (54)	− 183.0 (90)	1.7	147
Liquid helium He	− 271.4 (1.75)	− 268.9 (4.25)	4.5	18.1 (2.2 K)

N.B. This Table has been prepared to illustrate the general comparable thermophysical parameters for some liquid cryogens. There are many discrepancies in citations of such data in the literature. In particular, the thermophysical properties of liquid helium vary sharply with temperature over a small range. For this reason, the figures have often been rounded to whole numbers. Furthermore, the specific heats and thermal conductivities have been quoted for temperatures *close to* the melting point but not for any specific temperature. Data drawn from numerous and varied sources including Lange (1961), Touloukian et al. (1970), Weast (1971) and Vargaftik (1975).

considers that this will occur with liquid nitrogen, liquid oxygen, halocarbon 14 and halocarbon 13, and probably also with halocarbon 22 and propane. As indicated in Fig. 2.14 in §2.2.1, the highest cooling rate can often be obtained in the presence of nucleate boiling and, in fact, the *maximum nucleate boiling point* is that point which separates the phase of transition boiling from that of nucleate boiling (Fig. 2.14). At this point the heat transfer (expressed as heat removed per unit surface area of the specimen) is given by Rebhun (1972) as:

$$\frac{\dot{Q}}{A} = \frac{\pi}{24}\lambda_v \rho_v \left[\frac{g\sigma(\rho_1 - \rho_v)}{\rho_v^2}\right]^{0.25} \times \left[\frac{\rho_1}{\rho_1 - \rho_v}\right]^{0.5} \tag{2.10}$$

where

$\dot{Q}$ = heat removed
A = area
λ_v = latent heat of vapourisation
ρ_v = density of vapour
ρ_1 = density of liquid
g = acceleration due to gravity
σ = surface tension.

This equation assumes that the liquid is at its boiling point. Considerably greater heat fluxes can be obtained if the fluid is cooled below this point. One aim of rapid cooling techniques might, therefore, be to maximise nucleate boiling at a temperature relevant to biological specimens (e.g. *not* at the nucleate boiling point of liquid nitrogen at atmospheric pressure, 88 K, which is far too low to be of any significance). Bald (1984) has made a detailed theoretical study of the relative efficiency of cryogenic fluids in rapid cooling. He concludes that liquid nitrogen, maintained near its melting point of 61 K at a pressure in excess of the critical value of 33.5 bar (3.39×10^6 Pa) will produce the quickest cooling. This remains to be experimentally verified.

2.2.3b   The effects of specimen surfaces and shapes on heat transfer

Although heat conduction through the specimen and the necessity for heat transfer across the specimen/coolant boundary have been mentioned, the nature of the specimen surface has not been considered in detail. Frequently the large temperature difference between specimen and coolant leads to film boiling, with the retardation in cooling rate that this necessarily involves. It was reported by Cowley et al. (1961) that coating the surface of a specimen with an insulating material such as Vaseline or glycerol actually leads to increased cooling rates in liquid nitrogen. This effect was further considered by Rebhun (1972) who also studied the influence of adding particles or powders to the surface, which had previously been shown to increase cooling rates in liquid nitrogen (Luyet 1961; Cowley et al. 1961). Rebhun used insulating coatings such as glycerol, dibutyl phthalate, Vaseline, Formvar and a number of oils; and powders such as copper, carbon, protein, silica, pollen grains, spores, rosin and activated alumina. In general, insulating coatings and *non*-metallic powders improve the heat transfer when specimens are plunged into cryogens under conditions where nucleate boiling would not otherwise occur. The insulating coating appears to provide a layer between the specimen surface and the coolant such that the outer layer of the insulator is at, or below, the maximum nucleate boiling point, whereas the inner layer is at the temperature of the object. The net effect is that the additional surface heat transfer more than outweighs the extra thermal resistance imposed by the insulating layer and thus the specimen cools more quickly. The powders act by providing nucleating sites for the initiation of nucleate boiling at a higher temperature than normal. It should be noted that, although improvements of an order of magnitude or so may result when using cryogens that would not normally allow nucleate boiling within the relevant cooling range (e.g. liquid nitrogen), no improvement (or even a reduction in rate) occurs if the cryogen already permits nucleate boiling (e.g. propane). These techniques have not been widely applied to biological specimens but provide a useful insight into how mechanical/thermodynamic interactions can modify measured cooling rates. Other surface modifications (e.g. coating with PTFE; Umrath 1974b) have been used to enhance the 'wettability' of the specimen. Of more direct interest to the biologist is the fact, referred to earlier (§2.2.3a), that the initial cooling rate can be improved by maximising the exposure of the specimen surface to new, cold cryogen. One simple way of carrying this out is to have a deep bath of the coolant and project the specimen in at a moderate velocity ($1-5 \text{ m s}^{-1}$: too high injection velocities may mechanically damage or deform naked bulk specimens); this is one of the favourable aspects of the spray-freezing technique (§2.4.1c). The propane jet method (§2.4.1b) also maximises this feature by projecting the coolant at the specimen. An interesting alternative approach is to keep the coolant in constant motion (e.g. by ultrasonic means) or to vibrate the specimen at about 50 Hz as it is plunged into the coolant (Umrath 1974b).

Costello and Corless (1978) and Costello (1980) have given considerable attention to the size and shape of the specimen and its support in relation to cooling rate. With some supports Costello and Corless reported *improved* freezing at *lower* injection velocities and attributed this to an optimisation of the interaction between coolant and specimen. At high rates of immersion some designs of specimen support caused the coolant to splash away from the specimen, so retarding the cooling rate. The relationships between cryogen viscosity and surface tension and between specimen shape and entry velocity are paramount in determining the actual cooling rate. These relationships are ones that, with the exception of the work by Costello and Corless, have received rather little attention. The shape of the specimen can also influence cooling rate insofar as the relative surface area exposed to the coolant is concerned (Moor 1973). If a plate is cooled from one side only (half an infinite body) then the extent of any particular cooling rate into the specimen can be given by the expression of distance 'x'. x can, in fact, coincide with the limit of formation of ultracrystalline ice (so-called 'vitrification'). At all events, if the limit of such a cooling rate is the distance x in cooling a body from one side only, then it will be $4x$ if the body (platelet) can be simultaneously cooled from both sides; $6x$ if the specimen is cylindrical; and $7x$ if it is a sphere (Fig. 2.26). The theoretical advantage of using a specimen of cylindrical or spherical geometry for rapid cooling is clearly evident and has also been emphasised by Costello (1980), who showed that spherical samples of about 1.0 mm^3 cooled at rates of $<500 \text{ K s}^{-1}$ whereas such samples of 0.001 mm^3 had rates of $>50,000 \text{ K s}^{-1}$ under identical cooling conditions in liquid propane. The exception is that *very* thin samples of

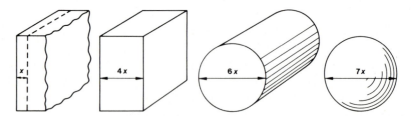

Fig. 2.26. The influence of the shape of an object on the maximum thickness of the well-frozen zone. For a half-infinite body (i.e. cooled from one side only) we may consider the depth of well-frozen material to be distance 'x'. Cooling from both sides of a platelet will produce a well-frozen zone of $4x$ while the values for a cylinder or sphere will be $6x$ or $7x$ respectively. (Adapted from Moor 1973.)

equivalent volume cooled even more rapidly: 11,900 K s⁻¹ for a 0.1 mm³ volume in a 20 μm-thick layer sandwiched between copper strips and cooled in propane. Despite the above comments, the balance between heat input to and heat withdrawal from a cooling specimen must be considered in relation to its overall cooling rate. This has led Pscheid et al. (1981) and Knoll et al. (1982) to stress the advantages of using a propane jet against a thin copper specimen support under which is the (thin) specimen which is, in turn, backed by a thermally insulating, plastic layer (§2.4.1b). Using this method, cooling rates of > 18,000 K s⁻¹ have been obtained.

2.2.4 Heat capacity

The heat capacity of a specimen is that quantity of heat required to increase its temperature by 1 K. Essentially, the heat capacity tells us how much heat

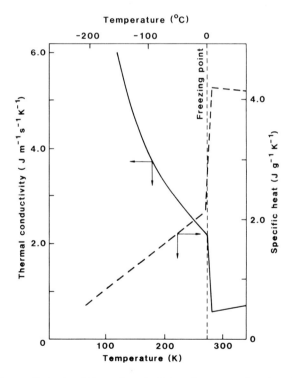

Fig. 2.27.   The specific heat and thermal conductivity of water as a function of temperature. (Redrawn from Bald and Fraser 1982.)

an object of specific composition and volume can 'store'. Materials of high heat capacity are able to 'hold' more heat than those of low heat capacity. Conversely, low heat capacity specimens will cool more quickly than high heat capacity ones of equivalent thermal conductivity. The *specific heat* of a substance is the heat capacity per unit mass. At constant pressure it is

TABLE 2.7

Specific heats of some liquids and solids at low temperatures

$(J g^{-1} K^{-1})$

Substance	Melting point °C (K)	Specific heat close to melting point
Water	0 (273)	4.2
Ice	0 (273)	2.1
Liquid nitrogen	− 210 (63)	2.06
Liquid helium	− 271.4 (1.75)	4.41
Ethanol	− 117 (156)	1.89
Freon 12	− 158 (115)	0.854
Freon 22	− 160 (113)	1.088
Isopentane	− 159.9 (113)	2.287
Propane	− 189.6 (84)	1.92
Ethane	− 183.5 (90)	2.27

	Specific heat at a temperature of (°C (K))				
	0 (273)	− 100 (173)	− 196 (77)	− 253 (20)	− 269 (4)
Plastic (PTFE)	1.04	0.69	0.33		
Sapphire		0.402	0.063	0.002	0.000008
Aluminium	0.89	0.77	0.37	0.00837	0.000261
Copper (crude)		0.7	0.33	0.010	0.000091
Copper (ultra-pure)	0.38	0.35	0.21	0.008	0.000091
Silver	0.23	0.23	0.146	0.0125	0.000124
Gold	0.126	0.121	0.096	0.016	0.00016
Titanium	0.54	0.418	0.23	0.006	0.000317
Stainless steel	0.47	0.39	0.17		
Glass (Pyrex)	0.71	0.5	0.22		
Quartz	0.75				

N.B. The data provided in this Table must be considered as *approximations*. There are many discrepancies in the literature and, hence, variations will be found. Furthermore, the cited temperatures have been selected to coincide with temperatures of interest to low temperature microscopists but the *specific heats* apply to temperatures *in the region of the temperatures specified* and are therefore not necessarily precise. Data drawn from numerous and varied sources including Lange (1961), Touloukian et al. (1970), Weast (1971) and Vargaftik (1975).

expressed in terms of specific heat (c_p) in J kg^{-1} K^{-1} and is itself temperature-dependent (see Fig. 2.3 in §2.1.1 for the heat capacity of water and ice and Fig. 2.27 for the specific heat and thermal conductivity of water and ice). Thus, the total heat capacity of a specimen derives from both its specific heat and its actual volume and this is the reason why, for example, thermocouples of identical composition show cooling rates that are proportional to volume (i.e. to heat capacity).

If we were able to specify the nature of the specimen that we wanted to freeze, then it would have high thermal conductivity, low density and low heat capacity. Unfortunately biologists are usually unable to modify the thermal conductivity and density and can only reduce the heat capacity by reducing the volume. It is, however, possible to consider the requisites for the cooling medium which, ideally, would have a high heat capacity as well as a high thermal conductivity. Some representative heat capacities of materials encountered in low temperature biology are cited in Table 2.7.

2.2.5   Cooling rates in liquid cryogens

The number of citations of cooling rates in the literature is extremely large and, as explained earlier, many of the citations are of little use for comparative purposes. However, Table 2.8 lists some of the rates that have been quoted when cooling in liquid cryogens. As will be seen from this table, actual measured cooling rates from different laboratories are not always in good agreement and it has not always been entirely clear which liquid coolant is 'best'. Rates have been measured using different sizes of thermocouple and, whereas improvements of 2.6 (Luyet and Kroener 1960, from 248 to 198 K), 5.8 (MacKenzie 1969), and 4.0 times (Umrath 1977) have been found when sub-cooled nitrogen is used instead of boiling nitrogen, the improvements recorded by Glover and Garvitch (1974) and Costello and Corless (1978) were only about 1.2–1.3 times. Absolute rates cited for liquid nitrogen at 77 K are: 39,000 K s^{-1} (Luyet and Kroener 1960, using a 25 μm thermocouple); 437 K s^{-1} (MacKenzie 1969, using a 250 μm thermocouple); 80 K s^{-1} (Robards 1979, using a 500 μm thermocouple); 1880 K s^{-1} (Glover and Garvitch 1974, using a 240 μm thermocouple) and 16,000 K s^{-1} (Costello and Corless 1978, using a 70 μm junction). In general, it appears that differences in rate are mainly ascribable to different thermocouple sizes and different immersion velocities but the comparative differences between the rates cited above for nitrogen at 77 K and at 63 K cannot be so dismissed. Costello and Corless (1978) discuss the problems arising from the thermo-

TABLE 2.8
Some comparative cooling rates ($K\ s^{-1}$)

Propane	Freon 22	Freon 12	Liquid nitrogen slush	Liquid nitrogen	Reference	Thermocouple diameter (μm)	Entry velocity (m s⁻¹)
19,100	9,000	6,580	1,700	800	Elder et al. 1982	300	1.4
98,000	66,000	47,000	21,000	16,000	Costello and Corless 1978	70	0.49
1,800	–	740	1,000	720	Zierold 1980	50	'dipped'
4,116	2,621	2,503	–	128	Schwabe and Terracio 1980	360	free-fall

All rates cited are between 273 and 173 K.

dynamic nature of sub-cooled nitrogen. Depending upon how the sub-cooling is effected, either by direct pumping as in Fig. 2.41 (§2.4.1a) or by indirect conduction as in Fig. 2.42, we may be dealing with a true mixture of solid and liquid nitrogen at 63 K ('slush') or we may simply have a sub-cooled liquid at some temperature close to (but not actually) 63 K. In either case there will be strong convective temperature gradients within the pool and some of the discrepancies in cooling rates probably result from this. Furthermore, the problems of immersing the specimen in a reproducible way in the sub-cooled nitrogen can also lead to inconsistencies that do not arise when the container need not be evacuated.

If we continue to examine absolute cooling rates from 273 to 173 K (Table 2.8), then Costello and Corless (1978) find that propane is significantly better (98,000 K s^{-1}) than any of the halocarbons (with halocarbon 22, 66,000 K s^{-1}, rather better than halocarbon 12). Rosenkranz (1975) calculated 273 to 133 K cooling rates for a 1.2 mm diameter water droplet and cited rates of 110, 460, 1410 and 2060 K s^{-1} for Freon 22, Freon 13, propane and sub-cooled nitrogen, respectively. Once again propane appears better than the halocarbons but this time not as good as sub-cooled nitrogen, although Costello and Corless (1978) have made some criticisms of the calculations and methodology of Rosenkranz. Glover and Garvitch (1974) also compared calculated and measured cooling rates under a variety of conditions. The measured rates always considerably exceeded the calculated rates, a fact that the authors attributed to the occurrence of forced convection during immersion, which they had not taken into account in their calculations. Using a 240 μm junction they obtained actual 273 to 173 K rates of 2,120 K s^{-1} (Freon 12), 1,880 K s^{-1} (nitrogen at 77 K), 2,320 K s^{-1} (pumped nitrogen) and 1,350 K s^{-1} (Freon 22). Rebhun (1972) cited his cooling rates in a different form, giving the final rate at 191 K. He obtained rates of over 5,000 K s^{-1} for propane and propylene, with lower rates for the halocarbons, except Genetron 23 (5,410 K s^{-1}). Once again, Freon 22 (3,976) proved better than Freon 12 (2,940). Umrath (1977) found propane and sub-cooled nitrogen to be better than Freon 22 and isopentane. Ethanol has been shown to be an extremely good cryogen down to about $-100°$C (173 K) at which point it becomes very viscous prior to freezing (Silvester et al. 1982) while these same authors also reported that ethane gave cooling rates, within 3.0 μl samples, approximately twice those obtained using propane. The virtues of ethanol as an efficient cryogen have also been described by Marchesse-Ragona (1984).

In summary, sub-cooled nitrogen does not uniformly produce the high rates that some authors have attributed to it. On the other hand, it is safe, easy to produce and does not leave a contamination of solid coolant on the specimen if this is transferred for storage in liquid nitrogen. The halocarbons are also relatively easy to use but have the disadvantages that they form solids at liquid nitrogen temperature; they may give contaminating peaks if the specimen is subsequently used for X-ray microanalysis (Echlin and Saubermann 1977; Gupta 1979); and they *may* have adverse health and environmental effects (Taylor and Harris 1970). Of the halocarbons, halocarbon 22 is better than 12 but the relatively unused halocarbon 23 has been found by Rebhun (1972) to give particularly high cooling rates. Umrath (1977) and Rosenkranz (1975) obtained very good rates with Freon 13 (Rebhun did not) but this is not a particularly easy cryogen to use in view of its high vapour pressure at room temperature (35 bar (3.5×10^6 Pa) at 25°C (298 K)). Costello and Corless (1978) discussed the particular virtues of Freon 13 as a cryogen and concluded that, using biological specimens of normal size for freezing work, the rates obtainable were less good than those obtained with Freon 22. Menold et al. (1976) carried out some experiments on freezing small droplets (approx. 2.0–20.0 mm diameter) of water in liquid nitrogen on Freon 13 at about 93 K. The cool-down rate from 273 to 173 K was measured using 50 μm wire thermocouples. The results show that rates of up to 100 K s^{-1} were obtained using liquid nitrogen, but rates in excess of 1,000 K s^{-1} resulted from quenching in Freon 13. The relationship between droplet size and cooling velocity was linear. In fact, the technique used was to re-cool ice spheres that had been warmed to exactly 273 K and hence the latent heat of crystallisation was not taken into account. However, the authors consider that this would not reduce the actual rates by more than a factor of two or three.

While not all past workers have experimented with liquid propane, the most recent work by Costello (1980), Elder et al. (1982) and in the authors' laboratories tends to reinforce the undoubted superiority of liquid propane as the best liquid cryogen (Table 2.8). This might be expected from its low melting point and high boiling point as well as its good thermal conductivity and specific heat. It must be concluded that, for the highest cooling rates in liquid cryogens, propane offers the best current possibility although the lesser-studied ethane may well be equally effective (Silvester et al. 1982). Propane has the added advantages that it is readily available and is relatively inexpensive. While pure propane melts at about 83 K, some commercial forms remain liquid at temperatures between 83 K and liquid nitrogen temperature (77 K) for prolonged periods. It has recently been shown (Jehl et al. 1981) that the addition of 20–50% isopentane allows propane to sub-cool below the temperature of liquid nitrogen (Fig. 2.28) so that the bath remains constantly liquid; the cooling properties are stated not to be adversely affected. The major criticism of the use of liquid propane is its potentially hazardous nature: it is highly inflammable (combustible at concentrations as low as 2% [20,000 ppm] in air); and it can be sub-cooled beyond the liquefaction point of oxygen, so allowing the condensation of oxygen to form a potentially explosive mixture (Stephenson 1954). Nevertheless, handled in small amounts using properly constructed apparatus and procedures, this should not be allowed to detract from the use of propane as a very efficient cryogen. Provision of a flammable gas sensor (Appendix 2) is a further sensible precaution to take in areas where propane is in use.

The confusion in the citation of cooling rates emphasises the need to produce an acceptable standard in this field. Because most biological specimens are heterogeneous and cool at a rate related to the thermal conductivity of the ice layer as it forms, it is probably simplest and most instructive to obtain an estimate of surface heat transfer properties simply by citing rates

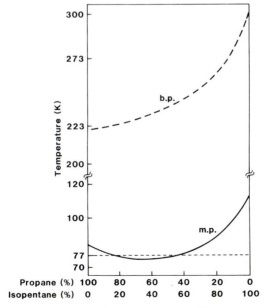

Fig. 2.28. Boiling points (**b.p.**) and melting points (**m.p.**) of different propane/isopentane mixtures. 30–40% isopentane in propane lowers the melting point below the temperature of liquid nitrogen. (Redrawn from Jehl et al. 1981.)

TABLE 2.9

Some criteria for the citation of cooling rates using thermocouples

(*i*) Thermocouple	Small size (25 μm junction would provide a good standard).
	Cite metals (Cu/Con is satisfactory for most purposes).
	Cite junction diameter and wire diameter.
	Cite how junction was made (better welded than soldered).
	Ensure wires are well insulated up to the junction, and carry out tests to ensure that heat is not transferred *via* wires.
	Standardise thermocouple and use the same standardised thermocouple for comparison.
(*ii*) Immersion	Immerse at a reproducible velocity and cite entry velocity and deceleration profile.
	Immerse to constant depth and cite depth.
	Cite nature of specimen immersion mechanism: whether specimen is rigidly held, vibrated, etc.
(*iii*) Cryogen	Cite volume of cryogen, depth of liquid, temperature and thermodynamic state.
	Measure and cite temperature variation within liquid pool.
	With freezing cryogens (e.g. halocarbons) ensure reproducible conditions (e.g. use a liquid pool at point of refreezing or a controlled temperature device to maintain the cryogen close to its freezing point).
	State whether cryogen was stirred or agitated in any way.
(*iv*) Specimen	If a specimen is used, cite:
	location of thermocouple in specimen;
	method of attachment of thermocouple to specimen;
	data relating to thermocouple as specified in (*i*) above;
	precise volume, dimensions and shape of specimen;
	orientation of specimen during immersion; nature of specimen (so far as possible if biological materials are being used);
	method of attachment of specimen to immersion apparatus.
(*v*) Presentation	Cite temperature range over which cooling velocity was calculated (273–173K is appropriate).
	Cite whether T vs t curve is linear; preferably show *graph* of T vs t or dT/dt vs t

of bare thermocouple response. The criteria provided in Table 2.9 would go some way towards allowing comparable data to be cited.

2.2.6   Heat transfer at cold surfaces

Those who have spilt liquid nitrogen onto their skin will know that the result need not be disastrous, despite the temperature of the coolant, provided that the liquid does not remain as a pool. This is largely because of the poor heat transfer characteristic resulting from film boiling. However, we would not think of placing our hand on the surface of a piece of metal even if it were far less cold than the temperature of liquid nitrogen, since a severe cold

TABLE 2.10

Thermal conductivities of some liquids and solids at low temperatures

$(J\ m^{-1}\ s^{-1}\ K^{-1})$

Substance	Melting point °C (K)	Thermal conductivity close to melting point
Water	0 (273)	0.5551
Ice	0 (273)	2.2
Liquid nitrogen	− 210 (63)	0.1322
Liquid helium	− 271.4 (1.75)	0.0181
Ethanol	− 117 (156)	0.2064
Freon 12	− 158 (115)	0.129
Freon 22	− 160 (113)	0.143
Isopentane	− 159.9 (113)	0.1
Propane	− 189.6 (84)	0.1
Ethane	− 183.5 (90)	0.24

Substance	Thermal conductivity at a temperature of (°C (K)):				
	0 (273)	− 100 (173)	− 196 (77)	− 253 (20)	− 269 (4)
Plastic (PTFE)	2.5	2.5	2.4	1.4	0.5
Sapphire	46	82	960	15,700	410
Aluminium	210	210	410		15,000
Copper (crude)	400	400	460	1,200	380
Copper (ultra-pure)	398	413	570	10,500	11,300
Silver	427	430	471	5,100	14,700
Gold	315	327	352	1,500	1,710
Titanium	22	25	34	28	5.76
Stainless steel	15	13	9.4	2.0	0.3
Glass (Pyrex)	1.0	0.9	0.55	0.15	0.1
Quartz	1.38				

N.B. The data provided in this Table must be considered as *approximations*. There are many discrepancies in the literature and, hence, variations will be found. Furthermore, the cited temperatures have been selected to coincide with temperatures of interest to low temperature microscopists but the *thermal conductivities* apply to temperatures *in the region of* the temperatures specified and are therefore not necessarily precise. Data drawn from numerous and varied sources including Lange (1961), Touloukian et al. (1970), Weast (1971) and Vargaftik (1975).

'burn' would result (see Appendix 1 on hazards in low temperature biology). This is a simple demonstration of the greater heat transfer across a solid interface than across a liquid one. This phenomenon is made use of in 'cold block' freezing (§2.4.2). The use of cold blocks for freezing biological specimens was first described by Eränko (1954), while Feder and Sidman (1958) and Van Harreveld and Crowell (1964) later constructed systems for efficient freezing in such a manner. A number of authors have subsequently described variations of the same technique (see §2.4.2).

Some metals increase considerably in thermal conductivity as the temperature is reduced. For example, the thermal conductivity of high purity copper is more than ten times greater at 10 K than at room temperature (Table 2.10). This characteristic may be compared with that of liquid cryogens, the absolute thermal conductivity of which is both lower (Table 2.6 in §2.2.3a) and *decreases* with falling temperature. Similarly, the heat capacity of metals is much higher than those of liquid cryogens (Tables 2.6 and 2.7). For these reasons, provided that a specimen can be brought rapidly into contact with the cold surface of a high conductivity metal, heat can be withdrawn at a very high rate indeed. It will be clear that the surface of the metal needs to be extremely smooth so that good surface contact can be achieved and must also be well protected from contamination (e.g. with condensing water vapour) prior to the freezing process if the full benefit of the improved heat transfer is to be obtained.

The cooling rate achievable with cold blocks is not, as is usual with liquid cryogens, limited by such factors as film boiling or other effects of that kind on heat transfer. Some comparisons of the cooling characteristics of different liquid cryogens and solid cold blocks have been restricted to thermal conductivity alone (which, for example, suggests both copper and sapphire as excellent materials to use as blocks for rapid cooling). However, this grossly oversimplifies the situation. It can be considered that, for thermal properties independent of temperature but with temperature varying with time, the important parameter is the thermal diffusivity (α) where:

$$\alpha = \frac{k}{\rho c_p} \tag{2.11}$$

where
$\quad k \qquad =$ thermal conductivity
$\quad \rho \qquad =$ density
$\quad c_p \qquad =$ specific heat.

However, when a warm specimen comes into contact with a cold surface the surface must necessarily warm up, thus changing α and, therefore, the cooling rate. Heuser et al. (1979) approached this problem by attempting to calculate the equilibrium temperature (T_o) of the boundary between the (solid) cold surface and the specimen. They calculated that the temperature (T) at distance x into a body at time t after initial contact will be:

$$T_x = T_i \, \mathrm{erf} \frac{x}{\sqrt{(4\alpha t)}} \qquad (2.12)$$

where

T_x	= temperature at distance x after time t
T_i	= initial temperature
x	= distance into body
α	= thermal diffusivity
t	= time
erf	= error function (Carslaw and Jaeger 1959).

Applying the boundary conditions for their model (Fig. 2.29), Heuser et al. derived:

$$T_o = \frac{A_1 T_1 + A_2 T_2}{A_1 + A_2} \qquad (2.13)$$

where

T_o	= equilibrium temperature of the boundary between the (solid) cold surface and the specimen
A	= $\sqrt{(k \rho c_p)}$
k	= thermal conductivity
ρ	= density
c_p	= specific heat.

It was argued that T_o is the only influence of the metal on what happens inside the specimen and, since this depends on A, then this combination of properties should be considered. Thus, for a low T_o, low T_1 and T_2 and high A_1 and A_2 are necessary.

It is instructive to consider the individual effect of each of the properties constituting A:

k (thermal conductivity) must clearly be as high as possible since this de-

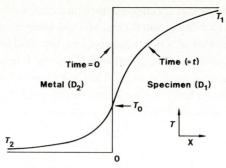

Fig. 2.29.   The theoretical distribution of temperatures in a metal freezing block and specimen just before (time = 0) and at any time after (time = t) the two make contact. It has been reasoned that, at the interface where the two make contact, the temperature at the surface of the metal block (which starts at T_2) must very soon become the same as the temperature of the surface of the tissue (which starts at T_1). This interface temperature may be called T_0. (Redrawn from Heuser et al. 1979).

termines the rate at which heat is conducted into the coolant (liquid or solid) and away from the specimen.

c_p (specific heat) determines the quantity of heat absorbed per unit temperature rise. It is desirable that a cold block should have high heat capacity so that it will absorb much heat without a large temperature rise which would reduce the temperature gradient (ΔT) and hence the cooling rate of the specimen (Eq. (2.3) in §2.2.3). Furthermore, the diffusion equation (α) predicts that *low* c_p will result in a high rate of temperature change which, for the coolant, would mean a high rate of temperature increase, which is clearly undesirable. Thus, c_p should be high for the coolant (block) but low for the specimen (although this is not usually a parameter that can be varied when dealing with biological materials).

ρ (density) of the coolant should also be as high as possible (thus cooling properties of solids are better than those of liquids which are better than gases) and, as with c_p, the density of the specimen would ideally be low.

Bald (1983) has re-evaluated the theoretical aspects of freezing using cold blocks, commenting that Eq. (2.13) above neglects the latent heat of fusion, which is released during freezing and is only valid provided that the density, specific heat and thermal conductivity of both the block and specimen remain constant. Using the method of Finite Elements, Bald has concluded that the ultimate in quick-freezing methods is obtained when using a high-purity *copper* block cooled to *19.1 K*, a temperature that will give cooling rates approximately 50% higher than if the block were initially cooled to 5.0 K. At all events, it remains clear that, in theory, cold blocks should give far higher cooling rates than liquid cryogens, a fact that is supported by the theoretical estimations of Jones (1984), although a more equivocal conclusion was reached by Kopstad and Elgsaeter (1982). Whether this method is, in fact, used must, therefore, depend primarily on the nature of the specimen to be frozen as well as what is subsequently to be done with it.

2.2.7 Comparison of cooling methods

From the above considerations the choice of the important parameters for a good cryogen can be made. As an effective method of rapid cooling, cold block-freezing clearly should have advantages over liquid quenching methods although it remains to be demonstrated that this theoretical advantage can be translated into practical fact. Cold block-freezing produces frozen specimens that have a relatively large area of *flat*, well-frozen (albeit only to a depth of 10–20 μm) tissue, so providing a good surface for subsequent cryoultramicrotomy (§4.3). Cooling using liquid cryogens also has advantages, not only on account of simplicity but also because, by moving either the specimen or the coolant, the temperature of the coolant at the interface can be maintained essentially static, thus maximising ΔT. Furthermore, because of the mechanical deformation or damage that must necessarily ensue, not all specimens lend themselves to the technique of being thrust against a metal surface and, finally, such a freezing method is essentially unilateral, i.e. cooling can only occur from one side of the specimen. Consequently, although heat withdrawal may be extremely good at one surface, the rate will rapidly diminish as the freezing front penetrates into the tissue. For very small specimens where better *bulk* freezing is required (as opposed to a very high rate at one surface) it may be better to contemplate spray-freezing (§2.4.1c), propane jet-freezing (§2.4.1b) or even simple quenching in a liquid bath. All of these methods improve the overall cooling capacity by withdrawing heat over the whole specimen surface.

In summary, when first contemplating how best to freeze a particular specimen, the following aspects should be considered. What is the basic information to be obtained from the specimen and what maximum size of ice crystal is compatible with this objective? There is, for example, no point in trying to produce 10 nm-diameter ice crystals if the specimen is to be studied in an SEM with an effective resolution far worse than this. How small can the specimen reasonably be made and is it 'solid' or could it be 'sprayed'? Is it possible to use well-frozen material at the periphery of the specimen or must deeper parts of the specimen also be studied? If a relatively large

(>0.1 mm in all dimensions) specimen block is to be well frozen through-out, then a cryoprotective is likely to be essential. Furthermore, such large specimens cannot be well frozen deeper than about 10–20 μm from the sur-face and there is therefore little point in utilising ultra-rapid methods. For such 'bulk' specimens, simply freezing in sub-cooled nitrogen or a halocar-bon bath will be equally as effective as other methods. Where the specimen can be made very small ($\ll$0.1 mm in at least one dimension), or where only the surface need be well frozen, then cold block-freezing, propane jets, plunging into propane or spray-freezing (if the specimen is 'sprayable') all serve equally well and the final choice will primarily depend on the nature of the specimen, the equipment to hand, subsequent processing steps and the information ultimately required.

2.3   Treatment of biological specimens prior to freezing

So far, we have almost exclusively considered model systems and cooling rates obtained using bare thermocouples or simple specimens. The situation becomes far more complicated when dealing with highly heterogeneous bio-logical specimens. Different cells of a tissue may freeze at different times and, depending on the nature of the contents of various intracellular com-partments, these may freeze with entirely different patterns of ice/eutectic formation. In addition, the low temperature biologist has a fundamental choice to make: is the specimen being frozen to provide the best preserva-tion of ultrastructure and retention of solutes in their *in vivo* positions; or is the aim to achieve the best viability of frozen cells after they have been thawed? Rarely are these two criteria of successful freezing met simulta-neously from the same cooling method. The following Section considers what happens when biological cells and tissues are cooled at different rates.

2.3.1   Cooling rates in relation to structure and viability

The acme of freezing biological specimens would be to do so in such a way that, without any pretreatment, the cells would remain potentially viable and would also maintain their *in vivo* ultrastructural relationships and solute gradients. Such a goal is at present not achievable for the vast major-ity of cells. It is usually necessary to make a choice between using a freezing protocol that changes the internal structure and solute concentrations of the cells but allows them to survive after thawing and a method that preserves 'good' cell structure even though the cell will not be viable if brought back to its normal growth temperature. The reasons for this distinction are relatively straightforward and arise from a consideration of what happens when cells are frozen at different rates (see Luyet 1966; Nei 1973; Farrant et al. 1977; Williams and Hodson 1978).

At slow (< 1.0 K s^{-1}) rates of cooling, crystallisation of ice commences in the external medium surrounding the cells and, by further slow crystal growth, water is withdrawn across the membranes of the cells so raising the intracellular concentration of solutes. This causes shrinkage and deformation of the cell structure (Figs. 2.30 and 2.31) but reduces the available free water within the cell so that, when the cell contents eventually freeze, they do so with the formation of small, non-injurious ice crystals. Such cells may survive and, if thawed under appropriate conditions, show high viability.

Intermediate rates of cooling (1.0 to 1,000 K s^{-1}) are not so slow that there is time for mass withdrawal of water across the cell membrane but are certainly not sufficiently rapid for the critical temperature range to be traversed so quickly that small, non-damaging crystals are produced. The consequence is that relatively large intracellular ice crystals form so that the cell is badly damaged morphologically (Figs. 2.30 and 2.31); solute relocation

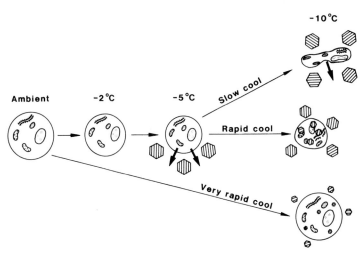

Fig. 2.30. Diagrammatic illustration of the effects of cooling cells at different rates. Slow cooling allows time for water withdrawal across the cell membrane with extracellular ice crystallisation; intermediate rates of cooling lead to large ice crystals both within and around cells; while very high cooling rates can produce extremely small intracellular ice crystals. (Adapted and redrawn from Mazur 1977.)

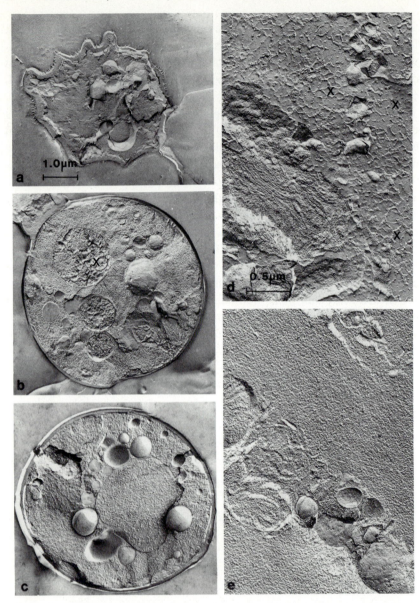

Fig. 2.31.   Micrographs illustrating cooling at different rates. (a–c) Yeast cells frozen at cooling rates of approximately 1.0 (a), 10 (b) and 1000 (c) K s^{-1}. The lowest rate produces a shrunken cell as a result of water withdrawal (see Fig. 2.30), while the intermediate rate leads to
some large intracellular ice crystals (**X**). The highest rate gives relatively good preservation *as
seen at this magnification*. (d–e) Isolated cardiac myocytes frozen by propane jet at about 8,000
(d) and 20,000 (e) K s^{-1}. The cell frozen at the lower rate (d) shows clear ice crystals (**X**) while
the more rapidly frozen myocyte (e) shows extremely fine ice grain structure. (Figs. 2.31d and
e reproduced, with permission, from Robards and Severs 1982.)

is considerable; membranes are disrupted; and the cell will certainly be dead when thawed. Such rates of cooling are of no use for either ultrastructural purposes or preservation of cell viability when applied to unprotected cells. Perhaps their only virtue is for the purposeful disruption of cell contents as a biochemical homogenisation method.

At very high rates of cooling ($> 1,000$ K s^{-1}) the ice crystal size within cells becomes much smaller (Figs. 2.30 and 2.31). If the critical temperature range can be traversed extremely rapidly ($> 10^4$ K s^{-1}) then it should be possible to freeze unprotected cells in such a way that they are both viable and also show good structural preservation. This has been achieved using ultra-rapid freezing techniques on suspensions of isolated cells (Moor 1973) but it is not possible for large specimens and tissues. For these, it is almost always necessary to protect the cells from freezing damage by the addition of a cryoprotectant (§2.3.2a).

The relationship between the structure and function of cells frozen at different rates has been discussed by a number of authors (references below). In relation to freeze-etching, Moor (1973) carried out experiments on the survival rate of yeast cells frozen at different cooling velocities (Fig. 2.32) and found that high viability could be achieved with either very low or very high cooling rates. Yeast, in fact, appears to be a rather tolerant organism

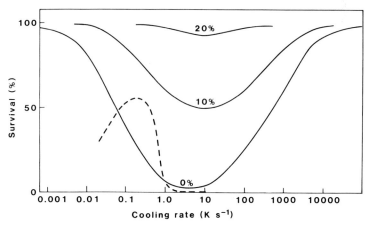

Fig. 2.32.   Survival of yeast cells in relation to cooling rate. The solid lines (taken from Moor 1973) demonstrate the cryoprotective effects of glycerol at different concentrations in enhancing viability of yeast cells after cooling at different rates. The dashed line (from Bank and Mazur 1973) shows yeast cell viability when cooled in distilled water and corresponds quite well with the equivalent curve from Moor's data. (Redrawn from data presented in Moor 1973 and Bank and Mazur 1973).

so far as freezing is concerned and it is much more difficult to obtain high viability at fast freezing rates with most other cells. Nei (1977, 1978) also investigated the structure and function of yeast cells together with that of mammalian cells and human erythrocytes. He found that the freezing pattern and post-thaw survival of cells varied with different cooling rates and that the optimal rates were different for yeast and red blood cells. He recorded the typical pattern of cell distortion at low cooling rates but good structural preservation with little ice crystal formation at high rates, particularly if cryoprotectants were used. The structure of erythrocyte membranes in relation to cooling rate has also been studied by Fujikawa (1981) while Leibo et al. (1978) studied ice formation as a function of cooling rate in mouse ova. The fact that different cells have different optimal cooling rates for survival has been emphasised by a number of workers, including Meryman (1974), Mazur (1977) and Farrant et al. (1977) (Fig. 2.33).

We shall return briefly to the subject of viability after thawing in §2.6. However, for a further consideration of the processes of freezing for ultrastructure research it should now be apparent that fast cooling rates will almost always be essential and that these may well *not* give cells that are potentially viable. Only a few authors have used slow-frozen cells for ultrastructural examination and this has usually been where the cells have been

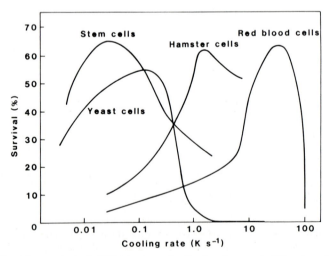

Fig. 2.33.   The survival of different types of cell when frozen at different cooling rates to −196°C. Each cell has a maximum survival at a different rate. (Redrawn from Mazur et al. 1970.)

known to remain viable (e.g. spermatozoa; Koehler 1970, 1972). As commented by Farrant et al. (1977), most cells which are able to recover on thawing are grossly shrunken at low temperatures, but since they are potentially functional they are of interest structurally. Further work in this area would be rewarding.

If it is decided that the best retention of cell structure can be achieved using high cooling rates, then the question of ice crystal size in relation to cooling rate in a particular subcellular location becomes of paramount importance. Some methods of freezing can produce extremely high cooling rates *provided that the specimen can be made small enough* (§2.4). If, however, the specimen is in any sense a 'bulk' specimen (>0.1 mm in all dimensions) then the cooling rate throughout the whole object will be inadequate to produce uniformly small ice crystals. This results primarily from the low thermal conductivity of liquid water and ice (§2.2.3; Fig. 2.24) and the consequent slow rate of heat withdrawal so that the critical range (effectively 273 to 173 K, see §2.1) cannot be traversed sufficiently rapidly to give small ice crystals. This is a fundamental feature of hydrated biological specimens and can only be surmounted by:

(*i*)   increasing the cooling rate (impossible in bulk specimens due to thermodynamic considerations; §2.2.3), or

(*ii*) by treating the specimen with a cryoprotectant (§2.3.2a) that will by some means reduce the ice crystal size for a given cool rate. It is this second method that becomes a necessity for many specimens.

2.3.2   *Principles of sampling and pretreatment*

The ideal method for freeze-fixation is simply to take the specimen and freeze it directly in such a way that it retains good morphology, ultrastructure, solute distribution and viability. For some isolated cellular specimens this goal can become a reality if optimal cooling methods are used (§2.4.1). However, most biological specimens have to receive some pretreatment which immediately poses problems relating to damage and artefacts. For instance, to freeze most *tissues* it is necessary to excise a small sample. This in itself leads to mechanical damage at the edges of the specimen. Even some of the fastest methods currently available for freezing only give good preservation to a depth of about 10 μm in hydrated tissue (Fig. 2.34; Table 2.11). If a cell is about 25 μm in diameter then clearly good freezing will not extend beyond the depth of mechanical damage. Furthermore, the necessity for

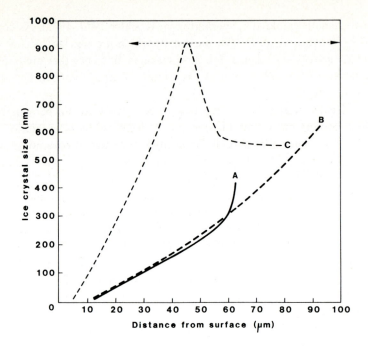

Fig. 2.34.   Ice crystal size at different depths from the surface of rapidly frozen cells. (A) Cold block-frozen pea root tip cells. (From Dempsey and Bullivant 1976a.) (B) Rat kidney cells examined by different low temperature techniques. (From Frederik and Busing 1981.) (C) Propane-frozen rat liver cells. (From Elder et al. 1982.) The significant feature of these curves is that, despite being from very different sources, they *all* show that small ice crystals are only found in the outer 10–20 μm of tissue. Elder et al. found that the peak in ice crystal size could occur over a relatively wide range of depths (fine horizontal dotted line) and it may well be that curves **A** and **B** would also have shown a maximum if the tissue had been examined at deeper levels.

specimen pretreatment with a cryoprotectant often leads to other consequential requirements. For example, if glycerol is used as a cryoprotective it will be found that many cells are impermeable to this compound and must be treated in some way (usually by fixation) to render them more permeable. Each of these steps moves freeze-fixation a little further from the ideal situation of direct freezing without pretreatment. However, for the majority of specimens they are necessities. Therefore the reasons for using them must be understood, the optimisation of their use must be sought and their capacity for producing artefacts must be constantly borne in mind (Table 2.12).

TABLE 2.11

Depth of ice crystal-free zone in tissue without cryoprotection

Depth (μm)	Method of quenching	Reference
10–15	Impact on Ag block (66 K) (1.0 m s^{-1})	Van Harreveld et al. (1974)
12	Impact on Cu block (77 K) (unspecified rate)	Dempsey and Bulivant (1976a,b)
15	Impact on Cu block (4 K) ($\leqslant$ 0.8 m s^{-1})	Heuser and Reese (1976)
20	LN$_2$ (77 K); motor-driven entry	Saetersdal et al. (1977)
5–8	LN$_2$ slush (65 K); 'rapid-entry'	Sevéus (1978)
25–30	Freon 22 (113 K); spring-driven entry	Handley et al. (1981)
10	Freon 22 (109 K)	Frederick and Busing (1981)
5–10	Freon 22 (109 K); compressed air-driven entry (0.6 m s^{-1})	Somlyo and Silcox (1979)
5–10	Propane (83 K), Freon 22 (113 K); gravity entry	Elder et al. (1982)
10	Propane jet (83 K)	Moor et al. (1976)
<1	Propane jet (83 K)	Pscheid et al. (1981)
500	Hyperbaric LN$_2$ at 2.1 × 10^8 Pa(77 K)	Moor et al. (1980)

Augmented and reproduced from Elder et al. (1982).

2.3.2a Cryoprotectants

The routine method of overcoming the problem of damaging ice crystal formation in specimens which have to be frozen in pieces larger than the critical size is the use of a cryoprotectant, with or without previous chemical fixation (see e.g. Nash 1966 and Skaer 1982 for reviews). Substances which allow cells to be frozen with the formation of smaller ice crystals and with less damage than would otherwise occur are now generally referred to as cryoprotectants. It has sometimes been sought to distinguish between *cryoprotectants* (additives used at low concentration where the objective is freeze-thaw survival) and *cryofixatives* (substances used to assist in the preservation of ultrastructural and physiological features in the solid hydrated state) (e.g. Franks 1977). This is not a useful distinction because the same compounds may serve either function. Furthermore, to attribute 'fixation' properties to cryoprotectives is misleading as no formation of covalent

TABLE 2.12
Possible artefacts as a consequence of sampling and specimen pretreatment

---

(1)   *Gravitation forces (centrifugation)*
  (*a*) Dislocation of cell components (organelles);
      especially in eucaryotic cells
  (*b*) Segregation (clustering) of membrane-intercalated particles (e.g. in bovine chro-
      maffin granules) by forced physical contact (Schuler et al. 1978)
  (*c*) Generation of vesicular or tubular membrane structures (e.g. mesosomes in procar-
      yotic cells) (Higgins et al. 1976)

(2)   *Mechanical stress during trimming or freezing*
  (*a*) Damage of superficial cell layers during trimming of the tissue block with a razor
      blade (consequences for the evaluation of fracture planes of specimens frozen on
      cold metal surfaces)
  (*b*) Specimen disruption during impact with a metal surface before freezing (cold block
      freezing) (Pinto da Silva and Kachar 1980)
  (*c*) Shear forces during pipetting of material and spray- or jet-freezing. Particularly
      strong shear forces can be expected during the emulsification necessary for the
      liquid paraffin-freezing technique (Buchheim 1972)

(3)   *Hydrostatic pressure*
  Lethal effect and structural damages as a consequence of the high hydrostatic load
  ($>2 \times 10^8$ Pa) necessary for high pressure-freezing techniques (Moor and Hoechli
  1970; Riehle and Hoechli 1973)

(4)   *Solute concentration*
  Evaporation of the surrounding (suspension) medium during trimming and mounting.
  Osmotic effects, changes of pH or ionic strength may provoke cell and/or organelle
  shrinkage or aggregation or membrane intercalated particles (Pinto da Silva 1972;
  Staehelin 1979; Raviola et al. 1980)

(5)   *Changes in the gas composition in the specimen environment*
  Anoxia (Raviola et al. 1980)

(6)   *Specimen temperature changes before freezing*
  Phase separation of membrane lipids (transition from the liquid crystalline to the gel
  state) and aggregation of membrane-intercalated particles during sampling and freezing as seen in native and isolated biological or artificial (model) membranes (Verkleij
  and Ververgaert 1975, 1978; Zingsheim and Plattner 1976; Maul 1979; Kachar et al.
  1980)

(7)   *Infiltration with cryoprotectants (without chemical fixation)*
  (*a*) Irreversible plasmolysis of cells (Richter 1968a,b; Richter and Sleytr 1970; Fineran
      1970c, 1978; Willison 1975, 1976; Willison and Brown 1979)
  (*b*) Cell or tissue degeneration during growth in glycerol-containing media (Fineran
      1978)
  (*c*) Changes in the form of organelles (e.g. swelling of mitochondria, vesiculation of
      ER cisternae) (Moor and Hoechli 1970; Moor 1971; Zingsheim and Plattner 1976)

(*d*) Ultrastructural changes in biological and artificial (model) membranes aggregation of membrane-intercalated particles (Plattner 1970, 1971; Mcl al. 1974; Sleytr 1974; Niedermeyer and Moor 1976; Groot and Leene ₁₇₁₇, Arancia et al. 1979)

(*e*) Masking of fine-structural details due to phase separation phenomena (as visualised by etching in most cryoprotected samples) (Staehelin and Bertaud 1971; Miller 1979)

(*f*) Changes in frequency of exposure of potential fracture planes (e.g. cell envelopes of Gram-negative bacteria) (Gilleland et al. 1973; Thornley and Sleytr 1974, 1978)

(*g*) Glycerol-induced membrane fusion (Chandler 1979)

(8) *Chemical fixation with glutaraldehyde*

(*a*) Slow fixation artefacts seen as changes in distribution of membrane-intercalated particles;
e.g. in systems which have a poor permeability for glutaraldehyde (fungal spores, mycelium) or systems which require fixation by perfusion (Schmalbruch 1980)

(*b*) Shrinkage of organelles (in plants) (Fineran 1970c, 1978)

(*c*) Reduction in frequency of membrane-orientated fracture planes (Furcht and Scott 1974; Nermut and Ward 1974)

(*d*) Changes in fracture behaviour of membrane constituents (e.g. changes in size 'morphology', frequency, and distribution 'partition coefficients' of membrane-intercalated particles (Dempsey et al. 1973; Staehelin 1973; Nermut and Ward 1974; Furcht and Scott 1975; Parish 1975; Bullivant 1978; Van Deurs and Luft 1979; Willison and Brown 1979; Satir and Satir 1979; Arancia et al. 1979; Kachar et al. 1980; Sleytr et al. 1981)

(*e*) Generation of particle-free membrane blisters (Hay and Hasty 1979; Shelton and Mowczko 1979)

(9) *Chemical fixation with osmium tetroxide*

(*a*) Reduction in frequency of membrane-orientated fracture planes (generally to a larger extent than with glutaraldehyde) (Nermut and Ward 1974)

(*b*) Destroys natural fracture planes of membranes (especially with membranes that contain significant quantities of unsaturated fatty acids) (Meyer and Winkelmann 1970; James and Branton 1971)

(*c*) Morphological changes of the nucleus of bacteria and generation of artificial membrane structures in bacteria (mesosomes) (Lickfeld 1968; Nanninga 1973; Fooke-Achterrath et al. 1974; Higgins et al. 1976; Ghosh and Nannings 1976)

(10) *Chemical fixation with glutaraldehyde and infiltration with crypoprotectant* (generally a combination of (7) and (8))

(*a*) Modification in the organization and partition of membrane-intercalated particles (Pinto da Silva and Miller 1975; Hasty and Hay 1977; Lefort-Tran et al. 1978)

(*b*) Masking of fine structural details due to phase separation phenomena (visualised by etching)

(11) *Treatment with non-penetrating cryoprotectants (e.g. polyvinylpyrrolidone, hydroxyethylstarch)*
Shrinkage of tissue (caused by a decrease of cytoplasmic water; 'osmotic dehydration') (Allen and Weatherbee 1979; Robards et al. 1980)

bonds or any fixation in a conventional sense is involved. Other terms have also been used (e.g. antifreeze agents) but cryoprotectant is probably the most useful and apt generic word. The purpose of such compounds is, literally, to protect the cells during the freezing process. The ways in which this can be accomplished vary but, in general, the useful effects of cryoprotectants are to: lower the equilibrium freezing point (T_f); lower the temperature of homogeneous ice nucleation (T_h) (increase the super-cooling capacity); raise the recrystallisation point (and hence reduce the critical cooling range (§2.1.1; Fig. 2.4) and the critical cooling rate (v_c)); lower the eutectic point; remove 'unbound' water; and remove nuclei for crystallisation. For useful general reviews on cryoprotection, see Meryman (1971) and Skaer (1982). Different cryoprotectants may have different combinations of these properties. In addition to its specific cryoprotection properties it is, of course, necessary that the cryoprotectant should have other virtues and, in particu-

TABLE 2.13

Properties of some cryoprotectants

Substance and formula	Molecular weight	Melting point ($^\circ$C (K))	Boiling point ($^\circ$C (K))
Penetrating			
Glycerol	92.11	20 (293)	290(563)
HOCH$_2$CH(OH)CH$_2$OH			
Dimethylsulphoxide	78.13	18.5(292)	189(462)
H$_3$COSCH$_3$			
Ethylene glycol (1,2 ethane diol)	62.07	$-11.5(262)$	198(471)
HOCH$_2$CH$_2$OH			
Non-penetrating			
Polyvinylpyrrolidone	44000–700000 K[a]		
Hydroxyethyl starch	450000		
Dextran	68500		
(C$_6$H$_{10}$O$_5$)$_n$			

[a] See Fig. 2.35.

lar, that it should not be unacceptably injurious to the cells nor cause excessive artefacts. *No* alien substance is completely without artefactual effects: its very presence and cryoprotective activity is bound to change the physiological and/or biochemical activity of the cell. All that can be done is to attempt to understand what the cryoprotectant is doing and thus to be able to interpret the results that arise from its use.

The range of cryoprotective substances is quite wide, although relatively few have been used for rapid freezing in ultrastructural work (see Table 2.13). A number of living organisms have developed extremely efficient natural cryoprotective substances and these could well benefit from further investigation to evaluate their use in biological studies (e.g. Baust and Edwards 1979; Kappen 1979; Zachariassen and Husby 1982; Baust et al. 1982 (a bibliography)). Some authors (e.g. Ananthanarayanan and Hew

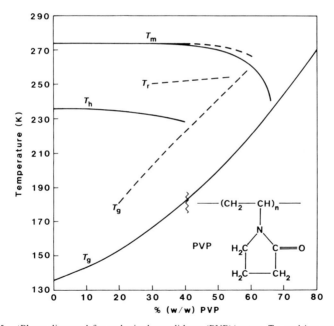

Fig. 2.35. 'Phase diagram' for polyvinylpyrrolidone (PVP)/water. T_m, melting temperature; T_h, temperature of homogeneous nucleation; T_g, glass transition temperature; T_r, recrystallisation temperature. Solid lines are for PVP of molecular weight (MW) 44,000; dashed lines are for MW 700,000. The lower limit for vitrification on the T_g curve for MW 44,000 is indicated by the vertical wavy line. The molecular structure of PVP is shown in the lower right corner of the diagram. (Redrawn from data provided in Rasmussen and Luyet 1970; MacKenzie and Rasmussen 1972 and Franks et al. 1977.)

1977) have attempted to emulate the effects of such natural cryoprotectants by synthesising polypeptides. If a distinction is to be made between different classes of cryoprotectant, then it is probably simplest to divide them into two groups: those which exert their effect by working *within* the cell (penetrating cryoprotectants such as glycerol (Fig. 2.10 in §2.1.2) and dimethyl sulphoxide); and those which operate *without* crossing the cell membrane (non-penetrating cryoprotectants such as polyvinylpyrrolidone (PVP) (Fig. 2.35) and hydroxyethyl starch). The different modes of action of these two classes of cryoprotectant have been discussed by McGann (1978).

Glycerol has been by far the most frequently used cryoprotectant for ultrastructural purposes. Its properties and cryoprotective capabilities have been well studied (Luyet et al. 1958; Luyet and Kroener 1966) and, although an increasingly long list of glycerol-induced artefacts is accumulating (Table 2.12), for many purposes it remains the cryoprotectant of choice. Although a 'penetrating' compound, the permeability of cell membranes to glycerol varies and prior fixation is sometimes required if the molecule is to get into cells (e.g. Richter 1968a). As with most cryoprotectants, glycerol must be used in relatively high concentrations ($>20\%$) to provide good cryoprotection. This means that, for example, osmotic effects must be taken into account when preparing processing solutions.

Dimethyl sulphoxide (DMSO) has also been widely studied (Rasmussen and MacKenzie 1968) and is well established as a cryoprotectant during the cryopreservation of cells in a viable state. However, it can have toxic effects on cell metabolism (Mathes and Hackenseller 1981), damages cell membranes and has been little used in ultrastructural work in comparison to glycerol. A useful study of its general toxicology was made by Wilson et al. (1965).

Other penetrating cryoprotectants have also been used, for example ethylene glycol (Richter 1968a,b) and ethanol (Richter 1968a,b; Schiller and Taugner 1980). However, for the majority of specimens for which a penetrating cryoprotectant is required, glycerol is the best substance to chose initially.

The non-penetrating compounds, such as polyvinylpyrrolidone (PVP) (Luyet and Rasmussen 1967; Luyet and Sager 1967; Rapatz and Luyet 1968; MacKenzie and Rasmussen 1972), hydroxyethyl starch (HES) (Allen and Weatherbee 1979; Körber and Scheiwe 1977) and dextran, had been well studied in their own right in relation to cryopreservation of cell viability before their use in ultrastructural studies was first investigated. Allen and Weatherbee (1980) compared the relative usefulness of HES and glycerol for the cryoprotection of erythrocyte ultrastructure. In a series of papers by Echlin, Franks and Skaer (Franks and Skaer 1976; Echlin et al. 1977; Franks et al. 1977; Skaer et al. 1977, 1978, 1979; Franks 1980; Skaer 1982), the virtues of these high molecular weight, hydrophilic polymers were explored. The method by which these cryoprotectants exert their action remains incompletely understood although it is thought that retardation of extracellular ice nucleation is one important aspect. Once again, it has been found that they need to be used at relatively high concentration ($>20\%$) at which quite pronounced osmotic effects can occur (Franks 1982; Wilson and Robards 1982). However, their usefulness in some situations is beyond question (see papers from Echlin, Franks and Skaer above and also Schiller et al. 1978, 1979) and, in particular, the fact that they do not penetrate the cell membrane (but see Barnard 1980, who describes endocytotic uptake) confers considerable advantages if material is to be subjected to microanalysis. Furthermore, these polymers provide a very satisfactory supporting matrix for cryosectioning (see, for example, Pihakaski and Sevéus 1980).

2.3.3 *Practical methods of sampling and pretreatment*

From a practical point of view it is convenient to distinguish between procedures suitable for suspensions of isolated cells or isolated cell components, and bulk specimens such as animal and plant tissues. The range of fixation schedules is very wide indeed and it is impossible, in this book, to refer to all methods that would be appropriate prior to cryofixation. The reader is referred to Glauert (1974) for a more complete discussion of practical methods of fixation.

2.3.3a *Pretreatment of isolated cells and cell components*

We may take isolated cells to be those cells which are naturally free living or have been cultured in this state. For the present purposes, these include procaryotic cells and some eucaryotic cells. Viruses can generally be dealt with using similar methods. Such material is best frozen as a small sample ($<1.0\ \mu$l) in suspension, or as a pellet, without cryoprotection using standard freezing methods (§2.4). Due to the low average water content, little ice crystal damage occurs within most procaryotic cells but extracellular ice crystals *can* cause distortions to cells. Extracellular ice crystals can also dislocate and disrupt highly hydrated slime, capsular material (Remsen and Watson 1972; Plattner et al. 1973), or cell appendages such as flagella or fimbriae (Sleytr 1978). These structures are often then trapped in the eutectic network surrounding the ice crystals. The average size of extracellular ice crystals can be reduced by freezing *dense* (close-packed) cell pellets, or by using a cryoprotectant.

Procaryotic cells can be prepared for freezing by following one of many different protocols. The cells are centrifuged down from the growth medium or taken directly from an agar surface. If a cryoprotectant is required, then the cells are resuspended in 20–30% buffered glycerol solution with a pH and a final ionic strength, equivalent to that of the growth environment (Lickfeld 1973). Infiltration with glycerol is performed at room temperature, at the growth temperature, or at 4°C. Recommended infiltration times vary from a few minutes up to a few hours and the correct time can often be judged by observing the complete reversal of plasmolysis that can be seen initially. In some instances it is even possible to *grow* cells on or in a medium containing 15–20% glycerol (Moor 1970).

Dormant and resting spores of fungi mostly have thick, impermeable walls and, consequently, severe ultrastructural changes can occur during chemical fixation procedures. These can be avoided, whenever possible, by selecting the best method of fixation. The protoplasts of dormant spores have a low water content and are therefore best frozen without cryoprotection as a dense pellet (Sleytr et al. 1969; Van Gool et al. 1970; Stocks and Hess 1970; Allen et al. 1971; Hess et al. 1972). To harvest spores, rinse the colonies with growth medium, water or, if previous experience has shown cryoprotection to be necessary, with a 20–30% solution of glycerol and then spin them down. Spores are sometimes difficult to 'wet' and this may make sedimentation difficult. This can often be overcome by adding a few drops of detergent (*Tween, Teepol*) to the suspension before centrifugation (Sassen et al. 1967).

If spores are to be kept from rehydrating during harvesting and pretreatment, they can be mixed as a dry powder with paraffin oil and frozen directly. The paraffin oil technique can also be applied to other specimens, such as spores of mosses (Schultz and Lehmann 1974) and pollen grains (Southworth and Branton 1971), which need to be kept dehydrated.

Although cryoprotection may not be strictly necessary, many workers have nevertheless suspended spores in a solution of glycerol before freezing (Stocks and Hess 1970; Brouchant and Demoulin 1971; Richmond and Pring 1971a,b; Cole 1973, 1975). Among other benefits, this leads to much smaller ice crystals in the extracellular solution. Impregnate with a 15–30% (v/v) aqueous solution of glycerol for between 30 min and 24 h at 4–20°C.

Depending on the incubation time, the temperature and the concentration of the glycerol solution, different rehydration or even germination states may be attained (Remsen et al. 1968; Stocks and Hess 1970).

Most studies on *fungal mycelia* have been made using chemically fixed and glycerinated specimens (Takeo et al. 1973; Takeo and Nishiura 1974). Fix with buffered 2–4% glutaraldehyde at room temperature for a few hours or at 4°C overnight. Vegetative fungal mycelia are best cut out and separated from the agar surface in suitably sized pieces (< 1.0 mm³) for direct mounting and freezing. For mounting prior to freezing, it is necessary to wet the pieces to provide enough adhesion to the specimen support. This is done by flooding the colony with water or growth medium, or by suspending the trimmed samples. Air inclusions are removed by briefly exposing the submerged mycelium to reduced atmospheric pressure in a desiccator. Fungi grown as agitated submerged cultures generally maintain a pellet-shaped colony structure during long incubation periods. Pellets of suitable size (< 1.0 mm³) can be mounted and then frozen, or frozen directly, while larger pellets can be trimmed down with a razor blade under a drop of growth medium on a dental wax plate.

Algal cells and *protozoa* usually have too high a water content to allow small ice crystals to be produced without cryoprotective measures when normal freezing methods are used, although spray-freezing and jet-freezing have been applied successfully to a number of algal and protozoan cells (Plattner et al. 1972, 1973; Lefort-Tran et al. 1978; Schuler et al. 1978). To minimise the risk of misinterpretation from artefact formation, such direct freezing methods should be tried before using pretreatments involving chemical fixation and cryoprotection). If no fast freezing methods are available or applicable, methods for pretreatment of algal cells and protozoa are similar to those already described for fungal material. Grow cells directly, or incubate them, in a cryoprotectant solution of 10–30% glycerol. Alternatively, fix the cells with glutaraldehyde (1–4%) prior to glycerination (Neushul 1970; Schwelitz et al. 1970; Werz and Keller 1970; Easterbrook 1971; Zerban et al. 1973; Bray et al. 1974; Pueschel 1977; Fineran 1978).

As mentioned earlier, the incubation time is a direct function of the diffusion constants for glycerol through various compartments and boundaries and can vary considerably from one cell type to another, especially in unfixed plant material. The rate of diffusion must be determined experimentally by studying cell morphology, by observing cell shrinkage or plasmolysis effects and by analysing the intracellular ice crystal size in relation to infiltration time. If the reaction and permeability of the specimen to glycerol is not known, it is safer to start with an appropriately chosen glutaraldehyde fixation (see Glauert 1974) before glycerol infiltration. For example, using *Micrasterias denticulata* Staehelin and Kiermayer (1970) first fixed cells for 10 min at room temperature in 1% phosphate buffered glutaraldehyde and then infiltrated them for 2 h with 30% glycerol in a 3% glutaraldehyde/phosphate buffer solution. Fixation with glutaraldehyde can be performed in the growth medium, a buffered solution, or in sea water (as described for the red alga *Palmaria palmata*; Pueshel 1977). After fixation, specimens are either transferred directly into a buffered 20–30% glycerol solution (Martinez-Paloma et al. 1976), or slowly infiltrated with 25–30% glycerol over a period varying from a few hours up to two days (Wunderlich and Speth 1972; Sattler and Staehelin 1974; Kitajima and Thompson 1977).

With *other isolated eucaryotic cells*, such as *tissue or monolayer cultures*, *spermatozoa* and *blood cells*, the same problems as those already discussed concerning fixation and cryoprotection become apparent if no fast freezing methods can be used directly (Plattner 1971; Plattner et al. 1973) and a variety of pretreatment procedures has been reported in the literature (Koehler 1970; Kouri 1971; Reith and Oftebro 1971; Pfenninger and Bunge 1974; Friend and Fawcett 1974; Flechon 1974). Again, diffusibility of glycerol into unfixed cells can vary considerably among similar cell types. For example, sheep red blood cells require several hours for proper infiltration whereas those of rat or humans reach an equilibrium state after a few minutes (Koehler 1970, 1972). For isolated cells which are sensitive to osmotic effects, 'complex' antifreeze media, containing hydrophilic substances with low osmotic activity, have been used successfully (Wecke 1968).

If specimens are cooled down before chemical fixation, then it is possible for phase separation to occur in the membrane lipids with concomitant development of areas devoid of intramembraneous particles as seen by freeze-fracture techniques (Tsien and Higgins 1974; Verkleij et al. 1975; Bayer et al. 1977). Thus if, for experimental purposes, pre-cooled samples are to be frozen, then structural comparisons with samples quenched from the normal growth temperature are desirable to avoid possible misinterpretation of membrane structure.

Isolated cell components can generally be frozen as a dense pellet or in suspension. With fast freezing methods such as 'sandwich' or spray-freezing, no cryoprotection of the specimen is necessary. Used with normal freezing methods, the ice crystal size can be minimised by adding between 5 and 30% glycerol to the suspending solution. Osmotically sensitive cell organelles, such as chloroplasts, vacuoles or mitochondria, are also often best prefixed with glutaraldehyde before infiltration with glycerol. Use fixation procedures similar to those for the cells and tissues from which the fraction is derived (Glauert 1974) or make up the fixative in the isolation medium. As discussed before, fixation and cryoprotection procedures should only be used if, by comparison with ultra-fast freezing methods, it can be shown that they do not produce unacceptable artefacts (Zingsheim and Plattner 1976). It is seldom realised that, with non-cryoprotected samples of membranes or cell organelles, the ice crystal size often does not affect the fine structure of membrane fracture faces and etched surfaces. If the material is suspended in very dilute buffer or distilled water, no solute/solvent phase separation can occur and even the formation of large ice crystals is not necessarily disturbing (Fineran 1970a; Southworth et al. 1975; Willison 1975, 1976).

2.3.3b   *Pretreatment of plant material*

Untreated, living plant material can only be successfully frozen, with very small ice crystals, if the water content is low enough. This only occurs in a few specimens, such as dormant spores (Schultz et al. 1973) or seeds and in cases of natural cold-hardening. As shown with pine needles (Leonhard and Sterling 1972), excellent cellular preservation is obtained when native cold-hardened tissue is frozen. However, there are considerable limitations to examining most non-cryoprotected living plant tissues (Fineran 1978).

In the absence of natural cryoprotection, it is possible to try to infiltrate living tissue with glycerol or to grow specimens in glycerol solutions. This second method has been particularly useful for studying root tips. For example, place *Allium cepa* bulbs with their bases in contact with 20% glycerol in tap water and grow them in the dark at 25°C for 4 days. From 0.5–1.0 cm long roots, cut out 200 μm thick, 1 mm long longitudinal sections for freezing (Moor 1964; Branton and Moor 1964). Similar studies on root tips of specimens grown in contact with glycerol have been reported by Northcote and Lewis (1968) and Hereward and Northcote (1973). They used germinated seeds of pea grown for 1–10 days on cotton wool soaked with 25% glycerol.

Fineran (1978) examined roots of *Avena sativa*, *Triticum vulgare* and *Zea mays*. Seeds are first germinated on moistened filter paper until the radicle emerges from the coleorhiza, at which stage they are placed on filter paper impregnated with 20% glycerol. Aeration is important, because roots immersed in glycerol often degenerate. To achieve optimal aeration, seeds are placed on top of narrow tubes of filter paper soaked in glycerol solution

(Fineran 1978). Sprouted seeds are incubated in 20% glycerol for 5–7 days, or for 3 days followed by 25% glycerol for about a further 3 days (Fineran 1970b, 1972, 1975). Root tips grown in contact with glycerol for 7 days show an essentially similar structure of their organelles, as compared with control roots, when examined by freeze-fracture replication or thin sectioning methods (Fineran 1970c). This does not hold for all plant material grown or incubated in glycerol solutions. Glycerol is osmotically active and when unfixed tissues are placed in osmotically unbalanced solutions they often plasmolyse.

The ability to metabolise glycerol is species-dependent and tissues of higher plants can suffer irreversible plasmolysis when exposed to glycerol. Even when cells deplasmolyse, serious changes in organelle morphology and membrane structure can occur (Moor and Hoechli 1970; Moor 1971). Controlled, slow infiltration with increasing concentrations of glycerol can prevent irreversible plasmolytic effects (Richter 1968a,b). A simple electronic apparatus for the slow admixture of glycerol has been described by Rottenburg and Richter (1969) but simple density gradient makers would serve the same purpose. Successful infiltration of unfixed pieces of tissue with cryoprotectant has been reported for leaf tissue (Hall 1967; Pyliotis et al. 1975), collenchyma strands (Chafe and Wardrop 1970; Hereward and Northcote 1973) and lichen (Ellis and Brown 1972). The method of choice strongly depends on the type of tissue and, consequently, many different methods of pretreatment involving different cryoprotectants, concentrations and infiltration procedures have been used (Fineran 1978).

Compared with glycerol, dimethyl sulphoxide (DMSO) is less suitable as a cryoprotective agent for living plant tissues or isolated cells. However, it is mentioned here because it may serve as a useful comparative cryoprotectant and, also, because the reader may find it recommended in some papers. It is often toxic at the concentrations needed to give adequate cryoprotection and it can also introduce specific artefacts (Pyliotis et al. 1975). Depending on the nature of the experiment, it is necessary to decide whether the potential physiological and structural effects induced by glycerol or DMSO can be tolerated. We recommend, in general, that glycerol is first used as the cryoprotectant whenever a new type of specimen is under study. With the possible exception of the application of non-penetrating cryoprotectants, there is no other way to induce frost-hardiness of tissue in the living state other than by cryoprotection. Thus, if cryoprotected living tissue is to be examined, a structural comparison of such samples with specimens that have been fixed with glutaraldehyde before infiltration with glycerol is advisable. This will help to distinguish the cryoprotectant-induced changes in cell structure. Structural studies on botanical material have also been reported where the formation of intracellular ice has not limited observations. Examples of specimens which have been frozen in their native state are: phloem (Johnson 1973, 1978), cellulose fibrils (Cox and Juniper 1973), tracheids (Puritch and Johnson 1973), water-conducting cells of moss (Hebant and Johnson 1976), roots (Robards et al. 1981) and germinated and ungerminated cereal grains (Buttrose 1971, 1973; Swift and Buttrose 1972, 1973; Barlow et al. 1973).

For tissues which cannot tolerate glycerol, or where unacceptably coarse structural alterations occur, mild chemical fixation prior to infiltration with glycerol is unavoidable. By comparison with other aldehydes tested, glutaraldehyde has been shown to be the most suitable fixative for both plant and animal material (Moor 1966). It increases the permeability of the plasmalemma and stabilises cell structure. A detailed description of the problems involved in processing botanical material for freeze-fracture replication procedures is given by Fineran (1978), who also discusses other structural changes introduced by glutaraldehyde fixation (such as the shrinkage of cell organelles). For some membranes, it has been shown that the general morphology is better preserved in glutaraldehyde-fixed tissue than in material grown or incubated in glycerol (Fineran 1978).

For fixation, glutaraldehyde is generally used in concentrations of 0.2–6% in buffered solutions of appropriate ionic strength and osmolarity for several hours. Alternatively, the fixative can be made up in glycerol solution, which allows simultaneous fixation and infiltration (Fineran 1978). Fineran (1978) found no marked differences in the quality of preservation of root tip cells between material fixed before glycerination and material that had been fixed and glycerinated simultaneously. For root tips, a fixation procedure of 3–6% glutaraldehyde in 0.025 M phosphate buffer for 1.5–2.0 h was used. Glutaraldehyde fixation prior to infiltration with glycerol has been used for a variety of botanical materials such as seeds (Barlow et al. 1973; Swift and Buttrose 1973), thick-walled tissue (Robards and Parish 1971), differentiating shoot cambium (Parish 1974), isolated protoplasts (Willison and Cocking 1975), infected tissue (Aist 1974), and tissue cultures (Sjölund and Smith 1974). Glutaraldehyde can be used successfully in concentrations as low as 0.2% (Sjölund and Smith 1974). Fixed specimens are usually infiltrated with 20–25% glycerol but in some procedures up to 50% glycerol has been used (Steere 1971).

In practice, the lowest concentration of glycerol capable of providing adequate cryoprotection should be used. To avoid the build-up of concentration gradients around specimens and to increase the rate of penetration until equilibrium is reached, samples should be agitated in the medium (e.g. by using a rotating tumbler). No general rule can be given about the optimal size of a sample for fixation and infiltration since this will depend strongly on the kind of tissue and the minimal size to which it can be trimmed without mechanical damage. The sample should not normally exceed a thickness of more than 2–3 mm at least in one direction and should preferably be *much* less than this. Similarly, although it is not possible to provide a pretreatment schedule that will be appropriate for all specimens, the information provided above will allow the beginner to make rational choices in the selection of all relevant parameters.

2.3.3c   Pretreatment of animal tissue

Since, with few exceptions, such as the *stratum corneum* of skin, the water content of animal tissues is high and even the best cooling methods only produce sufficiently small ice crystals in the peripheral 10–15 μm layer, cryoprotective pretreatments are again essential. As for many plant specimens, most animal tissues must be fixed before infiltration with glycerol to avoid necrosis or other structural changes induced by the cryoprotective agent.

Even now that chemical fixation before infiltration with cryoprotective agents such as glycerol or DMSO has become routine for animal tissue, such procedures must be seen as a compromise. Strictly speaking, they can only be considered 'safe', in terms of artefact formation if it is possible to obtain a comparison in the form of a sample that has received minimal chemical treatment. Even using methods that give good freezing in a peripheral zone, it is necessary to be aware that this well-frozen layer itself may have been damaged during the excision and/or trimming processes. Thus, as long as there are no methods available to give good freezing throughout pieces of untreated tissue, the use of chemical fixation in combination with cryoprotective agents has to be the compromise of choice.

As with plant material, brief fixation with formaldehyde or glutaraldehyde preserves the structure, renders the tissue permeable to glycerol (or other cryoprotectants) and allows infiltration without obvious shrinkage and plasmolysis.

If fixation is essential, it should be carried out as gently and as mildly as possible (Glauert 1974; Hayat 1981). Two main types of fixation procedure are possible. Firstly, small pieces of tissue (ideally not larger than 3.0 mm cubes and preferably much less) can be fixed immediately after they have been obtained (e.g. by biopsy) by immersing them in 2–3% buffered glutaraldehyde (Moor 1973; Southworth et al. 1975; Stolinski 1977; Willison and Brown 1979), or in Karnovsky's paraformaldehyde/glutaraldehyde fixative (Karnovsky 1965; Yee and Revel 1978). Alternatively, it is possible to fix the tissue *in vivo* by perfusion (Forssmann et al. 1975; Bullock 1979; Robards et al. 1981). This second method has the advantage that, for many tissues, better structural preservation is obtained if fixation starts while the animal is still alive. Preliminary treatments, such as sampling and glutaraldehyde fixation, are of great importance in determining whether fine structure is faithfully preserved. Furthermore, the correct conditions of pH, ionic strength and osmolarity, as determined for thin-sectioning methods (Glauert 1974), must be established and used.

Transfer fixed samples into a buffered 20–30% (v/v) solution of glycerol for between 1.0 and 5.0 h. Infiltration is generally best done at 4°C with frequent agitation to reduce the time necessary to reach equilibrium. Alternatively, the material can be transferred, after fixation, into a buffered 10% (v/v) solution of glycerol for 1.0–2.0 h and then, for the same time again, into a 25% glycerol solution. There has not been much systematic work published to show whether such stepwise infiltration of chemically fixed material improves structural preservation but we recommend it as a 'gentler' procedure which is less likely to lead to artefacts.

If artefacts associated with chemical fixation are to be assessed, then structural comparisons between unfixed and fixed/cryoprotected samples can be helpful. When tissues have not been stabilised by chemical fixation, it is preferable to infiltrate them with gradually increasing concentrations of cryoprotectant. This is usually carried out as rapidly as possible to minimise cryoprotectant-induced artefacts. If living tissue is infiltrated, special care must be given to providing the proper environmental conditions. For example Stolinski (1977) found it useful to buffer the cryoprotectant with a tissue culture solution to sustain tissue metabolism and viability until the sample was frozen. Especially for unfixed material, no general rule can be provided for the optimum infiltration time. Both the shortest practicable infiltration time and the lowest effective concentration of cryoprotectant must be established from studying morphological preservation. In practice Stolinski (1977) found, for example, that small cubes of rodent liver needed incubation periods of 1.0 h in 10% (v/v) glycerol, followed by 1.0 h in 25% (v/v) glycerol in a tissue culture solution. To minimise excessive tissue damage by osmotic shock, infiltration can be also performed using a continuous gradient.

Glycerol is generally superior to DMSO for infiltration of living tissue because it is less toxic at higher concentrations. Since even continuous infiltration of fresh animal tissue with the relatively non-toxic glycerol can result in gross structural alterations (Moor 1971; Echlin et al. 1977; Skaer et al. 1977), it is usually 'safer' to fix the material chemically before cryoprotectant treatment. This will particularly be the method of choice if no comparative studies on fixed and unfixed cryoprotected samples are to be carried out.

As well as penetrating cryoprotectants, non-penetrating compounds such as polyvinylpyrrolidone (PVP) and hydroxyethyl starch (HES) have been applied to animal tissues (Franks 1977; Schiller et al. 1978, 1979), although they can cause considerable cell shrinkage and consequent artefacts (Mazur 1970; Meryman 1971). They do, however, seem to have an important role to play in the range of cryoprotectants available. They are used at concentrations of between 20 and 30% made up in water or growth medium. As these solutions are osmotically active and as specimens to be treated with non-penetrating cryoprotectants are usually unfixed, exposure of the specimen to the cryoprotectant should be kept to a mimimum consistent with ensuring that the *whole* specimen surface is thoroughly coated with the cryoprotectant. At the concentrations used, the solutions are rather viscous and, therefore, some form of mechanical 'stirring', or other agitation, may be necessary.

2.3.3d   Ballistic methods of sampling and freezing

From much of what has gone before, it will be clear that not only should the cooling rate be maximised to ensure the 'best' freezing, but the method of obtaining the specimen prior to freezing will also have a strong bearing on the final quality of the preparation. The time interval between sampling and freezing should be as short as possible. One means of achieving this for samples from bulk material is the use of 'ballistic' methods of sampling: that is, to 'shoot' a sampler into or through the tissue to be examined. Where such methods have been employed, they have usually involved the rapidly sampled specimen being projected straight into the coolant, so that sampling and freezing occur within an extremely short time interval and as part of the same process. As an extreme of such methods may be instanced the hypodermic needle fired explosively by a modified 0.22 rifle that was used by Monroe et al. (1968) to acquire and freeze heart tissue (Fig. 2.36). Clearly, there are potential problems with such methods, not least those of

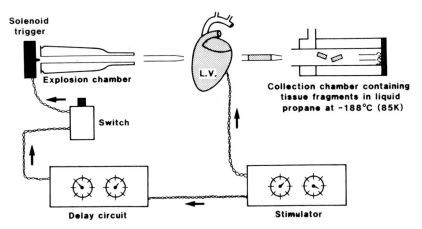

Fig. 2.36. Schematic diagram of the ballistic sampling technique used by Monroe et al. (1968) to collect and freeze cardiac tissue. The stimulus-inducing contraction in the isolated heart may be variably delayed to trigger the explosion, so driving the biopsy sampling projectile into the ventricle at any time during the contraction cycle. In the interval between leaving the heart and entering the propane the needle gets ahead of the tissue which thus cools rapidly in the liquid propane at $-188°C$ (85 K). (Redrawn from Monroe et al. 1968.)

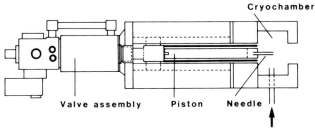

Fig. 2.37. A combined rapid sampling and freezing device designed by S.H. Chang and W.G. Mergner. This design supersedes a previous device (described by Chang et al. 1980). The four main component parts are illustrated. The 'cryogun' is connected to a supply of liquid nitrogen through flexible metal tubing. The trigger activates propulsion of the needle into the organ to be sampled; release of the trigger reverses the movement of the piston and brings about immediate withdrawal. A supply of cold nitrogen gas cools the insulated copper walls of the cryochamber so also cooling the needle and protecting the specimen from environmental changes; the slight positive pressure of the gas prevents the entry of moist air into the chamber. The needle is mounted on a PTFE piston and is designed to optimise cutting and removal of the tissue. Cutting is aided by rotation of the needle at 1200 rpm as it is projected forward. The propulsion mechanism is powered by dry nitrogen gas at 60–100 psi ($4.13–6.89 \times 10^5$ Pa). The temperature is monitored at the tip of the biopsy needle, within the copper mass of the cryochamber and within the cryochamber itself. Once the specimen has been obtained, it is removed using a special tool and is transferred to the chuck of a cryoultramicrotome where it can be sectioned. (Illustration redrawn from an original kindly provided by S.H. Chang and W.G. Mergner and reproduced with their permission).

Fig. 2.38.   Multiple sampling 'cryogun' described by Hearse et al. (1981) to obtain and freeze heart tissue. (Illustration provided by D.J. Hearse and reproduced with permission.)

percussion (pressure) shock to the cells and local frictional heating caused by the passage of the sample holder. However, it has been shown to be possible to extract a very small piece of tissue and rapidly freeze it with minimal damage.

Chang et al. (1980) have described a 'cryogun' (Fig. 2.37), which they have used for cryofixation of renal specimens prior to cryosectioning. This gun can 'cut' samples of 0.2–1.0 mm diameter and 0.5–1.0 mm long from the tissue to be sampled. The copper, needle-type specimen holder, cooled to liquid nitrogen temperature, is projected towards the specimen; the sample is cut out and immediately frozen by contact with the cold surface and is then retracted into the cooled body of the gun by a secondary spring. Using this device, the authors report improved cryofixation of tissue specimens with improved reliability of microanalytical data. There remains scope for further experimentation with such devices to minimise the possibility of post-mortem artefacts prior to freezing.

Hearse et al. (1981) employed a rather different approach in constructing a brass or stainless steel matrix of 70 cutters, each 2.0 mm × 2.0 mm in section (or 40 cutters of 4.0 mm × 4.0 mm) and 34 mm long, that was first explosively projected against the tissue to be sampled (heart) at 10^5 mm s^{-1}

(100 m s^{-1}) and then rapidly cooled to $-135°C$ (138 K) in a cold halocarbon (Fig. 2.38). While the cooling rate of such a device is low, it does have the advantage of sampling very quickly and of allowing minimal time between excision and freezing.

2.4 Freezing methods

A summary of rapid freezing methods is provided in Table 2.14 and a 'flow-diagram' for alternative preparation and freezing pathways is shown in Fig. 2.39. There are many publications reviewing rapid freezing methods, among which may be cited: Sitte (1979), Costello (1980), Elder et al. (1982), Escaig (1982), Plattner and Bachmann (1982), Plattner (1984) and Sitte (1984).

2.4.1 Freezing with liquid cryogens

The principles of freezing with liquid cryogens should now be clear but the selection of the most appropriate practical method and its proper application need to be carefully considered. Firstly, it must be stressed again that rapid cooling rates ($>1 \times 10^4$ K s^{-1}) can only be obtained throughout *extremely* small specimens ($\ll 100$ μm in one dimension) or within the surface 10–20 μm of larger ones. Consequently, if large (>0.1 mm in all dimensions) specimens are to be used, there is really very little point in applying ultra-rapid cooling methods because only the surface layer will cool quickly (as was demonstrated in Fig. 2.34 and Table 2.11 in §2.3.2). Such material will need to be cryoprotected and cryogens (such as liquid nitrogen at its equilibrium boiling point) that lead to severe film-boiling (§2.2.1) should be avoided. Liquid baths of sub-cooled nitrogen or halocarbons 12 or 22 should prove quite satisfactory. In addition, attempts should be made to ensure that the cooling is as consistent and reproducible as possible. This means that conditions should be recorded and standardised from run to run. One way of helping to achieve this is to use a cooling rate meter (Appendix 2), or some other device for monitoring cooling rates. This allows cooling conditions to be optimised for a specific method and also provides the possibility of checking that optimum rates are maintained from run to run.

As the next few sections demonstrate, liquid cryogens can be used in a variety of ways. However, there are some basic principles that apply for all applications. For example, the heat transfer characteristics of the cryogen

TABLE 2.14

A comparison of rapid freezing methods

Method	Description	Applications and advantages	Disadvantages
High pressure-freezing	Freezing at 2,000 bar (2.1×10^5 kPa) to sub-cool water so that the critical cooling range is much reduced and cooling rates of only 10^2 K s^{-1} are required for good freezing. Commercially available (Appendix 2)	Tissue blocks, suspensions, etc. Only method of obtaining well-frozen large (<1.0 mm) blocks of uncryo-protected hydrated tissue	Complex, expensive
Spray-freezing	Spraying fine droplets of suspensions, emulsions, etc. into liquid cryogens (e.g. propane). Equipment commercially available (Appendix 2; §2.4.1c)	Low viscosity solutions, suspensions, etc. Droplets ($\leqslant 0.1$ mm) or solid specimens ($\leqslant 0.05$ mm). Yields extremely good freezing results	Moderately complex apparatus required, although this can be made in any normal workshop. Only suitable for 'sprayable' specimens

Jet-freezing	Freezing specimens, usually sandwiched between thin metal plates, by impacting a jet of cold cryogen (usually propane) on one, or both, sides. Commercially available (Appendix 2; §2.4.1b)	Thin, flat, specimens such as cultures, monolayers, etc. Thickness of specimen $\leq$ 0.05 mm. Ideal for preparation of suitable specimens for freeze-fracturing	Complex and relatively expensive apparatus required. Jet may dislodge specimens
Plunge-freezing	Rapid immersion into liquid cryogen. The conditions of plunging must be optimised. Some appropriate apparatus available (Appendix 2; §2.4.1a)	Small tissue blocks. This technique is very versatile. Simple and inexpensive. Recommended as an initial procedure and in most circumstances where the highest cooling rates are unobtainable (because the specimen is too large).	Unless plunge conditions are optimised, poor cooling rates may be obtained. The surface may be rough and irregular and, therefore, the thin zone of well-frozen material may not be easy to examine by freeze-fracturing, cryosectioning etc.
Cold block-freezing	Rapid freezing by impacting a specimen onto the polished surface of a high thermal conductivity metal block at low temperature (77 or 4 K). Commercially available (Appendix 2; §2.4.2)	Gives highest heat transfer and yields excellent freezing in thin (10–20 μm) surface zone	Very complex and expensive apparatus required for routine use. Specimen may be deformed by contact with block. Liquid helium is expensive

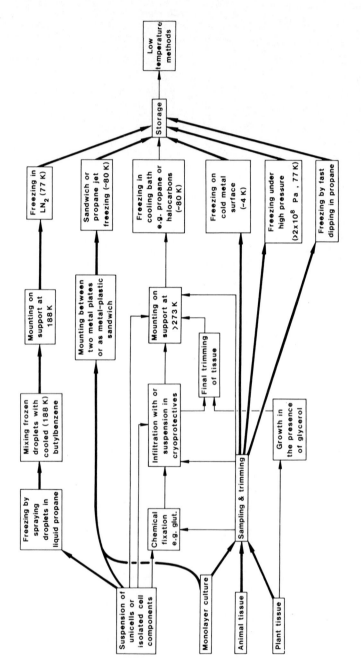

Fig. 2.39.   Scheme illustrating some of the many different pathways of biological specimen processing and freezing.

(and hence the cooling rate) are strongly dependent on the temperature difference between the specimen and coolant (ΔT) and the likelihood of film boiling is reduced if the temperature interval between the boiling point and the temperature of use is as large as possible. Therefore, liquid cryogens should generally be used at their lowest liquid temperature. Thus, even a difference of 10 K (e.g. 85–95 K) can have a significant effect on cooling rates obtained using liquid propane (Costello 1980; Fig. 2.40). In addition, it is always beneficial to present the specimen constantly with new coolant at its lowest temperature during the freezing process, either by plunging the specimen *through* the coolant or by 'squirting' the coolant over the speci-

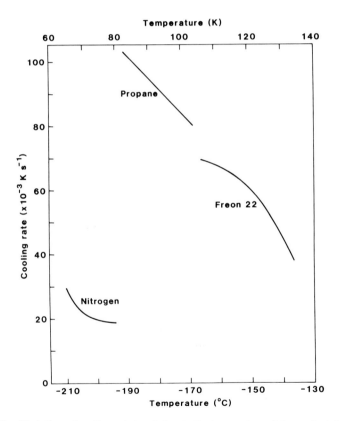

Fig. 2.40.   Variation of cooling rate in relation to the temperature of the coolant. Curves for propane, Freon 22 and liquid nitrogen are shown. A bare copper/Constantan thermocouple with a 70 μm diameter bead and 25 μm diameter wires was plunged 15 mm into the coolant at a rate of approximately 0.48 m s^{-1}. (Redrawn from Costello 1980.)

men. Other considerations will, of course, also apply but the following sections describe practical methods of achieving rapid cooling rates using cryogenic liquids that have proved satisfactory. Thus many potential coolants are eliminated on the grounds of their physical characteristics (such as too high a melting point, rapidly increasing viscosity as the freezing point is approached, too hazardous to use, or even too expensive). Furthermore, other coolants will also be ruled out because the complications associated with their use are not justified by the results obtained (liquid helium II appears to fall into this category although it also appears to be a poor coolant; §2.2.3a) or because they produce undesirable side effects (such as halogen contamination of samples by halocarbon cryogens prior to microanalysis) in relation to the particular experimental requirements. With such thoughts in mind, we can first look at the simplest and most widely used method of cooling: that of immersing the specimen directly in the liquid cryogen.

2.4.1a   Immersion method

The principles governing the selection of a suitable technique for rapid cooling by immersion should now be apparent (§2.2.5). Liquid nitrogen at its equilibrium boiling point is not a good coolant, but sub-cooled nitrogen, some halocarbons and cryogens such as propane or ethane may all be used. Current evidence suggests that, if the very fastest rates are to be obtained, then propane is the coolant of choice. We shall look at propane more closely later. However, we will first consider cooling methods where the requirement is that moderately sized specimens (perhaps cryoprotected pieces of tissue of about 1.0 mm^3) should be well-frozen under controlled conditions. In these circumstances, the very fastest cooling rates are not required and the use of either sub-cooled liquid nitrogen or one of the halocarbons condensed into a bath cooled by liquid nitrogen is recommended. Methods for cooling in liquid cryogens are reviewed, amongst others, by Umrath (1974b, 1975), Costello (1980) and Elder et al. (1982).

Sub-cooled liquid nitrogen. Liquid nitrogen is pumped down to the triple point of nitrogen (about 13.5 kPa, 1.01×10^2 Torr) and solidifies at a temperature of about 63 K. To obtain this transition, pour about 100–200 ml of liquid nitrogen into a well-insulated container (such as an expanded polystyrene cup) and evacuate this within a chamber (a desiccator can be used) pumped by a rotary pump through a relatively wide-bore tube (preferably $\geqslant 25$ mm in diameter; Fig. 2.41). Nitrogen 'ice' can form on the surface of the liquid, thus impeding further gas escape and causing the trapped

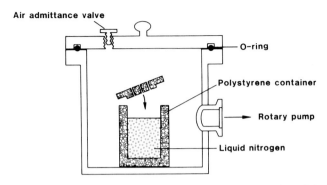

Fig. 2.41. A simple system for preparing liquid nitrogen 'slush' or sub-cooled nitrogen. A well-insulated (e.g. expanded polystyrene or polyurethane) container is placed in a small vacuum chamber. The chamber is evacuated through a relatively wide-bore tube (15–25 mm diameter). The liquid nitrogen firstly boils vigorously and then freezes. Air is readmitted quickly and the nitrogen remains sub-cooled for a considerable period, depending on the thermal properties of the container.

liquid to 'explode', so throwing liquid nitrogen over the interior of the chamber. This can be avoided by placing hollow drinking straws or 'anti-bump' granules in the container. Alternatively, the nitrogen can repeatedly be partially solidified and then reliquefied, by varying the vacuum, so allowing the whole pool to have sufficient time to reach the same temperature and freeze as an entity. As a further precaution, it is a good idea to place a perforated polystyrene lid over the container to trap any excessive splashing. When the nitrogen has solidified, the insulated container can be removed from the vacuum chamber.

As soon as air is leaked into the vacuum chamber, the nitrogen at 63 K starts to warm up leading, usually, to a mixture of liquid and solid nitrogen ('slushy' nitrogen): the mixture that is normally used for rapid cooling. Such a mixture is not in a state of stable equilibrium; there are strong temperature gradients and there may not be any solid nitrogen present, the pool simply being sub-cooled liquid. Consequently, there can be quite significant variation in cooling rates between runs when using this method. Quite apart from other considerations, the presence of solid nitrogen means that the heat of crystallisation is taken up before the whole pool becomes liquid (i.e. heat is accepted without warming the pool until the solid has all gone) and thus the thermodynamic considerations for the slush or for the sub-cooled liquid alone are rather different. If the sub-cooled nitrogen is contained in a well-insulated cup, then it will remain in the sub-cooled state for some period of time (a few minutes). However, as the temperature drifts upwards, the likelihood of significant film boiling during cooling becomes increasingly great and therefore the coolant should be used as soon as possible after removal from the vacuum apparatus.

A rather more convenient method of maintaining a liquid pool of sub-cooled nitrogen has been described by Umrath (1974a, 1975) in which an outer chamber of liquid nitrogen, which can be evacuated, surrounds an inner chamber which remains at atmospheric pressure (Fig. 2.42). Using this device, nitrogen can be maintained sub-cooled for prolonged periods, thus simplifying the consecutive freezing of a number of specimens. In all instances where sub-cooled nitrogen is used, it is advisable to check the temperature of the pool, either by noting the presence of solid nitrogen (63 K) or by using an electronic thermometer which has been calibrated for use at such low temperatures.

Halocarbons. The next most used group of liquid cryogens is the halocarbons, of which halocarbons 12 (CCl_2F_2) and 22 ($CHClF_2$) are the most popular. These gases become liquids at about 230–240 K and solids at about 114 K (Table 2.6; §2.2.3a) and they are therefore usually condensed into containers cooled by liquid nitrogen. The gases are available from suppliers (Appendix 2) in large cylinders or in small pressurised cans for laboratory use. The simplest way of obtaining the cooled liquid is to attach a piece of tubing to the container outlet and then to lead this tubing into the bottom of a container partially immersed in liquid nitrogen (Fig. 2.43). The con-

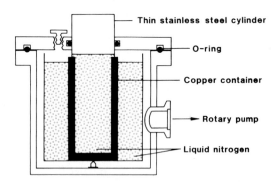

Fig. 2.42.   Apparatus for the continuous maintenance of sub-cooled liquid nitrogen (as described by Umrath 1974a). The outer vessel is evacuated, so sub-cooling the liquid nitrogen. This cools the liquid nitrogen in the inner container by conduction through the copper wall. Thus the inner vessel contains liquid nitrogen which is continuously maintained at low temperature (close to −210°C; 59 K) and can be used without the need for constantly breaking the vacuum as required for the apparatus illustrated in Fig. 2.41.

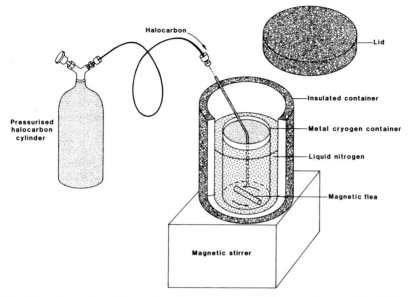

Fig. 2.43.   Preparation of halocarbons for use as cryogens. The pressurised gas is ducted from the cylinder and allowed to flow through a fine (approximately 1.0–2.0 mm diameter) tube into the metal cryogen container which is cooled by the surrounding liquid nitrogen. Gas flow must, initially, be rather slow so that there is time for the liquid to condense on the cold metal walls. Once a small 'pool' of liquid has condensed, the flow rate may be increased as the nozzle can be immersed in the liquid and further gas will condense directly. When sufficient cryogen has been condensed, the gas supply is turned off and the nozzle is removed. The cryogen is allowed to cool until it is close to its freezing point and the liquid is stirred using a magnetic stirrer or some other means to ensure that thermal gradients are minimised. Most cryogens that are used in this way will freeze solid when cooled with liquid nitrogen. This may be avoided either by adjusting (reducing) the level of the surrounding liquid nitrogen so that the heat withdrawal is insufficient to maintain the temperature below the freezing point or by inserting a small heater (a 'cartridge heater' (see Appendix 2) is suitable) into the pool of cryogen. An insulated lid is used to avoid excessive condensation of moisture into the cryogen.

tainer should ideally be made of a high-conductivity material such as copper. Gas is admitted until sufficient liquid has condensed in the container. Take care that liquid does not freeze in the end of the supply pipe, otherwise it may be blown out by increasing pressure. Subsequent cooling reduces the temperature of the halocarbon until, ultimately, it freezes. Immediately prior to using the cryogen, a high-thermal conductivity metal rod (brass or copper) is thrust into the solid halocarbon, thus causing the formation of a melted pool. As this pool begins to refreeze and is thus close to its freezing point, it is ready for use. As with sub-cooled nitrogen, cooling in halocarbons is not without practical problems. It is important that there should be a large enough liquid pool to allow an adequate depth of plunge of the specimen, but it is also important that all the liquid is very close to its freezing point. Using this method, there are bound to be thermal gradients and, again, it is difficult to use the same liquid pool to cool a number of specimens consecutively. Another consequence of plunging specimens into a pool that is re-freezing is that the specimen can become trapped in the solid cryogen. For these reasons it is useful if a system of maintaining the halocarbon at constant temperature is devised so that the excessive cooling by liquid nitrogen is offset by additional heat input. This can be achieved either by placing a small heater in the cryogen or by adjusting the contact area of the cryogen container with the surrounding liquid nitrogen. If this is done and some sort of stirring or agitation method is used to minimise temperature gradients within the pool (Fig. 2.43), the results will be far better and more reproducible. Some workers have achieved good results by very rapidly *raising* the container of cold halocarbon up to the specimen (e.g. Somlyo and Silcox 1979; Wendt-Gallitelli 1984). It is good practice to use halocarbons in a fume cupboard to avoid the build-up of high levels of vapour which may be toxic (Taylor and Harris 1970).

Propane. Propane (or other cryogens with similar properties, such as ethane) can be condensed in exactly the same manner as the halocarbons but, because of its lower freezing point, the gas flow should be slower. It is useful to partially cool the incoming propane gas by passing it through a copper coil immersed in liquid nitrogen (Fig. 2.44). *It cannot be stressed too strongly that propane is potentially an extremely hazardous chemical.* On the other hand, this should not cause its considerable virtues as a coolant to be ignored. *Under proper conditions*, there is no reason why propane should not be used both safely and routinely. The major hazard lies in the flammability of propane and the potentially explosive mixture that it may form with liquid oxygen which, with a boiling point of 90 K, may condense from the air into other liquids at lower temperatures. There are reports that ultrasonic devices may provide sufficient energy to cause liquid propane to explode and, therefore, they should not be used adjacent to such coolant baths. Clearly, simple precautions, such as avoiding naked flames, using spark-protected switches, using the minimum quantities required and minimising the time for the pool of liquid propane to 'stand around', will all help to minimise any risk. During the condensation phase there should be little risk of propane gas escaping provided that the flow rate from the supply is kept low. In addition, the relatively warm incoming gas causes the surrounding liquid nitrogen to boil and thus to form a protecting 'blanket'

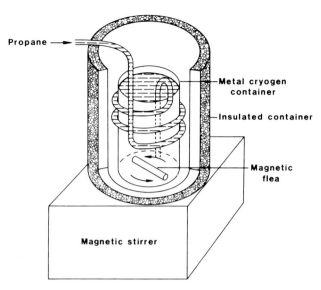

Fig. 2.44. Liquefaction of propane for use as a cryogen. The method is similar to that illustrated in Fig. 2.43 for halocarbons but, because the melting point of propane (approximately -190°C; 83 K) is lower, it is necessary to provide additional cooling for the incoming gas, usually by leading it through a coil of metal (copper) pipe that is immersed in the liquid nitrogen. Propane usually stays liquid for quite a long period when cooled with liquid nitrogen (especially if 'commercial', rather than pure, propane is used) and it can be maintained in the liquid state by the addition of 20–30% isopentane (Fig. 2.28; §2.2.5). Although not shown, an insulating lid is essential (as in Fig. 2.43). Cold propane should not be left standing for prolonged periods due to the danger of excessive amounts of liquid oxygen condensing into it, thus forming an explosive mixture (see text).

of inert nitrogen gas over the top of the propane container. Similarly, when the propane is close to its usable temperature ($\simeq 85$ K) its vapour pressure is so low and it is so far removed from its flash point (-104°C; 169 K) that the hazards are minimal. The major period for concern arises when the propane is to be disposed of: at this stage the propane must necessarily be warmed up; the vapour pressure will rise and the flash-point will be reached. In our experience, small quantities of liquid propane may be left safely to warm up and evaporate in a fume-cupboard provided that the fan motors and any other adjacent electrical equipment are spark-protected. Alternatively, the liquid may be carried outdoors and disposed of by pouring onto open ground well away from buildings, vehicles or any other object. Precautions should be taken by the operator to see that he stands up-wind of the propane as he pours it. It *is* possible to store the propane by immersing it completely in liquid nitrogen and allowing it to freeze prior to storage in a suitable liquid nitrogen-refrigerated vessel. However, this process should *not* be repeated due to the danger of liquid oxygen condensation into the propane during consecutive cycles. The problem of liquid oxygen condensation during normal use is probably small provided that the propane container is kept covered as much as possible and that the cryogen is not left standing around for excessively long periods. The constant flow of nitrogen gas over the mouth of the propane container also impedes the access of oxygen gas to the liquid. We can see, therefore, that propane is a perfectly reasonable cryogen to use and, indeed, it has some practical advantages apart from its superiority as a coolant described above (§2.2.5). For example, its low melting point means that it remains liquid in liquid nitrogen for long periods (thus alleviating the specimen-trapping problem that arises with the halocarbons). Furthermore, by the addition of up to 30% isopentane, the melting point of propane is reduced below that of liquid nitrogen although the coolant properties are said to remain unaffected (Jehl et al. 1981; Fig. 2.28 in §2.2.5).

Liquid helium II. Although liquid helium II is also a liquid coolant, its use and application is quite different from that of the cryogens described above, which are all used at liquid nitrogen temperature or above. Fernández-Morán (1960) has described the production of liquid helium II by rapidly lowering the pressure on liquid helium I to about 70 μm (9.3 Pa), which is equivalent to a temperature of 0.8 K. Because of the low heat of vapourisation of liquid helium, relatively large quantities must be used and the helium Dewar vessel must be surrounded by a liquid nitrogen-filled chamber. Furthermore, because the helium II must be maintained under vacuum, the specimen must be inserted via a vacuum lock. This method has not be widely used in recent times (but see Bullivant 1960) because it is not a very good coolant (§ 2.2.3a) and is very expensive. Reference to it is included here because its possible virtues are often quoted and it is important that the reader should understand the actual situation.

Methods of immersion in liquid cryogens. So far, only methods for obtaining liquid pools of cryogens have been described. However, equally important are the means whereby the specimen is plunged into the pool and the handling of specimens both before and after the freezing process. Most specimens need to be mounted on some form of support or carrier. Much will depend on what is to be done to the material once it has been frozen. For example, the requirements for mounting a specimen in a freeze-fracture unit are different from those for a cryoultramicrotome or for freeze-substitution or freeze-drying. Recommendations for specimen supports will be given in the Chapters relevant to the specific techniques. It is sufficient to comment here that the supports should have the minimum possible mass and should be of high thermal diffusivity. Conversely, the means of actually holding onto the specimen, together with its support, during freezing should allow the least possible heat conduction into the cooling specimen. In general, metal holders of low thermal conductivity and small cross-sectional area should be used. For example, very fine stainless steel tweezers are good for this purpose. Copper has probably been used more than any other metal for the specimen supports themselves and there are no obvious reasons for preferring gold or silver to copper (Table 2.10; §2.2.6)

Handley et al. (1981) drew attention to the critical importance of the thickness of any metal specimen support across which heat needs to be withdrawn during rapid cooling. As the effective time interval between cooling the outer and inner surfaces of such a support is proportional to the square of its thickness, it is clear that it should be as thin as possible. Unfortunately, metals with good heat transfer characteristics (e.g. gold and silver) are not very strong. Consequently, it is often not practically possible to use them as sheets less than 25–50 μm thick. Handley et al. used titanium which, while having thermal properties inferior to copper or silver (Table 2.10), has better tensile strength (Ti = 860 MPa; Cu = 211 MPa; Ag = 130 MPa) and was usable as foil envelopes only 4.0 μm thick. The time taken for heat to be transferred across a thin metal layer was calculated (Carslaw and Jaeger 1959) from:

$$t = \frac{0.36 \, d^2 \, \rho c_p}{k} \qquad (2.14)$$

where

t	= time interval for 100 K temperature decrease
d	= thickness of foil
ρ	= density
c_p	= specific heat
k	= thermal conductivity.

Fig. 2.45 illustrates the relationship between the thickness of copper or titanium sheets and the time taken for a 100 K temperature change to pass from one side of a sheet to the other. Using figures for thermal conductivity and specific heat at 25°C (298 K), Handley et al. (1981) found that it would only take 0.62 μs for a 100 K temperature change across a 4.0 μm thick titanium layer, equalling a rate of 1.6×10^8 K s^{-1}, which is well above the criti-

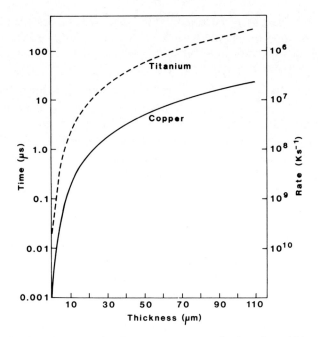

Fig. 2.45.   The time taken to conduct heat across copper or titanium sheets. This graph shows curves obtained by using Eq. (2.14) to calculate how long it takes for a 100 K temperature change to pass from one side of a metal sheet to the other. The better thermal conductivity of copper gives significantly shorter times although titanium can be used in thinner sheets.

cal cooling rate necessary for the formation of sufficiently small ice crystals and, therefore, not limiting to the ultimate specimen cooling rate that could be achieved by their methods. From Fig. 2.45 it can also be seen that the time taken to cool 100 K across a 50 μm titanium sheet would be about 1.0 ms (1.03×10^5 K s^{-1}) whereas across a copper sheet of equivalent thickness, the time would only be 7.6 μs (1.3×10^7 K s^{-1}): two orders of magnitude faster. Fig. 2.45 is thus useful in providing an estimate of the effect of the thickness of the specimen support on the highest cooling rate that can be obtained within the specimen itself.

The major requirements during the sampling and freezing stages are (ignoring the coolant characteristics):

(*i*)   the specimen should be damaged as little as possible, either by mechanical means or by desiccation;

(*ii*)   the specimen should be as small as possible;

(*iii*)   if mounted, the support should be of high thermal diffusivity and low mass;

(*iv*)   the specimen should not be allowed to pre-cool in cold gas above the liquid bath (Ryan and Purse (1984) have discussed this point) and

(*v*)   the specimen should be plunged into the coolant as rapidly as possible.

The first three points have already been discussed but the last two aspects require further consideration.

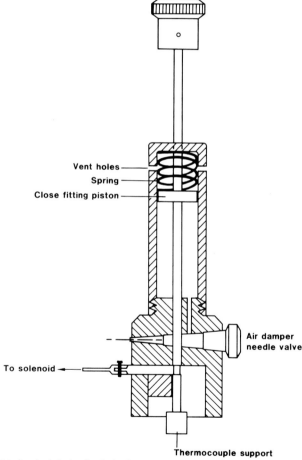

Fig. 2.46.   Mechanical device for 'injecting' specimens into liquid cryogens. The spring-powered injector is 'damped' by using a piston to force air out of the cylinder through an adjustable needle valve. (Redrawn from Glover and Garvitch 1974.)

When the specimen has been obtained and mounted on a support (if appropriate) it should be frozen as quickly as possible. Over any pool of very cold liquid there will be cold air which will pre-cool the specimen and may actually freeze it at very slow cooling rates if precautions are not taken to avoid this. Such precautions take two forms: firstly to minimise the layer of cold air above the bath, and secondly to move the specimen into the bath as rapidly as possible. In general, it is preferable to maintain the liquid level as close to the top of the container as possible and, except at the moment of freezing, to keep the container covered with an insulated lid. In this way

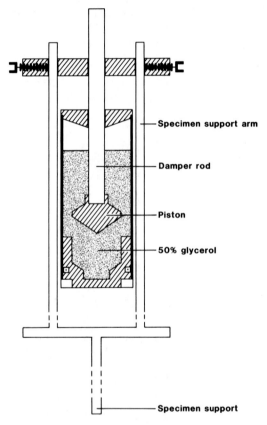

Fig. 2.47. A liquid damping system for specimen injectors. This system, as opposed to the air compression mechanism illustrated in Fig. 2.46, uses glycerol in a 'dash-pot' to damp the progress of the specimen rod into the cryogen. The specimen support and support arm are carried on a plate which is integral with the damper rod. When the specimen falls, the rate of movement is retarded as the piston moves through the glycerol. (Redrawn from Costello and Corless 1978.)

the coolant is kept at a low temperature and there is minimal depth of dense, cold gas above the liquid surface.

The method of inserting the specimen into the coolant is one that has only comparatively recently been given much attention. However, it does now seem clear that, within reason, higher rates of entry are advantageous (Costello 1980; Handley et al. 1981; Robards and Crosby 1983). Early techniques involved 'dipping' the specimen into the coolant bath. This method is *not* recommended: firstly, the entry rate is bound to be rather slow, and secondly such hand-held methods produce very inconsistent rates. Any serious work should be undertaken using some form of injector and/or stirrer mechanism that allows predetermined entry velocities to be obtained reliably and

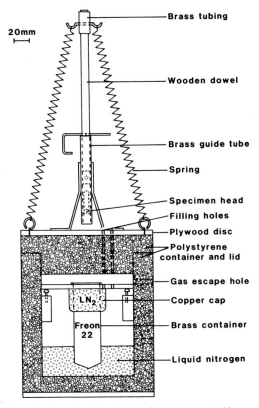

Fig. 2.48. Spring-assisted mechanism for injecting specimens at up to 10 m s⁻¹ into liquid cryogens. The cryogen used here is carefully cooled so that its surface is as cold as the bulk of the liquid; the descending specimen thus meets the coolant at its lowest temperature. (Redrawn from Handley et al. 1981.)

reproducibly. Many such devices have been described and some are illustrated in Figs. 2.46 to 2.50. Luyet and Gonzales (1951) used a 'guillotine' for rapid specimen insertion. Rebhun (1972) describes a small harpoon-like device that uses a rubber band to fire the specimen into the coolant. Glover and Garvitch (1974) built a mechanical injector (Fig. 2.46) that shot the specimen into the coolant at an average velocity of 2 m s^{-1} and simultaneously triggered the oscilloscope as it fired. Costello and Corless (1978) constructed another guillotine (based in part on the work of Heuser et al. (1976) for copper block freezing; Fig. 2.62; §2.4.2) that gave unassisted entry rates of 0.39 m s^{-1} and variable rates of 0.49–0.74 m s^{-1} when energised by rubber bands; it was damped using a glycerol-filled dash-pot (Fig. 2.47). Handley et al. (1981) produced rates up to 10 m s^{-1} with a spring-actuated device (Fig. 2.48); Echlin and Moreton (1976) described a free-fall injector giving an entry speed of 3 m s^{-1} and Bald and Robards (1978; see also Bald 1978a) also used a gravity-driven device giving a plunge velocity

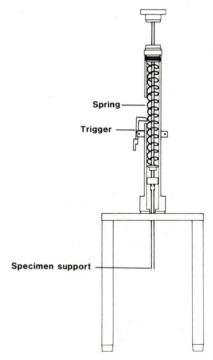

Fig. 2.49.   Simple spring-assisted mechanism for plunge-cooling specimens. Use of different spring strengths allows different entry velocities up to about 8.0 m s^{-1}. (Reproduced, with permission, from Robards and Crosby 1983.)

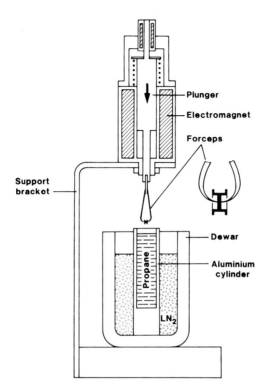

Fig. 2.50. Solenoid-operated rapid immersion device. The specimen is held between very fine 'pincers' (tweezers) to minimise thermal contact. Voltage variation of the electromagnet provides a variation in immersion velocity between 0.4 and 3.0 m s^{-1} with almost constant acceleration (200 m s^{-2} maximum). The stroke is 22 mm and the specimen is immersed 15 mm into the coolant. (Redrawn from Escaig 1982.)

of 5 m s^{-1} independently of springs, rubber bands, or other auxiliary systems. Numerous similar devices have been described. A simple spring-assisted injector giving rates between about 2.0 and 8.0 m s^{-1} is illustrated in Fig. 2.49 and an electric solenoid-driven plunger designed by Escaig is shown in Fig. 2.50. An elegant device for rapid plunging has been described by Sitte (1984) and is commercially available (Appendix 2). The specimen assembly should be 'held' so that the lowest possible pathway for heat transfer into the specimen is provided: fine stainless steel forceps or wire give sufficient mechanical rigidity with relatively low thermal conductivity (see Figs. 2.50 and 2.52 for examples of specimen holders). Summarising the available information, it is possible to construct a list of conditions that will

TABLE 2.15
Some criteria for efficient cooling in liquid cryogens

(*i*)	Specimens should be as small as possible
(*ii*)	Specimen supports should be as small as possible, of low mass and high thermal diffusivity
(*iii*)	For most purposes propane is the best liquid cryogen
(*iv*)	The coolant should be at its lowest liquid temperature
(*v*)	The coolant pool should be stirred to avoid temperature gradients
(*vi*)	The coolant vessel should be filled to the top (to minimise pre-cooling in the vapour phase)
(*vii*)	The entry rate of the specimen into the coolant should be high (>1.0 m s^{-1})
(*viii*)	There should be a sufficient depth of liquid to allow the specimen to keep moving at the initial entry velocity, at least over the early part of the critical cooling range (effectively 273–173 K)
(*ix*)	The specimen should be the first part of the injection assembly to contact the coolant
(*x*)	The specimen should be supported and 'held' using low mass, low thermal conductivity contacts

help to produce the most efficient cooling method available using the method of plunging into liquid cryogens (Table 2.15).

2.4.1b   Propane jets

In the previous section, the importance of moving the specimen relative to the cryogen was stressed. One of the important reasons for doing this is to ensure that the temperature of the coolant effectively remains the same during the whole cooling process. If this is not done, then the coolant adjacent to the specimen warms-up and may even boil, thus producing an insulating gas jacket. An alternative approach is to keep the specimen static but to project the coolant over it: this is the method of 'jet-freezing' which was first introduced by Moor et al. (1976) and subsequently described in greater detail by Müller et al. (1980). These authors again chose propane as the cryo-

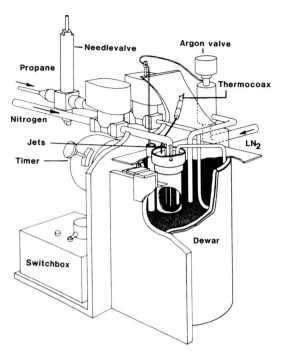

Fig. 2.51. Propane jet-freezing device designed by Müller et al. (1980). The purpose of the apparatus is to provide an opposed pair of high velocity jets of cold liquid propane, between which the specimen (see Fig. 2.52) can be thrust. This apparatus is commercially available (Appendix 2). (Reproduced, with permission, from Müller et al. 1980.)

gen. The principle of jet-freezing is simple but a specially constructed apparatus is necessary for its application (Fig. 2.51) and is commercially available (Appendix 2).

The propane is cooled down to working temperature (which should be < 93 K) in a vessel cooled by liquid nitrogen (Fig. 2.51). At the appropriate moment, the specimen (which is usually sandwiched between very thin copper plates; Fig. 2.52) is inserted between the jet nozzles. This is done using a special specimen holder. In the commercial equipment, this movement triggers pressurisation of the propane container which propels the cold liquid propane onto either side of the specimen. The propane must be at as low a temperature as possible prior to each run. This means that there must be some slight delay while the propane re-cools after consecutive runs. The atmosphere between the jets is bound to be cold and it is therefore essential that the specimen is brought extremely quickly between the jets and

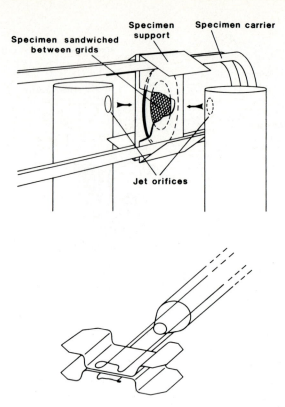

Fig. 2.52. Specimen holders for use with a propane jet-freezing device. A pair of ordinary electron microscope grids 'sandwich' the specimen. In the system shown in the upper diagram, the grids themselves are mounted between a pair of shaped copper supports, with central holes, so that the whole assembly can be transferred to a freeze-etching apparatus. Other arrangements may also be used as, for example, shown in the lower diagram. The two jet orifices must be exactly aligned horizontally opposite each other. The specimen holders are, in both the systems illustrated, approximately 3.0 mm across. (Upper diagram kindly provided by M. Müller and reproduced with his permission. Lower diagram redrawn from Robards and Severs 1982.)

that the propane stream is triggered immediately. If this is not done, there is a real risk of slow pre-cooling. The commercial equipment enables these criteria to be satisfied easily and routinely. Finally, when the jet has stopped, the very small, thermally sensitive, specimen will start to warm up quickly and it is therefore important to transfer it into a liquid nitrogen storage container as quickly as possible.

While the commercially available propane jet-freezing apparatus provides all the requirements for this technique, it is also possible for a well-equipped

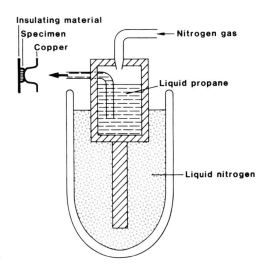

Fig. 2.53. Simple single-sided jet-freezing. Cold liquid propane is projected against one side of the specimen, the other side being made from a thermally insulating material to reduce heat input during freezing. (Redrawn from Pscheid et al. 1981.)

workshop to construct suitable equipment. There is no reason why this should not be accomplished successfully, but if it is tried it is imperative that due consideration is given to both the cryogenic and mechanical requirements of the apparatus. The points mentioned above must be met and features such as insulation around the jet orifices, exact alignment of jets and very quick response times to trigger the pressurisation must be optimised.

Pscheid et al. (1981) and Knoll et al. (1982) have used a rather simpler propane jet than the one described above to project a single stream of propane against *one* side of a specimen (Fig. 2.53). This equipment has the advantage that precise alignment and synchronisation of two jets is avoided and the actual 'gun' is much easier to construct (Fig. 2.53). It is therefore more easily applicable for use in laboratories where, either from choice or inability to purchase expensive equipment, the two-sided jet apparatus is not available. The performance of this system appears good: cooling rates of $> 18,000$ K s^{-1} have been recorded, which compare well with those obtained from two-sided jets. The explanation for this is partly that, in practice, *precise* synchronisation of double-sided jets is probably not possible and, in addition, the thermally insulating backing to the specimen in the single-jet system minimises heat input from this side while heat is being rapidly withdrawn from the other.

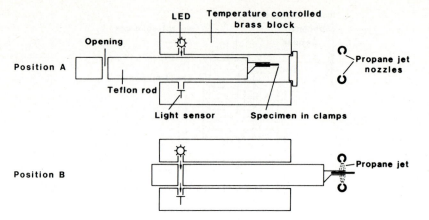

Fig. 2.54.   System for maintaining specimens at controlled, known temperature immediately prior to freezing. The arrangement illustrated here is designed for use with a propane jet device but could easily be adapted for other applications. When the specimen is thrust forward, the light from the light emitting diode (**LED**) falls onto the light sensor and thus triggers the jet of propane. (Redrawn from Van Venetië et al. 1981.)

Van Venetië et al. (1981) made a simple modification for the Müller et al. (1980)-type propane jet by arranging a specimen stage that would maintain temperatures around the specimen up to 330 K immediately prior to rapid cooling (Fig. 2.54). This allowed them to study liquid-solid phase transitions for some phospholipids. Although designed for use with a propane jet apparatus, a similar system could equally well be used with other rapid cooling methods.

While excellent results have been achieved using propane jet-freezing, there is little reason to believe that optimised plunging into liquid propane (as described in §2.4.1a) cannot produce equally acceptable results. The apparatus required is both simpler and cheaper and those starting in the field of cryobiology are advised to experiment with such systems before deciding whether they need to resort to more sophisticated devices.

2.4.1c   Spray-freezing

It has been repeatedly stressed that, to obtain the highest cooling velocities, the specimen size should be minimised. With emulsions and homogeneous suspensions of small particles, it is possible to freeze very small microdroplets (10–50 μm diameter) which will be representative of the bulk sample. The most effective way of producing such droplets and freezing them in the

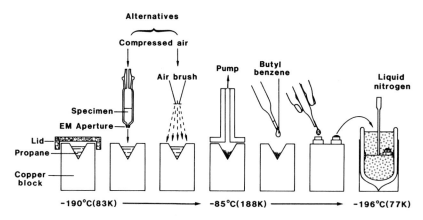

Fig. 2.55. The spray-freezing technique. Propane is condensed into a cold copper or brass block at about $-190°C$ (83 K). Using an air brush or a 'jet' sprayer, the liquid specimen (suspension, emulsion, etc.) is sprayed into the coolant. The temperature of the liquid propane is raised to about $-85°C$ and is then evaporated by connection to a rotary vacuum pump. The frozen droplets are then mixed into a 'slurry' with butyl benzene or some alternative substance having a melting point around -90 to $-100°C$ (§4.2.1; Table 4.1). Some of this slurry is transferred to the specimen holder, after which it is re-cooled to liquid nitrogen temperature. This technique has mainly been applied to freeze-etching (§6.2) but could equally well be used in other low temperature applications. A spray-freezing system is commercially available (Appendix 2). (Adapted from various illustrations published by the originators of this technique: Bachmann and Schmitt 1971.)

same process is by the use of an atomiser. This technique was pioneered by Bachmann, Plattner and their colleagues (Bachmann and Schmitt 1971; Bachmann and Schmitt-Fumian 1973; Plattner et al. 1973; Plattner and Bachmann 1982), primarily as a preliminary to freeze-fracture methods, although there is no reason why it should be so restricted. The freeze-fracture aspects of the method are dealt with more fully in Chapter 6 and only the freezing technique will be discussed here.

The apparatus is essentially rather simple (Fig. 2.55) and, while it can be relatively easily constructed in the average workshop, a fully equipped form is commercially available (Appendix 2). It comprises an artist's air brush which is used to spray a fine 'mist' of microdroplets (typically about 10 μm diameter) into a small container of liquid propane at about 83 K. The cooling rates are extremely high and, by small modifications to the size of the atomiser orifice, it is possible to spray cell fractions, micro-organisms and isolated cells.

The originators of the spray-freezing method used a Grapho air brush type I at pressures between 0.5 and 1.5 atmospheres (50–150 kPa); an Aerograph 'Super 63' Model A-504 retouching brush operated at pressures between 40 and 60 kPa has also been found suitable for atomising liquids (Lang et al. 1976) and other combinations would undoubtedly also be acceptable (Appendix 2). Ververgaert et al. (1973) modified the specimen loading arrangements so that volumes as low as 0.05 ml could be used. There must be enough suspending liquid to 'wrap' each isolated cell in a spray droplet and so the final ratio of specimen to suspending medium is important. This is usually about 1:1 (medium to pellet).

The small propane container is covered with a plate, in which is a hole slightly smaller than the top of the container: this prevents the ice, which forms from the air during spraying, falling into the cup (Fig. 2.55). Subsequently, the propane is evaporated, using the vacuum of a rotary pump, so leaving the rapidly frozen microdroplets in the cooled block and available for further processing.

The distance between the mouth of the spray gun and the coolant should be 50–100 mm and the sample (typically 0.2–0.3 ml) is sprayed at intervals of about 1.0 s to minimise warm-up of the coolant. Plattner et al. (1973) used the spray-freezing technique successfully to cool relatively large isolated cells: this was achieved by using low pressure (0.5 bar; 50 kPa) and, for *Paramecium* (30–40 μm diameter, 120–150 μm long), by using a 50 μm electron microscope aperture positioned in the mouth of the spray gun.

A small modification of the spray-freezing technique allows 'jet-freezing' whereby a continuous jet (rather than individual droplets) of suspension is propelled into the coolant (Fig. 2.55). A simpler 'gun' is required and the jet is formed by using a pressure of 1.0–2.5 bar (100–250 kPa) to propel the suspension through an electron microscope aperture of between 10 and 50 μm. It is necessary to move the jet about over the surface of the coolant as it is used so that undue warming of the coolant is minimised.

2.4.1d   High pressure freezing

As described in §2.1.2, when water is subjected to a pressure of about 2100 bar (2.1×10^8 Pa), its freezing point is lowered to about $-22°$C (251 K). The critical cooling rate for vitrification is then reduced to 2×10^4 K s^{-1} for water and to about 1×10^2 K s^{-1} for biological specimens. Thus, if it is possible to construct an apparatus that allows cells to be subjected to this high pressure and then rapidly cooled, a very significant improvement in freezing should be obtained. This principle was first elaborated in detail by Riehle

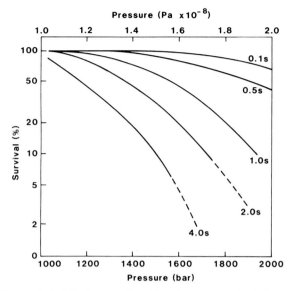

Fig. 2.56. The survival of *Euglena gracilis* subjected to pressure for different periods of time. (Redrawn from Riehle and Hoechli 1973.)

(1968a,b) who found that it was possible to produce very small (5–10 nm diameter) ice crystals when freezing 5% glycerol solutions. The crystallisation temperature fell by about 30 K and the speed at which the ice crystals formed decreased by a factor of 10^3.

One of the major problems in the design of a suitable apparatus is that most biological cells are very rapidly damaged by exposure to such high pressures (Moor and Hoechli 1970; Fig. 2.56) and, therefore, it is necessary to apply the pressure for only a millisecond or so before the cooling process commences. Riehle (1968a,b; Riehle and Hoechli 1973) described a suitable apparatus which was subsequently modified and has been used by Moor et al. (1980) to study the influence of high pressure freezing on mammalian nerve tissue. Although this apparatus is commercially available (Appendix 2), it is expensive; it is also unlikely to be within the capability of most laboratories to construct a copy of the published version. The technique is nevertheless potentially so important that its description is warranted here. It provides the only current or forseeable possibility of producing small (< 10 nm diameter) ice crystals at depths of up to 0.5 mm within a specimen.

The original version of the apparatus used a very thin-walled brass tube as the specimen holder while the later version described by Moor et al.

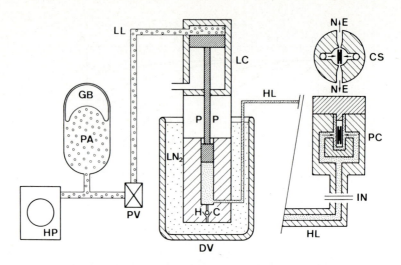

Fig. 2.57.   Diagram of a high pressure-freezing apparatus. **DV**, dewar flask; **GB**, gas balloon; **HC**, high pressure cylinder; **HL**, high pressure line; **HP**, hydraulic pump; **IN**, position of the injection chamber (see Fig. 2.58); **LC**, low pressure cylinder; **LL**, low pressure line; **PA**, pressure accumulator; **PC**, freezing chamber containing a specimen sandwich (Fig. 2.58); **PP**, pressure piston; **PV**, pressure valve. Small circles indicate oil and dots liquid nitrogen. The cross-section (**CS**) at the upper right is of the high pressure chamber showing the narrow apertures (**NE**) for gas escape. (Reproduced, with permission, from Moor et al. 1980.)

(1980) used a more conventional freeze-fracture specimen sandwich (Figs. 2.57 and 2.58). In this system, liquid nitrogen is used as both the coolant and the pressurising agent so that a pressure of 2100 bar (2.1×10^8 kPa) is maintained while 100 ml of liquid nitrogen is forced into the specimen chamber through two 0.3 mm diameter apertures within 0.5 s. Temperature fall (to 173 K) and pressure rise (to 2.1×10^8 kPa) occur simultaneously within 30 ms unless precautions are taken to delay cooling until pressurisation is complete. This is achieved by injecting a little warm isopropyl alcohol as a pressurising medium immediately prior to the liquid nitrogen. Thus, about 2.0 ml of isopropyl alcohol is forced into the specimen chamber as the pressure develops, after which the high pressure liquid nitrogen completes the cooling process. This systems allows the full pressure to be achieved and maintained prior to cooling but for a sufficiently short time so that, according to the authors, damage from pressurisation or contact with the alcohol is avoided. It is clear that, provided the technical complexities can be overcome and cells can be shown to withstand high pressure for the necessary time, this is a technique worthy of further study.

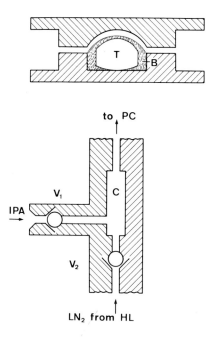

Fig. 2.58. Top, profile of a specimen sandwich for the high pressure freezing device (Fig. 2.57). **B**, droplet of buffer solution; **T**, tissue block. Below, injection chamber (**C**) for the uptake of isopropyl alcohol (**IPA**). This chamber is inserted into the high pressure line at its entry into the high pressure chamber (**PC** in Fig. 2.57). V_1 and V_2 are pressure valves for regulating the injection of isopropyl alcohol which precedes liquid nitrogen in order to delay the onset of cooling. (Reproduced, with permission, from Moor et al. 1980.)

2.4.1e Liquid paraffin method

Buchheim (1972, 1977) and Buchheim and Welsch (1978) have described a method of rapidly freezing small (< 30 μm diameter) globules by emulsifying them with liquid paraffin. Ice crystals of less than 50 nm diameter were reported. The technique involves mixing about 0.5 ml of the tissue or liquid with about 8.0 ml of viscous paraffin containing 5% glycerol mono-oleate, or 1–3% of a liquid polyglycerol ester, as an emulsifying agent. This mixture is then homogenised in a normal laboratory apparatus (e.g. Ultra-Turrax or Omni-Mixer, see Appendix 2 for manufacturers and suppliers). The intensity and duration of the homogenisation should be sufficient so that most globules are 5–10 μm in diameter, something that can easily be checked using a light microscope. Buchheim used small quanitities of the homogenate on standard freeze-fracture holders prior to freezing in Freon 22 but other, better, methods of cooling could obviously also be used. It is suggested that the reason for the production of very small ice crystals is that the differential thermal contraction of the two phases leads to the production of increased pressure within the small aqueous droplets and hence gain from the lowering of the freezing point so brought about (as for high pressure freezing; §2.4.1d).

2.4.1f   Droplet method

The principle of trying to reduce the specimen size so that high cooling rates can be achieved has been approached in a novel way by Menold and his colleagues (Menold et al. 1974, 1976; Menold and Lüttge 1979). This is the so-called 'droplet method' in which a droplet of a dispersion is positioned between the ends of two wires and is then frozen (Fig. 2.59). Such a system again allows maximum cooling rates to be achieved in relation to the size of the specimen.

Menold et al. (1976) describe a proprietary device that allows a pair of droplets to be positioned on the ends of wires so that they can be rapidly frozen. The wires should just be touching (Fig. 2.59) and should be of low thermal conductivity so that the specimen is not warmed by conduction once it has been frozen. It is recommended by the authors that freezing is carried out in liquid Freon 13 at approximately 93 K but it may be assumed that liquid propane would improve the cooling rate still further. Although this technique has primarily been used as a preliminary to freeze-fracture methods, there is no reason why the cooling procedure should not be used prior to other subsequent steps.

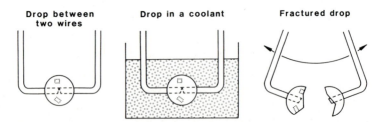

Fig. 2.59.   Droplet freezing method. A small droplet is suspended between the ends of two closely adjacent wires. High cooling rates are obtained because the droplet can be very small and has comparatively little metal 'supporting' material. This technique is usually used prior to freeze-fracturing (see §6.3.3a). (Redrawn from Menold et al. 1976.)

2.4.2 Freezing on cold surfaces

The improved heat transfer characteristics between hydrated biological specimens and solid surfaces of high thermal conductivity and diffusivity have been used by some workers in attempting to raise cooling rates (§2.2.6). If cold blocks are to be used, then it is sensible to chose the best combination of block materials and temperatures. For maximum heat transfer, ultra-pure copper cooled with liquid helium is clearly good (Table 2.10) but synthetic sapphire has even better thermal properties and, as pointed out by Meisner and Hagins (1978), could well be used in place of copper. (It might be added that there are no immediately obvious reasons for using gold or silver blocks in preference to copper; the thermal properties of copper are better than those of silver and gold at the relevant temperatures (Table 2.10) but it is essential that the copper is ultra-pure and that its surface is absolutely clean and untarnished, otherwise the beneficial properties will be greatly reduced.) For metals, the heat transfer characteristics improve dramatically with falling temperature. Thus it is generally preferable, for obtaining the fastest possible cooling rates, to use such cold blocks closer to the temperature of liquid helium than of liquid nitrogen. The work of Bald (1983) suggests that the optimum temperature for rapid cooling on a cold copper block is 19 K (§2.2.6).

Before deciding to use such a method, some thought should be given to the balance between advantages and disadvantages. High cooling velocities can undoubtedly be achieved but, as in all other cooling methods, the benefit will only be apparent within the first 10–20 μm depth from the surface of hydrated biological specimens (Table 2.11; §2.3.2). Furthermore, the heat extraction is from one side only and all the heat of the specimen (and its support) must move across this one boundary. For ideal cooling, it is essential that the wet, deformable specimen is brought extremely rapidly into very good, close contact with the surface of the cold block. This must inevitably mean that some degree of surface deformation is induced. The fact that the rapidly cooled surface is essentially flat can be most useful if the frozen specimen is to be sectioned on a cryoultramicrotome because it is then possible to cut sections within the plane of the well-frozen layer. All such considerations must be taken into account before deciding on the cooling method to be used. However, if it is concluded that cold block cooling is the technique to be followed, then a number of practical methods have been described.

The first practical use of cold block freezing was made by Eränkö (1954) and this was followed by experiments by Van Harreveld and Crowell (1964) and Van Harreveld et al. (1965), who give detailed practical instructions on the operation of such a method using sub-cooled liquid nitrogen to cool a silver block to about 67 K. Christensen (1971) used a similar system for freezing pancreas and liver prior to frozen thin sectioning. Van Harreveld et al. (1974) later described an improved version of their device which overcame some earlier difficulties (Fig. 2.60). In this apparatus, the highly polished silver surface is protected from condensation of moisture and gases by a flow of cold helium gas. The specimen passes through this gas layer sufficiently rapidly so that pre-cooling does not affect physiological process-

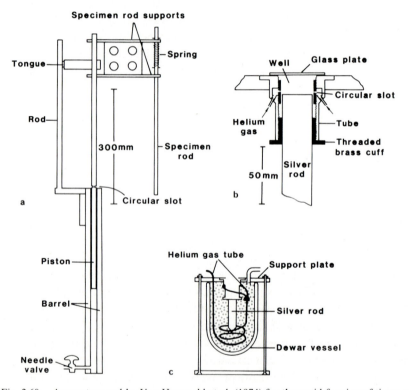

Fig. 2.60.   Apparatus used by Van Harreveld et al. (1974) for the rapid freezing of tissue on a cold block. Diagram (a) shows the general arrangement of the damped plunger device. Details of the cold block are shown in (b): the polished silver surface is cooled to liquid nitrogen temperature and purged with helium gas to avoid contamination; the glass protection plate is removed immediately prior to cooling. The general layout of the Dewar vessel and cold block is shown in (c). (Redrawn from Van Harreveld et al. 1974.)

es. The specimen (in the original description, a frog muscle) is attached to the end of a plunger device which moves rapidly towards the cold block, so thrusting the specimen against the cold surface. Van Harreveld et al. (1974) used a rate of 0.1 m s^{-1} with a 20 mm deep well immediately above the cold surface, this well being filled with the cold helium gas (Fig. 2.60).

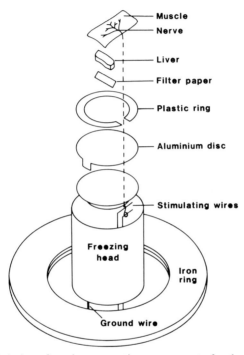

Fig. 2.61. Detailed view of specimen mounting arrangements for the cold block freezing experiments of Heuser et al. (1979). (For the purposes of this illustration, the freezing assembly is shown 'upside-down'. (In practice the specimen would face downwards.) The specimen (muscle) is seen on the freezing head, ready to be frozen. The tissue lies on a piece of liver to cushion the impact with the copper block and to raise its surface above the plastic ring. The liver sits on a slip of filter paper glued to an aluminium disc shaped to fit on the cold stage of the freeze-etching machine. The filter paper absorbs some of the Ringer's solution that clings to the liver and muscle. When the Ringer's solution freezes it bonds the liver and muscle tightly to the aluminium disc. The plastic ring prevents the muscle from being flattened beyond the reach of the knife in the freeze-etching unit. In this particular example the nerve can be stimulated immediately prior to impact through the aluminium wires which are attached to a stimulator. At the time of impact, the freezing head telescopes against the spring mounted inside it until the iron ring hits the electromagnet surrounding the copper block. The magnet prevents the freezing head from bouncing away from the copper block after impact. The iron ring is covered by a thin layer of foam rubber to cushion its impact with the magnet. (Redrawn from Heuser et al. 1979.)

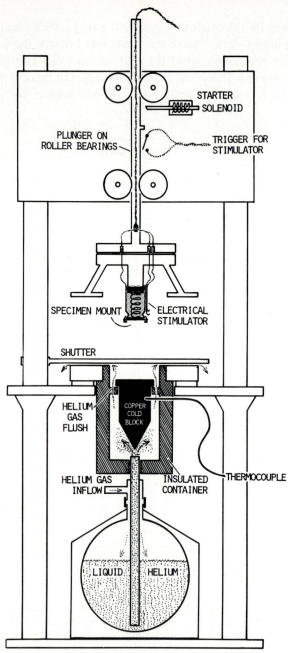

STARTER
SOLENOID

PLUNGER ON
ROLLER BEARINGS

TRIGGER FOR
STIMULATOR

SPECIMEN MOUNT

ELECTRICAL
STIMULATOR

SHUTTER

HELIUM
GAS
FLUSH

COPPER
COLD
BLOCK

HELIUM GAS
INFLOW

INSULATED
CONTAINER

THERMOCOUPLE

LIQUID   HELIUM

Fig. 2.62.   General arrangement of the cold block freezing device designed by Heuser et al. (1979) and now commercially available (Appendix 2). (Figure kindly supplied and reproduced by courtesy of J. Heuser.)

Such a depth was considered to be the minimum compatible with protecting the surface from contamination while not leading to unacceptable specimen pre-cooling.

Boyne (1979) used a modification of the Van Harreveld apparatus to freeze thin isolated stacks of the electric organ of a fish. He paid considerable attention to one of the major practical problems of cold block devices: optimising the force with which the specimen is thrust against the cold block. Excessive force deforms the specimen and, if no precautions are taken to avoid it, the freezing specimen may bounce away from the cold surface, so giving inconsistent cooling. To overcome these difficulties, it is usual to mount the specimen on a small piece of energy-absorbing material, such as foam rubber and/or even a thin slice of liver tissue (Fig. 2.61); this protects the specimen from being pushed too forcefully against the cold block. The problem of bounce has been overcome in the Heuser device (Fig. 2.62) by using a 'magnetic catch' that locks-on and prevents the specimen assembly from retracting once it has hit the cold block. Alternatively, Boyne (1979) eliminated bounce by using a hydraulic-pneumatic damping system on his plunger (see also Phillips and Boyne 1984).

Authors subsequent to Van Harreveld and his colleagues have described various modifications to the orginal cold block apparatus. Some (for example, Dempsey and Bullivant 1976a; Heath 1984), use extremely simple methods while others (e.g. Heuser et al. 1976, 1979) have described more sophisticated approaches.

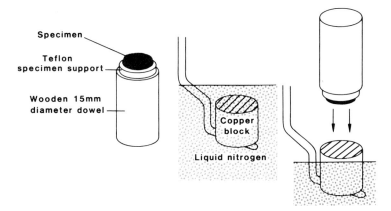

Fig. 2.63. A simple cold block freezing method. The copper block is cooled in liquid nitrogen from which, just prior to freezing, it is lifted so that its surface is *just* clear. The specimen is then firmly thrust against the highly polished surface of the cold copper block. (Redrawn from an original illustration in Dempsey and Bullivant 1976a.)

The Dempsey and Bullivant (1976a) technique must certainly be the simplest cold block method (Fig. 2.63) but the results demonstrated compare well with those achieved using more elaborate methods. The specimen is simply picked up with tweezers and quickly pushed against the surface of a cold, polished copper block that has just been lifted above the surface of liquid nitrogen in which it has been cooled. Even with such simple apparatus, the benefits of the higher rate of heat transfer across the cold copper surface at 77 K are realised. However, the depth of tissue with small ice crystals is again limited by the properties of the specimen to about 12 μm, a distance comparable to that obtained with liquid cryogens.

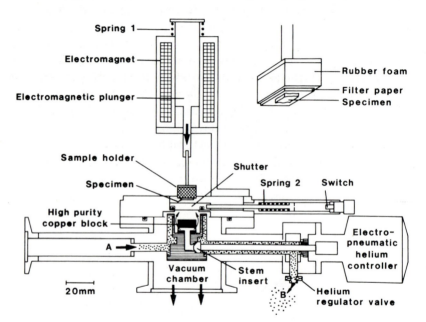

Fig. 2.64.   Cold block freezing device described by Escaig (1982). In this system a high purity copper block is fixed inside a chamber through which a regulated supply of liquid helium circulates (**A** to **B**). The specimen is mounted on the sample holder which is fixed to the tip of an electromagnetic plunger (see Fig. 2.50 in §2.4.1a). The vacuum chamber is pumped out to less than 1.3×10^{-2} Pa and the transfer line is connected to the device. Liquid helium then cools the device down to 6.0 K in less than 2.0 min. When the required temperature is reached, operation of the electropneumatic system opens the stem insert, so allowing cold helium gas into the vacuum chamber. When atmospheric pressure is reached, spring (**2**) pulls the shutter which, at the end of its travel, operates the switch. This then stops the electropneumatic system (which closes the stem insert and stops the flow of helium gas) and also activates the electromagnet (which brings the specimen into contact with the copper block). When the electromagnet ceases to be activated, spring (**1**) frees the specimen which is then stored in liquid nitrogen. (Illustration redrawn from an illustration kindly provided by J. Escaig.)

Heuser et al. (1976) (also Heuser et al. 1979), Escaig et al. (1977) (also Escaig 1982) and Sitte et al. (1977) are among those who have described more elaborate cold block freezing devices. The system devised by Heuser and his colleagues has produced some particularly elegant results and is now commercially available (see Appendix 2). It utilises a guided free-fall slider mechanism whereby the specimen is thrust against the copper block which is cooled with liquid helium. A shutter opens as the specimen falls, thus overcoming problems of condensation on the block face as well as of pre-cooling in a cold gas layer above the specimen. In the particular application of Heuser et al. (1979) the cold block device was used to freeze neuromuscular junctions of frog muscle where stimulation had been triggered by the closing of a microswitch as the specimen descended towards the cold surface. Thus, it was possible to control very precisely the time interval between stimulation and rapid freezing, a unique achievement.

The device of Escaig et al. (1977) (Fig. 2.64) also uses a copper block cooled with liquid helium. In the latest form of this apparatus (Escaig 1982), the specimen is projected against the block by means of an electromagnet and there is a shutter to protect the block from contamination.

The system of Sitte et al. (1977) provides a simple and neat method of using a copper cold block at liquid nitrogen temperature. In this device, the specimen is allowed to fall against the cold block. The authors point to the particular advantages of being able to obtain large areas of relatively flat, well-frozen tissue for cryoultramicrotomy.

Any of the commercially available cold block devices should give satisfactory results (suppliers are listed in Appendix 2) although the theoretical aspects of this technique are still being evaluated (Bald 1983).

2.4.3   Comparison of freezing methods

All the rapid freezing methods described above have been shown to produce excellent results, in terms of small ice crystal size, in some specimens. There have been arguments about the relative merits of the different methods but the information available tends to support the view that small specimens can be well-frozen by any of the methods while large ones will, with any method, only be well-frozen to a maximum depth of about 20 μm. This is a fundamental limitation imposed by poor heat transfer through hydrated biological specimens. Cryoprotection (§2.3.2a), or high pressure freezing (§2.4.1d), offer the only possibilities for decreasing ice crystal size in large specimens. The choice of freezing technique therefore depends more on the nature of the specimen and what is to be done with it once it has been frozen, than on the efficiency of the technique itself (Table 2.14; §2.4).

In the absence of any other criteria, it is best to start freezing tissues using rapid plunging methods. These can be established in any laboratory using simple and inexpensive apparatus. For many purposes they are be totally adequate. Spray-freezing is excellent for 'liquid' specimens and gives an extremely short time delay between 'sampling' and freezing whereas, in all the other methods, specimen sampling and mounting can be rather protracted. The droplet and liquid paraffin methods are also suitable for liquid specimens. Propane jets seem to offer advantages as much in terms of specimen handling (thin metal sandwiches) as in absolute cooling rates: this technique is thus particularly appropriate as a preparation for freeze-fracture methods. High pressure freezing *should* provide a very significant improvement in freezing quality but apparatus remains to be made generally available. Although cold blocks offer theoretical advantages over most of the other methods and some dramatic applications have been demonstrated (e.g. Heuser et al. 1979), their *general* usefulness has been less well documented (e.g. Table 2.11; §2.3.2). It hardly seems worthwhile using cold blocks unless at liquid helium temperature (Bald 1983) or unless the benefit of having a relative flat frozen layer (as for cryoultramicrotomy) outweighs the additional complexity. Alternatively, the requirement for a precise time course of freezing in relation to cellular activity may indicate the usefulness of cold block freezing. Thus, the individual microscopist must ultimately make his own choice of method but the most helpful advice is probably, initially, to 'keep it simple'!

2.5   *Subsequent processing and storage of specimens*

When the specimen has been frozen, it must be removed from the apparatus and then either used directly for further processing or stored for use at some future time. As stressed previously, once the specimen has been frozen it must always be handled in such a way that unnecessary temperature increases do not occur. In particular, unless the protocol of the technique demands it, the temperature should not be allowed to rise above about 90 K; the temperature above which it may be anticipated that some ice crystal growth may occur. Thus, it is important that pre-cooled tools are used and that transfers are accomplished as quickly as possible. These precautions are particularly relevant when using very small, thermally sensitive specimens.

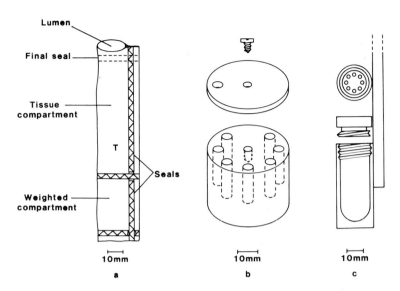

Fig. 2.65. Storage containers for specimens at low temperatures. (a) A sachet made from a plastic laminate (Appendix 2) which can be heat sealed while maintaining the tissue compartment under liquid nitrogen. (Redrawn from Appleton 1967.) (b) Simple container made from aluminium where a hole in the lid can be 'indexed' to align with storage compartments. (c) A 'dip-stick'-type storage container for use with liquid nitrogen Dewar vessels. (Redrawn from Krah et al. 1973.)

In many instances, the frozen specimen will be used immediately for further processing. It may well have been frozen on a specimen support that can be mounted directly onto a cryoultramicrotome or a freeze-fracture apparatus. However, it is often useful to be able to store a specimen for use at some future time. Although the principles of storage are simple, it is important that the technical aspects are fully considered so that no deleterious effects take place at this stage. For most purposes storage under liquid nitrogen is both safest and easiest. A number of suitable containers for storing specimens under liquid nitrogen have been recommended (Fig. 2.65). Appleton (1967) used a heat-sealed plastic/aluminium foil bag for cryoultramicrotomy specimens while a typical container for freeze-etch specimens is described by Krah et al. (1973) and Williamson (1984) has also described a useful system for storing specimens in liquid nitrogen. Rigler and Patton (1984) described another 'cryocell' storage container. The important features are that the specimen should be transferred into the container while under cold conditions (preferably under liquid nitrogen), that the container should itself be protected from warming while being transferred into the storage Dewar vessel and that, similarly, precautions should be taken to avoid heating the specimen when it is removed from the nitrogen and transferred to the next stage of whatever process is being used. It is also important that the containers should be indelibly and unambiguously labelled. Some felt-tip markers are stable under liquid nitrogen; engraving on metal is safe and permanent for containers that are frequently used; and some adhesive tapes are satisfactory for low temperature use.

It is possible to purchase ultra-low temperature deep freezes that maintain temperatures of approximately 170 K (Appendix 2). These may be safe for storing some specimens but they should be used with caution. Their lowest temperature is close to the expected temperature for ice recrystallisation and it is advised that they should only be considered 'safe' in this respect if rigorous tests on the particular specimens have been carried out.

2.6   Cooling and viability after thawing

The subject of cooling rates in relation to structure and viability has already been discussed in §2.3.1 but we should not leave the critical topic of freezing biological specimens without again briefly pausing to contemplate the interrelationships between cooling, structural preservation and viability. This topic has been well reviewed by a number of authors but the paper by Farrant et al. (1977) is particularly germane. These authors point to the probability of inducing toxic changes or structural alterations by using cryoprotectants at high enough concentrations to prevent ice crystal artefacts. Such cryoprotectants are usually only protective at low rates of cooling and may be positively harmful at higher rates. Lower concentrations of cryoprotectants ($<10\%$) may reduce ice crystal artefacts but they are unlikely to protect cells in a functional sense unless cooling rates are low. Thus, cells which are to survive upon thawing may be frozen by cooling very slowly or by using a two-step method; however, these cells will also be severely shrunken. The microscopist is presented with a choice: to cool cells as rapidly as possible, tolerating the presence of some intracellular ice which will almost certainly mean that the cells would not be viable after thawing, or to cool cells slowly, causing them to become shrunken but knowing that such cells may well show good viability on thawing. No general guidance can be given. A choice must made for each experiment carried out in relation to the type and quality of information required. It should, however, be noted that microscopists (and microanalysts) have, in general, preferred to work with cells that 'look good' rather than with those that can be demonstrated to survive the freezing process. This should act as a caution against the unheeding use of existing techniques and, perhaps, cause us to pause to think about exactly what we are trying to achieve in freezing the specimen before going on to use it in one of the methods described in the following Chapters.

References

Adrian, M., J. Dubochet, J. Lepault and A. McDowell (1984), Cryo-electron microscopy of viruses, Nature, Lond. *308*, 32.

Aist, J.R. (1974), A freeze-etch study of membranes in plasmodiophora-infected and non-infected cabbage root hairs, Can. J. Bot. *52*, 1441.

Allen, E.D. and L. Weatherbee (1979), Ultrastructure of red cells frozen with hydroxyethyl starch, J. Microscopy *117*, 381.

Allen, E.D. and L. Weatherbee (1980), Ultrastructural appearance of red cells frozen with glycerol for clinical use, Cryobiology *17*, 448.

Allen, J.V., W.M. Hess and D.J. Weber (1971), Ultrastructural investigations of dormant *Tilletia caries* teliospores, Mycologia *63*, 144.

Ananthanarayanan, V.S. and C.L. Hew (1977), A synthetic polypeptide with antifreeze activity, Nature, Lond. *268*, 560.

Appleton, T.C. (1967), Storage of frozen materials for cryostat sectioning and soluble-compound autoradiography, J. Microscopy *87*, 489.

Arancia, G., F. Rosati Valente and P. Trovalusci (1979), Effects of glutaraldehyde and glycerol on freeze-fractured *Escherichia coli*, J. Microscopy *118*, 161.

Arpaci, V. (1966), Conduction heat transfer (Addison-Wesley Publ. Co., London).

Aurich, F. and Th. Foster (1984), Temperature measurement during rapid cooling in μl-volumina using a fluorescence label. Cryo-Letters *5*, 231.

Bachmann, L. and W.W. Schmitt (1971), Improved cryofixation applicable to freeze-etching, Proc. natl. Acad. Sci. U.S.A. *68*, 2149.

Bachmann, L. and W.W. Schmitt-Fumian (1973), Spray-freezing and freeze- etching, in: Freeze-etching: techniques and applications, E.L. Benedetti and P. Favard, eds. (Soc. Fr. Microsc. Elect., Paris), p.73.

Bald, W.B. (1975), A proposed method for specifying the temperature history of cells during the rapid cooldown of plant specimens, J. exp. Bot. *26*, 103.

Bald, W.B. (1978a), A multi-purpose cryostat for freezing biological materials, Cryogenics *18*, 3.

Bald, W.B. (1978b), The thermal history of cells during the freezing of biological materials, Proc. 7th. Int. Cryogenic Eng. Conf., London, p.441.

Bald, W.B. (1979), A critical appraisal of the experimental techniques used in the freezing of biological materials, Proc. 15th Int. Congr. Refrig., Venice *3*, 225.

Bald, W.B. (1983), Optimising the cooling block for the quick-freeze method, J. Microscopy *131*, 11.

Bald, W.B. (1984), The relative efficiency of cryogenic fluids used in the rapid quench cooling of biological samples, J. Microscopy *134*, 261.

Bald, W.B. and A.B. Crowley (1979), On defining thermal history of cells during the freezing of biological materials, J. Microscopy *117*, 395.

Bald, W.B. and A.B. Crowley (1981), Modifications to the paper "On defining the thermal history of cells during the freezing of biological materials", J. Microscopy *122*, 331.

Bald, W.B. and J. Fraser (1982), Cryogenic surgery, Rep. Prog. Phys. *45*, 1381.

Bald, W.B. and A.W. Robards (1978), A device for the rapid freezing of biological specimens under precisely controlled and reproducible conditions, J. Microscopy *112*, 3.

Balko, B. and R.L. Berger (1968), Measurement and computation of thermojunction response times in the submillisecond range, Rev. scient. Instrum. *39*, 498.

Bank, H. and P. Mazur (1973), Visualization of freezing damage, J. Cell Biol. *57*, 729.

Barlow, K.K., M.S. Buttrose, D.H. Simmonds and M. Vesk (1973), The nature of the starch protein interface in wheat endosperm, Cereal Chem. *50*, 443.

Barnard, T. (1980), Ultrastructural effects of high molecular weight cryoprotectants dextran and polyvinyl pyrrolidone on liver and brown adipose tissue, J. Microscopy *120*, 93.

Baust, J.G. and J.S. Edwards (1979), Mechanisms of freezing tolerance in an antarctic midge, *Belgica antarctica*, Physiol. Entomol. *4*, 1.

Baust, J.G., R.E. Lee and R.A. King (1982), The physiology and biochemistry of low temperature tolerance in insects and other terrestrial arthropods: a bibliography, Cryo-Letters *3*, 191.

Bayer, M.E., M. Dolack and E. Houser (1977), Effects of lipid phase transition on freeze-cleaved envelope of *Escherichia coli*, J. Bacteriol. *129*, 1563.

Boutron, P. and A. Kaufmann (1979), Stability of the amorphous state in the system water-glycerol-ethylene glycol, Cryobiology *16*, 83.

Boyne, A.F. (1979), A gentle, bounce-free assembly for quick-freezing tissues for electron microscopy: application to isolated torpedine ray electrocyte stacks, J. Neurosci. Methods *1*, 353.

Branton, D. and H. Moor (1964), Fine structure in freeze-etched *Allium cepa* L. roots. J. Ultrastruct. Res. *11*, 401.

Bray, D.F., K. Nakamura, J.W. Costerton and E.B. Wagenaar (1974), Ultrastructure of *Chlamydomonas eugametos* as revealed by freeze-etching cell wall, plasmalemma and chloroplast membrane, J. Ultrastruct. Res. *97*, 125.

Brouchant, R. and V. Demoulin (1971), Ultrastructure de le paroi des basidiospores de *Lycoperdon* et de *Scleroderma*, Gasteromysetes comparee a celle de quelques autres spores de champignons, Protoplasma *72*, 179.

Brüggeller, P. and E. Mayer (1980), Complete vitrification in pure liquid water and dilute aqueous solutions, Nature, Lond. *288*, 569.

Buchheim, W. (1972), Zur Gefrierfixierung wässriger Lösungen, Naturwissenschaften *59*, 121.

Buchheim, W. (1977), Cryofixation of tissue without cryoprotectants, Naturwissenschaften *59*, 121.

Buchheim, W. and U. Welsch (1978), Freeze-etching of unglycerinated tissue dispersions by application of the oil emulsion technique, J. Microscopy *111*, 339.

Bullis, L.H. (1963), Vacuum-deposited thin-film thermocouples for accurate measurement of substrate surface temperature, J. scient. Instrum. *40*, 592.

Bullivant, S. (1960), The staining of thin sections of mouse pancreas prepared by the Fernández-Morán Helium II freeze-substitution method, J. Biophys. Biochem. Cytol. *8*, 639.

Bullivant, S. (1965), Freeze-substitution and supporting techniques, Lab. Invest. *14*, 440.

Bullivant, S. (1970), Present status of freezing techniques, in: Some biological techniques in electron microscopy, D.F. Parsons, ed. (Academic Press, New York), p. 101.

Bullivant, S. (1978), The structure of tight junctions, Proc. 9th Int. Congr. Electron Microscopy, Toronto *3*, 659.

Bullock, G.R. (1979), Preservation of myocardium for ultrastructural and enzymatic studies, in: Enzymes in cardiology: diagnosis and research, D.J. Hearse and J. de Leiris, eds. (Wiley-Interscience, Chichester), p. 461.

Burton, E.F. and W.F. Oliver (1935), Structure of ice at low temperatures, Proc. R. Soc. Lond. Ser. A *153*, 166.

Buttrose, M.S. (1971), Ultrastructure of barley aleurone cells as shown by freeze-etching, Planta *96*, 13.

Buttrose, M.S. (1973), Rapid water uptake and structural changes in imbibing seed tissues, Protoplasma *77*, 111.

Carslaw, H.S. and J.C. Jaeger (1959), Conduction of heat in solids (Oxford University Press).

Chafe, S.C. and A.B. Wardrop (1970), Microfibril orientation in plant cell walls, Planta *92*, 13.

Chandler, D.E. (1979), Quick freezing avoids specimen preparation artifacts in membrane fusion studies, in: Freeze-fracture: methods, artefacts and interpretations, J.E. Rash and C.S. Hudson, eds. (Raven Press, New York), p.81.

Chang, S.H., W.J. Mergner, R.E. Pendergrass, R.E. Bulger, I.K. Berezesky and B.F. Trump (1980), A rapid method of cryofixation of tissues *in situ* for ultracryomicrotomy, J. Histochem. Cytochem. *28* 47.

Christensen, A.K. (1971), Frozen thin sections of fresh tissue for electron microscopy, with a description of pancreas and liver, J. Cell Biol. *51*, 772.

Clark, J., P. Echlin, R. Moreton, A. Saubermann and P. Taylor (1976), Thin film thermocouples for use in scanning electron microscopy, Scanning Electron Microscopy 1976, *1*, 83.

Cole, G.T. (1973), A correlation between rodlet orientation and conidiogenesis in Hyphomycetes, Can. J. Bot. *51*, 2413.

Cole, G.T. (1975), The thallic mode of conidiogenesis in the Fungi imperfecti, Can. J. Bot. *53*, 2983.

Costello, M. J. (1980), Ultra-rapid freezing of biological samples, Scanning Electron Microscopy 1980, *2*, 361.

Costello, M.J. and J.M. Corless (1978), The direct measurement of temperature changes within freeze-fracture specimens during rapid quenching in liquid coolants, J. Microscopy *112*, 17.

Costello, M.J., R. Fetter and M. Höchli (1982), Simple procedures for evaluating the cryofixation of biological samples, J. Microscopy *125*, 125.

Cowley, C.W., W.J. Timson and J.A. Sawdye (1961), Ultra rapid cooling techniques in the freezing of biological materials, Biodynamica *8*, 317.

Cox, G. and B. Juniper (1973), Electron microscopy of cellulose in entire tissue, J. Microscopy *97*, 343.

De Quervain, M.R. (1975), Crystallization of water, a review, in: Freeze-drying and advanced food technology, S.A. Goldblith, L. Rey and W. W. Rothmayr, eds. (Academic Press, London, New York) p.3.

Dempsey, G.P. and S. Bullivant (1976a), Copper block method for freezing non cryoprotected tissue to produce ice crystal free regions for electron microscopy. I. Evaluation using freeze-substitution, J. Microscopy *106*, 251.

Dempsey, G.P. and S. Bullivant (1976b), Copper block method for freezing non cryoprotected tissue to produce ice crystal free regions for electron microscopy. II. Evaluation using freeze-fracture with a cryoultramicrotome, J. Microscopy *106*, 261.

Dempsey, G.P., S. Bullivant and W.B. Watkins (1973), Endothelial cell membranes: polarity of particles as seen by freeze-fracturing, Science *179*, 190.

Dowell, L.G. and A.P. Rinfret (1960), Low temperature forms of ice as studied by x-ray diffraction, Nature, Lond. *188*, 1144.

Dubochet, J. (1984), Electron microscopy of vitrified specimens, Proc. 8th Eur. Conf. Electron Microscopy, Budapest *2*, 1379.

Dubochet, J. and A.W. McDowall (1981), Vitrification of pure water for electron microscopy, J. Microscopy *124*, RP 3.

Dubochet, J., J. Lepault, R. Freeman, J.A. Berryman, and J.-C. Homo (1982), Electron microscopy of frozen water and aqueous solutions, J. Microscopy *128*, 219.

Easterbrook, K.B. (1971), Ultrastructure of *Crithidia fasciculata* – a freeze-etching study, Can. J. Microbiol. *17*, 277.

Echlin, P. and R. Moreton (1976), Low temperature techniques for scanning electron microscopy, Scanning Electron Microscopy 1976, 753.

Echlin, P. and A.J. Saubermann (1977), Preparation of biological specimens for X-ray microanalysis, Scanning Electron Microscopy 1977, *1*, 621.

Echlin, P., H.B. Skaer, B.O.C. Gardiner, F. Franks and M.H. Asquith (1977), Polymeric cryoprotectants in the preservation of biological ultrastructure. II. Physiological effects, J. Microscopy *110*, 239.

Elder, H.Y., C.C. Gray, A.G. Jardine, J.N. Chapman and W.H. Biddlecombe (1982), Optimum conditions for the cryoquenching of small tissue blocks in liquid coolants, J. Microscopy *126*, 45.

Ellis, E. A. and R. M. Brown (1972), Freeze-etch ultrastructure of *Parmelia caparata* (L) Ach., Trans. Amer. Microsc. Soc. *91*, 411.

Eränko, O. (1954), Quenching of tissue for freeze-drying, Acta Anat. *22*, 331.

Escaig, J. (1982), New instruments which facilitate rapid freezing at 83 K and 6 K, J. Microscopy *126*, 221.

Escaig, J., G. Geraud and G. Nicolas (1977), Congelation rapide de tissus biologiques. Mesure des temperatures et des vitesses de congelation par thermocouple en couche mince, C. R. Acad. Sci. Paris *284*, 2289.

Farrant, J., C.A. Walter, H. Lee, G.J. Morris, and K.J. Clarke (1977), Structural and functional aspects of biological freezing techniques, J. Microscopy *111*, 17.

Feder, N. and R.L. Sidman (1958), Methods and principles of fixation by freeze-substitution. J. Biophys. Biochem. Cytol. *4*, 593.

Fernández-Morán, H. (1960), Low temperature preparation techniques for electron microscopy of biological specimens based on rapid freezing with liquid helium II, Ann. N. Y. Acad. Sci. *85*, 689.

Fineran, B.A. (1970a), Evaluation of the form of vacuoles in thin sections and freeze-etch replicas of root tips, Protoplasma *70*, 457.

Fineran, B.A. (1970b), The effects of various pre-treatments on the freeze-etching of root tips, J. Microscopy *92*, 85.

Fineran, B.A. (1971), Ultrastructure of vacuolar inclusions in root tips, Protoplasma *71*, 1.

Fineran, B.A. (1972), Fracture faces of tonoplast in root tips after various conditions of pretreatment prior to freeze-etching. J. Microscopy *96*, 333.

Fineran, B.A. (1973a), Association between endoplasmic reticulum and vacuoles in frozen root tips, J. Ultrastruct. Res. *43*, 75.

Fineran, B.A. (1973b), Organization of the Golgi apparatus in frozen etched root tips, Cytobiology *8*, 175.

Fineran, B.A. (1975), Freeze-etching and the ultrastructural morphology of root tip cells, Phytomorphology *25*, 398.

Fineran, B.A. (1978), Freeze-etching, in: Electron microscopy and cytochemistry of plant cells, J.L. Hall, ed. (Elsevier/North-Holland, Amsterdam), p. 279.

Flechon, J.E. (1974), Freeze-fracturing of rabbit spermatozoa, J. Microscopie *19*, 59.

Fletcher, N.F. (1970), The chemical physics of ice (Cambridge University Press).

Fooke-Achterrath, M., K.G. Lickfield, V.M. Reusch, U. Aebi, U. Tschope and B. Menge (1974), Close-to-life preservation of *Staphylococcus aureus* mesosomes for transmission electron microscopy, J. Ultrastruct. Res. *49*, 270.

Forsmann, W.G., J. Metz and D. Heinrich (1975), Gap junctions in the Hemotrichorial placenta of the rat, J. Ultrastruct. Res. *53*, 374.

Franks, F. (1977), Biological freezing and cryofixation. J. Microscopy *111*, 3.

Franks, F. (1980), Physical, biochemical and physiological effects of low temperatures and freezing: modification by water soluble polymers, Scanning Electron Microscopy 1980, *2*, 349.

Franks, F. (1982), Apparent osmotic activities of water soluble polymers used as cryoprotectants, Cryo-Letters *3*, 115.

Franks, F. and H.le B. Skaer (1976), Aqueous glasses as matrices in freeze-fracture electron microscopy, Nature, Lond. *262*, 323.

Franks, F., M.H. Asquith, C.C. Hammond, H.le B. Skaer and P. Echlin (1977), Polymeric cryoprotectants in the preservation of biological ultrastructure. I. Low temperature states of aqueous solutions of hydrophilic polymers, J. Microscopy *110*, 223.

Frederik, P.M. and W.M. Busing (1981), Strong evidence against section thawing whilst cutting on the cryo-ultratome, J. Microscopy *122*, 217.

Friend, D.S. and D.W. Fawcett (1974), Membrane differentiations in freeze-fractured mammalian sperm, J. Cell Biol. *63*, 641.

Fujikawa, S. (1981), The effect of various cooling rates on the membrane ultrastructure of frozen human erythrocytes and its relation to the extent of haemolysis after thawing, J. Cell Sci. *49*, 369.

Furcht, L.T. and R.E. Scott (1974), Influence of cell cycle and cell movement on the distribution of intra-membrane particles in contact-inhibited and transformed cells, Exp. Cell Res. *88*, 311.

Furcht, L.T. and R.E. Scott (1975), Modulation of the distribution of plasma membrane intramembranous particles in contact-inhibited and transformed cells, Biochim. Biophys. Acta *401*, 213.

Ghosh, B.K. and N. Nanninga (1976), Polymorphism of the mesosome in *Bacillus licheniformis* (749c and 749): influence of chemical fixation monitored by freeze-etching, J. Ultrastruct. Res. *56*, 107.

Gilleland, H.E., J.D. Stinnett, I.L. Roth and R.G. Eagon (1973), Freeze-etch study of *Pseudomonas aeruginosa*: localization within the cell wall of an ethylenediaminetetraacetate extractable component, J. Bacteriol. *113*, 417.

Glauert, A.M. (1974), Fixation, dehydration and embedding of biological specimens, in: Practical methods in electron microscopy, Vol. 3, A.M. Glauert, ed. (North-Holland, Amsterdam).

Glover, A.J. and Z.S. Garvitch (1974), The freezing rate of freeze-etch specimens for electron microscopy, Cryobiology *11*, 248.

Groot, C. de and W. Leene (1979), The influence of cryoprotectants, temperature, divalent cations and serum proteins on the structure of the plasma membrane in rabbit peripheral blood lymphocytes, Eur. J. Cell Biol. *19*, 19.

Gulik Krzwicki, T. and M.J. Costello (1978), Use of low temperature X-ray diffraction to evaluate freezing methods in freeze-fracture electron microscopy, J. Microscopy *112*, 103.

Gupta, B.L. (1979), Electron microprobe X-ray analysis of frozen hydrated sections with new information on fluid transporting epithelia, in: Microbeam analysis in biology, C.P. Lechene and R.R. Warner, eds. (Academic Press, London) p. 375.

Hall, J.A. (1966), The measurement of temperature (Chapman and Hall, London).

Hall, D.M. (1967), Wax microchannels in the epidermis of white clover, Science *158*, 505.

Handley, D.A., J.T. Alexander and S. Chien (1981), Design and use of a simple device for rapid quench freezing of biological samples, J. Microscopy *121*, 273.

Haselden, O.G. (1971), Cryogenic fundamentals, (Academic Press, New York).

Hasty, D.L. and E.D. Hay (1977), Freeze-fracture studies of the developing cell surfaces. II. Particle-free membrane blisters on glutaraldehyde fixed corneal fibroblasts are artefacts, J. Cell Biol. *78*, 756.

Hay, E.D. and D.L. Hasty (1979), Extrusion of particle-free membrane blisters during glutaraldehyde fixation, in: Freeze-fracture: methods, artefacts and interpretations, J.E. Rash and C.S. Hudson, eds. (Raven Press, New York), p. 59.

Hayat, M.A. (1981), Fixation for electron microscopy (Academic Press, New York).

Hearse, D.J., D.M. Yellon, D.A. Chappell, R.K.H. Wyse and G.R. Ball (1981), A high velocity impact device for obtaining multiple, contiguous, myocardial biopsies, J. Mol. Cell. Cardiol. *13*, 197.

Heath, I.B. (1984), A simple and inexpensive liquid helium 'slam freezing' device, J. Microscopy *135*, 75.

Hebant, C. and R.P.C. Johnson (1976), Ultrastructural features of freeze-etched water conducting cells in *Polytrichum*, Cytobiologie *13*, 354.

Heide, H.-G. and E. Zeitler (1984), Water in cryomicroscopy, Proc. 8th Eur. Reg. Conf. Electron Microscopy, Budapest *2*, 1388.

Hereward, F.V. and D.H. Northcote (1973), Freeze-fracture planes of the plasmalemma of some higher plants revealed by freeze-etch, J. Cell Sci. *13*, 621.

Hess, W.M., J.L. Bushnell and D.J. Weber (1972), Surface structures and unidentified organelles of *Lycoperdon perlatum* Pers. basidiospores, Can. J. Microbiol. *18*, 270.

Heuser, J.E. and T.S. Reese (1976), Freeze substitution applied to the study of quick frozen synapses, J. Cell Biol. *70*, 357a.

Heuser, J.E., T.S. Reese and D.M. Landis (1976), Preservation of synaptic structure by rapid freezing, Cold Spring Harbor Symp. Quant. Biol. *40*, 17.

Heuser, J.E., T.S. Reese, M.J. Dennis, Y. Jan, L. Jan and L. Evans (1979), Synaptic vesicle exocytosis captured by quick freezing and correlated with quantal transmitter release, J. Cell Biol. *81*, 275.

Heverly, J.R. (1949), Supercooling and crystallisation, Trans. Amer. Geophys. Union *30*, 205.

Higgins, M.L., H.C. Tsien and L. Daneo-Moore (1976), Organisation of mesosomes in fixed and unfixed cells, J. Bacteriol. *127*, 1519.

James, R. and D. Branton (1971), Correlation between saturation of membrane fatty acids and presence of membrane fracture faces after osmium fixation, Biochim. Biophys. Acta *233*, 504.

Jehl, B., R. Bauer, A. Dörge and R. Rick (1981), The use of propane/isopentane mixtures for rapid freezing of biological specimens, J. Microscopy *123*, 307.

Johnson, R.P.C. (1973), Filaments but no membranous transcellular strands in sieve pores in freeze-etched translocating phloem, Nature, Lond. *244*, 464.

Johnson, R.P.C. (1978), The microscopy of P-protein filaments in freeze- etched sieve pores, Planta *143*, 191.

Jones, G.J. (1984), On estimating freezing times during tissue rapid freezing, J. Microscopy *136*, 349.

Kachar, B., J.A. Serrano and P. Pinto da Silva (1980), Particle displacement in epithelial cell membranes of rat prostate and pancreas induced by routine low temperature fixation, Cell Biol. Int. Reports *4*, 347.

Kappen, M.M. (1979), Protective agents with respect of freezing tolerance of leaves. Investigations with the halophyte *Halimione portulacoides*, Oecologia *38*, 303.

Karnovsky, M.J. (1965), A formaldehyde-glutaraldehyde fixative of high osmolality for use in electron microscopy, J. Cell Biol. *27*, 137A.

Kitajima, Y. and G.A. Thompson (1977), Tetrahymena strives to maintain fluidity interrelationships of all its membranes constant, J. Cell Biol. *72*, 744.

Knoll, G., G. Oebel, and H. Plattner (1982), A simple sandwich-cryogen-jet procedure with high cooling rates for cryofixation of biological materials in the native state, Protoplasma *111*, 161.

Koehler, J.A. (1970), Freeze etching study of rabbit spermatozoa with particular reference to head structures, J. Ultrastruct. Res. *33*, 598.

Koehler, J.A. (1972), Freeze etching technique, in: Principles and techniques of electron microscopy, Vol. 2., M.A. Hayat, ed. (Van Nostrand Reinhold, New York), p. 53.

Kopstad, G. and A. Elgsaeter (1982), Theoretical analysis of specimen cooling rate during impact freezing and liquid-jet freezing of freeze-etch specimens, Biophys. J. *40*, 163.

Körber, C and M.W. Scheiwe (1977), Effects of hydroxyethyl starch on sodium-water phase diagram and its influence on the freezing process of cells, Cryobiology *14*, 705.

Kouri, J. (1971), Ultrastructure of BHK 21 cells as revealed by freeze-etching and fixation methods, J. Microscopie *11*, 331.

Krah, S., L.A. Staehelin and E. Nettersheim (1973), A new type of storage container for freeze-etch specimens, J. Microscopy *99*, 349.

Lane, L.B. (1925), Freezing points of glycerol and its aqueous solutions, Ind. Eng. Chem. *17*, 924.

Lang, R.D.A., P. Crosby and A.W. Robards (1976), An inexpensive spray-freezing unit for preparing specimens for freeze-etching, J. Microscopy, *108*, 101.

Lange, N.A. (1961), Handbook of chemistry, 10th edition, (McGraw Hill, New York).

Lefort-Tran, M., T. Gulik, H. Plattner, J. Beissen and W. Weissner (1978), Influence of cryofixation procedures on organisation and partition of intramembrane particles, Proc. 9th Int. Congr. Electron Microscopy, Toronto *2*, 146.

Leibo, S.P., J.J. McGrath and E.G. Cravalho (1978), Microscopic observation of intracellular ice formation in unfertilized mouse ova as a function of cooling rate, Cryobiology *15*, 257.

Leonhard, R. and C. Sterling, (1972), Freeze-etched surfaces in potato starch, J. Ultrastruct. Res. *39*, 85.

Lickfeld, K. G. (1968), Der frostgerätzte Bakterienkern. Ein Beitrag zur Klärung seiner Tertiärstruktur, Z. Zellforsch. *88*, 560.

Lickfeld, K.G. (1973), Die gefrierätzung von Bakterien, in: Methodensammlung der Elektronenmikroskopie, G. Schimmel and W. Vogell, eds. (Wiss. Verlags. Ges., Stuttgart) p. 1.

Lonsdale, K. (1958), The structure of ice, Proc. R. Soc. Lond. Ser A. *247*, 424.

Lusena, C.V. (1960), Ice propagation in glycerol solutions at temperatures below $-40°C$, Ann. N. Y. Acad. Sci. *85*, 541.

Luyet, B.J. (1960), On various phase transitions occurring in aqueous solutions at low temperatures, Ann. N. Y. Acad. Sci. *85*, 549.

Luyet, B.J. (1961), A method for increasing the cooling rate in refrigeration by immersion in liquid nitrogen or in other boiling baths. Biodynamica *8*, 331.

Luyet, B.J. (1966), Anatomy of the freezing process in physical systems, in: Cryobiology, H.T. Merymman, ed. (Academic Press, New York), p. 115.

Luyet, B.J. (1970), Physical changes occurring in frozen solutions during rewarming and melting, in: The frozen cell, G.E.W. Wolstenholme and M. O'Connor eds. (J. and A. Churchill, London), p. 27.

Luyet, B.J. and F. Gonzales (1951), Recording ultra rapid changes in temperature, Refrig. Eng. *5*, 191.

Luyet, B.J. and C. Kroener, (1960), The highest obtainable cooling velocities. Proc. 4th. Ann. Meeting Amer. Biophys. Soc., Philadelphia, L7.

Luyet, B.J. and C. Kroener (1966), The temperature of the "glass transition" in aqueous solutions of glycerol and ethylene glycol. Biodynamica *10*, 33.

Luyet, B.J. and D.H. Rasmussen (1967), Study by differential thermal analysis of the temperatures of instability in rapidly cooled solutions of polyvinylpyrrolidone, Biodynamica *10*, 137.

Luyet, B.J. and D.H. Rasmussen (1968), Study by differential thermal analysis of the temperatures of instability of rapidly cooled solutions of glycerol, ethylene glycol, sucrose and glucose, Biodynamica *10*, 167.

Luyet, B.J. and D. Sager (1967), On the existence of two temperature ranges of instability in rapidly cooled solutions of polyvinylpyrrolidone, Biodynamica *10*, 133.

Luyet, B.J., C. Kroener and G. Rapatz (1958), Detection of heat of recrystallisation in glycerol-water mixtures, Biodynamica *8*, 73.

MacKenzie, A.P. (1969), Apparatus for the partial freezing of liquid nitrogen for the rapid cooling of cells and tissues, Biodynamica *10*, 341.

MacKenzie, A.P. and D.H. Rasmussen (1972), Interactions in the water-PVP system at low temperatures, in: Water structure at the water-polymer interface, H.H.G. Jellinek, ed. (Plenum Publishing Corporation, New York), p. 146.

Marchesse-Ragona, S.P. (1984), Ethanol, an efficient coolant for rapid freezing of biological material, J. Microscopy *134*, 169.

Martinez-Paloma, A., P. Pinto Da Silva and B. Chavez (1976), Membrane structure of *Entamoeba histolytica* – fine structure of freeze-fracture membranes, J. Ultrastruct. Res. *54*, 148.

Marto, P.J., J.A. Moulson and M.D. Maymard (1968), Nucleate pool boiling of nitrogen with different surface conditions, J. Heat Transfer *90*, 437.

Mathes, G. and H.A. Hackenseller (1981), Correlations between purity of dimethyl sulfoxide and survival after freezing and thawing, Cryo-Letters *2*, 389.

Maul, G.G. (1979), Temperature dependent changes in intramembrane particle distribution, in: Freeze-fracture: methods, artefacts and interpretations, J.E. Rash and C.S. Hudson, eds. (Raven Press, New York) p. 37.

Mayer, E. and P. Brüggeller (1982), Vitrification of pure liquid water by high pressure jet freezing, Nature, Lond. *298*, 715.

Mazur P. (1966), Physical and chemical basis of injury in single-celled micro-organisms subjected to freezing and thawing. In: Cryobiology, H.T. Meryman, ed. (Academic Press, London).

Mazur, P. (1970), Cryobiology – the freezing of biological systems, Science *168*, 939.

Mazur, P. (1977), The role of intracellular freezing in the death of cells cooled at supraoptimal rates, Cryobiology *14*, 251.

Mazur, P., S.P. Leibo, J. Farrant, E.H.Y. Chu, M.G. Hanna and L.H. Smith (1970), Interactions of cooling rate, warming rate and protective additive on the survival of frozen mammalian cells, in: The frozen cell, G.E.W. Wolstenholme and M. O'Connor, eds. (Churchill, London), p. 69.

McGann, L.E. (1978), Differing actions of penetrating and nonpenetrating cryoprotective agents, Cryobiology *15*, 382.

McIntyre, J.A., N.B. Gilula and M.J. Karnovsky (1974), Cryoprotectant induced redistribution of intramembranous particles in mouse lymphocytes, J. Cell Biol. *60*, 192.

McMillan, J.A. and S.C. Los (1965), Vitreous ice: irreversible transformations during warmup, Nature, Lond. *206*, 806.

Meisner, J. and W.A. Hagins (1978), Fast freezing of thin tissues by thermal conduction into sapphire crystals at 77 Kelvin, Biophys. J. *21*, 149a.

Menold, R. and G. Lüttge (1979), Freeze-etching of dispersions without contamination of the fracture faces, Microscopica Acta 81, 317.Menold, R., B. Lüttge and W. Kaiser (1974), Zum Einfrieren von Suspensionen für elektronenmikroskopische Untersuchungen. Colloid Polymer Sci. *252*, 530.

Menold, R., G. Lüttge, and W. Kaiser (1976), Freeze-fracturing: a new method for investigation of dispersions by electron microscopy, Adv. Colloid Interface Sci. *5*, 281.

Meryman, H.T. (1966), Review of biological freezing, in: Cryobiology, H.T. Meryman, ed. (Academic Press, New York), p. 1.

Meryman, H. T. (1971), Cryoprotective agents, Cryobiology *8*, 173.

Meryman, H.T. (1974), Freezing injury and its prevention in living cells, Ann. Rev. Biophys. Bioeng. *3*, 341.

Meyer, H.W. and H. Winkelmann (1970), Die Darstellung von Lipiden bei der gefrierätzpräparation und ihre Beziehung zur Gefrieranalyse biologischer Membranen, Exp. Pathol. *4*, 47.

Miller, K.R. (1979), Artefacts associated with deep etching technique, in: Freeze-fracture: methods, artefacts and interpretations, J.E. Rash and C.S. Hudson, eds. (Raven Press, New York), p. 31.

Mishima, L., L.D. Calvert and E. Whalley (1984), 'Melting ice' I at 77K and 10 kBar: a new method of making amorphous solids, Nature, Lond. *310*, 393.

Monroe, R.G., W.J. Gamble, C.G. La Farge, R. Gamboa, C.L. Morgan, A. Rosenthal and S. Bullivant (1968), Myocardial ultrastructure in systole and diastole using ballistic cryofixation, J. Ultrastruct. Res. *22*, 22.

Moor, H. (1964), Die Gefrierfixation lebender Zellen und ihre Anwendung in der Elektronenmikroskopie, Z. Zellforsch. *62*, 546.

Moor, H. (1970), Die Gefrierätztechnik, in: Methodensammlung der Elektronenmikroskopie, G. Schimmel and W. Vogell, eds. (Wissenchaftliche Verlagsanstallt, Stuttgart), p. 1.

Moor, H. (1971), Recent progress in the freeze-etching technique, Phil. Trans. R. Soc. Ser B *261*, 121.

Moor, H. (1973), Etching and related problems, in: Freeze-etching, techniques and applications, E.L. Benedetti and P. Favard, eds. (Soc. Française de Microscopie Electronique, Paris), p. 21.

Moor, H. and M. Hoechli (1970), The influence of high pressure freezing on living cells, Proc. 7th Int. Congr. Electron Microscopy, Grenoble *1*, 445.

Moor, H., J. Kistler and M. Müller (1976), Freezing in a propane jet, Experientia *32*, 805.

Moor, H., G. Bellin, C. Sandri and K. Akert (1980), The influence of high pressure freezing on mammalian nerve tissue, Cell Tissue Res. *209*, 201.

Müller, M., N. Meister and H. Moor (1980), Freezing in a propane jet and its application in freeze-fracturing, Mikroskopie *36*, 129.

Nanninga, N. (1973), Freeze-fracturing of microorganisms. Physical and chemical fixation of *Bacillus subtilis*, in: Freeze-etching, techniques and applications, E.L. Benedetti and P. Favard, eds. (Soc. Française de Microscopie Electronique, Paris), p. 151.

Nash, T. (1966), Chemical constitution and physical properties of compounds able to protect living cells against damage due to freezing and thawing, in: Cryobiology, H.T. Meryman, ed. (Academic Press, New York), p. 179.

Nei, T. (1973), Growth of ice crystals in frozen specimens, J. Microscopy *99*, 227.

Nei, T. (1977), Freezing patterns and post-thaw survival of yeast cells, Cryobiology *14*, 711.

Nei, T. (1978), Structure and function of frozen cells: freezing patterns and post-thaw survival, J. Microscopy *112*, 197.

Nermut, M.V. and B.J. Ward (1974), Effect of fixatives on fracture plane in red blood cells, J. Microscopy *102*, 29.

Neushul, M. (1970), A freeze-etching study of the red alga *Porphyridium*, Amer. J. Bot. *57*, 1231.

Niedermeyer, W. and H. Moor (1976), Effect of glycerol on structure of membranes: a freeze-etch study, Proc. 6th Europ. Congr. Electron Microscopy, Jerusalem *2*, 108.

Northcote, D.H. and D.R. Lewis (1968), Freeze-etched surfaces of membranes and organelles in the cells of pea root tips, J. Cell Sci. *3*, 199.

Parish, G.R. (1974), Seasonal variation in the membrane structure of differentiating shoot cambial cells demonstrated by freeze-etching, Cytobiologie *9*, 131.

Parish, G.R. (1975), Changes of particle frequency in freeze-etched erythrocyte membranes after fixation, J. Microscopy *104*, 245.

Pfenninger, K.H. and R.P. Bunge (1974), Freeze-fracturing of nerve growth cones and young fibres. A study of developing plasma membrane, J. Cell Biol. *63*, 180.

Phillips, T.E. and A.F. Boyne (1984), Liquid nitrogen-based quick freezing: experiences with bounce-free delivery of cholinergic nerve terminals to a metal surface, J. Electron Microsc. Tech. *1*, 9.

Pihakaski, K. and L. Seveus (1980), High polymeric protective additives in cryoultramicrotomy of plants. I. PVP infusion of fixed plant tissues. Cryo-Letters *1*, 494.

Pinto da Silva, P. (1972), Translational mobility of membrane intercalated particles of human erythrocyte ghosts. pH dependent, reversible aggregation, J. Cell Biol. *53*, 777.

Pinto da Silva, P. and B. Kachar (1980), Quick freezing versus chemical fixation: capture and identification of membrane fusion intermediates, Cell Biol. Int. Reports *4*, 625.

Pinto da Silva, P. and R. G. Miller (1975), Membrane particles on fracture faces of frozen myelin, Proc. Natl. Acad. Sci. U.S.A. *72*, 4046.

Plattner, H. (1970), A study on the interpretation of freeze-etched animal tissues and cell organelles, Mikroskopie *26*, 233.

Plattner, H. (1971), Bull spermatozoa – a reinvestigation by freeze-etching using widely different cryofixation procedures, J. Submicrosc. Cytol. *3*,19

Plattner, H. (1984), Ultrarapid freezing and freeze-etching of cell monolayers and suspensions, Proc. 8th Europ. Reg. Conf. Electron Microscopy, Budapest *3*, 1731.

Plattner, H. and L. Bachmann (1982), Cryofixation: a tool in biological ultrastructural research, Int. Rev. Cytol. *79*, 237.

Plattner, H., W.M. Fischer, W.W. Schmitt and L. Bachmann (1972), Freeze-etching of cells without cryoprotectants, J. Cell Biol. *53*, 116.

Plattner, H., W.W. Schmitt-Fumian and L. Bachmann (1973), Cryofixation of single cells by spray freezing, in: Freeze-etching, techniques and applications, E.L. Benedetti and P. Favard, eds. (Soc. Française de Microscopie Electronique, Paris), p. 81.

Pryde, J.A. and G.O. Jones (1964), Properties of vitreous water, Nature, Lond. *170*, 685.

Pscheid, P., C. Schudt and H. Plattner (1981), Cryofixation of monolayer cell cultures for freeze-fracturing without chemical pretreatments, J. Microscopy *121*, 149.

Pueschel, C.M. (1977), A freeze-etch study of the ultrastructure of red algal pit plugs, Protoplasma *91*, 15.

Puritch, G.S. and R.P.C. Johnson (1973), The structure of freeze-etched bordered-pit membranes of Abies grandis, Wood Sci. Technol. *7*, 256.

Pyliotis, A.N., D.J. Goodchild and L.H. Grimme (1975), The regreening of nitrogen deficient *Chlorella fusca*. II. Structural changes during synchronous regreening, Arch. Microbiol. *103*, 259.

Rapatz, G.L. and B.J. Luyet (1968), Combined effects of freezing rates and various protective agents on the preservation of human erythrocytes, Cryobiology *4*, 215.

Rapatz, G.L., L.J. Menz and B.J. Luyet (1966), Anatomy of the freezing process in biological materials, in: Cryobiology, H.T. Meryman, ed. (Academic Press, New York), p. 139.

Rasmussen, D.H. (1982), Ice formation in aqueous systems, J. Microscopy *128*, 167.

Rasmussen, D.H. and B.J. Luyet (1969), Complementary study of some non-equilibrium phase transitions in frozen solutions of glycerol, ethylene glycol, glucose and sucrose, Biodynamica *10*, 319.

Rasmussen, D.H. and B.J. Luyet (1970), Contribution to the establishment of the temperature-concentration curves of homogeneous nucleation in solutions of some cryoprotective agents, Biodynamica *11*, 33.

Rasmussen, D.H. and A.P. MacKenzie (1968), The glass transition in amorphous water. Application of the measurements to problems arising in cryobiology, J. Phys. Chem. *75*, 976.

Rasmussen, D.H. and A.P. MacKenzie (1968), Phase diagram for the system water/dimethyl sulphoxide. Nature, Lond. *220*, 1315.

Rasmussen, D.H. and A.P. Mackenzie (1972), Effects of solutes on ice-solution interfacial free energy: calculation from measured homogeneous nucleation temperatures, in: Water structure at the water-polymer interface, H.H.G. Jellinek, ed. (Plenum Press, New York), p. 126.

Raviola, E., D.A. Goodenough and G. Raviola (1980), Structure of rapidly frozen gap junctions, J. Cell Biol. *87*, 273.

Rebhun, L.I. (1972), Freeze-substitution and freeze-drying, in: Principles and techniques of electron microscopy, M.A. Hayat, ed. (Van Nostrand Reinhold, New York), p. 3.

Reith, A. and R. Oftebro (1971), Structure of HeLa cells after treatment with glycerol as revealed by freeze-etching and electron microscopic methods, Exp. Cell Res. *66*, 385.

Remsen, C.C. and S.W. Watson (1972), Freeze-etching of bacteria, Int. Rev. Cytol. *33*, 253.

Remsen, C.C., S.W. Watson, J.B. Waterbury and H.G. Trüper (1968), Fine structure of *Ectothiorhodospira mobilis* Pelsh, J. Bacteriol. *95*, 2374.

Rey, L.R. (1960), Thermal analysis of eutectics in freezing solutions, Ann. N. Y. Acad. Sci. 510.

Richmond, D. V. and R. J. Pring (1971a), Fine structure of *Botrytis fabae Sardina conidia*, Ann. Bot. *35*, 175.

Richmond, D. V. and R. J. Pring (1971b), Fine structure of germinating *Botrytis fabae Sardina conidia*, Ann. Bot. *35*, 493.

Richter, H. (1968a), Die Reaktion hochpermeabler Pflanzenzellen auf die Gefrierschutzstoffe (Glyzerin, Athylenglykol, Dimethylsulfoxid), Protoplasma *65*, 155.

Richter, H. (1968b), Low-temperature resistance of *Campanula*-cells treated with glycerol, Protoplasma *66*, 63.

Richter, H. and U. Sleytr (1970), Gefrieratzung des Assimilations Parenchyms von *Asparagus sprengeri* Regel, Mikroskopie *26*, 329.

Riehle, U. (1968a), Uber die Vitrifizierung verdunnter wassriger Losungen, Doctoral thesis (Juris Druck and Verlag, Zurich).

Riehle, U. (1968b), Fast freezing of organic specimens for electron microscopy, Chem. Ing. Tech. *40*, 213.

Riehle, U. and M. Hoechli (1973), The theory and technique of high pressure freezing, in: Freeze-etching, techniques and applications, E.L. Benedetti and P. Favard, eds. (Soc. Française Microscopie Electronique, Paris), p. 31.

Rigler, M.W. and J.S. Patton (1984), The cryocell. A simple inexpensive cryogenic storage device for microscopic specimens, J. Microscopy, *134*, 335.

Rinderer, L. and F. Haenseler (1959), Heat transfer in superfluid helium, in: Proc. 10th Congr. Refrig. (Pergamon, London)*1*, 243.

Robards, A.W. (1980), A microprocessor controlled rapid cooling rate meter, Cryo-Letters *1*, 384.

Robards, A.W. and P. Crosby (1979), A comprehensive freezing, fracturing and coating system for low temperature SEM, Scanning Electron Microscopy 1979, *2*, 325.

Robards, A.W. and P. Crosby (1983), Optimisation of plunge freezing: linear relationship between cooling rate and entry velocity into liquid propane, Cryo-Letters *4*, 23.

Robards, A.W. and G.R. Parish (1971), Preparation and mechanical requirements for freeze-etching thick walled plant tissue, J. Microscopy *93*, 61.

Robards, A.W. and N.J. Severs (1982), A comparison between cooling rates achieved using a propane jet device and plunging into liquid propane, Cryo-Letters *2*, 135.

Robards, A.W., U.B. Sleytr, G.R. Bullock, P.D. Sibbons and K.H. Quine (1980), Thin plate cleavage device for Leybold- Heraeus Bioetch 2005 freeze-etching unit, Proc. 7th Europ. Reg. Conf. Electron Microscopy, The Hague *2*, 740.

Robards, A.W., G.R. Bullock, M.A. Goodall and P.D. Sibbons (1981), Computer assisted analysis of freeze-fractured membranes following exposure to different temperatures, in: Effects of low temperatures on biological membranes, G.J. Morris and A. Clarke, eds. (Academic Press, London), p. 219.

Rosenkranz, J. (1975), Course of temperature variation in an object during freeze-etching procedure, Arzneim Forsch. *25*, 454.

Rottenberg, W. and H. Richter (1969), Automatische Glyzerinbehandlung pflanzlicher Dauergewebszellen fur die Gefrieratzung, Mikroskopie *25*, 313.

Ryan, K.R. and D.H. Purse (1984), Rapid freezing: specimen supports and cold gas layers. J. Microscopy *136*, RP5.

Saetersdal, T.S., J. Røli, R. Myklebust and H. Engedal (1977), Preservation of shock-frozen myocardial tissue as shown by cryoultramicrotomy and freeze-fracture studies, J. Microscopy *111*, 297.

Sassen, M.M.A., C.C. Remsen and W.M. Hess (1967), Fine structure of *Penicillium megasporum* conidiospores, Protoplasma *64*, 75.

Satir, B.H. and P. Satir (1979), Partitioning of intramembrane particles during freeze-fracture procedures, in: Freeze-fracture: methods, artefacts and interpretations, J.E. Rash and C.S. Hudson, eds. (Raven Press, New York), p. 43.

Sattler, C.A. and L.A. Staehelin (1974), Ciliary membrane differentiations in *Tetrehymena pyriformis*, J. Cell Biol. *62*, 473.

Schenk, H. (1959), Heat transfer engineering (Prentice-Hall, London).

Schiller, A. and R. Taugner (1980), Freeze-fracturing and deep etching with the volatile cryoprotectant ethanol reveals true membrane surfaces of kidney structures, Cell Tissue Res. *210*, 57.

Schiller, A., R. Taugner and E. Rix (1978), Die Anwendung kryoprotectiver Substanzen in der Gefrierbruchtechnik, Mikroskopie *34*, 19.

Schiller, A., U. Sonnhof and R. Taugner (1979), Tissue function-compatible cryoprotectants in cryoultramicrotomy and freeze fracturing, Mikroskopie *35*, 23.

Schmalbruch, H. (1980), Delayed fixation alters pattern of intramembrane particles in mammalian muscle fibres, J. Ultrastruct. Res. *70*, 15.

Schuler, G., H. Plattner, W. Aberer and H. Winkler (1978), Particle segregation in chromaffin granule membranes by forced physical contact, Biochim. Biophys. Acta *513*, 244.

Schultz, D. and H. Lehmann (1974), Use of freeze-etching technique to demonstrate the ultrastructure of dormant cells, Proc. 8th. Int. Congr. Electron Microscopy, Canberra *2*, 712.

Schultz, D., H.V. Neidhart, E. Perner, J. Jaenicke and J. Sommer (1973), Verbesserte Darstellung der Feinstruktur ruhender Pflanzenzellen, Protoplasma *78*, 41.

Schwabe, K.G. and L. Terracio (1980), Ultrastructural and thermocouple evaluation of rapid freezing techniques, Cryobiology *17*, 571.

Schwelitz, F.D., W.R. Evans, H.H. Mollenhauer and R.A. Dilley (1970), The fine structure of the pellicle of *Euglena gracilis* as revealed by freeze-etching, Protoplasma *69*, 341.

Sevéus, L. (1978), Preparation of biological material for X-ray microanalysis of diffusible solutes. 1. Rapid freezing of biological tissue in nitrogen slush and preparation of ultra-thin frozen sections in the absence of trough liquid, J. Microscopy *112*, 269.

Shelton, E. and W.E. Mowczko (1979), Scanning electron microscopy of membrane blisters produced by glutaraldehyde fixation and stabilised by post fixation in osmium tetroxide, in: Freeze-fracture: methods, artefacts and interpretations, J.E. Rash and C.S. Hudson, eds. (Raven Press, New York), p. 67.

Shepard, M.L., C.S. Goldstein and F.H. Cocks (1976), The H_2O-NaCl-glycerol phase diagram and its application to cryobiology, Cryobiology *13*, 9.

Silvester, N.R., S. Marchese-Ragona and D.N. Johnston (1982), The relative efficiency of various fluids in the rapid freezing of protozoa, J. Microscopy *128*, 175.

Sitte, H. (1979), Cryofixation of biological material without pretreatment, Mikroskopie *35*, 14.

Sitte, H. (1984), Equipment for cryofixation, cryoultramicrotomy and cryosubstitution in biomedical TEM-routine, Zeiss Magazine for Electron Microscopists *3*, 25.

Sitte, H., H. Fell, W. Hobli, H. Kleber and K. Neumann (1977), Fast freezing device, J. Microscopy *111*, 35.

Sjöland, R.D. and D.A. Smith (1974), Freeze-fracture studies of photosynthetic deficient supergranal chloroplasts in tissue cultures containing virus-like particles, J. Cell Biol. *60*, 285.

Skaer, H.le B. (1982), Chemical cryoprotection for structural studies, J. Microscopy *125*, 137.

Skaer, H.le B., F. Franks, M.H. Asquith and P. Echlin (1977), Polymeric cryoprotectants in the preservation of biological ultrastructure. III. Morphological aspects, J. Microscopy *110*, 257.

Skaer, H.le B., F. Franks and P. Echlin (1978), Nonpenetrating polymeric cryofixatives for ultrastructural and analytical studies of biological tissues, Cryobiology *15*, 589.

Skaer, H.le B., F. Franks and P. Echlin (1979), Freeze-fracture studies of ice recrystallization in tissues quenched in aqueous solutions of polyvinylpyrrolidone, Cryo-Letters *1*, 61.

Sleytr, U.B. (1974), Freeze-fracturing at liquid helium temperature for freeze-etching, Proc. 8th Int. Congr. Electron Microscopy, Canberra *2*, 30.

Sleytr, U.B. (1978) Regular arrays of macromolecules on bacterial cell walls – structure, chemistry, assembly and function, Int. Rev. Cytol. *53*, 1.

Sleytr, U.B., H. Adam and H. Klaushofer (1969), Die Feinstruktur der Konidien von *Aspergillus niger*, V. Tiegh., dargestellt mit Hilfe der Gefrierätztechnik, Mikroskopie *25*, 320.

Sleytr, U.B., H. Groesz and W. Umrath (1981), Interpretation of morphological data obtained by freeze-fracturing at very low temperatures, Acta Histochemica *23*, Suppl., S29.

Somlyo, A.V. and J. Silcox (1979), Cryoultramicrotomy for electron probe analysis, in: Microbeam analysis in biology, C.P. Lechene and R.R. Warner, eds. (Academic Press, New York) p. 535.

Southworth, D. and D. Branton (1971), Freeze-etched pollen walls of *Artemesia pycnocephala* and *Lilium humboldt*, J. Cell Sci. *9*, 193.

Southworth, D., K. Fisher and D. Branton (1975), Principles of freeze-fracturing and etching, in: Techniques in biochemical and biophysical morphology, D. Glick and R.M. Androsenbaum, eds. (Wiley Interscience, New York, London), *2*, 247.

Staehelin, L.A. (1973), Analysis and critical evaluation of the information contained in freeze-etch micrographs, in: Freeze-etching, techniques and applications, E.L. Benedetti and P. Favard, eds. (Soc. Française de Microscopie Electronique, Paris), p. 223.

Staehelin, L.A. (1979), A simple guide to the evaluation of the quality of a freeze-fracture replica, in: Freeze-fracture: methods, artefacts and interpretations, J.E. Rash and C.S. Hudson, eds. (Raven Press, New York), p. 11.

Staehelin, L.A. and W.S. Bertaud (1971), Temperature and contamination dependent freeze-etch images of frozen water and glycerol solutions, J. Ultrastruct. Res. *37*, 146.

Staehelin, L.A. and O. Kiermayer (1970), Membrane differentiation in the Golgi complex of *Micrasterias denticulata* Bréb. Visualised by freeze-etching, J. Cell Sci. *7*, 787.

Steere, R.L. (1971), Retention of three dimensional contours by replicas of freeze-fracture specimens, Proc. 29th Ann. Meeting EMSA, p.242.

Stephenson, J.L. (1954), Caution in the use of liquid propane for freezing biological specimens, Nature, Lond. *174*, 235.

Stephenson, J.L. (1956), Ice crystal growth during the rapid freezing of tissues, J. biophys. biochem. Cytol. *2*, 45.

Stephenson, J.L. (1960), Ice crystal formation in biological materials during rapid freezing, Ann. N. Y. Acad. Sci. *85*, 535.

Stocks, D.L. and W.M. Hess (1970), Ultrastructure of dormant and germinated basidiospores of a species of *Psilocybe*, Mycolgia *62*, 176.

Stolinski, C. (1977), Freeze-fracture replication in biological research development, current practice and future prospects, Micron *8*, 87.

Swift, J.G. and M.S. Buttrose (1972), Freeze-etch studies of protein bodies in wheat scutellum, J. Ultrastruct. Res. *40*, 378.

Swift, J.G. and M.S. Buttrose (1973), Protein bodies, lipid layers and amyloplasts in freeze-etched pea cotyledons, Planta *109*, 61.

Takeo, K. and M. Nishiura (1974), Ultrastructure of polymorphic mucor as observed by means of freeze-etching, Arch. Microbiol. *98*, 175.

Takeo, K., I. Uesaka and K. Uehira (1973), Fine structure of *Cryptococcus neoformans* grown in vitro as observed by freeze-etching, J. Bact. *113*, 1442.

Taylor, G.J. and W.S. Harris (1970), Cardiac toxicity of aerosol propellants, J. Amer. Med. Ass. *214*, 81.

Thornley, M.J. and U.B. Sleytr (1974), Freeze-etching of outer membranes of *Pseudomonas* and *Acinetobacter*, Arch. Microbiol. *100*, 149.

Touloukian, Y.S. et al (1970), Thermophysical properties of matter, Vols.*1–13* (IFI/Plenum, New York, Washington).

Tsien, H.C. and M.L. Higgins (1974), Effect of temperature on distribution of membrane particles in *Streptococcus faecalis* as seen by the freeze-fracture technique, J. Bacteriol. *118*, 725.

Umrath, W. (1974a), Cooling bath for rapid freezing in electron microscopy, J. Microscopy *101*, 103.

Umrath, W. (1974b), Gefrierschock in Kuhlbadern fur die Elektronenmikroskopie, Beitr. elektronenmikroskop. Direktabb. Oferfl. *7*, 103.

Umrath, W. (1975), Rapid freezing in open cooling baths, Arnzeim Forsch. *25*, 450.

Umrath, W. (1977), Prinzipien der Kryopraparationsmethoden – Gefriertrocknung und Gefrieratzung, Mikroskopie *33*, 11.

Ushiyama, M., E.G. Cravalho and R.L. Levin (1979), Volumetric changes in yeast cells during freezing at constant cooling rates, J. Membr. Biol. *6*, 112.

Van Deurs, B. and J.H. Luft (1979), Effects of glutaraldehyde fixation on the structure of tight junctions: a quantitative freeze-fracture analysis, J. Ultrastruct. Res. *68*, 160.

Van Gool, A. P., J. Meyer and R. Lambert (1970), The fine structure of frozen etched *Fusarium conidiospores*, J. Microscopie *9*, 653.

Van Harreveld, A. and J. Crowell (1964), Electron microscopy after rapid freezing on a metal surface and substitution fixation, Anat. Rec. *149*, 381.

Van Harreveld, A. and J. Trubatch (1979), Progress of fusion during rapid freezing for electron microscopy, J. Microscopy *115*, 243.

Van Harreveld, A., J. Crowell and S.K. Malhotra (1965), A study of extracellular space in central nervous tissue by freeze-substitution. J. Cell Biol. *25*, 117.

Van Harreveld, A., J. Trubatch and J. Steiner (1974), Rapid freezing and electron microscopy for the arrest of physiological processes, J. Microscopy *100*, 189.

Van Venetië, R., W.J. Hage, J.G. Bluemink and A.J. Verkleij (1981), Propane jet freezing – a valid ultra-rapid freezing method for the preservation of temperature dependent lipid phases, J. Microscopy *123*, 287.

Vargaftik, N.B. (1975), Tables on the thermophysical properties of liquids and gases in normal and dissociated states. 2nd. Edn. (Hempisphere, Washington, London).

Venrooij, G.E.P.M., A.M.H.J. Aertsen, W.M.A. Hax, P.H.J.T. Ververgaert, J.J. Verhoeven and H.A. Van der Vost (1975), Freeze-etching: freezing velocity and crystal size at different locations in samples, Cryobiology *12*, 46.

Verkleij, A.J., P.H. Ververgaert, R.A. Prins and L.M.G. van Golde (1975), Lipid phase transitions of the strictly anaerobic bacteria *Veillonella parvula* and *Anaerovibrio lipolytica*, J. Bacteriol. *124*, 1522.

Verkleij, A.J. and P.H. Ververgaert (1975), Architecture of biological and artificial membranes as visualised by freeze-etching, Ann. Rev. Phys. Chem. *26*, 101.

Verkleij, A.J. and P.H. Ververgaert (1978), Freeze-fracture morphology of biological membranes, Biochim. Biophys. Acta *515*, 303.

Ververgaert, P.H.J.T., A.J. Verkleij, J.J. Verhoeven and P.F. Elbers (1973), Spray-freezing of liposomes. Biochim. Biophys. Acta *311*, 651.

Weast, R., ed. (1971 and other editions), CRC handbook of chemistry and physics, (Chemical Rubber Company, Ohio).

Wecke, J. (1968), Vergleichende elektronenmikroskopische Untersuchungen an Membranstrukturen von latent lebenden Organismen verschiedener systematischer Stellung zur Frage der Unit-Membran-Konzeptes, Inauguraldissertation, (Freie Universität, Berlin).

Wendt-Gallitelli, M.F. (1984), Importance of freezing tissue in defined functional states for reproducible X-ray microanalysis in heart muscle, Proc. 8th Europ. Reg. Conf. Electron Microscopy, Budapest *3*, 1677.

Werz, G. and G. Keller (1970), Die Struktur des Golgi-Apparates bei gefriergeätzten Dunaliella-Zellen, Protoplasma 69, 351.

Westwater, J.W. (1959), Boiling heat transfer, American AEEE Scientist *47*, 427.

White, G.K. (1979), Experimental techniques in low temperature physics (Clarendon Press, Oxford).

Williams, L. and S. Hodson (1978), Quench cooling and ice crystal formation in biological tissues, Cryobiology *15*, 323.

Williamson, F.A. (1984), Storage of small samples under liquid nitrogen, J. Microscopy *134*, 125.

Willison, J.H.M. (1975), Plant cell wall microfibril deposition revealed by freeze-fractured plasmalemma not treated with glycerol, Planta *126*, 93.

Willison, J.H.M. (1976), An examination of the relationship between freeze-fractured plasmalemma and cell wall microfibrils, Protoplasma *88*, 187.

Willison, J.H.M. and R.M. Brown (1979), Pretreatment artefacts in plant cells, in: Freeze-fracture: methods, artefacts and interpretations, J.E. Rash and C.S. Hudson, eds. (Raven Press, New York), p. 51.

Willison, J.H.M. and E.C. Cocking (1975), Microfibril synthesis at the surface of isolated Tobacco mesophyll protoplasts, a freeze-etch study, Protoplasma *84*, 147.

Willson, J.E., D.E. Brown and E.K. Timmens (1965), A toxicologic study of dimethyl sulfoxide, Toxicology appl. Pharmacol. *7*, 104.

Wilson, A.J. and A.W. Robards (1982), Some experiences in the use of a polymeric cryoprotectant in the freezing of plant tissue, J. Microscopy *125*, 287.

Wunderlich, F. and V. Speth (1972), Membranes in *Tetrahymena*. I. The cortical pattern, J. Ultrastruct. Res. *41*, 258.

Yannas, I. (1968), Vitrification temperature of water, Science *160*, 298.

Yee, A.G. and J. Revel (1978), Loss and reappearance of gap junctions in regenerate liver, J. Cell Biol. *78*, 554.

Zachariassen, K.E. and J.A. Husby (1982), Antifreeze effect of thermal hysteresis agents protects highly supercooled insects, Nature, Lond. *298*, 865.

Zerban, H., M. Wehner and G. Werz (1973), Uber die Feinstruktur des Zellkerns von Acetabularia nach Gefrieratzung, Planta *114*, 239.

Zierold, K. (1980), Cryofixation of tissue specimens studied by cooling rate measurements and scanning electron microscopy, Microscopica Acta *83*, 25.

Zingsheim, H.P. and H. Plattner (1976), Electron microscopy methods in biology, in: Methods in membrane biology, Vol. 7, E.D. Korn, ed. (Plenum Publishing Company, New York) p.1.

Chapter 3

Direct viewing and analysis
of cold specimens

3.1 Principles and problems

The most obvious and direct way of dealing with a biological specimen once it has been frozen is to look at it while still in this hydrated frozen state. This, however, is not without its attendant problems. While bulk, frozen specimens are now readily and routinely observed in the scanning electron microscope (SEM), both the preparation and examination of thin frozen specimens required for transmission electron microscopy (TEM) still present very substantial difficulties.

It is technically very demanding to produce thin specimens from bulk frozen material and, moreover, when such samples are irradiated in an electron microscope, they can be all too easily irreversibly damaged by the effects of heating, by radiolysis of ice or by other direct or indirect results of electron bombardment. A major reason for putting cold or frozen specimens into electron microscopes is to make use of the potentially improved retention of diffusible components in their *in vivo* positions so that different microanalytical techniques can be applied. It is *not* the purpose of this book to deal with microanalytical methods, which are fully discussed elsewhere in this series (Chandler 1977). However, the preparation of specimens for such use must obviously take into account the requirements of the final observational and microanalytical processes. Thus, while we will not discuss microanalysis *per se* in any detail, the implication that many of the preparatory techniques will be used with this goal in mind must necessarily condition our assessment of the experimental procedures used as we move from the native biological structure towards the hostile environment of an electron microscope column.

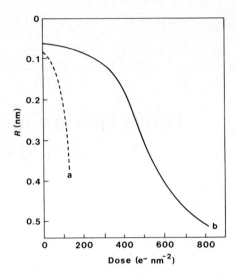

Fig. 3.1.   Radiation damage in crystals of L-valine at different temperatures in relation to electron dose. The value of R indicates the maximum resolution remaining in the diffraction pattern: R is the inverse of the average spatial frequency corresponding to the highest diffraction orders visible in the two axial directions of the pattern. Curve (a) represents data recorded at a beam voltage of 60 kV with the L-valine at room temperature while (b) was recorded with a beam voltage of 65 kV at liquid helium temperature. The lower temperature clearly confers some protection on the specimen. This result is typical of a number of similar experiments although earlier work had suggested greater protection of the specimen at low temperatures than indicated here. (Abridged and redrawn from Lamvik et al. 1983.)

Although most of this book is concerned with the production, and subsequent utilisation, of hydrated frozen specimens, there are also good reasons for using cold stages in electron microscopes so that other specimens (not necessarily hydrated) can be viewed and analysed at low temperatures. In particular, it has been shown that radiation damage is reduced when specimens are maintained at low temperature instead of at room temperature (e.g. Cosslett 1978; Glaeser and Taylor 1978; Taylor 1978; Dubochet et al. 1982; Talmon 1982a,b; Fig. 3.1), although the actual *quantitative* reduction remains a matter of debate (see below). A number of authors have pointed to the extremely low electron dose that will bring about irreversible changes in biological specimens at 'normal' temperatures in a microscope column (e.g. Kellenberger 1982). The radiation dose to the specimen is measured as the number of electrons (n_0) incident on one square nanometre. n_0 then has units of e$^-$ nm^{-2} where 6.25 e$^-$ nm^{-2} is equivalent to a dose of 1 coulomb per square metre (C m^{-2}). Misell (1978) provides a useful discussion of electron dose in relation to radiation damage. Taylor (1978) has commented that the 'safe' dose for high resolution studies on unstained biological specimens is less than about 100 e$^-$ nm^{-2} (and, for some specimens, less than 10 e$^-$ nm^{-2}) while Cosslett has stated that most biological specimens will be damaged with an input energy of between 1 and 10 C m^{-2} (approximately 6 to 60 e$^-$ nm^{-2}). When it is realised that the normal electron dose simply to record an image (i.e. ignoring the time taken for finding the right position on the specimen, focusing, etc.) at a magnification of 4000 times is about 20 e$^-$ nm^{-2} (Glaeser and Taylor 1978), the magnitude of the problem is clearly seen. There has been considerable debate about the actual degree of protection from radiation damage conferred by the use of low temperatures. In part this has arisen because of the difficulties inherent in ensuring that the actual specimen in an electron microscope is held at a specific, very low temperature (e.g. 4.2 K). Knapeck and Dubochet (1980) suggested that radiation damage was reduced by a factor of 30 to 300 when the specimen was observed at 4.2 K in a microscope using a liquid helium super-conducting lens. However, subsequent reports (e.g. Lamvik et al. 1983) have rated the protective value of low temperatures much more conservatively with an improvement factor of between four and six. Discussions of this matter are contained in Talmon (1982a,b) and Dubochet et al. (1982) (see also Knapeck et al. 1984). We shall return to this problem in §3.4.1 and §3.4.2 but suffice it to say that, for some high resolution purposes, even a gain of a factor of two would be sufficient to warrant the use of low temperature stages and, therefore, the practical methods of handling and viewing any specimen at low temperature fall within the ambit of this book.

While the direct insertion of a cold specimen onto the low temperature stage of an electron microscope is conceptually one of the simplest ways of treating a frozen specimen, even this is not without its difficulties. The specimen must be kept cold while it is moved from its place of cooling into the microscope and, furthermore, it must be protected from the accumulation of contamination by condensing gases. In addition, few specimens can be viewed immediately without additional processing. This Chapter describes the different means of preparing specimens for observation and analysis at low temperatures, methods for the protection of specimens during transfer and, finally, cold stages for both scanning and transmission electron microscopes.

3.1.1   Types of specimen

The types of specimen viewed at low temperatures may be broadly divided into four main categories (Fig. 3.2):

(1)   frozen-hydrated – with the possibility for partial or complete freeze-drying at some stage during the preparation pathway;

(2)   frozen-fractured – a simple modification of (1) that allows internal surfaces to be exposed;

(3)   frozen-section – thin pieces of frozen-hydrated or frozen-dried tissue that can be viewed by transmission imaging, and

(4)   frozen thin-layer – which are not necessarily hydrated but which are observed at low temperature to minimise radiation damage.

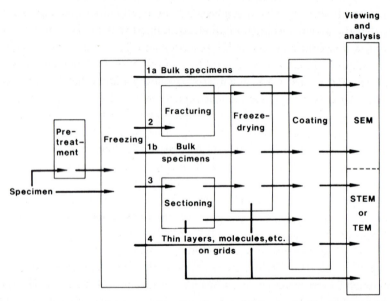

Fig. 3.2.   Diagrammatic illustration of preparation possibilities for the direct observation of biological specimens using low temperature methods. All specimens may be frozen directly or after pretreatment (depending on the nature of the specimen and the requirements of the experiment). (1a) Bulk specimens, after coating, may be viewed directly on the cold stage of an SEM. Alternatively (1b) they can be frozen-dried and then transferred to the SEM for viewing at ambient temperature. (2) Bulk specimens can also be fractured open, perhaps partially frozen-dried (freeze-etched) to reveal further structure, and then coated before viewing in the SEM. (3) Specimens for viewing in the transmission mode must be prepared so that they are thin enough for electrons to traverse. This can be done by cryosectioning possibly followed by freeze-drying and/or coating. (4) Thin layers are usually prepared directly and, again after the possibility of coating, are viewed directly.

3.2 Instrumental requirements for low temperature electron microscopy

This is primarily a book of *practical* methods in electron microscopy. However, it has also been part of the authors' philosophy in this, still emergent, field of low temperatures that the background and underlying principles of the use of specific techniques and procedures should be fully understood so that sensible and correct decisions can be made in selecting appropriate methods. This approach becomes especially important when discussing equipment. Thus, while it is possible to provide a detailed statement on how to use a cryoultramicrotome or how best to freeze a particular type of specimen, the same approach cannot be employed when dealing with, for example, cold stages in electron microscopes. For freeze-sectioning (see Chapter 4) a relatively high degree of operator knowledge and training is necessary, whereas with cold stages, provided that the right equipment is available, the operator has little scope for individual variation and merely has to follow the specific instructions of the designer. In this section we shall, therefore, be more concerned with the basic *principles* underlying the design of the stages as illustrated with a few selected examples; see Butler and Hale (1981) for a full discussion of cold stages. It is certainly *not* the intention of this book to provide the information necessary for the reader to produce his own cold stage or transfer device! This is a specialist area which relatively few individuals will have either the aptitude, facilities or inclination to enter. The purpose of what follows is therefore to provide the basic information and fundamentals that will apply to *any* system and then to look at one or two examples of how these have been put into practice.

3.2.1 Microscopes

The principle requirements for cold stages in electron microscopes are that they should easily maintain the specimen at the lowest required temperature, that they should be mechanically and thermally stable and that the specimen should be protected from contamination by molecules condensing from the residual gas in the vacuum system. If microanalysis is to be carried out, then the cold stage should also be constructed so that minimum background radiation is generated.

The technology required for the construction and use of cold stages to protect specimens from the effects of radiation damage is, to quite a large extent, the same as that required for purposes of observation and analysis, although there may be substantial differences in detail. Therefore, with the present exception of liquid helium-cooled stages, it is possible to discuss the technological requirements generally, irrespective of the purposes that the specific cold stage will ultimately serve. It should, however, be noted that, whereas for high resolution molecular studies the prime requirement is for low temperature and high spatial resolution, for X-ray microanalysis the emphasis is on good X-ray performance (thus precluding massive sample holders) and high temperature stability so that specimen drift is minimised over the relatively long counting period. For these reasons low temperature stages for X-ray use need to be more delicate than those for high resolution structural studies.

The actual temperature of the cold stage will depend primarily on whether liquid nitrogen or liquid helium is the cryogenic source. For SEMs, liquid nitrogen (or adiabatic expansion of nitrogen – the Joule-Thomson effect) has been adequate for most purposes but, for high resolution TEMs or scanning transmission electron microscopes (STEMs), helium-cooled stages have advantages in terms of increased protection from radiation damage. Liquid helium-cooled stages are *very* specialised pieces of equipment and they will not be considered further here. The majority of cold stages have used nitrogen as a coolant and operate down to a lowest temperature of about 63 K (the temperature of sub-cooled nitrogen). It is now generally accepted that cold stages for biological work should be designed to operate at as low a temperature as possible (even if this is not always required). Some early SEM cold stages, including commercial designs, with a minimum working temperature of only about 140 K are no longer considered adequate: any cold stage should be capable of being cooled to < 110 K.

SEM and TEM cold stages differ considerably in their construction. SEM stages tend to be quite large and capable of maintaining a relatively bulky specimen at low temperature, whereas TEM stages need 'less' cooling but much better thermal and mechanical stability and precision. The parameter determining the lowest operating temperature for the specimen is the ratio between the rate of heat input to the specimen and heat withdrawal from it. Heat input to the specimen mainly arises from beam heating (which can normally be reduced to an acceptable level; see §3.4.1) and from heat radiated from surrounding surfaces. It is usually possible to minimise this source of heat by using well-positioned cold shrouds. The rate of heat withdrawal depends upon the temperature of the cooling source, its temperature difference from the specimen, and the magnitude of the thermal resistances

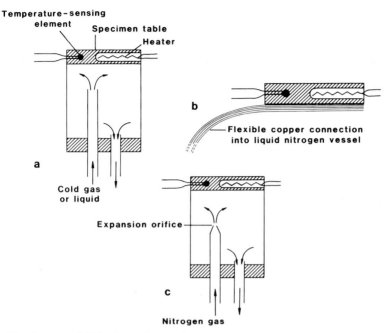

Fig. 3.3. Some possibilities for cooling specimen stages in scanning electron microscopes and similar equipment. (a) Direct supply of piped liquid or gaseous nitrogen. This method can produce temperatures close to $-196°C$ (77 K). A temperature-sensing element (thermocouple, thermistor or platinum resistance element) is usually incorporated so that the temperature can be accurately monitored and also so that, using a heater, the temperature of the stage can actually be controlled. (b) A flexible metallic connection can be used as an alternative to pipes. The metal chosen is almost always copper because of its high thermal conductivity. The stage cannot be cooled to such a low temperature as with directly piped nitrogen but there may be advantages of ease of manipulation of the stage and lack of mechanical disturbance. (c) The Joule-Thomson refrigerator makes use of the principle of the cooling effect produced by the adiabatic expansion of a compressed gas. This system can also produce temperatures very close to liquid nitrogen temperature.

between the specimen and the coolant. To provide the best conduction of heat from the specimen it is necessary to have a conduction pathway of large cross-sectional area, minimum length, and high thermal diffusivity, and with a minimum number of mechanical junctions. Thus, a most efficient cooling system is to have sub-cooled nitrogen passing through copper tubing to the underside of a copper specimen support. It is then possible to maintain the specimen temperature close to that of the coolant (Fig. 3.3). However, mechanical constraints on specimen movement, and other factors, such as the possibility of vibration arising from flowing liquid, often preclude direct liquid cooling. Cold nitrogen gas can also be used although this again necessitates the use of piping between the coolant (a container of vapourising liquid nitrogen) and the specimen stage. For these reasons, many designers of cold stages (particularly for SEMs) have preferred to use flexible, solid metallic conduction pathways between the specimen and the coolant. These provide improved flexibility of stage movement and reduce problems of vibration. If such solid connections are used, then it is necessary to maximise the cross-sectional area while retaining flexibility. Stranded fine copper thread is ideal and preferable to braided copper (Robards and Crosby 1979). It is obviously necessary to have mechanical junctions in such a 'solid' pathway but they should be kept to a minimum because it is possible to 'lose' up to 10 K cooling capacity across each junction (Taylor and Burgess 1977). The flexible copper wire must be *very* tightly clamped at its ends and solder should only be used to a minimum extent as a space filler (its thermal conductivity is much worse than that of pure copper). Apart from solid or piped conduction pathways using liquid or gaseous nitrogen as the cooling source, some systems have used the Joule-Thomson refrigerator effect (e.g. Pawley and Norton 1978). This operates on the principle of cooling through the adiabatic expansion of a gas (nitrogen) under pressure and can provide temperatures very close to that of liquid nitrogen.

While the specimen stage may be capable of good thermal performance, it is, of course, the temperature of the specimen itself that is critical. It is therefore most important that thermal contact between the specimen and the cooled specimen stage is as good as possible. This means that mechanical connections should all be very close ('tight') and high thermal conductivity metals (preferably copper) should be used. If all the criteria mentioned here are taken into account, then it should be possible to keep a specimen very close to the coolant temperature (i.e. within a few Kelvin for liquid/gas piped systems and for the Joule-Thomson refrigerator; within 10–20 K for solid conducting pathways). For many reasons it will not always be desirable to keep the specimen at the lowest possible temperature on the cold stage in the microscope. In consequence it is usually advisable to have a heater in the cold stage so that, by balancing heat input with heat extraction, it is possible to maintain any desired temperature. Clearly, there must also be a temperature monitoring device (thermocouple or resistance element) in the stage so that the actual temperature can be determined and so that, if required, the temperature can be automatically controlled using a suitable electronic device to balance heating and cooling and thus achieve the actual temperature demanded.

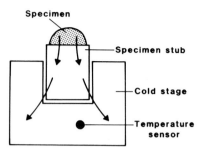

Fig. 3.4. Heat flow through a specimen and cold stage. Heat is withdrawn through the specimen towards the cooling source. It is important to position the temperature-sensing element as close as possible to the specimen itself. Any mechanical junctions in the thermal pathway will introduce a loss of cooling capacity.

As the specimen will be very cold, it will act as an efficient cold trap for condensable molecules in the vacuum system (especially water), against which it should be protected. The usual method of achieving this is to position one or more cold plates (anticontaminators) close to the specimen and, preferably, making a large solid angle with it (see §5.2.1). This plate should be at least 10–20 K colder than the specimen itself. From this it will be apparent that, if the specimen stage is operated at the lowest possible temperature, then the probability of contamination is greatly increased since the anticontaminator is unlikely to be cooler than the specimen and, therefore, will have a much reduced effect. The basic principles of a simple EM cold stage are illustrated in Fig. 3.4.

3.2.2 Transfer devices

If a specimen is to be viewed at low temperature in an EM, then it is usually cooled externally before being put into the microscope. This means that a very cold specimen has to be moved from its original cooling site, through various preparative procedures, into the EM through the air (except in a relatively few instruments where the specimen is cooled in the EM itself or in a chamber attached directly to it). The specimen must be protected against both excessive warming and contamination with condensing vapours during this transfer. It is usual, therefore, to protect the specimen during such transfer stages by use of a 'transfer device'. Together, the transfer device and the specimen mount should allow the specimen to be both maintained at low temperature and protected from contamination. The requirement for low temperature can be met either by provision of a

constant cooling source or by relying on the thermal capacity of a previously cooled block. Contamination is avoided either by providing a good
vacuum environment (with low partial pressure of condensable molecules)
and/or by shrouding the specimen with a cold, protective surface. The cold
shroud may either be cooled continuously or rely on thermal capacity. If
the specimen is well shrouded with low temperature surfaces and is supported on low thermal conductivity mounts, then its own heat gain will be
very much reduced. A variety of transfer devices has been constructed (see
§3.3.7 and §3.4.7), some with continuous evacuation and liquid nitrogen
cooling, others relying on cold shielding alone. The physical demands are
clear, and it is simply a matter of ensuring that they are met within the geometric demands of the microscope and accessory units being used.

3.2.3   Other ancillary equipment

So far, we have discussed the freezing of specimens and their transfer onto
the cold stage of an electron microscope. In addition, there are other intermediate processes that may have to be accommodated, notably fracturing
of bulk specimens, sublimation of ice and coating of specimens with an electrically conducting layer. Once again, the precise methods and equipment
will vary but the principles are clear. The requirement for maintenance of
low temperature with low contamination is a paramount consideration.

The frozen specimen, if it is a 'bulk' object, may be crudely fractured
under liquid nitrogen (§6.3.3c), cleaved or sectioned in a special chamber (or
on a cryoultramicrotome), or manipulated while under observation in a
SEM.

Ice can be sublimed by raising the temperature of the whole block, either
in a special chamber or on the SEM cold stage, or by surface heating using
a radiant heater. Heating of the whole specimen is the method used in
freeze-etching (§6.5) and is capable of close control of bulk temperature.
However, it can be slow for large specimens and raises the whole block to
approximately the same temperature. It also means that the surface will be
the *coldest* part of the specimen (Fig. 3.5). If a radiant heater is used, then
it may be difficult to obtain such a precise control over sublimation rates
(although Talmon (1980) has provided some useful calculations relating
heater current to rate of ice removal), but surface ice (hoar frost) can be
removed easily and quickly; only the *surface* is warmed appreciably, the remainder of the specimen staying cold. Sublimation of ice is preferably carried out in an ancillary chamber rather than in the EM itself, although there

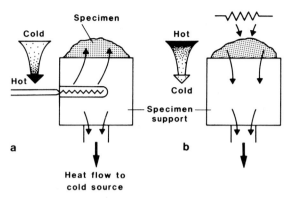

Fig. 3.5. Thermal consequences of subliming ice from specimens either by warming the whole specimen (a) or by radiant heat to the the specimen surface (b). (a) The freeze-etching approach warms the whole specimen block to the etching temperature. The specimen surface, being furthest from the heating source, remains as the coldest site along the temperature gradient. (b) If a radiant heater is used, it applies heat to the surface, the remainder of the specimen still being cooled. Thus, in this situation, the surface is the warmest part of the gradient. Either method of etching can be used but the thermal consequences should be taken into consideration in relation to the requirements of the specimen.

are situations where the very purpose of the experiment is to observe the behaviour of subliming ice or to 'etch' to a very precise level. In such cases, sublimation must be performed within the microscope but, in other circumstances, it is better to protect the microscope column against contamination.

Coating of a low temperature specimen, with gold, platinum or carbon for example, is a process that has occasioned misgivings in relation to probable specimen heating. However, it is now clear that a variety of thin film deposition techniques can be used without leading to undue damage through heating or other effects. Evaporative methods were the first to be used (Echlin 1975, 1978b, 1981; Pawley and Norton 1978; Echlin et al. 1980) and remain necessary if carbon is to be deposited (although ion beam deposition appears practicable; Franks et al. 1980). More recently it has been shown that diode sputter-coating with gold can also be accomplished with minimal heating (Robards et al. 1981). Using estimates of the maximum possible heat load to specimens when sputter-coated with gold using the equipment illustrated in Fig. 3.11 (§3.3.4), it was concluded that the surface temperature could not increase by more than 6 K (2 K W^{-1} energy input). Although they have not, as yet, been routinely applied to low temperature specimens, Penning sputtering (Jakopić et al. 1978; Peters 1980) and ion-beam deposition (Franks et al. 1980) would undoubtedly also provide acceptable layers without damage. In fact, it may well be found that these methods will provide fine grain films that will allow higher resolution studies on low temperature specimens. The subject of grain structure in relation to deposition conditions during sputter-coating and ion-beam coating has been investigated by Echlin et al. (1982) and Kemmenoe and Bullock (1983) and, while again not directly concerned with low temperature specimens, some useful relevant conclusions arise. In general it is preferable to use the lowest possible voltage and current consistent with depositing an adequate film within a reasonable period in order to minimise the heat arising from the dispersion of energy within the specimen.

Evaporative and sputtering techniques have their own relative advantages and disadvantages. Evaporation is carried out in a high vacuum, hence facilitating contamination protection, but avoidance of 'shadowing effects' may be necessary. Sputtering is simpler and may give a more even coating but, as it is carried out at low vacuum (0.1 Torr, 13.3 Pa), it is important to ensure protection from contamination. Despite the relative merits and demerits, it is a fact that both methods have been operated successfully on low temperature specimens. Evaporation of carbon can provide problems because the energy input needs to be so much higher than for metals, but both carbon arc evaporation and resistance heating of carbon fibre 'string' (Vesely and Woodisse 1982; Peters 1984; Smits et al. 1984) have been employed successfully. The problems of heat damage during specimen coating obviously become more formidable for specimens of low thermal capacity, such as those for observation in the transmission mode, than for bulk SEM specimens. Nevertheless, it appears that, provided the physical principles of specimen heating are appreciated and the practical application is appropriately selected, significant heat damage can be avoided.

3.3   Frozen bulk specimens

By bulk specimens we generally mean those that are too thick to be traversed by electrons. Such specimens will normally be viewed in an SEM and may be required for either observation and/or analysis. There has been a number of useful papers in this area, among which the most general include: Nei (1974), Nei et al. (1973, 1974), Fuchs and Lindemann (1975), Koch (1975), Marshall (1975, 1977, 1980), Echlin and Moreton (1976), Echlin (1978a), Fuchs et al. (1978b), Pawley and Norton (1978), Robards and Crosby (1979) and Zeirold (1979).

3.3.1   Frozen-hydrated bulk specimens

Conceptually the simplest of all low temperature techniques for EM is that of taking a specimen, freezing it, and then looking at it while it is maintained in the frozen state on the cold stage of a SEM. When suitable equipment is available, such a method now provides a simple and rapid means of preparing specimens with minimal damage. For example, it is possible to take delicate structures such as small flowers, mushroom fruiting bodies or insect larvae (to give just a few typical examples) and to have them on the cold stage of a SEM within a few minutes of starting operations. Quite often the surfaces that are to be viewed will be covered with a film of water that will have to be removed (by sublimation of ice) before the true surface can be seen. With most scanning electron microscope specimens it is normally necessary to provide a conducting film over the final frozen surface before it is irradiated by the electron beam. For bulk specimens the preparation

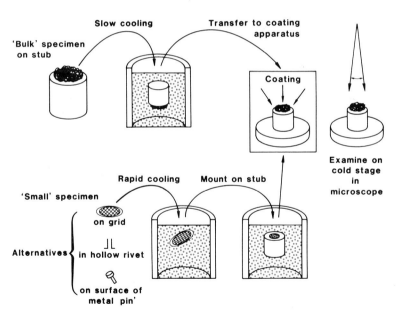

Fig. 3.6.   Alternative methods of preparing frozen specimens for low temperature scanning electron microscopy. If large specimens are used, the cooling rate is of little significance. Therefore the whole specimen can be frozen on a relatively large stub (for example, using sub-cooled nitrogen). However, if high cooling rates are required, the specimen is first frozen on a small support (such as a 'pin', grid or rivet) before being attached, under liquid nitrogen, to a larger stub.

route is thus clear. Unless the specimen is very small (<0.1 mm in at least one dimension) there will be little point in trying to freeze it very rapidly. Furthermore, the actual resolution that is attained in the SEM should be considered in relation to the probable ice crystal size developed in the specimen. In most circumstances it is perfectly adequate to cool the relatively large specimens in sub-cooled nitrogen or in a halocarbon bath (§2.4.1a) and, indeed, the specimen may be mounted on the SEM stub itself for this purpose. If the specimen is much smaller or, perhaps, if a microbial suspension or cell fraction is to be examined, then it may be thought preferable to use a rapid cooling method. The specimen should then be frozen on a support of low thermal mass that can subsequently be mechanically attached to the cold SEM stub (Fig. 3.6).

Once the frozen specimen is on the stub, it needs to be protected against two main damaging conditions: firstly it should not be warmed unintentionally above about 160 K (which would cause the inadvertant sublimation of ice, see §5.2.1); secondly, because it is at low temperature, it will act as a good trap for condensing molecules (e.g. water and hydrocarbon oils) that will contaminate the surface. The means of avoiding these pitfalls are clear: the specimen should be maintained (by continuous cooling or by being mounted on a support of adequate thermal capacity) at a temperature below 160 K; and it should be protected against contamination either by surrounding it with even colder surfaces and/or by evacuating the vessel in which it is contained so that the partial pressures of condensible gases are drastically reduced. Different experimental systems have used different combinations of these two solutions (§5.2.1).

If a cryogen with a melting point warmer than the temperature of the cold stage is used (e.g. a halocarbon), then this may stay as a solid residual contaminant on the specimen surface. This may then either be removed by warming the specimen under controlled conditions so that sublimation of the solid cryogen occurs, or the problem may be circumvented by using a cryogen of low melting point (e.g. propane).

If the sublimation of water (or contaminating cryogen) from the specimen surface is required, then some means of doing this under controlled conditions is necessary. Two possibilities exist: firstly, the whole temperature of the specimen may be raised to >170 K so that freeze-drying in vacuum will start to take place (exactly as in freeze-etching; §5.2.1 and §6.5). This can, in principle, be carried out either on the cold stage of an EM or in an accessory chamber. Alternatively, surface ice may be sublimed away using a radiant heater. The 'freeze-etching' approach has the advantage that it is controllable and that it may be effected in either the microscope column or an accessory chamber. On the other hand, it can be rather slow because the temperature of the whole bulk specimen on the stub must be raised substantially (Fig. 3.5). Radiant heating is less controllable, although empirical determinations can lead to reproducible results and Talmon (1980) has made some calculations relating radiant energy to the amount of sublimation achieved, and is not carried out in the microscope. It is relatively fast as the *bulk* temperature of the specimen is effectively unchanged. An advantage of sublimation in the microscope is that the surface can be watched as the ice disappears. However, as a general rule the sublimation of excessive amounts of additional contamination into electron-optical columns is not recommended and, if this is done, it is essential that the microscope should have effective cold traps close to the source of contamination. This discussion relates to hydrated frozen specimens; that is, those containing frozen water. The physical behaviour of ice at different temperatures under vacuum is now well understood (§5.2.1; §6.5) and the operational parameters for the design of equipment may thus be determined. However, low temperature techniques, including low temperature SEM, may often be called on for the processing and observation of non-aqueous systems. For such specimens, the behaviour of cold material under vacuum may be quite different from that of ice. For example, it may be that a component of the specimen is an oil of low melting point and high vapour pressure, in which case even very low operational temperatures may be inadequate to prevent sublimation into the vacuum. The only remedy would be to use a system with sufficient cooling capacity (possibly using liquid helium) to reduce the vapour pressure of the relevant oil below the point at which it would sublime into the vacuum system. The theoretical calculations would be as for water/ice, but substituting the appropriate parameters for the oil.

Once the required, possibly etched, surface has been revealed, it is usually necessary to coat it with a conducting layer prior to viewing. Frozen-hydrated specimens usually show serious charging if not so coated. This subject has been fully discussed (Nei et al. 1973; Brombach 1975; Marshall 1975; Echlin, 1978b; Fuchs et al. 1978b). Brombach (1975) and Marshall (1975) both described the phenomenon of surface charging, which was evaluated more fully in the paper of Fuchs et al. (1978a). While it is to be expected that pure ice (with a DC conductivity of approximately 1–10 nohms m^{-1}; Hobbs 1974) will show substantial charging, the presence of electrolytes in frozen biological samples increases the conductivity by some orders of magnitude and thus reduces the charging problem considerably. When a specimen is freeze-dried, its conductivity is reduced and it charges more readily than the equivalent fully hydrated material, and consequently charging can be seen to increase as etching progresses. In general, it has been found that quite high accelerating potentials can be used (up to 30 keV) so long as the beam current is kept very low (Echlin 1975; Woods and Ledbetter 1976; Echlin 1978a,b,c; Echlin 1981). Fuchs et al. (1978a) showed that an adequate conducting film deposited on the specimen surface effectively eliminates *surface* charging but does not necessarily reduce *space* charging. These authors used electrodes to collect electron currents at different positions around model specimens in an EM and came to the conclusion that charging need not be a problem so long as the electron range (mean depth of penetration) is greater than about half the specimen thickness and the specimen is mounted on an effective rear 'electrode' (earth contact). They also noted that the irradiation of bulk samples with a narrow electron beam leads to a space charge field close to the surface that reduces the effective depth for analysis (excitation volume) and, furthermore, that charge storage develops very quickly, even in coated specimens. Quite clearly, thick bulk specimens will normally need to be coated with a conducting layer to inhibit charging. A number of possibilities exists for accomplishing this, including resistance heating, diode and triode sputter-coating, ion-beam sputtering, electron-beam coating and Penning sputtering. These techniques have all been well reviewed in the SEM literature (Echlin 1978a,c; Echlin et al. 1980) but only the first two have been used to any significant extent in low temperature applications.

To evaporate metals, they must be heated above their boiling point (see also the previous discussion in §3.2.3). Consequently, evaporative coatings necessarily involve the generation of high temperatures (Table 6.2; §6.4.1b) locally around the source. Nevertheless, evaporation of gold, gold alloys, chromium and carbon onto frozen surfaces without unduly deleterious effects has been demonstrated in a number of laboratories. Sputter-coating has the advantages of requiring only a low vacuum (0.1 Torr, 13.33 Pa) and also gives a more even coating over static, uneven surfaces than does resistance evaporation. It has been shown both theoretically and in practice (Robards and Crosby, 1979; Robards et al. 1981; §3.3.6) not only that diode sputter-coating is acceptable for providing conducting films on frozen specimens but also that the heat input to the specimen can be very low indeed. An advantage of resistance heating is that it is usually carried out in a high vacuum ($< 1 \times 10^{-5}$ Torr; 1.33×10^{-3} Pa) and the theoretical possibility for contamination of the specimen is thus reduced. However, properly designed sputtering systems working at higher pressures than resistance evaporators have also been used in the total absence of contamination (Robards and Crosby 1979; Beckett et al. 1982). (In fact, to our knowledge, K. Oates was the first microscopist to demonstrate (in about 1977) that the sputter-coating technique could be satisfactorily applied to low temperature specimens; personal communication.) Other methods of coating at low temperatures, such as ion-beam sputtering, can certainly be used although they have not yet been applied routinely. For most purposes gold, gold alloy or platinum layers have proved satisfactory for observational work, while carbon, beryllium or chromium have been favoured for microanalytical determinations. If the system uses a sputter-coater, then the carbon may have to be evaporated in a poor vacuum ($> 5 \times 10^{-4}$ Torr; 6.67×10^{-2} Pa). This is not necessarily a disadvantage because there are more collisions of the carbon atoms with residual gas molecules, leading to a less directional coating (see Willison and Rowe 1980 for these techniques). Using carbon fibre 'string' as the evaporation source, it is simple to produce a carbon layer on a frozen specimen in a vacuum of 10^{-3} to 10^{-4} Torr (1.33×10^{-1} to 1.33×10^{-2} Pa).

Having been coated, the specimen now simply has to be inserted into the microscope and onto the cold stage: it must again be borne in mind that the same stringent conditions of protection from contamination during transfer must be applied.

3.3.2 Frozen-fractured bulk specimens

Relatively few specimens are such that the surface of interest is directly revealed in a SEM without further treatment. More usually, it is an internal surface of which views are required and it is then necessary to cut or fracture the specimen before coating it and looking at it in the SEM. Where possible it is best that the *frozen* specimen should be cut or fractured because largely *clean* fracture planes through the frozen material will then be produced. If the surface is revealed by exposing it when the tissue or sample is unfrozen, then deformation can occur and, if the specimen is wet, some etching will be necessary to show the real surface. Furthermore, debris will often contaminate this exposed surface.

To produce a fracture through the frozen specimen, many possibilities exist (Fig. 3.7) and each worker has to consider what is most suitable for the particular material in hand. Whether the specimen is 'cut' or 'cleaved' at low temperature, we may expect that a brittle fracture will result (§6.3.1). There may, of course, be specific reasons for fracturing at warmer tempera-

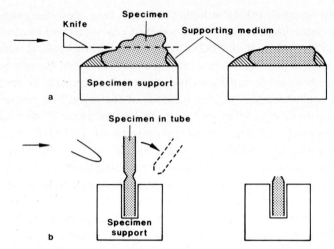

Fig. 3.7.   Possibilities for producing fractures through frozen SEM specimens. (a) Using some sort of knife system (cooled microtome, scalpel, razor blade) a fracture can be produced through the specimen while it is maintained cold. Alternatively (b), a simple fracture can be induced, for example by breaking apart pairs of rivets that have been frozen together or by cross-fracturing a tube containing the specimen.

tures (for example, to look at temperature-related changes in an emulsion) but even then it is usually required that the specimen is subsequently maintained at the lowest practical temperature consistent with the absence of contamination. Once the fractured surface has been exposed, it may be etched by subliming ice away so that biological structures are revealed, and the specimen is then ready for coating and viewing.

3.3.3   Preparation

The methods of preliminary preparation and freezing of specimens have been described in Chapter 2. It remains to be decided whether the specimen should be frozen on a relatively large specimen block (stub) or on a much smaller support which can subsequently be mounted onto a larger carrier. These are choices for the individual user to make, bearing in mind the ultimate objective that is to be achieved. In a typical system (Fig. 3.8), the whole specimen, on its stub, is cooled in sub-cooled nitrogen. Whenever possible it is preferable that the stub has good thermal conductivity and a high heat capacity: copper is the usual choice. It is important that the specimen is in as good contact with the metallic stub, or holder, as possible. Before cooling it should be 'buttressed' with the tissue solution or, if appropriate, with a material such as polyvinylpyrrolidone (PVP), 'Tissue-TeK' (Appendix 2) or a similar medium. The two criteria of good thermal contact with the stub and good mechanical stability are critical to the attainment of satisfactory results.

When the specimen is frozen on a large block or stub, it must be appreciated that, while the specimen itself may cool to the temperature of the cryogen quite rapidly, the stub can take very much longer to achieve an equilibrium temperature. Thus, it must be ensured that the specimen/stub combination is left in the coolant for a sufficiently long time for this equilibration to take place. This can easily mean as much as a minute or so in the coolant. It is best to test the time taken for each different type of stub to cool down by carrying out some test runs with a thermocouple positioned at the centre of the supporting block.

With the frozen specimen on its stub or carrier, we can go on to the next stage. This may well involve inserting the carrier into a transfer device (see §3.3.7). For the moment, however, we shall concern ourselves with methods for revealing faces for observation and analysis, i.e. fracturing and/or etching.

3.3.4 Fracturing

Many frozen specimens have surfaces that are not of direct interest to the observer, who requires some internal plane. It is possible either to fracture the specimen (usually under liquid nitrogen) before mounting onto a specimen stub, or to carefully 'section' the specimen using a cryoultramicrotome so that a relatively smooth, frozen surface is produced. The first method requires no further comment, since it is simple and can easily be adapted to the particular requirements of different specimens (Fig. 3.7; §3.3.2). Cryoultramicrotomy is described in the next Chapter and is used in the same way to prepare smooth surfaces for low temperature SEM as to prepare thin frozen specimens. In either of these two methods the final specimen ends up on the stub to be put on the SEM cold stage.

When a special accessory chamber, either remote from or attached to the SEM (a 'dedicated system'), is used, the frozen specimen is fractured in the chamber with a needle, knife or microtome. The system shown in Fig. 3.8 provides a variety of simple devices that allow the specimen to be relatively crudely fractured. This chamber is part of an integrated system that enables the specimen to be moved, in a transfer device, from the freezing chamber to the fracturing/coating chamber and subsequently to the microscope itself.

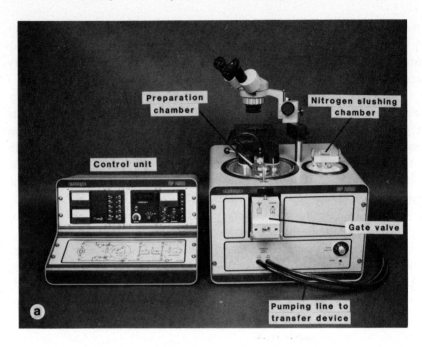

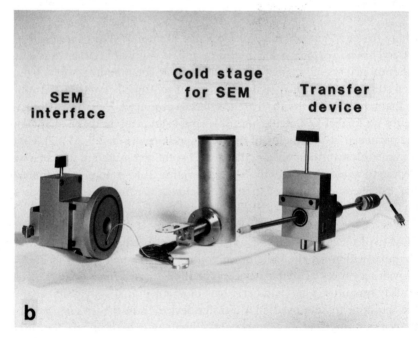

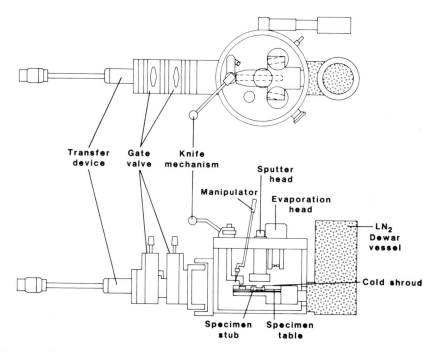

Fig. 3.8. The EMscope SP2000 Sputter-Cryo System (see Appendix 2) for low temperature scanning electron microscopy (see also Fig. 3.11). (a) Illustration of whole Sputter-Cryo System with control unit on the left and the preparation unit on the right. (b) Components of the system. (c) Schematic diagram of the system. (Illustrations by courtesy of EMscope laboratories, Ltd.)

Similar systems have been built-up using equipment already available in the laboratory. For example, a number of workers (Saubermann and Echlin 1975; Fuchs and Lindemann 1975) have used a transfer device to move the frozen specimen onto the cold stage of a vacuum or freeze-etch apparatus where fracturing and coating can easily be accomplished (Fig. 3.9). Some dedicated systems have been constructed with rather more precise (but nevertheless simple) microtomes. An example is illustrated in Fig. 3.10, in which the chamber is actually attached to the microscope itself. In some special circumstances it may be useful to observe the specimen in the microscope while it is being fractured. Again, this causes no special problems beyond the need to provide suitable vacuum lead-throughs to the micromanipulators or other fracturing devices (Hayes and Koch 1975; Koch 1975).

Whatever means is used to fracture the specimen, it is important that the fracturing tool is as cold as, if not colder than, the specimen itself. If a microtome is used, then it is usual to provide a positive means of cooling the blade (see Figs. 3.9 and 3.10) but if simple needle or scalpel blades are employed then they can be cooled by 'parking' them in close contact with some convenient cold surface in the system (e.g. in the apparatus in Figs. 3.8 and 3.11 it is possible to cool the needles by placing them in specially machined recesses in the cold table). It should also be remembered that any material that is fractured or otherwise broken away from the main body of the cold specimen will probably be in poorer thermal contact with the cold sink and will therefore warm up, with the consequent possibility of water sublimation and condensation onto the adjacent, colder, specimen surface

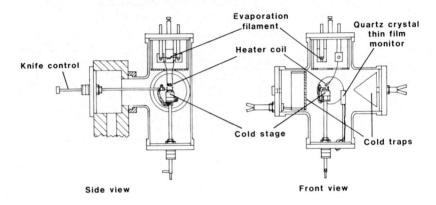

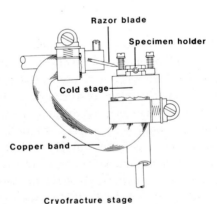

Fig. 3.9. A method of using the Balzers BAE 120 vacuum evaporator (see Appendix 2) as a fracturing and coating apparatus for preparing specimens for low temperature scanning electron microscopy. These illustrations show the cold stage, cooled microtome knife and evaporative coating arrangements. A large cold trap minimises specimen contamination. (Relabelled and reproduced, with permission, from Saubermann and Echlin 1975.)

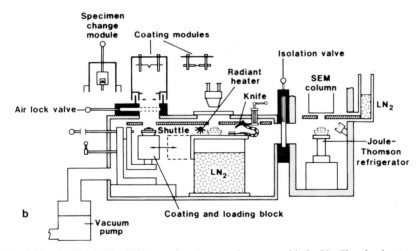

Fig. 3.10. (a) The AMR 1000A scanning electron microscope with the Bio-Chamber low temperature preparation unit attached (see Appendix 2). (b) Diagrammatic representation of the Bio-Chamber, showing the high vacuum system, evaporative coating system, the 'shuttle' system, and the Joule-Thomson refrigerator in the SEM. (Fig. 3.10b redrawn from Pawley and Norton 1978.)

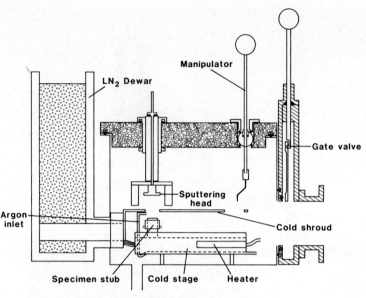

Fig. 3.11.   The fracturing/coating chamber of the Sputter-Cryo system designed by Robards and Crosby (1979). This system, as opposed to the equipment illustrated in Fig. 3.10, is designed to operate away from the SEM itself. There is thus a gate valve so that the specimen can be inserted and removed via a vacuum-lock. The specimen is positioned on a temperature-controlled cold stage where it can be cleaved, fractured or otherwise manipulated. It is then moved along the cold stage so that it can be sputter-coated with a metal (gold or platinum) or, alternatively, coated with evaporated carbon (not illustrated). The same principles as those described here are used in the EMscope SP2000 system (Fig. 3.8). (Redrawn from Robards and Crosby 1979.)

(this subject is more fully discussed in §6.5.1). Care should be taken to try to avoid this problem, either by attempting to move the fractured fragments as far away from the specimen surface as possible and/or by providing as much cold-trapping surface around the specimen as possible (e.g. as in Fig. 3.11).

Having successfully produced the required specimen surface, it may either be coated and viewed directly or it may be 'etched' so that the surface topography is accentuated (by lowering the ice level) or so that, if it has not been fractured at all, any obscuring surface ice is removed.

3.3.5   Sublimation

Sublimation of ice from bulk specimens is almost always carried out under vacuum although there is no reason in principle why, as for thin sections, it should not also be achieved through careful control of temperature under a dry inert gas (e.g. nitrogen) at atmospheric pressure. The theory of ice sublimation under vacuum is fully discussed in Chapter 5, from which it will be seen that etching is effected, and controlled, by manipulating the *temperature* of the specimen (ice) surface under vacuum. For bulk low temperature specimens, ice sublimation can therefore be brought about by:

(*i*)   raising the bulk temperature of the specimen above about 170 K while under high vacuum (strictly, in a vacuum with a low partial pressure of water; see §5.2.1a) on a temperature-controlled stage such as in the accessory chambers shown in Figs. 3.8 and 3.9;

(*ii*)   raising the bulk temperature of the specimen on its stage in the microscope; or,

(*iii*) etching the surface with a radiant heater as provided in the apparatus shown in Figs. 3.8 and 3.10.

Deeper etching is usually necessary for low temperature SEM than for freeze-etching, although this depends on what is required to be revealed and must be determined empirically for most specimens. (Data provided in Table 5.2 in §5.2.1a allow a preliminary estimate of etching times and temperatures to be made). After etching by warming the whole bulk specimen, the block should be recooled so that etching ceases and the surface is stable. When fracturing and/or etching have taken place, the specimen is ready for viewing but must usually be first coated with a conducting film. If etching has been effected in the microscope, then the specimen will have to be retrieved so that it can be coated. Some operators like to watch the specimen as ice sublimes from its surface in the microscope. Such specimens will, of course, charge badly as the surface will not be coated or, if a coating had already been deposited, this will become disrupted by the etching process. If this method is used (and contamination of the microscope column in this way is not recommended), then the specimen (on its stage) should be rapidly re-cooled once the desired degree of etching has been achieved, before it is withdrawn for coating.

3.3.6   Coating

As described previously (§3.2.3), coating is usually achieved either by resistance heating or by sputtering methods. These two approaches are represented in the apparatus in Figs. 3.10 and 3.11. During coating, the specimen

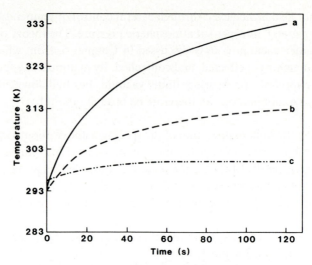

Fig. 3.12.   Typical heating curves during sputter-coating. (a) Specimen heating during sputter-coating using 40 mA at 1.2 kV. (b) Specimen heating during sputter-coating using 20 mA at 1.0 kV. (c) Specimen heating during sputter-coating using an experimental quadrupole head operating with 25 mA at 0.44 kV as described by Robards et al. (1981). This is similar to the system incorporated into the EMscope SP2000 Sputter-Cryo System (Fig. 3.8) and shows that specimen heating during sputter-coating is an insignificant problem as long as the operating parameters are chosen correctly and the coating head is properly designed. (Drawn from data presented in Robards et al. 1981.)

should be maintained at the lowest possible temperature consistent with the capacity of the system to protect it from contamination. It will be seen that the Bio-Chamber (Fig. 3.10; Pawley and Norton 1978) uses a high vacuum whereas the Sputter-Cryo (Fig. 3.8; Robards and Crosby 1979) works at low vacuum. However, the critical point is whether or not there are any *condensable* molecules in the residual vacuum and, in this respect, the Sputter-Cryo provides a very low contamination environment because it is rarely opened to the atmosphere (only for occasional servicing) and is constantly purged with dry argon gas. Consequently, while the total pressure is relatively high, the partial pressure of condensable molecules (particularly H_2O) is extremely low. Furthermore, the cold shrouds in both systems serve as additional anticontamination surfaces. As well as these types of purpose-designed system, other authors have successfully interfaced transfer devices with, for example, freeze-etching units so that the frozen specimen can be transferred into that apparatus to be coated (Saubermann and Echlin 1975; Fuchs and Lindemann, 1975). The salient points to be remembered for coating at low temperature are that the specimen should be at a reasonable distance from the heat source (radiant heat load falls off in relation to the inverse square of the distance), that the input power for evaporation or sputtering should be kept as low as possible, and that it is better to minimise heat input by coating for a long time at low deposition rates rather than for a short time at high rates (Fig. 3.12; see §6.4.1b for further discussion of this topic). If evaporative techniques are used, then some means of ensuring that the specimen is evenly coated should be sought (such as moving it during coating). It should be mentioned that a disadvantage of coating with carbon for low temperature work is that the thin carbon film often cracks, so leading to impaired thermal and electrical conductivity.

3.3.7   Transfer

To accomplish the different steps necessary in preparing a specimen for low temperature SEM, the specimen must be moved to different work positions. This is done either within a dedicated accessory chamber attached to the SEM (Fig. 3.10), or by using a portable transfer device. The first method has the advantage of simplicity and, perhaps, speed in operation. The second can be more versatile and does not require the accessory chamber to be attached to the SEM. There do not appear to be overwhelming advantages or disadvantages to either method. If the cold specimen is to be moved external to the vacuum of the microscope column, then a protecting transfer device is essential. A very simple type of device is illustrated in Fig. 3.13 (Saubermann and Echlin 1975). This merely relies on the cold-trapping capacity of a relatively high thermal capacity tube into which the specimen is withdrawn. Provided that the tube is reasonably long, most condensable molecules will be trapped on the walls of the tube before they can arrive at the specimen. A rather more elaborate system is shown in Fig. 3.14 in which the specimen is retracted into a cold-shrouded 'garage' and is also contained within an evacuated chamber. The transfer device carries its own

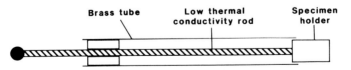

Fig. 3.13.   A simple transfer device for cold SEM specimens. The tube is first cooled with liquid nitrogen so that, when the specimen is retracted into it, potentially contaminating vapour molecules (such as water) are trapped on the cold metal inner wall. (After a design described by Saubermann and Echlin 1975.)

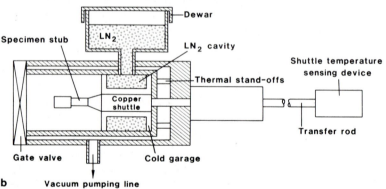

Fig. 3.14.   A cooled transfer device for frozen SEM specimens. (a) Photograph showing the specimen holder, gate valve and Dewar vessel. (b) Diagram illustrating the components of the transfer device. When enclosed in the evacuated device, the specimen is constantly cooled. Its temperature is monitored at all times. (Photograph supplied and reproduced by courtesy of Cambridge Scientific Instruments, Ltd. (Appendix 2). Diagram redrawn and reproduced with permission from Cambridge Scientific Instruments Ltd.)

liquid nitrogen supply and thus maintains a very low temperature throughout the change from one work position to another. Yet another approach is embodied in the apparatus illustrated in Fig. 3.15, which again has an eva-

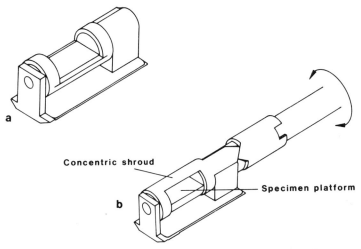

Fig. 3.15. A 'self-shrouding' stub as used in the EMscope SP2000 Sputter-Cryo system. (Diagram redrawn and reproduced with permission from EMscope Laboratories Ltd.)

cuated chamber (Fig. 3.8) but with the specimen stub carrying its own high thermal capacity, automatically closing cold shroud. These examples represent typical approaches and constructions; there are many others described in the literature (e.g. Aldrian and Windisch 1984). What is important is that, once the specimen has been frozen, its temperature should never rise so much that ice sublimes (unintentionally) or so that unwarranted ice recrystallisation takes place. A good working rule is that the *surface* of the specimen should not inadvertently rise above 140 K. It is thus necessary to be able to monitor the complete temperature history of a specimen from freezing to viewing and to see that no untoward temperature excursions have occurred. Contamination during transfer is minimised by the provision of cold shrouds and by ensuring that the vacuum vessels are always back-filled with a dry inert gas (usually argon or nitrogen) rather than with air, and also by evacuating 'dead volumes' between adjacent vacuum valves before these are opened to allow continuity between evacuated compartments.

3.3.8 Viewing and analysis

The basic requirement of the EM is that it has a cold stage and a cold anti-contaminator plate. The cold stage should be temperature-controlled and capable of going to as low a temperature as possible (certainly < 100 K). In operation, the anticontaminator should preferably be kept 20–30 K

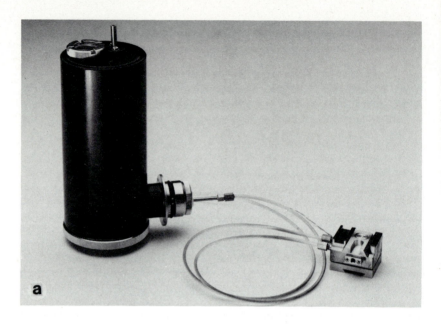

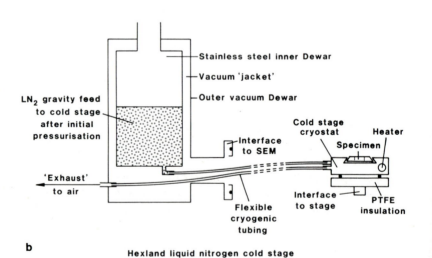

Hexland liquid nitrogen cold stage

Fig. 3.16.   The Hexland cold stage for an SEM using piped liquid nitrogen (see Appendix 2). (a) The Hexland cold stage. (b) Diagrammatic illustration of the cold stage, showing the direct liquid nitrogen piping and the dovetail slot in the stage to accept the specimen mounts. (Photograph supplied and reproduced by courtesy of EM Technology – Hexland Ltd. Diagram redrawn from an original supplied by EM Technology – Hexland Ltd.)

colder than the specimen itself. For SEM work, the cold stage can be kept rather simple and, provided that the 'fit' of the specimen stub into the stage is good and secure, there should not be a problem in temperature attainment and control. Three forms of cold stage are illustrated in Fig. 3.3 in §3.2.1; one uses a stranded copper cooling pathway (as in the system illustrated in Figs. 3.8 and 3.11), another uses a piped system (Fig. 3.16), and the third type uses the Joule-Thomson refrigerator principle (Fig. 3.10). Other commercial forms of heating/cooling stages are also available and may be suitable for low temperature applications, although some of these are more complex (and hence more expensive) than is required and may not have the ultimate low temperature capability required for hydrated specimens. It is important that the cold stage is mounted on thermally insulating (e.g. nylon) spacers to separate it from the main body of the SEM stage. However, some provision for maintaining an electrically conducting pathway from the specimen to instrument earth must be made. A fine wire of low heat transfer capacity but high electrical conductance is adequate.

If microanalysis is contemplated, then the normal instrumental modifications to minimise background counts should be performed (see Chandler 1977). Surfaces close to the specimen which are likely to generate spurious X-rays should be coated with carbon or fabricated from beryllium. It should be remembered, however, that carbon is not a good heat conductor and any carbon shrouding should take this into account. For example, it would be preferable to coat the top of a copper specimen stub with a thin film or 'wafer' of carbon (or, even better if available, beryllium) rather than to fabricate the whole stub from carbon as might be done for analysis at ambient temperatures.

3.4 Frozen thin specimens

Surface signals are the predominant ones retrieved from bulk specimens (§3.3). At low temperatures, such samples have almost always been studied in SEMs. However, when it comes to thin specimens, we have the option of low temperature conventional TEM, or STEM in an instrument that is either basically a TEM or an SEM. If an SEM cold stage is adapted for transmission work, then the problems are not too severe as there is generally plenty of space for necessary components and the engineering is relatively straightforward and similar to that required for bulk specimens (indeed, the same stage can often be adapted for either mode of operation). When it comes to thin specimens for observation in true TEM or STEM instruments, then the problems are all magnified by an order of magnitude or so. The specimen is much smaller, has a lower thermal capacity, and its temperature is therefore much more liable to be changed, especially during transfer. The specimen mount must also necessarily be smaller and this means that it is common to mount the specimen into a holder which is itself transferred into the microscope (Fig. 3.6; §3.3.1).

3.4.1   Frozen thin sections

The preparation of frozen thin sections is described in the next Chapter and, for the purposes of the present discussion, it is sufficient to assume that we have produced a section which is essentially a fragment of biological tissue 'embedded' in a matrix of ice. Reference to tables of sublimation rates (Fig. 5.3 and Table 5.2 in §5.2.1a) show how very quickly such a section would completely freeze-dry at temperatures much above 160 K, and so the maintenance of a *very* low temperature during all processing and observational stages is of paramount concern. This may be more difficult than is immediately obvious. Whereas a bulk frozen specimen can be mounted on a large, high thermal capacity, high thermal conductivity metal block, the thin sections must be mounted on a grid (or similar supporting structure) where much of the frozen material will not be in contact with an efficient heat transfer element (e.g. a copper grid bar) and may, in fact, be supported on a plastic film with poor heat transfer characteristics. There is always a rather poor heat conduction pathway through the ice of the specimen to the nearest cold sink. A second criterion for the effective use of frozen thin sections is, therefore, that the sections should be mounted in the best possible thermal contact with the ultimate cooling source, this mounting being consistent, of course, with the requirements of the particular viewing/analytical methods to be used. If these two main criteria are fulfilled (lowest operational temperature and best possible thermal contact with the substrate), then the stages of coating the section with a conducting film (if necessary) and transferring it into the microscope can be accomplished without damage. The environment within the microscope is, however, far more hostile. In a high vacuum the specimen is subjected to beam intensities greater than 300 $C\ m^{-2}$ which may lead both to direct heating and also to the radiolysis of ice. As shown in the discussion in §3.3.1, however, problems of charging in thin sections are less than for bulk frozen specimens (Echlin 1975, 1978b, 1981; Fuchs et al. 1978a). This means that many workers have found it unnecessary to provide a conductive coating for thin frozen sections prior to viewing and analysis.

Talmon and his colleagues (Talmon and Thomas 1977a,b, 1978, 1979; Talmon et al. 1979; Talmon 1982a) have considered in detail the effects of beam heating and radiolysis of ice. Using a model for a 'moderately thick specimen' (a red blood cell mounted on a metalised polymer film), Talmon and Thomas (1977b) determined the theoretical temperature rise under a range of different operating conditions. It was calculated that a STEM probe current of < 10 nA in a 20 nm diameter beam would raise the specimen temperature by about 5 K, the temperature increase being proportional to the probe current if all other variables were maintained constant. The maximum surface temperature rise of a specimen on the cold stage of a SEM depends on beam current, incident beam energy, ratio of sample area to scanned area, and the temperature and thermal diffusivity of the sample. The temperature change varies directly in relation to beam current and is inversely proportional to specimen thermal conductivity. Some representative data provided by Talmon and Thomas (1977b) are provided in Table 3.1 and Fig. 3.17. It can easily be seen (Fig. 3.17) that water loss by sublimation is very sensitive to surface temperature, and the beam conditions should therefore be selected to minimise temperature rise (see also §5.2.1 and §6.5).

Talmon and Thomas went on (1978) to calculate heating temperature profiles in moderately thick SEM/STEM specimens on a cold stage. For thin specimens, one temperature maximum in the profile was found near the specimen surface and this maximum shifted away from the surface as the temperature increased. For thicker specimens (those approaching the maximum thickness for electron penetration), two maxima were found: one close to the surface, the other close to the sample/substrate interface. These results are interesting and potentially important, particularly in relation to X-ray microanalysis where (as opposed to using the secondary electron mode for image formation) the temperature at the main site of X-ray emission is of some consequence.

The conclusion of this and other work on beam heating is that it is possible to minimise temperature increase in the specimen by operating with the lowest practical beam current and using well-designed cold stages to ensure the lowest possible sample temperature. However, radiation damage may occur, mostly arising from ionisation processes that lead to bond breakage, cross-linking, mass loss and free radical formation (Talmon 1982a). These processes proceed independently of temperature although it has been clearly demonstrated that the *effects* of radiation damage are reduced if the

TABLE 3.1
Maximum surface temperature rise of thin and thick sections

| Beam voltage | Thickness | Conditions | | ΔT (fullscan) | ΔT (spot mode) |
| | | Beam current | Spot size | | |
(kV)				(K)	(K)
25	100 nm	80 pA	20 nm	<1	1
25	100 nm	6.0 nA	0.1 μm	<1	8
25	2.0 μm	80 pA	20 nm	<1	<1
25	2.0 μm	6.0 nA	0.1 μm	<1	4
100	2.0 μm	1.0 nA	20 nm	<1	<1
100	2.0 μm	50 nA	0.1 μm	<1	15

From Talmon and Thomas (1977b).

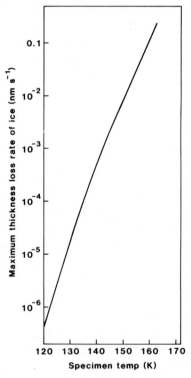

Fig. 3.17.   Thickness loss of a frozen hydrated specimen as a function of temperature. Water loss through only one face is assumed. (Redrawn from Talmon and Thomas 1977b.)

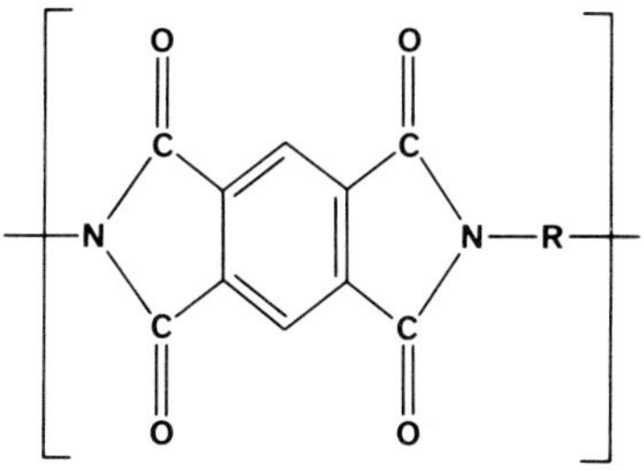

Fig. 3.18. The general molecular structure of polyimides. (Redrawn from Talmon et al. 1979.)

specimen temperature is lowered. The explanation of this appears to reside in the reduced mobility of the products of radiation damage lessening the chances of subsequent deleterious interactions (Cosslett 1978). Under the conditions within an electron microscope, ice can be readily radiolysed by an impinging electron beam. This effect may well be far more important than direct heating of the specimen and becomes increasingly pronounced when high electron doses are required as, for example, for high magnifica- tions or for microanalysis (Talmon et al. 1979; Talmon 1982a,b). Talmon et al. (1979) showed that free radicals produced from the radiolysis of ice (OH$^{\cdot}$, H$^{\cdot}$ and HO$_2^{\cdot}$radicals) significantly etched Formvar films used as sub- strates for transmission microscopy. They concluded that more resistant, hydrophilic polymers, such as the polyimides (Fig. 3.18), are preferable and demonstrated a reduction in etching when a 20–40 nm thick polyimide film was used in place of Formvar. Carbon films and carbon films coated with SiO$_2$ were also resistant to etching by the radiolytic products of ice.

Finally, in relation to the radiolysis of ice, it should be noted that not only may the specimen be etched by the free radicals and ionised molecular frag- ments, but there is also the mass loss of the ice itself to be taken into ac- count. This loss increases in proportion to the beam density and is therefore at a maximum when magnifications are high, when (in a scanning mode) a small area is scanned, and when microanalysis is being carried out. An example provided from theoretical calculations by Talmon et al. (1979) is illustrative. Using a beam current of 0.1 nA at 100 keV (10 nC) to scan an area of 0.1×0.1 μm on a 1.0 μm thick frozen-hydrated section for 100 s, the beam heating would be quite negligible but 1.5×10^{-14} g of ice would be lost through radiolysis. However, assuming that the specimen has the density of water, it would only have a mass of 1.0×10^{-14} g ! Although the etching effect may have been exaggerated in this example, it can be seen that most of the water would be lost, and this greatly limits the application of quantitative analysis. Although in the example provided 100 keV electrons were used, lower voltage electrons (as in SEMs) would lose more energy to the specimen and the anticipated mass loss per coulomb of impinging electrons is likely to be greater in an SEM than in a STEM. As stated earlier, even though the radiolysis itself is temperature-independent, the subsequent diffusion of the reaction products is not and it is therefore still beneficial to maintain the specimen at the lowest possible temperature.

The above comments concerning possible sublimation or radiolysis of ice lead on to a major matter of concern in all low temperature work with thin specimens: how can we tell that the specimen is still frozen-hydrated and not freeze-dried? For bulk specimens this is a relatively insignificant question, but for thin specimens it is a paramount consideration. In a detailed discussion of this problem, Saubermann and Echlin (1975) outlined the following possible methods of detecting whether ice has been lost:

(*i*)   By monitoring the specimen temperature and the surrounding total pressure and partial pressure for water during preparation, examination and warm-up. In this way it is possible to detect changes in the temperature and partial pressure of water vapour that would indicate that sublimation is occurring. When the, presumptively hydrated, specimen is warmed-up there should be a corresponding increase in the partial pressure of water and the total pressure of the residual gas in the vacuum system. (It should be borne in mind, however, that the small change in partial pressure of water brought about by warming a frozen thin section would, in practice, be almost or completely undetectable.)

(*ii*)   The determination of section mass by measuring 'continuum' radiation during and after analysis will reveal no significant change if the specimen has remained fully hydrated.

(*iii*)   If the thin specimen is in an instrument equipped with both TEM (STEM) and SEM imaging, then it is possible to observe directly changes between the frozen-hydrated and freeze-dried states.

(*iv*)   Similarly, structural changes are seen between frozen-hydrated and freeze-dried specimens.

(*v*)   Sections shrink during warm-up if they are still frozen-hydrated.

Two further methods were proposed by Dorge et al. (1975):

(*vi*)   A comparison of peak:background ratios of the elements being analysed should show a dramatic increase as the specimen changes from the frozen-hydrated to the freeze-dried state.

(*vii*)　If the specimen is deliberately freeze-dried in the microscope, there should be a disappearance of the oxygen peak from the residual gas spectrum.

A further possibility is the 'electron transmission method', which allows mass-thickness determination by monitoring either the transmitted electron beam (Halloran et al. 1978) or the zero energy-loss electrons (Hosoi et al. 1981; Leapman et al. 1984)

Clearly, it is not possible to apply all these criteria to every observation on frozen-hydrated specimens. Nevertheless, it is of the utmost importance to know whether the specimen is fully frozen-hydrated, partially freeze-dried or completely freeze-dried. Only by giving this matter detailed consideration, in conjunction with a precise knowledge of the relevant temperatures, total pressures and partial pressures of surrounding gases, is it possible to have any confidence in the interpretation of data, whether structural or microanalytical. Now that the requirements for preparing and viewing thin frozen sections are better understood, some remarkable results are starting to emerge (e.g. Chang et al. 1983).

3.4.2　Frozen thin layers

In view of the very clear correlation between the effects of radiation damage and temperature (Cosslett 1978; Glaeser and Taylor 1978), those who wish to work at the highest resolution on biological specimens will necessarily need to use low temperature stages. Indeed, the ultimate aim is to work at liquid helium temperature although this may necessitate complex and costly equipment, possibly involving superconducting lenses (Fernández-Morán 1966; Heide and Urban 1972; Hörl 1974; Dietrich 1978; Butler and Hale 1981; Lefranc et al. 1982). It is not the intention to discuss here the *reasons* for making observations on specimens in the form of thin layers at low temperatures: these are fully discussed elsewhere (e.g. Dubochet et al. 1982). Suffice it to say that thin layers (of the order of 100 nm thick) can be frozen sufficiently rapidly so that good structural preservation is achieved. This is particularly true when observing protein crystals, for example, where the water content may be only about 50%. As with thin frozen sections, it is critical that the specimen is maintained in the fully frozen-hydrated state (unless it is intended to be partially or completely freeze-dried). Specimens must therefore either be produced directly at the correct thickness for obser-

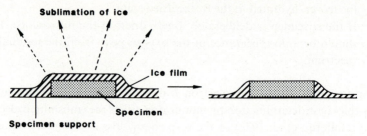

Fig. 3.19. Controlled sublimation of ice to reveal the specimen while maintaining it in a frozen-hydrated state.

vation or they must be reduced to working thickness by controlled sublimation of ice (Fig. 3.19). Furthermore, once the required thin hydrated layer has been produced, it must be treated most carefully to ensure that no additional water loss occurs. This protection can be achieved by combining the right conditions of temperature in relation to vacuum and by physically protecting the specimen on both sides by thin films of plastic and/or carbon (§3.4.4). Frozen thin specimens have also been observed on a specially constructed cold stage in the high voltage electron microscope (Fotino and Giddings 1984).

The production of frozen-hydrated thin layers of specimens such as viruses, protein crystals and cell membranes does not involve any chemical treatment and specimens are obtained from which very high resolution information can be retrieved. There have been some remarkable demonstrations of the applications of such techniques (Taylor 1978; Lepault et al. 1983; Adrian et al. 1984; Dubochet 1984; Lepault 1984) and, while the technical demands of specimen preparation and microscope cold stage construction are complex, they are, nevertheless, now quite well understood. With the exception of the preparation steps, subsequent treatment of thin frozen layers is essentially the same as for frozen-hydrated thin sections, including the requirement for adequate protection during transfer stages. Finally, it is worth commenting that, although it is imperative that specimens are kept cold if high resolution (< 0.34 nm cited by Taylor 1978) is to be achieved, few of the present cold stages currently available for TEMs have such a good mechanical stability as those for use at normal ambient temperature (Taylor 1978).

3.4.3 Preparation

Thin specimens usually end up on some form of grid, with or without a supporting film. As usual, the grid is selected on the basis of its thermal properties, mesh size, whether it will interfere with X-ray analysis, and other criteria. The specimen on the grid is often sandwiched between plastic films which serve both to support it and also to reduce water loss by sublimation. Thin specimens are usually of one of two types: frozen suspensions, liquids and the like, or frozen thin sections. Thin sections will be dealt with more fully in the next Chapter although the technology for processing the sections once they have been obtained is essentially similar to that for thin layers. Thus, we shall look at this topic first.

3.4.4   Thin layers

Thin layers may take the form of liquids in which components such as virus particles, membraneous fragments, subcellular fractions, and protein crystals may be suspended or with which other liquids may form an emulsion. In other words, it is possible to produce a layer of such materials so that it can be frozen very rapidly indeed and then observed in the transmission mode without further treatment (apart, perhaps, from some sublimation of ice). It is usual to freeze such specimens directly on their supporting coated grid. As the specimen is so small, freezing can be extremely good. For example, Dubochet and McDowall (1981) and Dubochet et al. (1982) have shown that if copper grids, coated with a thin, hydrophilic carbon film, are allowed to fall through a mist of microdroplets from a nebulizer into liquid ethane or liquid propane (Fig. 3.20), water layers $< 1.0 \ \mu m$ thick are truly vitrified (as judged from electron diffraction). This is a recent development that will undoubtedly be more widely explored using specimens other than simple liquids alone.

Taylor and Glaeser (1976), Taylor (1978) and Talmon et al. (1979) are among those who have provided details for the satisfactory production of frozen thin layers. The third paper is particularly useful in describing the relevant practical details. The liquid sample is trapped between two thin films supported by copper grids. For reasons already given (§3.4.1), Talmon et al. very much prefer polyimide films about 20–40 nm thick, whereas Taylor and Glaeser (1976) have used SiO-coated carbon films for their work. Willison and Rowe (1980) give details for preparing silicon monoxide films. Polyimides have good heat, radiation and chemical resistance and are much better in these respects than, for example, Formvar. Thin films of polyimide can be produced by dipping a glass microscope slide length-wise and vertically into a 0.75% (w/v) solution of the prepolymer (the poly(amide-acid)), drying for 10 min at 90°C and then curing at 300°C for 3.5 h to obtain the

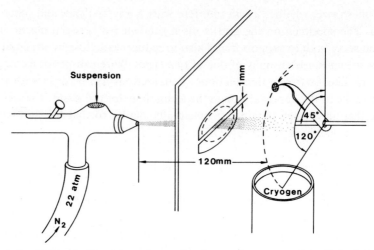

Fig. 3.20.   Schematic illustration of spray-freezing apparatus as used by Dubochet et al. (1982) to produce vitrified water. 50% of the atomised droplets have a volume less than 30 μm^3 and these are deposited on the 200 mesh coated grid at a frequency of about 1 droplet per square. The time between deposition and freezing is about 0.12 s and the rate of entry into the cryogen (ethane) is about 2.0 m s^{-1}. (Redrawn from Dubochet et al. 1982.)

final polyimide. The coated glass slides are dipped into 12% hydrofluoric acid for 5–10 s so that the films can easily be floated off onto a water surface. Such films are very fragile and are put onto grids using the spade method described by Hall (1966). This involves making a grid spade by attaching a piece of 200 mesh foil, about 50 mm × 50 mm, to a wire handle which is also bent to surround and support the mesh itself. The grids are placed on the floating film and then gently picked up by raising the spade from below the film and grids. Talmon et al. have used either 'Skybond 705' (Monsanto) or 'P13N' (Ciba-Geigy) as their original material for casting polyimide films (see Appendix 2 for list of suppliers). These are both supplied as 15–20% solutions of the prepolymer, either of which is then diluted in a 3:1 mixture of *n*-methyl pyrrolidone and xylene to provide the final solution into which the microscope slides are dipped.

To prepare a thin layer of liquid for EM, a 400 mesh, film-covered grid is placed, film side uppermost, on a sheet of filter paper; a microdroplet of the required liquid is placed on the film and then a 100 mesh, coated grid is positioned, film side downwards, over the droplet (taking care, if necessary, to align the bars of the two grids). This squeezes some liquid out, so leaving a thin film between the coated grids (Fig. 3.21). It is important not to allow liquid to get onto the outside surfaces of the grids or the specimen will be too thick when frozen. To avoid excessive premature drying through evaporation (especially if particularly volatile components are present) this mounting process should take place in a glove box in which there is a saturated atmosphere of the relevant liquid. The sandwich is then rapidly frozen by one of the methods described in §2.4.1. Talmon et al. (1979) plunged their specimens into sub-cooled nitrogen but propane would give higher cooling rates, particularly if mixed with isopentane so that it does not form a solid layer when the specimen is transferred into liquid nitrogen. Although the method described here is simply one specific technique, it provides acceptable solutions to all the fundamental problems that need to be overcome in preparing this sort of specimen. Other support films could be used (such as hydrophilic carbon films) but the principles and constraints will be the same. The next stage is to pick the specimen up from the liquid nitrogen in which it will have been placed after freezing and mount it for transfer to the next processing operation.

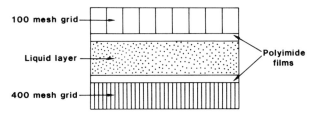

Fig. 3.21.   Preparation of a thin layer for freezing. (Redrawn from Talmon et al. 1979.)

3.4.5   Cryosections

Cryosections, once mounted firmly onto a grid (§4.3.2c) can be treated in essentially the same way as thin films. Some workers prefer to place an additional plastic film over the frozen section. This can be accomplished using the 'loop' method or by using a second, 'matched', coated grid. Although the polyimide technique of Talmon et al. (1979) clearly has merits, many workers still use Formvar or Collodion films for this application. Production of these is straightforward and described in detail in Chapter 3 of Willison and Rowe (1980). Nylon films have also been used (Saubermann and Echlin 1975).

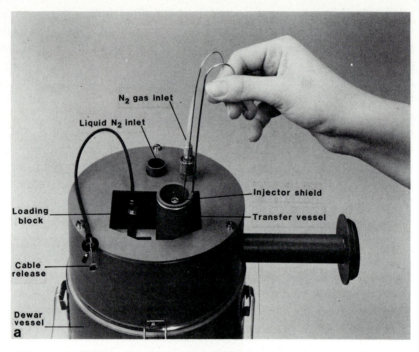

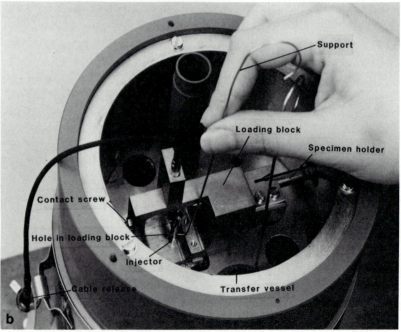

3.4.6 Coating

In many instances coating is not required for thin specimens, since the problem of charging is not the same as for bulk specimens where the vast majority of the impinging electrons are trapped by the specimen. However, deposition of a conducting layer, such as aluminium or carbon, may be required, and then the precautions already described in §3.3.6 are necessary. The problems of obtaining a uniform coating are less acute (the specimen is flatter) but the probability of heat damage is significantly increased. The specimen must be maintained at its lowest possible temperature while on the cold stage. Once again, it should be recalled that it is much better to use low deposition rates (low energy input) for longer periods rather than high rates for shorter times (see Fig. 3.12).

3.4.7 Transfer

If specimens are to be observed in transmission mode in an SEM, then the transfer devices described in §3.3.7 will, with small modifications, be suitable. However, the situation becomes more complicated if a true transmis-

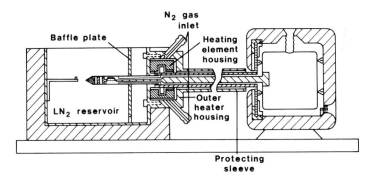

Fig. 3.23. Cold stage transfer module described by Perlov et al. (1983) for the JEOL JEM 100CX electron microscope (see also Fig. 3.28). (Redrawn from Perlov et al. 1983.)

Fig. 3.22. Cryotransfer unit PW6361 for the Philips 400 TEM (see Appendix 2). This is part of a complete system (see also Fig. 3.25) for the transfer and viewing of frozen thin specimens. This system allows the frozen specimen to be brought, at liquid nitrogen temperature, into the transfer vessel where it can be loaded into a cooled specimen holder in a dry nitrogen atmosphere. The specimen holder is then transferred into the microscope under the protection of a double mechanical shielding system (Fig. 3.25). (Reproduced by courtesy of Philips Electron Optics.)

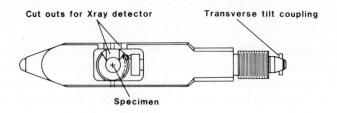

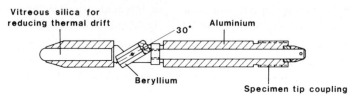

Fig. 3.24.   Double-tilt specimen tip for the Philips EM400 cold stage as manufactured by Gatan Inc. (Redrawn and reproduced by courtesy of Gatan Inc.; see Appendix 2.)

sion instrument is used; not least because the low thermal capacity of the specimen means that extra precautions must be taken to ensure that undue warming does not occur during transfer. One practical problem has been that, while a number of manufacturers have provided cold stages for transmission microscopes, few have also supplied suitable transfer devices. The situation is now gradually changing, with some well-designed systems, such

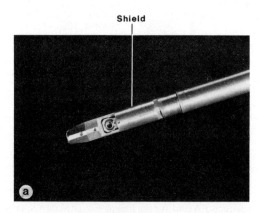

Fig. 3.25.   (a–c) Transmission holder PW6599/00 for the Philips Cryosystem (Appendix 2). The specimen can be loaded into this holder using the injector system illustrated in Fig. 3.22. This side-entry holder for the EM 400 series of Philips microscopes allows for viewing temperatures down to −195°C and is cooled by the circulation of cold, dry nitrogen gas. Temperature monitoring is incorporated and there is a double mechanical shroud to avoid contamination during transfer. (Reproduced by courtesy of Philips Electron Optics.)

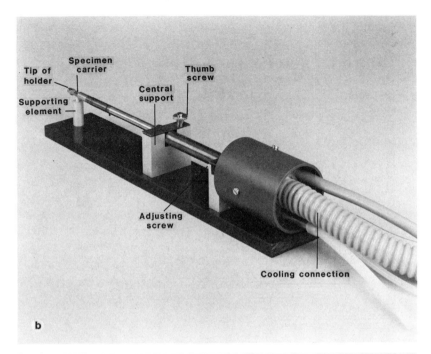

b

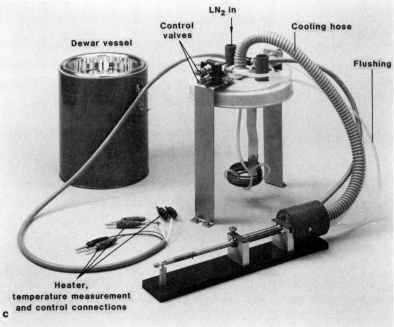

c

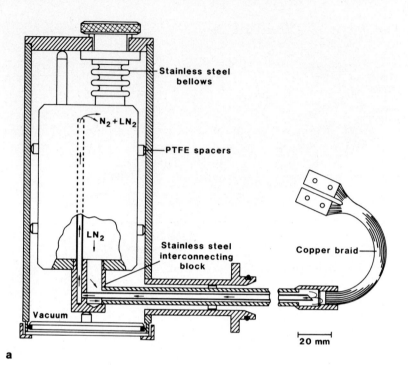

a

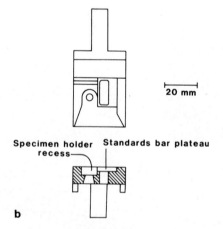

b

Fig. 3.26.   Modification of a low temperature stage for scanning electron microscopy to provide for transmission use. (a) General arrangement of Dewar vessel and cold stage. (b) Plan and vertical section of cold stage. (Redrawn from Taylor and Burgess 1977.)

as that available for the Philips 400 (Fig. 3.22; Hax and Lichtenegger 1982; see also Tatlock et al. 1984) and for Zeiss TEMs (Hezel 1984), but many workers have found it necessary to construct their own accessory to fulfil this function. In addition to the necessity of maintaining low specimen temperature during transfer, the restricted space available in a TEM means that the mechanical design has to be neater and, furthermore, the higher instrumental resolution makes additional demands on stage stability. A detailed description of a transfer device for TEM work has been provided by Hutchison et al. (1978a,b), while another example of a transfer device for TEM work is illustrated in Fig. 3.23.

3.4.8 Viewing and analysis

A number of manufacturers now provide liquid nitrogen-cooled stages for transmission instruments (e.g. Figs. 3.24 and 3.25) and others have been described in the literature (see Butler and Hale 1981). For transmission viewing in an SEM, the problems are relatively easily overcome, merely requiring a suitable transmission holder attached to a slightly modified SEM cold stage (e.g. Fig. 3.26). As such stages are almost always required for use in conjunction with microanalysis, precautions must be taken to provide a low X-ray background environment, and a beam current meter (Faraday cage) and space for standards (see Chandler 1977) are also required.

The higher resolution of a true TEM or STEM instrument requires a major step forward in the construction of a suitable cold stage (see Butler and Hale 1981). Not only must the temperature be maintained low enough but the mechanical stability (in relation to varying temperature) must also be of a high order. It is usually necessary for stages to be allowed to equilibrate at a particular temperature for a period of time so that they are stable before observation begins. To produce such stages requires specialised knowledge and precision engineering for successful construction, and such details fall outside the scope of this book. However, such stages are now being produced commercially, together with appropriate transfer devices, so that a complete low temperature pathway from freezing the specimen to observation and analysis in the microscope can be pursued.

The Philips integrated low temperature transfer device and cold stage for the EM 400 provides an excellent example of a commercially available system capable of routine use for observational studies and/or analytical ones (Figs. 3.22 and 3.25). Recently, a number of other stages for work at high resolution have also been described. Hayward and Glaeser (1980) published

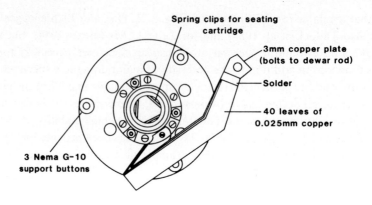

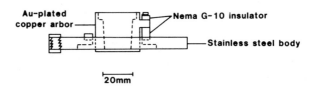

Fig. 3.27.   A high resolution cold stage for the JEOL 100B and 100C transmission electron
microscopes. (Redrawn from Hayward and Glaeser 1980.)

a design for a high resolution cold stage compatible with a JEOL microscope (Fig. 3.27). This liquid nitrogen-cooled stage only achieves a temperature of 183 K but provides a resolution of at least 0.6 nm as demonstrated
on purple membranes of bacteria. Perlov et al. (1983) have also produced
a transfer module and a simple cooling holder for use with a JEOL JEM
100C TEM. This allows frost-free transfer during which the temperature of
the sample does not exceed 120 K. The specimen may be held at a temperature between 100 and 450 K during observation while retaining a resolution
of 1.5 nm (Fig. 3.28). Nicholson et al. (1982) have designed and built a stage
that is cooled to 83 K using nitrogen gas, the minimum specimen temperature achievable being 108 K. This stage has a drift rate of 0.2–0.3 nm s^{-1}
and, being designed for microanalytical work, contributes less than 7% to
Brehmsstrahlung (X-ray 'white' radiation or 'continuum' radiation) at voltages between 9.5 and 14.5 keV (Fig. 3.29).

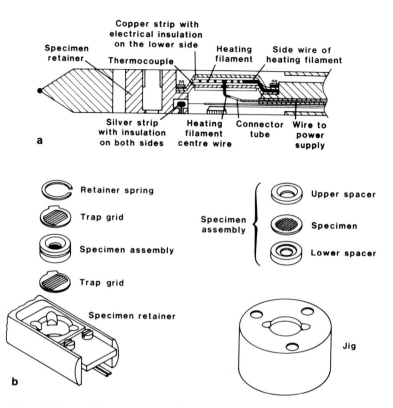

Fig. 3.28. (a) The modified tip section of the JEOL cooling holder with a heating bridge installed as described by Perlov et al. (1983) and used in conjunction with a transfer device (Fig. 3.23). (Redrawn from Perlov et al. 1983.) (b) The specimen assembly and its jig for the type of holder illustrated in (a). (Redrawn from Talmon et al. 1979.)

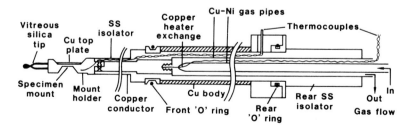

Fig. 3.29. Low temperature specimen holder designed for microanalytical work in conjunction with a JEOL microscope by Nicholson et al. (1982). (Redrawn from Nicholson et al. 1982.)

Now that both SEM and TEM cold stages and accessory devices are becoming more widely available it is inevitable that applications will be broadened as more users have the opportunity to use low temperature techniques to investigate their particular material. However, in concluding this Chapter it is, perhaps, as well to stress that the final choice of system and operating specifications ultimately rests with the individual user who should, therefore, have a clear understanding of what is to be achieved and what operational conditions are necessary to facilitate the particular study. Thus, to provide an example, although a commercial cold stage may be available that will cool to 183 K, this will be totally unsuitable for frozen-hydrated specimens (they would rapidly freeze-dry). On the other hand, such a stage may be very suitable for reducing radiation damage in *dry* specimens viewed at low temperatures. The ultimate decision lies with the user!

References

Adrian, M., J. Dubochet, J. Lepault and A.W. McDowall (1984), Cryo-electron microscopy of viruses, Nature, Lond. *308*, 32.

Aldrian, A.F. and G.A. Windisch (1984), A versatile cryo-transfer system for SEM, Proc. 8th Eur. Reg. Conf. Electron Microscopy, Budapest *1*, 99.

Beckett, A., R. Porter and N.D. Read (1982), Low temperature scanning electron microscopy of fungal material, J. Microscopy *125*, 193.

Brombach, J.D. (1975), Electron beam X-ray microanalysis of frozen biological bulk specimens below 130 K. II. The electrical charging of the sample in quantitative analysis, J. Microscopie Biol. Cell *22*, 233.

Butler, E.P. and K.F. Hale (1981), Dynamic experiments in the electron microscope, in: Practical methods in electron microscopy, Vol. 9, A.M. Glauert, ed. (Elsevier/North-Holland, Amsterdam).

Chandler, J.A. (1977), X-ray analysis in the electron microscope, in: Practical methods in electron microscopy, Vol. 5, A.M. Glauert, ed. (Elsevier/North-Holland, Amsterdam).

Chang, J.J., A.W. McDowall, J. Lepault, R. Freeman, C.A. Walter and J. Dubochet (1983), Freezing, sectioning and observation artefacts of frozen hydrated sections for electron microscopy, J. Microscopy *132*, 109.

Cosslett, V.E. (1978), Radiation damage in the high resolution microscopy of biological materials: a review, J. Microscopy *113*, 113.

Dietrich, I. (1978), Superconducting lenses, Proc. 9th Int. Congr. Electron Microscopy, Toronto *3*, 173.

Dorge, A., R. Rick, K. Gehring, J. Mason and K. Thurau (1975), Preparation and applicability of freeze-dried sections in the microprobe analysis of biological soft tissue, J. Microscopie Biol. Cell, *22*, 205.

Dubochet, J. (1984), Electron microscopy of vitrified specimens, Proc. 8th Eur. Reg. Conf. Electron Microscopy, Budapest *2*, 1379.

Dubochet, J. and A.W. McDowall (1981), Vitrification of pure water for electron microscopy, J. Microscopy *124*, RP3.

Dubochet, J., J. Lepault, R. Freeman, J.A. Berriman and J.-C. Homo (1982), Electron microscopy of frozen water and aqueous solutions, J. Microscopy *128*, 219.

Echlin, P. (1975), Sputter coating techniques for scanning electron microscopy, Scanning Electron Microscopy 1975, *1*, 217.

Echlin, P. (1978a), Low temperature biological scanning electron microscopy, in: Advanced techniques in biological electron microscopy, J.K. Koehler, ed. (Springer-Verlag, Berlin), p. 89.

Echlin, P. (1978b), Coating techniques for scanning electron microscopy and X-ray microanalysis, Scanning Electron Microscopy 1978, *1*, 109.

Echlin, P. (1978c), Low temperature scanning electron microscopy: a review, J. Microscopy *112*, 47.

Echlin, P. (1981), Recent advances in specimen coating techniques, Scanning Electron Microscopy 1981, *1*, 79.

Echlin, P. and R.B. Moreton (1976), Low temperature techniques for scanning electron microscopy, Scanning Electron Microscopy 1976, *1*, 753.

Echlin, P., A.N. Broers and W. Gee (1980), Improved resolution of sputter-coated metal films, Scanning Electron Microscopy 1980, *1*, 163.

Echlin, P., B. Chapman, L. Stoter, W. Gee and A. Burgess (1982), Low voltage sputter coating, Scanning Electron Microscopy 1982, *1*, 29.

Fernández-Móran, H. (1966), High resolution electron microscopy with superconducting lenses at liquid helium temperatures, Proc. natl. Acad. Sci. U.S.A. *56*, 801.

Fotino, M. and T.H. Giddings, Jr. (1984), Ultrastructural imaging of whole cells by fast freezing and cryo-HVEM, Proc. 8th Eur. Reg. Conf. Electron Microscopy, Budapest *3*, 1823.

Franks, F., C.S. Clay and G.W. Peace (1980), Ion beam thin film deposition, Scanning Electron Microscopy 1980, *1*, 155.

Fuchs, W. and B. Lindemann (1975), Electron beam X-ray microanalysis of frozen biological bulk specimens below 130 K. I. Instrumentation and specimen preparation, J. Microscopie Biol. Cell. *22*, 227.

Fuchs, W., J.D. Brombach and W. Trosch (1978a), Charging effect in electron irradiated ice, J. Microscopy *112*, 63.

Fuchs, W., B. Lindemann, J.D. Brombach and W. Trosch (1978b), Instrumentation and specimen preparation for electron beam X-ray microanalysis of frozen hydrated bulk specimens, J. Microscopy *112*, 75.

Glaeser, R.M and K.A. Taylor (1978), Radiation damage relative to transmission electron microscopy of biological specimens at low temperature: a review, J. Microscopy *112*, 127.

Hall, C.E. (1966), Introduction to electron microscopy, 2nd Edn. (McGraw-Hill, New York), p. 284.

Halloran, B.P., R.G. Kirk and A.R. Spurr (1978), Quantitative electron probe microanalysis of biological thin sections: the use of STEM for measurement of local mass thickness, Ultramicroscopy *3*, 175.

Hayes, T.L and G. Koch (1975), Some problems associated with low temperature manipulation in the scanning electron microscope, Scanning Electron Microscopy 1975, *1*, 35.

Hayward, S.B. and R.M. Glaeser (1980), High resolution cold stage for the Jeol 100B and 100C electron microscopes, Ultramicroscopy *5*, 3.

Hax, W.M.A. and S. Lichtenegger (1982), Transfer, observation and analysis of frozen hydrated specimens, J. Microscopy *126*, 275.

Heide, H.G. and K. Urban (1972), A novel specimen stage permitting high resolution electron microscopy at low temperatures, J. Phys. E: scient. Instrum. *5*, 803.

Hezel, U.B. (1984), A high-performance cryotransfer system and cryostage for transmission electron microscopy, Zeiss Magazine for Electron Microscopists *3*, 23.

Hobbs, P.V. (1974), Ice physics (Clarendon Press, Oxford).

Hörl, E.M. (1974), A versatile liquid helium cooling stage for the transmission electron microscope, Proc. 8th Int. Congr. Electron Microscopy, Canberra *1*, 1.

Hosoi, J., T. Oikawa, M. Inoue, Y. Kokubo and K. Hama (1981), Measurement of partial specific thickness (net thickness) of critical-point-dried cultured fibroblasts by energy analysis, Ultramicroscopy *7*, 147.

Hutchinson, T.E., W.A.P. Nicholson, B.W. Robertson, R.P. Ferrier and W.H. Biddlecombe (1977), A new low temperature specimen stage and transfer system for the JEOL JEM 100C, Developments in electron microscopy, D.L. Misell, ed., Ins. Phys. Conf. Ser. *36*, 49.

Hutchinson, T.E., D.E. Johnson and A.P. MacKenzie (1978), Instrumentation for direct observation of frozen hydrated specimens in the electron microscope, Ultramicroscopy *3*, 315.

Jakopić, E., A. Brunegger, R. Essl and G. Windisch (1978), A sputter source for electron microscopic preparation, Proc. 9th Int. Congr. Electron Microscopy, Toronto *1*, 150.

Kellenberger, E. (1982), Radiation damage in biological materials in perspective with other limitations, Proc. 10th Int. Congr. Electron Microscopy, Hamburg *1*, 33.

Kemmenoe, B.H. and G.R. Bullock (1983), Structure analysis of sputter-coated and ion-beam sputter-coated films: a comparative study, J. Microscopy *132*, 153.

Knapeck, E. and J. Dubochet (1980), Beam damage to organic material is considerably reduced in cryo-electron microscopy, J. mol. Biol. *141*, 147.

Knapeck, E., H. Formanek, G. Lefranc and I. Dietrich (1984), New aspects for the interpretation of cryoprotection illustrated on L-valine, Proc. 8th Eur. Reg. Conf. Electron Microscopy, Budapest *2*, 1395.

Koch, G.R. (1975), Preparation and examination of specimens at low temperatures, in: Principles and techniques of scanning electron microscopy. Biological Applications, Vol. 4, M.A. Hayat, ed. (Van Nostrand Reinhold, New York), p.1.

Lamvik, M.K., D.A. Kopf and J.D. Robertson (1983), Radiation damage in L-valine at liquid helium temperature, Nature, Lond. *301*, 332.

Leapman, R.D., C.E. Fiori and C.R. Swyt (1984), Mass thickness determination by electron energy loss for quantitative X-ray microanalysis in biology, J. Microscopy *133*, 239.

Lefranc, G., E. Knapek and I. Dietrich (1982), Construction principles for stable superconducting lens systems as required for obtaining high resolution images, Proc. 10th Int. Congr. Electron Microscopy, Hamburg *1*, 391.

Lepault, J. (1984), Frozen suspensions, Proc. 8th Eur. Reg. Conf. Electron Microscopy, Budapest *2*, 1413.

Lepault, J., F.P. Booy and J. Dubochet (1983), Frozen biological suspensions, J. Microscopy *129*, 89.

Marshall, A.T. (1975), Electron probe X-ray microanalysis in: Principles and techniques of scanning electron microscopy, Vol. 2, M. A. Hayat, ed. (Van Nostrand Reinhold, New York), p. 103.

Marshall, A.T. (1977), Electron probe X-ray microanalysis of frozen hydrated biological specimens, Microscopica Acta *79*, 254.

Marshall, A.T. (1980), Quantitative X-ray microanalysis of frozen hydrated bulk biological specimens, Scanning Electron Microscopy 1980, *2*, 335.

Misell, D.L. (1978), Image enhancement and interpretation, in: Practical methods in electron microscopy, Vol. 7, A.M. Glauert, ed. (Elsevier/North-Holland Biomedical Press, Amsterdam).

Nei, T. (1974), Cryotechniques, in: Principles and techniques of scanning electron microscopy. Biological applications, Vol. 1, M.A. Hayat, ed. (Van Nostrand Reinhold, New York), p. 113.

Nei, T., H. Yotsumoto, Y. Hasegawa and Y. Nagasawa (1973), Direct observations of frozen specimens with a scanning electron microscope, J. Electron Microscopy *22*, 185.

Nei, T., H. Yotsumoto, Y. Hasegawa and M. Hasegawa (1974), Freeze-etching in scanning electron microscopy, J. Electron Microscopy *23*, 137.

Nicholson, W.A.P., W.H. Biddlecombe and H.Y. Elder (1982), Low X-ray background low temperature specimen stage for biological microanalysis in the transmission electron microscope, J. Microscopy *126*, 307.

Perlov, G., Y. Talmon and A.H. Falls (1983), An improved transfer module and variable temperature control for a simple commercial cooling holder, Ultramicroscopy *11*, 183.

Peters, K.-R. (1980), Penning sputtering of ultra thin metal films for high resolution electron microscopy, Scanning Electron Microscopy 1980, *1*, 113.

Peters, K.-R. (1984), Precise and reproducible deposition of thin and ultrathin carbon films by flash evaporation of carbon yarn in high vacuum, J. Microscopy *133*, 17.

Pawley, J.B. and J.T. Norton (1978), A chamber attached to the scanning electron microscope for fracturing and coating frozen biological samples, J. Microscopy *112*, 169.

Robards, A.W. and P. Crosby (1979), A comprehensive freezing, fracturing and coating system for low temperature scanning electron microscopy, Scanning Electron Microscopy 1979, *2*, 325.

Robards, A.W., A.J. Wilson and P. Crosby (1981), Specimen heating during sputter coating, J. Microscopy *124*, 143.

Saubermann, A.J. and P. Echlin (1975), The preparation, examination and analysis of frozen hydrated tissue sections by scanning transmission electron microscopy and X-ray microanalysis, J. Microscopy *105*, 155.

Smits, H.Th.J., A.L.H. Stols and A.M. Stadhouders (1984), A carbon fibre evaporator to coat frozen hydrated cryo-sections, Proc. 8th Eur. Reg. Conf. Electron Microscopy, Budapest *3*, 1687.

Talmon, Y. (1980), Rate of sublimation of ice by radiative heating in freeze-etching, Proc. 38th Ann. Meeting EMSA, San Francisco, p. 618.

Talmon, Y. (1982a), Thermal and radiation damage to frozen hydrated specimens, J. Microscopy *125*, 227.

Talmon, Y. (1982b), Frozen hydrated specimens, Proc. 10th Int. Congr. Electron Microscopy, Hamburg *1*, 25.

Talmon, Y. and E.L. Thomas (1977a), Temperature rise and sublimation of water from thin frozen hydrated specimens in cold stage microscopy, Scanning Electron Microscopy 1977, *1*, 265.

Talmon, Y. and E.L. Thomas (1977b), Beam heating of a moderately thick cold stage specimen in the SEM/STEM, J. Microscopy *111*, 151.

Talmon, Y. and E.L. Thomas (1978), Electron beam heating temperature profiles in moderately thick cold stage STEM/SEM specimens, J. Microscopy *113*, 69.

Talmon, Y. and E.L. Thomas (1979), Open system microthermometry – a technique for the measurement of local specimen temperature in the electron microscope, J. mater. Sci. *14*, 1697.

Talmon, Y., H.T. Davis, L.E. Scriven and E.L. Thomas (1979), Cold-stage microscopy system for fast frozen lipids, Rev. scient. Instrum. *50*, 698.

Tatlock, G.J., T.J. Hurd, R.E.W. Banfield and P.J. Rushton (1984), A vacuum cryo-transfer holder for a Philips TEM, Proc. 8th Eur. Reg. Conf. Electron Microscopy, Budapest *1*, 97.

Taylor, K.A. (1978), Structure determinations of frozen, hydrated, crystalline biological specimens, J. Microscopy *112*, 115.

Taylor, P.G. and A. Burgess (1977), Cold stage for electron probe microanalyser, J. Microscopy *111*, 51.

Taylor, K.A. and R.M. Glaeser (1976), Electron microscopy of frozen hydrated biological specimens, J. Ultrastruct. Res. *55*, 448.

Vesely, D. and G. Woodisse (1982), Carbon coating with carbon fibre filament, Proc. R. microsc. Soc. *17*, 137.

Willison, J.H.M. and A.J. Rowe (1980), Replica, shadowing and freeze-etching techniques, in: Practical methods in electron microscopy, Vol. 8, A.M. Glauert, ed. (Elsevier/North-Holland, Amsterdam).

Woods, P.S. and M.C. Ledbetter (1976), Cell organelles at uncoated cryofractured surfaces as viewed with the scanning electron microscope, J. Cell Sci. *21*, 47.

Zierold, K. (1979), Scanning electron microscopy and X-ray microanalysis of cryofractured cells and tissues, Beitr. Elektronenmikroskop. Direktabb. Oberfl. 12.

Chapter 4

Freeze-sectioning

4.1 Principles and problems

In principle, the idea of cutting thin frozen sections from biological tissue is simple. In practice it turns out to be rather more difficult. Since the earliest attempts to cut thin frozen sections (Fernández-Morán 1952; Bernhard 1965; Bernhard and Leduc 1967; Appleton 1969, 1972a,b,c; Christensen 1969, 1971; Dollhopf et al. 1972), the rate of publication of papers on this subject has increased dramatically. This interest reflects the realisation that frozen thin sections potentially offer great rewards in terms of ultrastructural preservation and maintenance of the *in vivo* distribution and concentration of solutes. However, thin frozen sections *are* difficult to obtain, difficult to maintain fully hydrated and uncontaminated, and difficult to view without problems arising from radiation damage and other sources. Were it not for the very real and unique opportunities available, one wonders whether such a technique would be pursued at all. Various terms have been used in the literature to describe this technique: we shall call it '*cryoultramicrotomy*', following the practice of Reid (1974) in an earlier volume of this series. This is a *very* rapidly changing field and, therefore, only the most recent developments will be described: except where their importance and/or utility have been retained, equipment and methods which are now (even after only a relatively few years) of historical interest only will not be dealt with.

The theory of cryosectioning is, as yet, by no means totally understood. At simplest, it should be possible to differentiate between a *cutting* and a *fracturing* process, in the knowledge that hydrated biological tissues are likely to contain substances that will show cleavage/cutting phenomena that are strongly affected by temperature. Saubermann (1980) has discussed such matters in detail. He cites the melt-freeze theory of Thornburg and Mengers (1957), who suggested that the local pressure exerted by the advancing knife caused melting of the ice remaining in the tissue as a result of melting point depression. However, this concept was based on the, possibly erroneous, assumption that the physical properties of frozen biological tissue blocks are similar to those of frozen pure water. It was estimated by Thornburg and Mengers that the energy produced during sectioning would be about 2×10^4 ergs (2×10^{-3} J) although subsequent direct measurements by Saubermann et al. (1977) indicated that the energy of cutting 0.5 μm thick cryosections ranged between about 1300 ergs (1.3×10^{-4} J) at 243 K ($-30°$C) to 5000 ergs (5×10^{-4} J) at 193 K ($-80°$C). It was calculated that this *could* provide a melting zone of between 0.4 and 1.6 nm. A number of authors have turned their attention to this problem because it could clearly be quite crucial. If melting takes place during sectioning (or at any other time), then solute redistribution may be possible and the whole authenticity of the technique for the localisation of soluble components is thrown open to doubt. The concensus appears to be that melting either does not occur or, if it does, it is to such a very limited degree as to be insignificant, and is therefore not a limitation to the use of the method (Hodson and Marshall 1972; Appleton 1974; Saubermann et al. 1977; Frederik and Busing 1981; Frederik et al. 1982a; Karp et al. 1982).

As an alternative theory, Saubermann (1980) has drawn the analogy with metal machining, suggesting that the main difference is simply that, whereas in cryosectioning it is the 'chip' that one wants to retain, in metal machining it is the chip that is removed. Using this comparison, the knife is considered to be forced into the frozen block, causing compression and stress at the knife edge. The force is eventually released by rupture along planes of least resistance with additional relief through plastic deformation which depends greatly on the ductility of the specimen. If the block is brittle, it is suggested that the compressive force causes rupture along the shear angle prior to the movement of the chip over the knife, so leading to a discontinuous chip as opposed to the continuous chip (section) that is produced from less brittle, more ductile material (Fig. 4.1). As the state of ductility/brittleness of frozen specimens depends both on the temperature and on the nature of the specimen itself including, particularly, the size of the ice crystals (see Sleytr and Robards 1977; see also §6.3.1), the importance of selecting the appropriate preparation/sectioning conditions is self-evident. One of the most useful contributions to ease of cryosectioning is to produce a specimen within

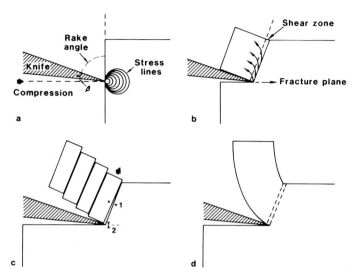

Fig. 4.1. (a) Schematic diagram of metal machining model of cryosectioning. The knife causes stress and compression in the block along the direction shown by the heavy arrow. The rake angle of the knife is defined as the angle between a line perpendicular to the cut surface and the top surface of the knife. The cutting angle is the sum of the knife angle (α) and the clearance angle (β). The general distribution of stress from a knife edge is shown by the stress lines. (b) The compressive forces are relieved by fracture or rupture along the fracture plane which is the path of least resistance. This may be between crystal planes but most probably follows the same fracture planes as frequently observed in freeze-fractured material. Some relief of the compressive forces occurs through plastic deformation in the shear zone. The amount of plastic deformation depends greatly on the ductility of the tissue. Plastic flow, from the compressive forces, tends to occur in the direction shown by the arrows. (c) A discontinuous chip is formed if the tissue is brittle, non-compressible, and does not undergo plastic deformation in the shear zone. Instead, stress from the knife is relieved by fracture along the two planes shown by the small arrows (**1** and **2**). Slippage along fractures in the direction shown by the heavy arrow causes the chip sequence to overlap. Such sections would not be useful for X-ray analysis. (d) A continuous chip is formed when plastic deformation can occur in the shear zone. In this case, the tissue is sufficiently ductile to promote plastic flow over the knife surface. The bending forces imparted by the knife cause the chip to curl. (Redrawn from Saubermann 1980.)

which, either locally or in general, the ice crystals are small. Conversely, blocks containing very large ice crystals are usually difficult, if not impossible, to cryosection. The explanation for many of the conflicting reports of success or failure, even using the same tissue, is that they undoubtedly stem, at least in part, from differences in ice crystal size within the blocks (see §2.1.2, and Fig. 2.34 in §2.3.2).

It will be clear that the temperature at which sections are cut will greatly affect the nature of the cutting/fracturing process. Saubermann (1979) showed that continuous chips from mouse liver could be obtained with lower 'work' at 243 K ($-30°C$) compared with the greater work needed to produce discontinuous chips at 193 K ($-80°C$). Such findings point to the possibility of improving sectioning by cutting at warmer temperatures (e.g. 243 K; $-30°C$), although it can also be argued that ice recrystallisation is more likely to occur and any melting zone would be extended. However, the work needed at the higher (warmer) temperature is less and the likelihood of inducing thermal damage is therefore also less ! Partly for the reasons given above, Saubermann et al. (1981a,b) advocated sectioning at temperatures closer to 240 K ($-33°C$) than 200 K ($-73°C$); a suggestion that has met with strong opposition from some other workers. For example, it has been suggested that Saubermann did not see ice recrystallisation because the crystals in their specimens were already so large that additional growth would be extremely slow; only very small crystals (such as might be produced at the surface of a large block) would grow rapidly at the warmer temperatures (Sevéus 1980). More work remains to be done before the theoretical basis of variations in the physical characteristics of the specimen at different temperatures in relation to both sectioning and ice transformation phenomena are properly understood. Nevertheless, enough is known and agreed to allow us to go some way towards producing and sectioning frozen specimens under optimised conditions.

The critical temperature during the cryosectioning process is clearly that at the block face itself. This, in turn, is largely dependent on the temperature of the surrounding atmosphere. Saubermann (1980) stresses the importance of achieving a stable thermal equilibrium between the block face and the knife, and that consequently there is little point in separately cooling the specimen and the knife. However, most cryomicrotomists find it preferable to have full control over both knife and specimen cooling. Such considerations have important implications for the design of cryoultramicrotomes. Similarly, it has been found that thin (<100 nm) sections need to be cut at lower temperatures than thickers ones (Sevéus and Kindel 1974; Saubermann 1980). Thus it is apparent that any cryoultramicrotome should at least have a means of obtaining a good range of specimen temperature control so that the operator can experiment with the best conditions for the particular specimen in hand.

Consideration should be given to the knife angle and rake angle (see Fig. 4.1) when sectioning (Reid 1974). Saubermann (1980) recommends a low knife angle and a high rake angle to minimise the tendency to bend the brittle specimen. The chip can be further constrained from bending as it moves across the knife by use of an anti-roll plate (§4.3.2a), or by the ingenious vacuum method of Appleton (1974) described later (§4.3.2b). Saubermann (1980) states that an advantage of metal knives for low temperature sectioning is that they have smaller knife angles than glass or diamond knives and can thus provide a lower clearance angle and high rake angle. There is the further advantage that metal knives have better thermal stability than glass ones.

The rate of heat input to the specimen is directly proportional to the cutting speed used. If heat input is too rapid the specimen face will warm up. It is therefore necessary to create an acceptable balance between cutting speed, block face temperature and section thickness. Given that the section thickness is usually determined by the requirements of the operator, it is sensible, if consistent results are to be obtained, to maintain a constant cutting speed while varying the block face temperature, which is determined by the balance between the thermal properties of the specimen together with the temperature of the specimen holder and the surrounding atmosphere, to give optimum sectioning properties. However, the hardness of the specimen also varies with temperature and cutting speed may have to be varied to help compensate for this temperature-dependent variation in hardness.

4.2 Equipment

It will be clear that the basic requirement for frozen thin sectioning is an ultramicrotome that will operate with the specimen area at a very low temperature. 'Cryostats' (a poor term for conventional microtomes in cold boxes) have long been available for preparing sections with thicknesses down to about 10 μm and are in widespread routine use for light microscopical investigations. The more rigorous requirements of thin frozen sections have meant that, over a decade or so, cryoultramicrotomes have evolved from slightly modified conventional ultramicrotomes to specially designed and engineered apparatus. As the specimen has to be protected at all stages, the cryoultramicrotome cannot be considered alone: it must be 'interfaced' with both the initial freezing process and with subsequent procedures such as specimen coating and viewing. This usually necessitates some sort of transfer device (see §3.2.2). An important aspect that should also not be overlooked is that the success or failure of many difficult techniques, and cryoultramicrotomy is certainly no exception, can be determined by the accessory tools that are available. It is necessary to have a range of forceps, clamps, grid supports and so on that have been selected and/or modified to work at the specific temperatures chosen. Frozen thin sections are often allowed to freeze-dry prior to subsequent handling and some provision must also be made for this process to be carried out under controlled conditions.

4.2.1   Basic requirements of cryoultramicrotomes

The cryoultramicrotome should have good temperature control of the specimen and knife (either independently or by cooling the whole surrounding atmosphere, see §4.2.2) to temperatures considerably lower than about 170 K ($-100°C$); the specimen area should be protected from water vapour, and hence condensation; and the usual facilities of an ultramicrotome (adjustable cutting speed, knife angle, section thickness, etc.) should be available and operable down to the lowest achievable temperature of the equipment. The manner in which the specimen is mounted is important, because it is often required that the smallest possible specimen is frozen (to provide highest cooling rates) whereas thermal and mechanical stability during sectioning demand that the specimen must be mounted firmly onto some form of substantial support having good thermal conductivity and heat capacity. It

TABLE 4.1

Some low temperature 'glues' and other substances used during cryoultramicrotomy

Liquid	Boiling point °C (K)	Melting point[a] °C (K)	Vapour pressure[b] Torr (Pa) (K)
Toluene $C_6H_5CH_3$	110.6 (384)	-95 (178)	755 (1.0×10^5) (110)
n-Butyl benzene $C_6H_5(CH_2)_3CH_3$	183.3 (456)	-88 (185)	755 (1.0×10^5) (139)
Isopentane $(CH_3)_2CHCH_2CH_3$	27.9 (301)	-159.9 (113)	–
n-Heptane $CH_3(CH_2)_5CH_3$	98.4 (372)	-90.6 (183)	4×10^{-3} (5.33×10^{-1}) (173)
DMSO CH_3OSCH_3	192 (465)	18.0 (292)	–

[a] Temperatures given for 760 mm pressure.
[b] Vapour pressures given together with relevant temperature.

is not generally possible to clamp frozen specimens directly because ice is too brittle. Therefore, material is usually either frozen on a very small support which can then be mounted onto the microtome arm, or the specimen is frozen alone and subsequently stuck to a chuck, or other support, using a special low temperature 'glue' (§4.3.1b; §4.3.2c; Table 4.1).

Cryoultramicrotomes are of two basically different types: either the whole ultramicrotome is placed in a cold box (the conventional 'cryostat' approach), or just the specimen/knife area of the ultramicrotome is enclosed so that this region can be kept at the appropriate selected temperature. Both forms are commercially available and both the instruments and the methods of operating them have often been modified over the past few years. The descriptions provided here predominatly refer only to the latest versions. Whichever form of cryoultramicrotome is used, it should be possible to transfer the frozen sections to a suitable area for freeze-drying, if this is required, and it should also be possible to take up the frozen sections and to transfer them into other ancillary equipment, without allowing them either to freeze-dry or to become contaminated.

4.2.2 Cryoultramicrotomes

The two types of cryoultramicrotome (cryostat or specific specimen area cooling) have their own particular merits and demerits. The advantage of the cryostat approach is that it provides a large working volume at a stable low temperature around the specimen. It is easy to keep the many tools, vials and chemicals in the large cold box. However, such equipment is often very uncomfortable to use. The operator must usually stand, bent-over, with his/her hands thrust down into the cold chamber, to mount the specimen and to make adjustments. Furthermore, change in specimen temperature (particularly warming) can take an extremely long time because the whole mass of the microtome must be brought to the required temperature. The early cryosectioning attachments for specimen area cooling on conventional ultramicrotomes were also often less than satisfactory in many respects, but more recent versions appear to have overcome many of the previous problems, such as the maintenance of temperature stability. Moreover, the ultramicrotomes can be used in the conventional sitting position: an important consideration during the painstaking processes involved. Cryostat systems tend to use more liquid nitrogen than the attachments which have smaller volumes to cool. In principle, in either cryostat or attachment, cooling can be achieved by using cold nitrogen gas ducted continuously or intermittently into the chamber, or by evaporators in the case of cryostats. Specimen mount and knife holder can be warmed independently, using heaters, so that their bulk temperature can be varied from that of the surrounding atmosphere, and the temperature of incoming nitrogen gas can be adjusted by passing the cold gas over some form of heater. Various combinations of these cooling methods have been used in different designs of cryoultramicrotome (see below). The critical considerations are that the specimen block face and the knife should achieve the desired temperatures and that these temperatures should be very stable. Early studies often stressed the importance of differences between apparent specimen and knife temperatures. More recently the trend has been towards using the same temperature for both. This simplifies control systems because it means that only the temperature of the surrounding atmosphere needs to be accurately and stably controlled. When cooling is effected by enclosing just the specimen/ knife region within a small chamber, there is the additional advantage that all the mechanical parts of the microtome remain at ambient temperature. This allows a low consumption of coolant and better operation of the fine controls of the microtome. However, the chamber must be continuously cooled with a flow of dry, cold nitrogen gas so that no humid air can reach the cooled parts. Also, there must be no opening below the top of the chamber through which the cold nitrogen gas can 'overflow', since this would leave the upper area unprotected by dry gas and, therefore, at risk from 'frosting-up' by condensing water vapour. Two alternative solutions to these problems have been found (§4.2.2b).

It has already been noted that a versatile arrangement must be available for mounting the specimen. Thermal and mechanical stability should be combined with good thermal conductivity once the specimen has been mounted. Saubermann (1980) has emphasised the advantages of metal knives when used at low temperatures, partly due to their better thermal conductivity and partly to their small knife angle. He notes that razor blades, which can have a knife angle of $<25°$, can be used to cut sections down to 0.5 μm.

4.2.2a   Cryostat ultramicrotomes

Typical of this approach is the Slee Type TUL cryoultramicrotome (Fig. 4.2; see Appendix 2 for supplier). Details of using such a device have been given by Appleton (1974, 1978) and Gupta (1979). The top-opening cold box contains a Porter-Blum ultramicrotome; this box can be cooled down

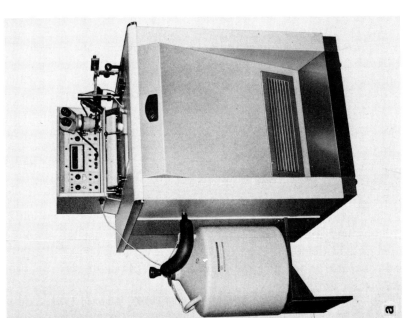

Fig. 4.2. (a) General view of the Slee TUL Cryostat Cryo-Ultramicrotome. This type of cryoultramicrotome embodies a large, temperature-controlled chamber within which the microtome is housed. Compressor-cooling of the chamber is augmented by liquid nitrogen cooling when the microtome is in use. (b) Internal view of the cryostat chamber showing the microtome, freeze-drier, remote controls, standby refrigerator and storage space for specimens and work tools. (Courtesy of Slee Medical Equipment Ltd.)

to about 220 K ($-53°C$) using an electrically driven compressor system and is further cooled by liquid nitrogen which is injected into the work chamber under electronic control. This provides a vibration-free and reliable system capable of operating down to 170 K ($-103°C$). It is necessary for the ultramicrotome to be specially prepared to operate satisfactorily at low temperatures (something not required for cryoattachments to ultramicrotomes). Section thicknesses between 25 and 500 nm can be set and cut at a range of speeds (1–12 mm s^{-1}) using this electronically controlled, motor-driven microtome. Specimen feed is obtained through a mechanical lever system. It is possible to heat the knife, on its 'universal' stage, so that it is warmer than the ambient temperature within the cabinet. Unique to this system is the vacuum probe device described by Appleton (1974, 1978) which allows ribbons of frozen sections to be gently drawn out from the knife edge (Fig. 4.12 in §4.3.2b). Within the main work chamber is a separate, small compartment for storage and/or freeze-drying of sections. This has its own temperature control arrangements, allowing temperatures as low as 120 K ($-153°C$) to be used for storage or temperatures between 250 and 150 K (-23 and $-123°C$) to be maintained for freeze-drying. A special set of handling tools is provided with the equipment.

4.2.2b   Ultramicrotome cryoattachments

This approach to cryoultramicrotomy has been more popular than the cryostat system over the past few years. In part, this is because a number of (ultramicrotome) manufacturers have produced commercially available 'kits' which enable a conventional ultramicrotome to be used at low temperatures. Interestingly, and fortunately, the three major systems currently available (from Dupont-Sorvall, LKB Produkter AB and Reichert-Jung; Appendix 2) embody a number of different design features which allow useful comparisons to be made. It is very significant, and a reflection of the still evolving state of the art, that numerous papers have been published which describe modifications to commercially available apparatus. Thus, while the basic equipment can now be bought, the newcomer to cryoultramicrotomy must still be willing, not only to experiment to find the best preparation methods for the particular biological material in hand, but also to spend some time in adapting the equipment to facilitate the different processes that are to be carried out.

The *Dupont-Sorvall* equipment (Hagler et al. 1980; Barnard 1982; Biddlecombe et al. 1982) is a derivative of the Christensen (1967, 1971) system in which liquid nitrogen is fed into an insulated chamber around both the knife and the specimen. The specimen is supported on a 'bridge' that hangs over the wall of the cryochamber; this is one of the two most common ways of keeping the specimen and knife cold while the rest of the microtome is at ambient temperature (Fig. 4.3). The latest version (the FS-1000) is designed for operation with the Sorvall MT-5000 ultramicrotome (Fig. 4.4). It comprises a freezing chamber, a modified specimen mount, a low temperature controller, which senses and maintains the preset temperature of both the specimen and the knife, and an insulated liquid nitrogen delivery system. The stage of the FS-1000 is a thermally insulated aluminium chamber containing a knife holder which accepts glass, diamond or razor blade knives (Fig. 4.5).' The specimen holder accepts a variety of specimen 'pins' (Fig. 4.5).

The controller initiates the delivery of the correct amount of liquid nitrogen to maintain the preset temperatures in the chamber. A phase separator in the supply line ensures that only liquid gas is allowed into the chamber.

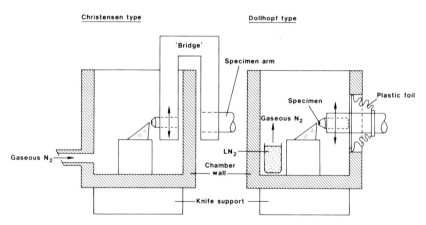

Fig. 4.3. Comparison of two methods of thermally insulating the cold specimen in a cryoultramicrotome from the main body of the equipment. The 'Christensen type' (Christensen 1967, 1969) uses a 'bridge', providing a poor heat conduction pathway, so that the main part of the specimen arm is thermally isolated from the specimen itself. This type is represented in (for example) the Reichert-Jung FC4 cryosystem and the Sorvall FS-1000 cryosectioning system. The alternative 'Dollhopf type' (Dollhopf et al. 1972) has a continuous, low thermal conductivity, specimen arm that enters the cryochamber through a hole in the rear wall. This hole is sealed with a thin, flexible plastic film which thus allows movement of the microtome arm. The LKB Cryokit 14800 employs this method. (This figure redrawn from an original illustration kindly provided by H. Sitte, Universität des Saarlandes, F.R.G.)

a

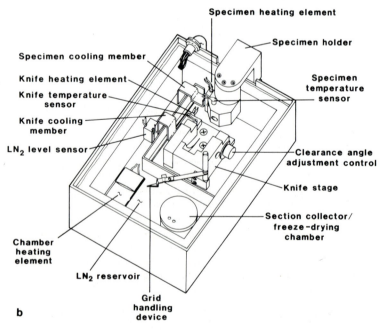

Specimen heating element

Specimen holder

Specimen cooling member

Knife heating element

Knife temperature sensor

Knife cooling member

LN₂ level sensor

Specimen temperature sensor

Clearance angle adjustment control

Knife stage

Section collector/ freeze-drying chamber

Chamber heating element

LN₂ reservoir

Grid handling device

b

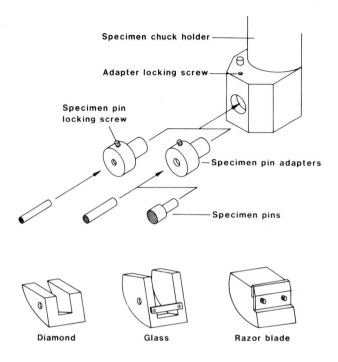

Specimen chuck holder

Adapter locking screw

Specimen pin
locking screw

Specimen pin adapters

Specimen pins

| Diamond | Glass | Razor blade |

Knife holders

Fig. 4.5. Specimen and knife handling facilities for the Sorvall FS1000 cryosectioning system. A variety of specimen pins can be used to mount specimens prior to freezing. The low heat capacity pins can be attached to the pin adapters, under liquid nitrogen, and then quickly transferred into the cryochamber and inserted into the cooled specimen chuck holder. The FS1000 allows the use of diamond, glass or metal (razor blade) knives by using different holders. (Redrawn from an original kindly provided by Biomedical Products Division, Dupont (U.K.) Ltd.)

Entry of moist, surrounding air is completely eliminated through a constant flow of nitrogen gas generated by a heating element in the bottom of the chamber.

Accessories supplied with the Sorvall system include: a grid holder/freeze-drying chamber; different sized specimen pins and pin holders; a grid

Fig. 4.4. (a) Sorvall FS1000 Cryosectioning system. The attachment of a cold-box to the Sorvall MT5000 ultramicrotome, together with liquid nitrogen supply and control unit, is illustrated here. (b) Diagrammatic illustration of the cryochamber of the FS1000 as seen from above (see also Fig. 4.5). (Fig. 4.4a kindly supplied by Biomedical Products Division, Dupont (U.K.) Ltd. and reproduced with permission. Fig. 4.4b redrawn from an original diagram provided by Dupont U.K. Ltd.)

handling device; knife holders; film support rings; a cryochamber cover with three doors and various other cryotools. This system allows two ranges of chamber temperature to be selected: -120 to $-160°C$ (153–113 K) or -20 to $-80°C$ (253–193 K).

The *LKB Cryokit 14800* attachment, together with the LKB 14860 cryotools, represents one of the best known cryoultramicrotomy systems. Numerous authors have published results from this equipment, including Sevéus (1979, 1980), who has been concerned with its commercial development. The cryochamber encloses the whole specimen/knife area and gives sufficient space for other ancillary items to be included (Fig. 4.6). Two liquid nitrogen reservoirs are provided (one serving the specimen head, the other the knife stage) and thus the knife and specimen stage temperatures can be independently and automatically controlled by varying the coolant level. The specimen head can achieve temperatures down to 100 K ($-173°C$) while the knife cools to about 120 K ($-153°C$). There is automatic control between these temperatures and temperatures approaching 273 K ($0°C$). The specimen holder is rapidly interchangeable, so several holders can be pre-cooled and inserted into the chamber. A rotatable specimen holder allows trimming to be carried out. Either diamond or glass knives can be used, spare knives also being housed within the chamber. Using a liquid nitrogen 'booster' it is possible to evaporate some liquid nitrogen from within the cryochamber, thus providing the anticondensation advantages of purging with dry nitrogen gas.

The *Reichert-Jung Cryochamber FC 4* (Sitte et al. 1979) was specifically developed for use with the Reichert OmU4 Ultracut ultramicrotome (Sitte et al. 1979, 1980; Sitte 1982, 1984). It represents an extremely advanced approach to cryoultramicrotomy. The walls of the 'cryochamber' are of aluminium, the side walls being cooled directly by about 1.0 litre of liquid nitrogen stored in an adjacent 'twin-tank' (Fig. 4.7). This tank, together with the cooled chamber, is insulated by surrounding styrofoam which is itself enclosed within an aluminium jacket that is continuously held at 293 K ($20°C$). More than 75% of the nitrogen gas escaping from the supply tank is warmed up to room temperature and hence causes minimal disturbance of ultramicrotome function. A small ($<25\%$) amount of the cold gas is guided by aluminium plates to the bottom of the chamber and provides a continuous, smooth and laminar flow of gas from the bottom to the top of the open chamber. The specimen is fixed on a bridge which is connected to the specimen arm of the Ultracut and enters the chamber through the open top (Fig. 4.7b). All parts of the specimen bridge exposed to the room

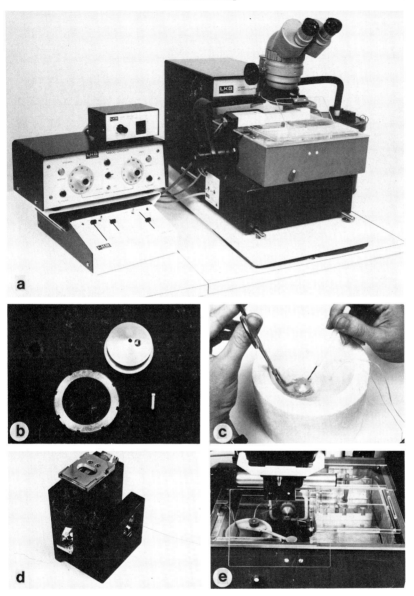

Fig. 4.6.   LKB 14800 Cryokit cryosectioning system. (a) General view of LKB 14800 Cryokit attached to the Ultrotome V. (b) Specimen holder and locking ring together with one of the silver 'pins' on which specimens are frozen. (c) The specimen, on its pin, is secured to the specimen holder under liquid nitrogen before transfer to the microtome. (d) The grid carrier holds the grid in a horizontal plane, close to the knife edge, while sectioning. The sections can thus easily be manoeuvred onto the grid. (e) The general layout of the specimen, knife and grid holder in the cryochamber is shown in this illustration. (Illustrations by courtesy of LKB-Produkter AB.)

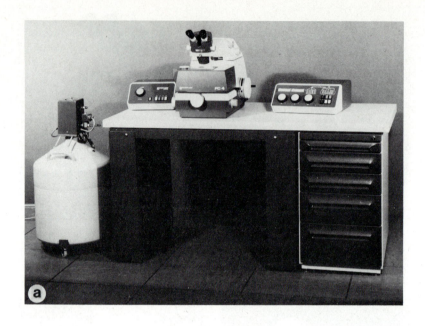

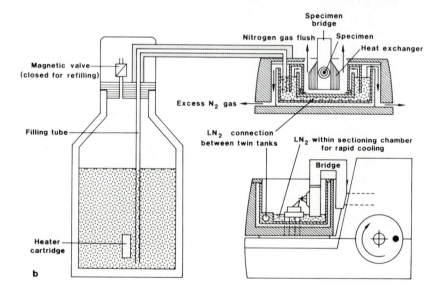

atmosphere are separately heated to about 293 K (20°C), so preventing frosting of these parts or any other thermal effect. The specimen holder, knife holder and preparation plate each contain a temperature sensor and a heating element. The preparation plate is used for flattening specimens and/or for freeze-drying. Since the atmosphere within the chamber is cooled down by the cold aluminium walls, together with the cold nitrogen gas flow from bottom (77 K; −196°C) to top (123 K; −150°C), specimen, knife and preparation plate reach temperatures of about 100 K (−173°C). By partially filling the chamber with liquid nitrogen, a further reduction in temperature (or more rapid initial cool-down) can be achieved. The heaters can be used to warm the three separate components rapidly to up to 390 K (+117°C), so eliminating films of water condensation that would otherwise occur during more gradual warming to lower temperatures.

Any temperature of the three main components (knife, specimen and preparation plate) between 100 K (−173°C) and 273 K (0°C) can be accurately maintained (±5°C) over prolonged periods and the chamber temperature can be held to a constant value ±1°C at any given level. Since the walls of the chamber are, under any operational conditions, colder than the other components, any condensation will occur on the walls, so leaving the important areas free of contamination.

The FC 4 chamber is easily and quickly mounted onto the Ultracut knife support. There is an entry system so that sections can be transferred out of the cryochamber and onto other equipment. Cryogen consumption is low (3.0 l N_2 h^{-1}) even at the lowest temperatures and the twin cryogen tanks are automatically refilled from the main Dewar.

The standard knife holder system allows the subsequent use of two precooled glass knives or one glass and one special diamond knife, as well as one stainless steel knife for trimming. Adaptors allow the use of special dia-

---

Fig. 4.7.   Reichert-Jung Ultracut ultramicrotome with FC4 cryoattachment. (a) Photograph illustrating general layout of the ultramicrotome, FC4 cryochamber, supporting table and liquid nitrogen Dewar vessel.(b) Diagrammatic illustration of the main components of the FC4 cryoattachment. Liquid nitrogen is supplied, under pressure from the Dewar vessel, to the twin tanks surrounding the specimen/knife area of the ultramicrotome. Rapid cooling can be obtained by 'overfilling' the tanks so that some liquid nitrogen spills into the cryochamber. Once working temperature has been achieved, it can be maintained constant within precise limits (see text). (Fig. 4.7a reproduced by courtesy of Reichert-Jung (U.K.) Ltd. Fig. 4.7b redrawn from an original illustration provided by H. Sitte, Universität des Saarlandes, F.R.G. and reproduced by courtesy of H. Sitte and Reichert-Jung.)

mond knives. The clearance angle can be adjusted between 0 and 9° while ±10° eucentric adjustment of the knife around the vertical axis is also possible. A number of optical or acoustic warning signals exists to prevent system malfunctions or to avoid loss of coolant, etc. While the details provided here only serve to indicate the main features of the FC 4 cryoattachment, it can be seen that this gives a very versatile instrument that can be used with relative ease over prolonged periods.

The cryoultramicrotomes described above each have their own virtues and specific features: all have been used successfully for the preparation of frozen-hydrated thin sections. However, the success or failure of cryoultramicrotomy depends very greatly on:

(*i*)   the nature of the specimen itself,
(*ii*)   the manner in which it is prepared for sectioning,
(*iii*) the physical parameters that are selected for the sectioning process, and
(*iv*)   the skill of, and/or investment of time by the operator, which must be considerably more than for conventional ultramicrotomy.

It is these factors that will be discussed below.

4.3   Sectioning

4.3.1   Specimen preparation for sectioning

The basic methods of specimen preparation prior to ultra-thin frozen sectioning have already been described in Chapter 2 and it is important that the reader should have a good understanding of these.

4.3.1a   Pretreatment

The choice of whether or not to fix the specimen, or to pretreat it in any other way, resides with the experimenter and depends on the nature of the information that is required. In the early days of cryoultramicrotomy, fixation, cryoprotection, encapsulation (of the specimen), use of trough liquids, and post-staining were all commonly advocated. With the recognition of the particular importance of cryoultramicrotomy in the localisation of diffusible elements and the application of histo- and cytochemical methods, the trend has been much more towards minimal use of pretreatments and any other 'wet' processes which would detract from the effectiveness of such techniques. It may be decided that the best course of action *is* to fix the tissue, in which case one of the conventional recipes can be used (Glauert 1974; Hayat 1981). Because cryoultramicrotomy is often used in an attempt to retain otherwise diffusible substances within the specimen, it has sometimes been suggested that vapour fixation (e.g. with formaldehyde fumes; Bernhard and Viron 1971) is preferable to wet fixation prior to cryoultramicrotomy. The individual must assess the requirements of the particular preparation method and chose accordingly. After fixation, some cryoprotectant treatment may be used. Indeed, fixation may often be necessitated *because* cryoprotection is required (§2.3.2a). Glycerol is the most commonly used cryoprotectant (§ 2.3.2a) but infiltration with sucrose (20% w/v) gives advantages in terms of better 'sectionability'.

4.3.1b   Mounting specimens for freezing

Two main considerations apply in chosing how best to mount the specimen for freezing. Firstly, the specimen needs to be mounted firmly onto the microtome specimen support and, secondly, the opportunity may be taken (particularly if a penetrating cryoprotectant was not used during the earlier processing stages) to confer some cryoprotection by 'embedding' or 'encapsulating' the specimen in a suitable medium. Appleton (1978) used small brass or copper specimen holders in the end of which a small, rough-surfaced cup had been machined (Fig. 4.8; see also Fig. 4.5). Tissues often adhere to the

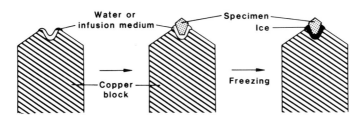

Fig. 4.8.   Methods of mounting specimens for sectioning. Simple means of attaching a specimen to a metal support prior to freezing. The water or surrounding medium (for mounting media see Appendix 2) provides good adhesion of the tissue block to the support, once frozen. The same principle can be used for mounting specimens onto other supports (e.g. see also Figs. 4.5 and 4.6). If frozen specimens are to be attached to supports, then low temperature glues may be necessary (§4.2.1; Table 4.1). (Redrawn from an original diagram in Roland 1978.)

holder by their inherent moisture but a thin application of a suitable medium such as Tissue-Tek (Appendix 2) or simply a 6% solution of methyl cellulose (Sevéus 1979) can help although, as cautioned by Appleton, any tissue actually in contact with the mounting medium should be regarded with suspicion when interpreting structural or analytical information. The virtues of polyvinylpyrrolidone (PVP) or hydroxyethyl starch (HES) as both cryoprotectant and mountant have been reported by Echlin et al. (1977, 1982). A 25% solution acts as a good supporting medium, confers cryoprotection in suppressing growth of ice crystals, and, importantly in the current context, sections well (see also Biddlecombe et al. 1982). Thus, tissues embedded in a matrix of PVP often prove to be more amenable to sectioning than tissues frozen without such treatment. Other 'encapsulating' media that have been used include 1% bovine serum albumin subsequently cross-linked with glutaraldehyde; and 20% gelatin which may need to be infiltrated for a considerable period (e.g. 20 min at 37°C followed by 1 week at 4°C). The decision for the user again depends on the nature of the specimen and what is required of it, but a sensible first attempt would be to mount some specimens simply using 6% methyl cellulose and some others using PVP or HES as an embedding matrix.

4.3.1c   Freezing specimens

The method of freezing used by Appleton (1978), in which quite large (1.0 mm³) tissue blocks within a substantial metal holder are cooled by plunging into sub-cooled nitrogen, will certainly not provide ultra-rapid cooling rates. The balance has to be drawn between obtaining a suitable large, undamaged, tissue block and obtaining a high enough cooling rate to suppress formation of large ice crystals. As stated above, cryoprotectives will, therefore, often be necessary. The requirements are more fully discussed in §2.3.2a. Sevéus (1979, 1980) has described the well-established method of freezing specimens on the ends of small silver 'pins'. She has used 'solid' nitrogen at 65 K ($-208°C$) and states that the outer 6–8 μm of the specimen is essentially free of observable ice. Once again, the absolute cooling rate is unlikely to be high and the undulating nature of the surface, together with the mechanical damage that will usually have taken place at the surface itself, means that it is difficult, if not impossible, to take complete advantage of the very small ice crystals in the superficial zone. Gupta (1979) has used rather similar methods, dropping specimens on 'pins' or 'microchucks' into halocarbon 13 (Freon 13). Many workers have preferred not to use halocarbons because they can contaminate the specimen (possibly leading to spurious peaks during microanalysis) and may also remain as liquid surface layers over the rest of the specimen during sectioning. Nevertheless, Gupta considers that the virtues of Freon 13 outweigh the disadvantages so long as excess cryogen is shaken from the tissue blocks after freezing. If the specimen is of any significant size (> 0.1 mm in all dimensions), and particularly if it is mounted on a specimen support, then it is best simply to freeze it by plunging into sub-cooled nitrogen (§2.4.1a). Only if the specimens are *very* small is it worth using the ultra-rapid cooling methods described in §2.4.

Sitte et al. (1979, 1980) have given the matter of freezing prior to cryo-ultramicrotomy special attention and conclude that the vast improvement in 'sectionability' obtained with very small ice crystals is a goal well worth pursuing. For this reason they have advocated the use of carefully controlled cold block-freezing (§2.4.2), rather than plunge-cooling, prior to cryosectioning. This not only yields very small ice crystals within the surface zone but also provides a relatively flat surface that can be oriented parallel to the knife edge, so making it possible to cut quite large areas of section through well-frozen cells. If the tissue surface is very carefully oriented as parallel as possible to the surface of the polished metal block (metal 'mirror') prior to freezing, then deformation is reduced to a minimum (Fig. 4.9). Cold block freezing also leads to dimensional changes which are made worse by incomplete relief of shear stress caused by dry sectioning. This technique is relatively complex and the beginner would do well to experiment with simple plunge-cooling methods until he or she understands well enough what is involved before embarking on cold block-freezing.

It may be advantageous to freeze the specimen alone (i.e. without a support) so that very high cooling rates can be obtained, such as when spray-freezing cells. In these circumstances the frozen cells will need to be 'stuck' to the specimen support (while, of course, maintaining low temperature). This requirement has led to the search for suitable low temperature 'glues', i.e. substances with a melting point lower than the temperature at which specimens will be mounted (typically lower than about 80°C; 193 K), with a significantly higher boiling point and with a relatively low vapour pressure at the temperatures to be used (Table 4.1; §4.2.1). The use of all such low temperature glues must be viewed with particular caution when microanalysis is subsequently to be carried out because they may contribute either to specific elemental peaks (as is the case with the halocarbons) and/or to the continuum radiation. In practice, substances with melting points around

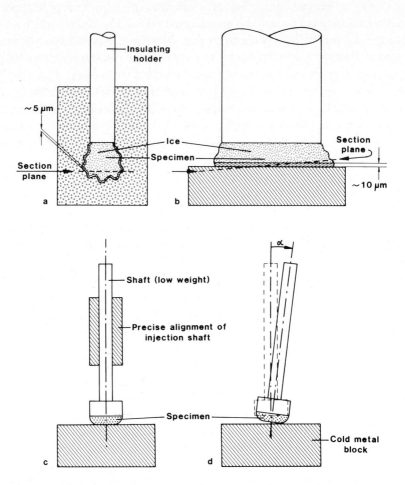

Fig. 4.9.   Comparison of plunge-freezing and cold block-freezing in preparation for cryosectioning. (a) Tissue blocks frozen in liquid cryogens only have a superficial layer of 5–10 μm thickness that is well-frozen with small ice crystals. This is the region from which cryosections are cut most easily. However, because the edge of the block is likely to be irregular, cryosections will be incomplete and/or contain large areas of poorly frozen material (see §2.3.1). (b) If the specimen is frozen against a cold metal block, then a relatively large, flat area of well-frozen tissue is obtained so that cryosections, even if not *exactly* parallel to the surface, will contain large areas with very small ice crystals in them (see §2.4.2). (c) To be able to produce large, flat areas of well-frozen tissue, it is necessary to ensure that the direction of specimen travel is exactly normal to the cold block surface, otherwise (d) only a very small region of well-frozen tissue results. (Redrawn from an original illustration kindly provided by H. Sitte and reproduced with permission.)

$-100°$C (170 K) have proved useful. For example, *n*-heptane (Steinbrecht and Zierold 1982, 1984) has a melting point of $-90.6°$C (182 K) and a boiling point of $+98.4°$C (372 K) together with a vapour pressure of 4×10^{-3} Torr (5.33×10^{-1} Pa) at $-100°$C (compared with 1×10^{-5} Torr (1.33×10^{-3} Pa) for ice at the same temperature). Thus, specimens can be 'mounted' with the 'glue' at temperatures warmer than $-90°$C and will become stuck to the support as the temperature is lowered. Furthermore, the vapour pressure – which is higher than that of water – allows freeze-drying of sections without risk of solvent extraction of the specimen. Steinbrecht and Zierold reported that *n*-heptane sections satisfactorily at temperatures between -115 and $-130°$C (158 and 143 K). Some other low temperature glues are listed in Table 4.1 in §4.2.1.

4.3.1d   *Mounting specimens on the ultramicrotome*

When the specimen has been frozen on its holder, it is transferred to the chuck of the cryoultramicrotome. This should present no special problems although, of course, it is always essential that the specimen is not allowed to warm up inadvertently once it has been frozen. The simplest method is usually to transfer the specimen, protected within a small insulated container of liquid nitrogen, into the cold chamber of the ultramicrotome from where it can be picked up, using pre-cooled tools, and inserted into the chuck. Because ice is so brittle, it is not normally possible to clamp frozen specimens directly. However, if there is a need to fill a space between a frozen specimen, or a specimen holder, and its chuck, then thin, soft, indium film (Appendix 2) is very suitable for ensuring good thermal contact.

Although some workers have preferred not to trim frozen specimens, most appear to believe this to be necessary, particularly if ribbons of sections are to be produced. Because of the need to maintain a low specimen temperature, trimming is usually carried out with the specimen mounted on the microtome arm. Appleton (1978) recommends trimming in much the same way as for conventional ultramicrotomy (see Reid 1974). The leading and trailing block edges are made parallel to the knife edge, while the lateral edges are trimmed to produce an approximately square face. Trimming can be accomplished by rotating the knife holder and then 'cutting' a pyramid with approximately $30°$ sides. It is, of course, true that trimming often removes much of the best frozen parts of the specimen. If cold block-freezing has been used, then trimming should be confined to the edges of the block so that the well-frozen face is left intact. In some microtomes (e.g. the Rei-chert-Jung FC 4; Sitte et al. 1979, 1980), a special steel alloy 'knife' with an elliptical edge is used specifically for trimming (Fig. 4.10). An alternative method of shaping the block is to use a pre-cooled metal file and gently to

Fig. 4.10.   Knife holder for the Reichert-Jung FC4-Cryosystem. (a) General arrangement of the knife holder in relation to the specimen in the cryochamber. (b) Three knives are mounted: on the left, a special Diatome (Appendix 2) diamond knife for dry sectioning; in the centre, an eliptical metal trimming edge; and, on the right, a normal glass knife. (Reproduced by cour-tesy of Reichert-Jung.)

abrade opposing faces until the required configuration is obtained (Biddle-combe et al. 1982). An important point that can easily be overlooked is that the specimen should be allowed to come to thermal equilibration with its holder/chuck before sectioning is commenced. This may take a considerable time (30 min–1.0 h) but, if not allowed to take place, sectioning is erratic and not under proper control.

4.3.2   Dry sectioning

4.3.2a   Cutting the sections

Whereas the general principles of sectioning apply irrespective of the type of cryoultramicrotome being used, there are inevitably substantial differ-ences in details of operation from one instrument to another. Therefore, only general procedures will be described here, and the reader will need to refer to the relevant manual of the microtome being used for details of parti-cular operational conditions.

For most work involving *thin* frozen sections, temperatures in the region of the specimen/knife of the order of $-80°C$ (193 K) and lower are required. For the LKB 14800 Cryokit the following standardised sectioning condi-tions are suggested (Sevéus 1980):

(*i*)   specimen temperature $-140°C$ (133 K);
(*ii*)   chamber temperature between -100 and $-120°C$ (173 and 153 K);
(*iii*) knife temperature between -100 and $-120°C$ (173 and 153 K);
(*iv*)   knife angle 40°, clearance angle 4°; and
(*v*)   sectioning speed as low as possible, usually 1–2 mm s^{-1}.

Only trial and error will determine *precisely* the best conditions for any par-ticular specimen. Under such conditions the sections appear transparent, glossy and hard, resembling cellophane.

Steel knives are preferred for cutting sections of frozen material down to about 0.5 μm thickness, after which glass or diamond knives are necessary. Special diamond knives are available for cryosectioning from Diatome (see Appendix 2), while Reichert Jung AG (see Appendix 2) make knives of har-dened stainless steel which are suitable for low temperature work.

It is difficult to cut frozen sections to a precise predetermined thickness. Although interference colours can often (if not always) be seen, they do not provide the good indication of thickness that is available during conven-tional ultramicrotomy. Also the colours do not indicate actual thicknesses equivalent to those seen during wet sectioning of resin-embedded blocks because of differences in refractive index.

Frozen thin sections, and particularly those that are cut in the 'thicker' range (>0.2 μm), often show a tendency to curl as they are cut. This has led to many suggestions for the use of anti-roll plates (Fig. 4.11) or other devices to preclude this from happening. The simplest type of mechanism is similar to that used for conventional cryotomy and consists of a glass coverslip positioned very close to the knife edge (Fig. 4.11d). Hellström and Sjöström (1974) describe the construction and use of such a plate. Dörge et al. (1978) also used a small glass anti-roll plate which was manoeuvred with a micromanipulator so that it was 5.0 μm away from the cutting edge, so forming a small gap through which the sections could pass. Boll et al.

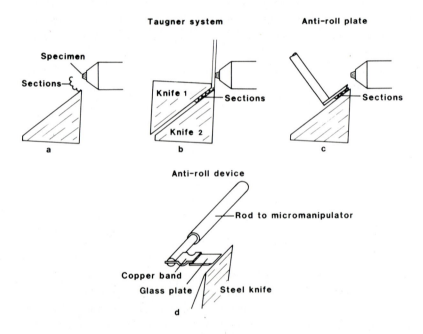

Fig. 4.11.   Cryosection anti-roll systems. (a) When cryosections are cut dry, they tend to curl away from the knife edge, so making it difficult or impossible to pick them up easily as ribbons. Different systems for preventing this rolling have been devised. Some are illustrated here; an alternative method is shown in Fig. 4.12. (b) The 'Taugner' system (see Boll et al. 1975) uses the slit between two closely adjacent glass knives to direct the ribbon of sections as it develops. (c) A glass coverslip makes an excellent anti-roll plate when attached to a support arm such as that shown in (d) (Fig. 4.11d redrawn from Rick et al. 1982.)

(1975) devised a simple system for stretching sections and preventing them from rolling by feeding them as they are cut between the adjacent edges of two carefully aligned glass knives (Fig. 4.11b). Schiller and Taugner (1979) actually coated the cut surface of a frozen tissue block with metal (platinum) and carbon prior to cutting each section. They found that this made the handling of the sections easier, the advantages including:

(*i*)   enhancement of contrast without the need for staining;

(*ii*)   stabilisation of the sections for subsequent freeze-drying, histo-chemistry, TEM observation, or analysis without further coating;

(*iii*) protection of the sections against artefacts during autoradiography; and

(*iv*)   allowing the production of useful surfaces for bulk specimen micro-analysis.

Frozen sections can be extremely difficult to handle because they often become electrostatically charged, in which case they may well be subject to strong repulsive forces when attempts are made to place them onto grids. Various methods have been tried to overcome this problem (e.g. Nicholson 1978). Fine metal needles are better than hairs for handling the specimen (Biddlecombe et al. 1982). Antistatic 'pistols' (Appendix 2) have sometimes proved to be helpful, and Sitte has successfully incorporated a Simco-singlespike (Appendix 2) into the chamber of his cryoultramicrotome to re-duce problems arising from static electricity. Sevéus (1980) refers to coating the knife with an evaporated carbon film that has electrical continuity with the frame of the microtome. This is said to be useful although the thickness of the carbon layer appears to be critical.

4.3.2b   Section collection

Once the sections have been cut, they are collected, transferred to a grid (or other support) and made to adhere firmly to this 'carrier'.

Appleton (1973) introduced the method of vacuum collection of ribbons of sections at temperatures of about −60 to −70°C (193 to 203 K) (Fig. 4.12). Although this accessory is standard on the Slee cryoultramicrotome, a number of workers have adapted it to be used on other systems. It is im-portant that the vacuum 'pull' is both even and reproducible. The leading edge of the first section to be cut is gently 'nudged', using a mounted hair or a cold micro-needle, towards the mouth of the vacuum tube. As successive sections are cut, the vacuum tube is gradually and gently retracted from the knife, thus drawing out the ribbon. When sufficient sections have been cut, they are lowered onto a support positioned immediately below the ribbon.

The vacuum collection technique cannot be used at temperatures much below $-70°C$ (203 K), and so at lower temperatures the sections must be collected singly. They can be picked up using a mounted hair (e.g. an eyelash) or thin metal needle and transferred to the grid which is usually positioned, for convenience, as close to the knife edge as possible. A method of picking up dry sections that has been found useful by many workers is that

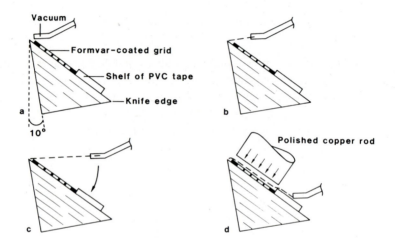

Fig. 4.12.   The vacuum method of producing a ribbon of dry cryosections as devised by Appleton (1973). (a) A weak vacuum (using a 'fish-tank' pump or similar device) is applied through the flattened orifice of a serum needle which is positioned very close to the knife edge. A Formvar-coated grid is supported just below the knife edge using a 'shelf' of plastic tape. (b) As the first section is cut it is very gently 'induced', if necessary using a mounted eye-lash, to enter into the opening of the serum needle. The vacuum should be *just* sufficient to 'hold' the section without drawing it further into the needle. (c) As sectioning proceeds, the needle is moved, using a micro-manipulator, step-by-step away from the knife edge, so maintaining a continuous, extended ribbon of sections. When enough sections have been cut, the micro-manipulator is used to lower the needle, and hence the sections, onto the waiting grid. (d) A highly polished, cold, copper rod is then used to press the sections into close contact with the grid, after which the mounted sections are either freeze-dried or used directly for frozen-hydrated observation or analysis. (e) (opposite) Illustration showing cryosectioning in progress using the vacuum collection method. (Diagrams redrawn from Appleton (1973) and Fig. 4.12e reproduced with the permission of T.C. Appleton, North-Holland Publishing Company and Slee Medical Equipment Ltd.)

of using a droplet of freezing sucrose (Tokuyasu 1973; Fig. 4.13). Using a small syringe, hair or platinum loop, a droplet (0.5–1.0 mm^3) of saturated (or approximately 2.0 M) sucrose is inserted into the cold environment of the sectioning chamber. As the droplet cools it becomes very viscous, at which stage it is touched against the section(s), which adhere to it and can be lifted away from the knife edge. This procedure requires some practice to be carried out successfully: if the droplets are still liquid they may freeze onto the knife edge! When the droplet is removed from the cryochamber, it melts and the sections expand; the droplet with sections is then put onto a coated grid which is floated, sections downwards, on water to remove the

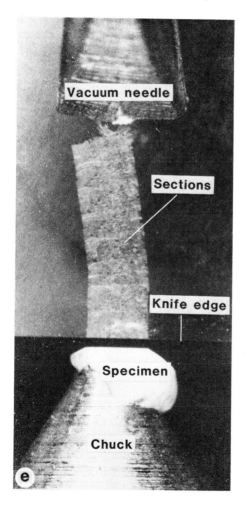

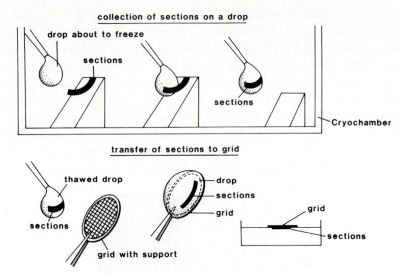

Fig. 4.13.   Freezing sucrose droplet method of picking up dry cryosections as devised by Tokuyasu (1973). A droplet of about 2.0 M sucrose, just as it is about to freeze, is brought into contact with the cryosections, which adhere to it and can be lifted from the knife edge. The frozen droplet is then transferred to an EM grid, allowed to thaw so that the sections flatten onto the grid, and the sections are then washed free from sucrose by floating the grid, section side downwards, on distilled water (see text). (Redrawn from Roland 1978.)

sucrose. Sometimes sections from unfixed specimens become dispersed during this process due to surface tension effects. This may be overcome by adding 0.5–2% gelatin to the sucrose solution and subsequently washing the sections through decreasing concentrations of sucrose (Tokuyasu and Singer 1976; see also Tokuyasu 1980). While this method does involve the contact of liquids with the sections, it provides a useful intermediate pathway between totally dry sectioning (and collection) and other 'wet' methods which use trough liquids (§4.3.3).

4.3.2c   Section mounting

While coated grids are the commonest supports for frozen sections, many other systems have also been used. These usually comprise a small metal collar across which a suitable plastic film can be suspended (see, for example, Fig. 3.28 in §3.4.8). The precise size and shape depends upon the mounting arrangements on the stage of the particular microscope. It is particularly important that supporting films should be made hydrophilic (by glow discharge or other means; see Willison and Rowe 1980) prior to the collection of frozen sections.

Frozen sections do not adhere well to mounting films or grids and must usually be firmly pressed against their supporting layer. This must obviously be done with the sections maintained at low temperature. The most common method is to press the sections against the support using the highly polished end of a cold metal (copper) or plastic (nylon) rod (Christensen 1971). This is done either by hand or using a simple system to ensure that the pressure applied is consistent and even. For example, Sevéus (1980) recommends leaving the sections under a chilled weight in a press for a period of 1 h prior to further processing. It is often useful to 'press' the sections between *two* coated grids (e.g. Biddlecombe et al. 1982) so that, when separated, the sections are well attached to one of the grids. Tvedt et al. (1984) have produced a device (Fig. 4.14) for this purpose. Other devices for pressing sections have also been described, such as the one illustrated in Fig. 4.15.

An alternative method of ensuring good contact between sections and their substrate is to float them on a drop of cold (−100°C; 173 K) isopentane on their supporting films. This serves the two-fold purpose of removing folds from the sections and helping to provide good contact with the substrate when the temperature is reduced again (Saubermann et al. 1977).

n-Butyl benzene (Saubermann and Echlin 1975) or toluene (Somlyo et al. 1982) can be used as low temperature 'glues' to 'stick' sections to their supporting film. These glues have already been discussed in §4.3.1c (see also Table 4.1 in §4.2.1 for other substances having similar properties).

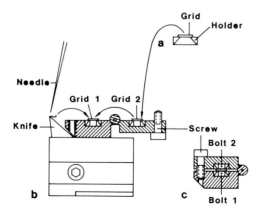

Fig. 4.14. 'Press' device for cryosections as devised by Tvedt et al. (1984). (Redrawn after Tvedt et al. 1984.)

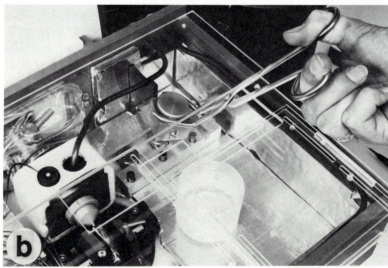

Fig. 4.15. 'Pressing' cryosections using the tools provided with the LKB 14800 Cryokit. (a) The grid carrier is used to transport the grid, with sections, to the press assembly. (b) The cold specimen press is placed onto the grid. This system has gold, Teflon (PTFE) and brass pressing surfaces provided. (Illustrations by courtesy of LKB-Produkter, AB.)

A second plastic (e.g. Formvar) film can be placed *over* the sections once they have been mounted onto their support to ensure greater stability. This does not appear to interfere significantly with any subsequent freeze-drying. Once the sections have been made to adhere firmly to the support, they may either be freeze-dried (§ 5.5.2d) or transferred directly onto the cold stage of a microscope for transmission viewing and/or analysis.

Cryosections are often freeze-dried before viewing and/or analysis at ambient temperature (Frederik et al. 1984). The drying process is conveniently carried out on a temperature-controlled platform in the microtome cold chamber. The theory of freeze-drying is discussed in Chapter 5 and its application to cryosections in §5.5.2d. Zingsheim (1984) has made a study of sublimation rates of ice in a cryoultramicrotome and has concluded that the danger of even partial (unintentional) dehydration of 0.5–1.0 μm thick sections is less serious than commonly assumed.

It has generally been believed that frozen-hydrated thin sections show very little intrinsic contrast in transmission electron microscopes. However, the recent work of Dubochet and his colleagues (Dubochet et al. 1982; Chang et al. 1983; McDowall et al. 1983; Dubochet and McDowall 1984) has indicated that quite remarkable contrast *can* be observed within such sections. Much remains to be done before we are fully able to appreciate the mechanisms of contrast formation in viewing frozen-hydrated sections.

4.3.3 Wet sectioning

Any liquid coming into contact with frozen sections must be regarded with suspicion because of the probable induction of artefacts, including the production of contaminating elements prior to microanalysis and, most importantly, the probable loss or relocation of diffusible components of the specimen. Such considerations have turned many workers against using *any* liquids during cryosectioning. Microanalytical or cytochemical studies are now carried out almost exclusively using 'dry' sections throughout. Furthermore, wet sectioning can *usually* only be done at relatively high temperatures (> $-60°C$; 213 K). Despite these restrictions, the morphological appearance of sections cut using trough liquids can be much better than that of dry sections and the sectioning process itself is easier. It is for the individual worker to make a choice after careful consideration of all the requirements involved.

Dimethylsulphoxide (DMSO) is the most commonly used trough liquid for cutting frozen wet sections (Bernhard 1971) but Sjöström and Thornell

(1975) concluded that this led to the extraction of elements. DMSO is also known to be toxic to some cellular systems and can inhibit some enzymes (e.g. acid phosphatases). Nevertheless, DMSO is a convenient choice for someone contemplating a first attempt at cryoultramicrotomy using a trough liquid. A 60% solution of DMSO freezes at about $-60°C$ (213 K).

Alternatives to DMSO are available. Cyclohexane can be used at temperatures below $-130°C$ as a trough liquid (Hodson and Marshall 1970, 1972). Ethylene glycol, methylene glycol, halocarbons, isopentane and even liquid nitrogen have all been used as trough liquids with varying degrees of success. However, in general, their use at temperatures below the probable recrystallisation point for ice (i.e. lower than about $-80°C$; 193 K) is problematical and not recommended for the novice in this field. Thus, while the use of trough liquids may lead to easier sectioning, will prevent the 'rolling' of sections, and can produce pleasing morphological details, the current concensus – coupled with improvements in cryoultramicrotomy apparatus and techniques – tends to suggest that they should only be used where their presence is unavoidable.

4.3.3a   Contrast enhancement

Contrast enhancement (staining) is used for morphological, but not microanalytical, studies and is usually only applied to 'wet' sections. Either positive or negative staining methods can be employed. The freshly cut sections should not be allowed to dry out. Floating grids on a solution of 0.2–4.0% sodium silicotungstate or 1–2% neutralised potassium phosphotungstate at pH 7.2 for about 10–30 s at 35–40°C (308–313 K) provides good negative staining. The precise time depends on the nature of the specimen and must be established by experimentation. Positive staining can be effected using lead citrate (e.g. Reynolds 1963; 5.0 s–1.0 min) or uranyl acetate (1.0–5.0 min) in the conventional manner (Lewis and Knight 1977) but for much shorter times than for embedded material as the stains have more direct access to the structures with which they react. Tsuji (1974) has outlined a method for simultaneous section collection and staining using a droplet of silicotungstate. After staining, the grids should be rinsed briefly by floating on clean water and then carefully dried on a piece of filter paper. Staining using osmium tetroxide vapour has also been suggested and may be preferable if redistribution of intracellular components is to be minimised, although it does not work if sections are coated with a developed nuclear emulsion.

4.4   Applications of freeze-sectioning

Frozen thin sections can serve as the starting point for a number of important morphological and analytical applications (Table 4.2). The basic rationale is that rapid freezing will have retained both ultrastructural detail and soluble compound distribution in a more natural state than other preparation methods. In interpreting the results from frozen thin sections it is, therefore, of critical importance that the changes that may have occurred during the preparation procedure, such as ice crystal formation, are both noted and understood.

TABLE 4.2
Typical applications of ultrathin cryosections

Tissue/Cell	*Technique*	*Application*	*Reference*
Various	Protein A-gold labelling	Antigen localisation	Roth (1982)
Bean cotyledons	Immunoferritin labelling	Phaseolin localisation	Baumgartner et al. (1981)
Human lymphoid tissue	Double immuno-enzymatic labelling	Antigen localisation	Mason and Woolston (1982)
Frog skin	X-ray analysis	Cellular electrolyte concentration	Rick et al. (1979)
Rabbit cardiac muscle	X-ray analysis	Organelle electrolyte concentration	Hagler et al. (1983)
Catalase crystals	Low temperature TEM	Structural effects of freezing	Chang et al. (1983)
Various	Autoradiography and X-ray analysis	Localisation of diffusible substances	Baker and Appleton (1976)
Mammalian gut epithelium	Autoradiography	Localisation of diffusible substances	Johnson and Bronk (1979)
Influenza virus antigens	Immunolabelling of sections	Antigen localisation	Beesley and Campbell (1984)

Cryosectioning has produced some excellent images of chemically untreated cells (e.g. Frederik et al. 1982b, 1984; Chang et al. 1983; McDowall et al. 1983) but it is in the general area of cytochemistry and microanalysis that the technique has a particularly significant contribution to make.

The application of cytochemical methods following frozen thin sectioning falls outside the scope of the present book. The reader is referred to one or more of the increasing number of papers reporting the use of cytochemical (including immunocytochemical) applications following cryoultramicrotomy (e.g. in Bullock and Petrusz 1982, 1983; Polak and Varndell 1984; Table 4.2). Frozen thin sections are also ideal starting points for the use of autoradiography to localise radiolabelled compounds within cells (Williams 1977; Table 4.2).

However many frozen thin sections are used for determination of structure, it has to be admitted that their real usefulness resides in the analytical information that they are potentially able to provide. It is therefore necessary to adopt rigorous criteria to ensure that any elemental or soluble compound redistribution is detected so that false interpretations will be avoided. It is useful to prepare 'standards', such as a gelatin solution containing dissolved salts (Chandler 1977), which can be frozen and sectioned using the same procedures as used for the biological tissue so that ion concentrations can be determined (see also Roomans and Sevéus 1977; Somlyo and Silcox 1979; Zierold 1982a,b, 1984; Gupta and Hall 1984). Appleton (1974, 1978) has shown how it is possible to prepare frozen salt solutions so that any possibility of elemental diffusion is avoided.

References

Appleton, T.C. (1969), The possibilities of locating soluble labelled compounds by electron microscope autoradiography, in: Autoradiography of diffusible substances, L.J. Roth and W.E. Stumpf, eds. (Academic Press, New York), p. 301.

Appleton, T.C. (1972a), "Dry" frozen sections of unfixed and unembedded biological material for X-ray microanalysis of naturally occurring diffusible electrolytes: the cryostat approach, J. Microscopie *13*, 144.

Appleton, T.C. (1972b), "Dry" ultra-thin frozen sections for electron microscopy and X-ray microanalysis: the cryostat approach, Micron *3*, 101.

Appleton, T.C. (1972c), Transmission microscopy and analysis of frozen sections, Proc. 5th. Eur. Reg. Conf. Electron Microscopy, Manchester, p. 56.

Appleton, T.C. (1973), Cryoultramicrotomy, possible applications in cytochemistry, in: Electron microscopy and cytochemistry, E. Wisse, W.Th. Daems, I. Molenaar and P. van Duijn, eds. (North-Holland, Amsterdam), p. 229.

Appleton, T.C. (1974), A cryostat approach to ultra-thin frozen sections for electron microscopy; a morphological and X-ray analytical study, J. Microscopy *100*, 49.

Appleton, T.C. (1978), The contribution of cryo-ultramicrotomy to X-ray analysis in biology, in: Electron probe microanalysis in biology, D.A. Erasmus, ed. (Chapman and Hall, London), p. 148.

Baker, J.R.J. and T.C. Appleton (1976), A technique for electron microscope autoradiography (and X-ray microanalysis) of diffusible substances using freeze-dried fresh frozen sections, J. Microscopy *108*, 307.

Barnard, T. (1982), Thin frozen dried cryosections and biological X-ray microanalysis, J. Microscopy *126*, 317.

Baumgartner, B., K.T. Tokuyasu and M.J. Chrispeels (1981), Immunocytochemical localisation of reserve protein in the endoplasmic reticulum of developing bean (*Phaseolus vulgaris*) cotyledons, Planta *150*, 419.

Beesley, J.E. and D.A. Campbell (1984), The use of ultrathin cryosections for localisation of influenza virus antigens in infected vero cell cultures. Histochemistry *80*, 497.

Bernhard, W. (1965), Ultramicrotomie a basse temperature, Ann. Biol. *4*, 5.

Bernhard, W. (1971), Improved techniques for the preparation of ultrathin frozen sections, J. Cell Biol. *49*, 731.

Bernhard, W. and E.H. Leduc (1967), Ultrathin frozen sections. I. Methods and ultrastructural preservation, J. Cell Biol. *34* 757.

Bernhard, W. and A. Viron (1971), Improved techniques for the preparation of ultrathin frozen sections, J. Cell Biol. *51*, 732.

Biddlecombe, W.H., D.M. Jenkinson, S.A. McWilliams, W.A.P. Nicholson, H.Y. Elder and D.W. Dempster (1982), Preparation of cryosections with a modified Sorvall MT2B ultramicrotome and cryoattachment, J. Microscopy *126*, 63.

Boll, H.U., J. Eschwey, G. Reb and R. Taugner (1975), A section stretching apparatus for ultracryotomy, Experientia *31*, 618.

Bullock, G.R. and P. Petrusz (eds.) (1982), Techniques in immunocytochemistry, Vol. 1 (Academic Press, London).

Bullock, G.R. and P. Petrusz (eds.) (1983), Techniques in immunocytochemistry, Vol. 2 (Academic Press, London).

Chandler, J.A. (1977), X-ray microanalysis in the electron microscope, in: Practical methods in electron microscopy, Vol. 5, A.M. Glauert, ed. (Elsevier/North-Holland, Amsterdam).

Chang J.-J., A.W. McDowall, J. Lepault, R. Freeman, C.A. Walter and J. Dubochet (1983), Freezing, sectioning and observation artefacts of frozen hydrated sections for electron microscopy, J. Microscopy *132*, 109.

Christensen, A.K. (1967), A simple way to cut frozen thin sections of tissue at liquid nitrogen temperatures, Anat. Rec. *157*, 227.

Christensen, A.K. (1969), A way to prepare frozen thin sections of fresh tissue for electron microscopy, in: Autoradiography of diffusible substances, L.J. Roth and W.E. Stumpf, eds. (Academic Press, New York), p. 349.

Christensen, A.K. (1971), Frozen thin sections of fresh tissue for electron microscopy with description of pancreas and liver, J. Cell Biol. *51*, 772.

Dollhopf, F.L., G. Lechner, K. Neumann and H. Sitte (1972), The cryoultramicrotome Reichert, J. Microscopie *13*, 152.

Dörge, A., R. Rick, K. Gehring and K. Thurau (1978), Preparation of freeze-dried cryosections for quantitative X-ray microanalysis of electrolytes in biological soft tissues. Pfleugers Arch. Eur. J. Physiol. *373*, 85.

Dubochet, J. and A.W. McDowall (1984), Cryoultramicrotomy: study of ice crystals and freezing damage, Proc. 8th Eur. Reg. Conf. Electron Microscopy, Budapest *2*, 1407.

Dubochet, J., A.W. McDowall, R. Freeman and J. Lepault (1982), Cryoprotection on organic specimens, Proc. 10th Int. Congr. Electron Microscopy, Hamburg *1*, 19.

Echlin, P., H.le B. Skaer, B.O.C. Gardiner, F. Franks and M.H. Asquith (1977), Polymeric cryoprotectants in the preservation of biological ultrastructure. II. Physiological effects, J. Microscopy *110*, 239.

Echlin, P., C.E. Lai and T.L. Hayes (1982), Low temperature X-ray microanalysis of the differentiating vascular tissue in root tips of *Lemna minor* L., J. Microscopy *126*, 285.

Fernández-Morán, H. (1952), Application of ultrathin freezing sectioning technique to the study of cell structures with the electron microscope, Arch. Fisik, *4*, 471.

Frederik, P.M. and W.M. Busing (1981), Ice crystal damage in frozen thin sections – freezing effects and their restoration, J. Microscopy *121*, 191.

Frederik, P.M., W.M. Busing and A. Persson (1982a), Concerning the nature of the cryosectioning process, J. Microscopy *125*, 167.

Frederik, P.M., W.M. Busing and W.H.A. Hax (1982b), Frozen, hydrated and drying thin cryosections observed in STEM, J. Microscopy *126*, RP1.

Frederik, P.M., W.M. Busing and W.H.A. Hax (1984), Observations on frozen hydrated and drying thin cryosections, Proc. 8th Eur. Reg. Conf. Electron Microscopy, Budapest *2*, 1411.

Glauert, A.M. (1974), Fixation, dehydration and embedding of biological samples, in: Practical methods in electron microscopy, Vol. 3, A.M. Glauert, ed. (North-Holland, Amsterdam).

Gupta, B.L. (1979), The electron microprobe X-ray analysis of frozen hydrated sections with new information on fluid transporting epithelia, in: Microbeam analysis in biology, C.P. Lechene and R.R. Warner, eds. (Academic Press, London), p. 375.

Gupta, B.L. and T.A. Hall (1984), X-ray microanalysis of frozen-hydrated cryosections: problems and results, Proc. 8th. Eur. Reg. Conf. Electron Microscopy, Budapest *3*, 1665.

Hagler, H.K., K.P. Burton, C.A. Greico, L.E. Lopez and L.M. Buja (1980), Techniques for cryosectioning and X-ray microanalysis in the study of normal and injured myocardium, Scanning Electron Microscopy 1980, *2*, 493.

Hagler, H.K., L.E. Lopez, J.S. Flores, R.J. Lundswick and L.M. Buja (1983), Standards for quantitative energy dispersive X-ray microanalysis of biological cryosections: validation and application to studies of myocardium, J. Microscopy *131*, 221.

Hall, T.A. and B.L. Gupta (1982), Quantification for the X-ray microanalysis of cryosections, J. Microscopy *126*, 333.

Hayat, M.A. (1981), Fixation for electron microscopy (Academic Press, New York).

Hellström, S. and M. Sjöström (1974), An anti-roll plate for flattening of "dry" ultra-thin frozen sections, J. Microscopy *102*, 197.

Hodson, S. and J. Marshall (1970), Ultramicrotomy – a technique for cutting ultrathin sections of unfixed frozen biological tissues for electron microscopy, J. Microscopy *91*, 105.

Hodson, S. and J. Marshall (1972), Evidence against through-section sawing whilst cutting on the ultramicrotome, J. Microscopy *95*, 459.

Johnson, I.T. and J.R. Bronk (1979), Electron microscope autoradiography of a diffusible intracellular constituent, using freeze-dried frozen sections of mammalian intestinal epithelium, J. Microscopy *115*, 187.

Karp, R.D., J.C. Silcox and A.V. Somlyo (1982), Cryoultramicrotomy: evidence against melting and the use of a low temperature cement for specimen orientation, J. Microscopy *126*, 157.

Lewis, P.R. and D.P. Knight (1977), Staining methods for sectioned material, in: Practical methods in electron microscopy, Vol. 5, A.M. Glauert, ed. (Elsevier/North-Holland Biomedical Press, Amsterdam).

Mason, D.Y. and R.E. Woolston (1982), Double immunoenzymatic labelling, in: Techniques in immunocytochemistry, Vol. 1, G.R. Bullock and P. Petrusz, eds. (Academic Press Inc., London), p. 135.

McDowall, A.W., J.-J. Chang, R. Freeman, J. Lepault, C.A. Walter and J. Dubochet (1983), Electron microscopy of frozen hydrated sections of vitreous ice and vitrified biological samples, J. Microscopy *131*, 1.

Polak, J.M. and I.M. Varndell (eds.) (1984), Immunolabelling for electron microscopy (Elsevier, Amsterdam).

Nicholson, P.W. (1978), A device for static elimination in ultramicrotomy, Stain technol. *53*, 237.

Reid, N. (1974), Ultramicrotomy, in: Practical methods in electron microscopy, Vol. 3, A.M. Glauert, ed. (North-Holland Publishing Co., Amsterdam).

Reynolds, E.S. (1963), The use of lead citrate at high pH as an electron-opaque stain in electron microscopy, J. Cell Biol. *17*, 208.

Rick, R., A. Dorge, K. Gehring, R. Bauer and K. Thurau (1979), Quantitative determination of cellular electrolyte concentrations in thin freeze-dried cryosections using energy-dispersive X-ray microanalysis, in: Microbeam analysis in biology, C.P. Lechene and R.R. Warner, eds. (Academic Press Inc., London), p. 517.

Rick, R., A. Dorge and K. Thurau (1982), Quantitative analysis of electrolytes in frozen dried sections, J. Microscopy *125*, 239.

Roland, J.-C. (1978), General preparation and staining of thin sections, in: Electron microscopy and cytochemistry of plant cells, J.L. Hall, ed. (Elsevier/North-Holland, Amsterdam), p. 1.

Roomans, G.M. and L.A. Sevéus (1977), Preparation of thin cryosectioned standards for quantitative microprobe analysis, J. Submicrosc. Cytol. *9*, 31.

Roth, J. (1982), The protein A-gold (pAg) technique – a qualitative and quantitative approach for antigen localisation on thin sections, in: Techniques in immunocytochemistry, Vol. 1, G.R. Bullock and P. Petrusz, eds. (Academic Press Inc., London), p. 107.

Saubermann, A.J. (1979), General considerations of X-ray microanalysis of frozen hydrated tissue sections, Scanning Electron Microscopy 1979, *2*, 607.

Saubermann, A.J. (1980), Application of cryosectioning to X-ray microanalysis of biological tissue, Scanning Electron Microscopy 1980, *2*, 421.

Saubermann, A.J. and P. Echlin (1975), The preparation, examination and analysis of frozen hydrated tissue sections by scanning transmission electron microscopy and X-ray microanalysis, J. Microscopy *105*, 155.

Saubermann, A.J., W.D. Riley and R. Beeuwkes (1977), Cutting work in thick section cryoultramicrotomy, J. Microscopy *111*, 39.

Saubermann, A.J., P. Echlin, P.D. Peters and R. Beeuwkes (1981a), Application of scanning electron microscopy to X-ray analysis of frozen hydrated sections. I. Specimen handling techniques, J. Cell Biol. *88*, 257.

Saubermann, A.J., R. Beeuwkes and P.D. Peters (1981b), Application of scanning electron microscopy to X-ray analysis of frozen hydrated sections. II. Analysis of standard solutions and artificial electrolyte gradients, J. Cell Biol. *88*, 268.

Schiller, A. and R. Taugner (1979), Cryo-ultramicrotomy after coating the cut surface of the tissue block with carbon and metals before sectioning, Histochemie *62*, 153.

Sevéus, L. (1979), The subcellular distribution of electrolytes. A methodological study using cryoultramicrotomy and X-ray microanalysis. Thesis, University of Stockholm, Sweden.

Sevéus, L. (1980), Cryoultramicrotomy as a preparation method for X-ray microanalysis, Scanning Electron Microscopy 1980, *4*, 161.

Sevéus, L. and L. Kindel (1974), Dry cryo-sectioning of human and animal tissue at a very low temperature (133 K), 8th Int. Congr. Electron Microscopy, Canberra *2*, 52.

Sitte, H. (1982), Instrumentation for cryosectioning, Proc. 10th Int. Congr. Electron Microscopy, Hamburg *1*, 9.

Sitte, H. (1984), Equipment for cryofixation, cryoultramicrotomy and cryosubstitution in biomedical TEM-routine, Zeiss Magazine for Electron Microscopists *3*, 25.

Sitte, H., K. Neumann, H. Hässig, H. Kleber and G. Kappl (1979), FC4-cryochamber for the Reichert-ultramicrotome UmU4-Ultracut, Arzneim. Forsch. *29*, 1809.

Sitte, H., K. Neumann, H. Hässig, H. Kleber and G. Kappl (1980), FC4-cryochamber for Reichert-Om U4-ultramicrotome Ultracut, Proc. 7th Eur. Reg. Conf. Electron Microscopy, The Hague *1*, 540.

Sjöström, M. and L.-E. Thornell (1975), Preparing sections of skeletal muscle for transmission electron analytical microscopy (TEAM) of diffusible elements, J. Microscopy *103*, 101.

Sleytr, U.B. and A.W. Robards (1977), Plastic deformation during freeze-cleavage: a review, J. Microscopy *110*, 1.

Somlyo, A.V. and J. Silcox (1979), Cryoultramicrotomy for electron probe analysis, in: Microbeam analysis in biology, C.P. Lechene and R.R. Warner, eds. (Academic Press, London), p.535.

Somlyo, A.P., H. Shuman and A.V. Somlyo (1982), X-ray mapping, electron energy loss analysis and quantitative electron probe analysis in biology, Proc. 10th Int. Congr. Electron Microscopy, Hamburg *1*, 143.

Steinbrecht, R.A. and K. Zierold (1982), Cryo-embedding of small frozen specimens for cryoultramicrotomy, Proc. 10th Int. Congr. Electron Microscopy, Hamburg *3*, 183.

Steinbrecht, R.A. and K. Zierold (1984), A cryoembedding method for cutting ultrathin cryosections from small frozen specimens, J. Microscopy *136*, 69.

Thornburg, W. and P.E. Mengers (1957), An analysis of frozen section techniques. I. Sectioning of fresh-frozen tissue. J. Histochem. Cytochem. *5*, 47.

Tokuyasu, K.T. (1973), A technique for ultramicrotomy of cell suspension and tissues, J. Cell Biol. *57*, 551.

Tokuyasu, K.T. (1980), Immunocytochemistry of ultrathin frozen sections. Histochem. J. *12*, 381.

Tokuyasu, K.T. and S.J. Singer (1976), Improved procedures for immunoferritin labeling of ultrathin frozen sections, J. Cell Biol. *71*, 894.

Tsuji, S. (1984), Cryoultramicrotomy: simultaneous transference, "ionic fixation" and negative staining of fresh frozen sections by means of a silico-tungstic acid droplet, Proc. 8th Eur. Reg. Conf. Electron Microscopy, Budapest *3*, 1691.

Tvedt, K.E., G. Kopstad and O.A. Haugen (1984), A section press and low elemental support for enhanced preparation of freeze-dried cryosections, J. Microscopy *133*, 285.

Williams, M.A. (1977), Autoradiography and immunocytochemistry, in: Practical methods in electron microscopy, Vol. 6, A.M. Glauert, ed. (Elsevier/North-Holland, Amsterdam).

Willison, J.H.M. and A.J. Rowe (1980), Replica, shadowing and freeze-etching techniques, in: Practical methods in electron microscopy, Vol. 8, A.M. Glauert, ed. (Elsevier/North-Holland, Amsterdam).

Zierold, K. (1982a), Preparation of biological cryosections for analytical electron microscopy, Ultramicroscopy *10*, 45.

Zierold, K. (1982b), Cryopreparation of mammalian tissues for X-ray microanalysis in STEM, J. Microscopy *125*, 149.

Zierold, K. (1984), X-ray microanalysis of freeze-dried cryosections: advantages and limitations, Proc. 8th Eur. Reg. Conf. Electron Microscopy, Budapest *3*, 1695.

Zingsheim, H.P. (1984), Sublimation rates of ice in a cryo-ultramicrotome, J. Microscopy *133*, 307.

Chapter 5

Freeze-drying

5.1 Introduction

As discussed in Chapter 2, water is an important structural component of most biological systems. Thus, for the examination of non-frozen biological material in an electron microscope this water must be removed. For example, if a soft specimen (e.g. a piece of tissue) is transferred into the vacuum of a SEM, the tissue water firstly boils. Subsequently extraction of the latent heat of evaporation causes the remaining water to freeze. Surface tension forces produced during evaporation, as well as the ice crystals generated during freezing, cause severe distortions and displacements in most specimens. Although environmental cells (Butler and Hale 1981), in combination with high voltage electron microscopy, provide another possibility for the examination of hydrated unfrozen specimens they are expected to be of only limited use in biology.

In electron microscopy the following drying methods are used:

(*i*)   Evaporative-drying from distilled water or volatile buffers.

(*ii*)   Evaporative-drying from (highly) volatile solvents which have replaced the specimen water in a previous preparation step.

(*iii*)   Freeze-drying of hydrated specimens (lyophilisation).

(*iv*)   Freeze-drying of specimens substituted with non-aqueous volatile solvents.

(*v*)   Critical point-drying (after replacing the specimen water with an organic solvent) from CO_2 or a suitable halocarbon (e.g. Freon 13).

Many reviews, monographs and comprehensive papers have been published on dehydration procedures for biological specimens, including Goldblith et al. (1975), Nermut (1977), Umrath (1983) and McKenzie (1965).

As illustrated in the phase diagram for water (Fig. 5.1), there are three principal physical routes by which specimen water can be removed. Air-drying (Route 1) involves passing across the boundary between the liquid and vapour phases. In Route 2, the freeze-drying procedure, the specimen water is first converted into the solid state by freezing (crossing the boundary line between liquid and solid states) and is subsequently sublimed (traversing the boundary between solid and vapour phases) by reducing the partial pressure of water over the specimen below the saturation vapour pressure of the ice in the specimen. As can be seen from the diagram, the critical point-drying route for water (Route 3) is of no practical interest since the specimen would be pressure cooked (critical point of water: 374°C (647 K) and 218 atm (2.21×10^7 Pa)) under these conditions. Consequently, for critical point-drying procedures, water must first be replaced by a solvent (e.g. liquid CO_2 or a halocarbon) whose critical point lies within a less damaging range.

Studies comparing air-drying with other methods for removing water have convincingly demonstrated that air-drying is of very limited value (Boyde 1974a; Humphreys 1975; Nermut 1977; Boyde and Franc 1981). With the possible exception of very resistant and rigid specimens (e.g. inorganic mineralised tissue, mature wood, pollen grains, exoskeletons of insects) most native (unfixed) specimens shrink severely during air-drying.

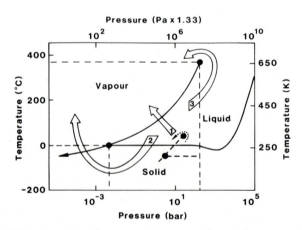

Fig. 5.1.   The pressure-temperature phase diagram for water. This shows the different possible ways of moving from the liquid phase to the vapour phase. Route 1 is air-drying; Route 2 is freeze-drying; and Route 3 is critical point-drying. The dashed line shows the equivalent phase diagram for carbon dioxide, including the position of the critical point, superimposed on that for water. (See also Fig. 1.1.)

These drying artefacts cannot be avoided, whether the surrounding phase is air, dry air, cool air, hot air or any other gas or combination of gases (Boyde 1980). The shrinkage which is seen as a macroscopic phenomenon during *dehydration* with solvents is nothing compared with the enormous force which is exerted at the air/water interface during *drying* by surface tension, which is 73 dynes cm^{-1} (7.3×10^{-2} N m^{-1}) at 18°C for water. Anderson (1951) estimated that the surface tension forces during air-drying of specimens from water are of the order of one ton per square inch (160 kg cm^{-2}). Some improvement is obtained by drying from solvents which have a lower surface tension (e.g. diethyl ether, 17 dynes cm^{-1} (1.7×10^{-2} N m^{-1}); acetone, 24 dynes cm^{-1} (2.4×10^{-2} N m^{-1})). In addition, the high volatility of these organic solvents (see Table 5.1a and b and Fig. 5.14 in §5.4.1) reduces the time taken for the air/liquid interface to pass through the specimen, thus resulting in less shrinkage. The structural preservation of specimens can be improved by chemical fixation before exchanging the water for a liquid of lower surface tension. However, for most specimens the surface tension forces at the air/liquid interface are so great that the resulting artefacts (see §5.6) will still be unacceptable (Boyde 1980).

TABLE 5.1a
Boiling point and vapour pressure of liquids used for air-drying

Liquid	Formula	Boiling point[a]		Vapour pressure[b]		
		°C	K	mm Hg (Torr)	bar	Pa
Distilled water	H_2O	100	373	17.5	0.0233	2.33×10^3
Ethyl alcohol	C_2H_5OH	78.4	351.6	44.0	0.0585	5.87×10^3
Chloroform	$CHCl_3$	61.4	334.6	160.5	0.2133	2.14×10^4
Isopentane	$(CH_3)_2CHCH_2CH_3$	27.9	301.0	432.4	0.5746	5.76×10^4
Diethyl ether	$CH_3CH_2OCH_2CH_3$	34.5	307.7	446.9	0.5939	5.96×10^4
Halocarbon 11	CCl_3F	23.7	296.9	641.3	0.8523	8.55×10^4
Halocarbon 13	$CClF_3$	−81.4	191.8	24624.0	32.7240	3.28×10^6

[a] Citations of boiling points for some liquids vary from source to source. Small variations are of little significance in the present context and, for this reason, boiling point citations have been rounded to 0.1°C. Boiling points are given at 760 mm pressure (1.01×10^5 Pa).
[b] Vapour pressures cited at 20°C (293 K).
Results during air-drying improve with increasing vapour pressure of the liquid (from chloroform to halocarbon 13). Adapted and expanded from Klein and Stockem (1976).

TABLE 5.1b

Boiling and melting points of some liquids

Liquid	Formula	Boiling point		Melting point	
		$^\circ$C	K	$^\circ$C	K
Carbon dioxide	CO_2	(Sublimes)		-78.3	194.9
Distilled water	H_2O	100	373.2	0	273
Ethyl alcohol	C_2H_5OH	78.4	351.6	-114.7	158.5
Chloroform	$CHCl_3$	61.4	334.6	-6.5	266.7
Isopentane	$(CH_3)_2CHCH_2CH_3$	27.9	301.0	-159.9	113.3
Methanol	CH_3OH	64.7	337.9	-97.9	175.3
DMSO	CH_3OSCH_3	192	465.2	18.4	291.6
Acetone	C_3H_6O	56.1	329.3	-96.5	176.7
Diethyl ether	$CH_3CH_2OCH_2CH_3$	34.5	307.7	-116.2	157.0
Halocarbon 11	CCl_3F	23.8	296.9	-111.0	162.0
Halocarbon 13	$CClF_3$	-81.4	191.8	-181.0	92.2
Diethylamine	$C_4H_{11}N$	55.5	328.7	-48	225.2
Benzene	C_6H_6	80.1	353.3	5.6	278.8
Dimethylformamide	C_3H_7ON	70.5	343.7	-61	212.2
Glycerol	$HOCH_2CH(OH)CH_2OH$	290	563.2	20	293.2
Amylacetate	$CH_3CO_2(CH_2)_4CH_3$	149.3	422.5	-70.8	202.4
Isoamylacetate	$CH_3COOCH_2CH_2CH(CH_2)_2$	140	413.2	-78.5	194.7
Halocarbon 113	$CCl_2F\text{-}CCl_2F$	48	321.2	-36.4	236.8
Camphene	$C_{10}H_{16}$	160	433.2	45.0	318.2
Tertiary butanol	$(CH_3)_3COH$	82.2	355.4	11.0	284.2

Temperature citations for some liquids vary from source to source. For this reason, citations are restricted here to 1.0 or 0.1°C. This has little significance for their use in freeze-drying, freeze-substitution and related processes. Boiling points and melting points are cited at 760 mm pressure (1.01×10^5 Pa).

To avoid the effects of surface tension, the specimen cannot be dried directly from the liquid phase. As can be seen from Fig. 5.1, freeze-drying or critical point-drying (from suitable solvents) are alternative procedures. Boyde and his collaborators have demonstrated, in a series of very convincing experiments, that frozen-dried or critical point-dried specimens show much less shrinkage than air-dried material (Boyde 1980; Boyde and Franc 1981). This Chapter describes in detail methods for freeze-drying both hydrated specimens and specimens substituted with non-aqueous volatile solutions.

5.2   Basic principles of freeze-drying

An enormous stimulus for technical developments and the analysis of parameters involved in the freeze-drying process has come from the pharmaceu-
tical and food industries (for reviews see Goldblith et al. 1975; Mellor 1978).

Boyde (1978a) has defined freeze-drying for electron microscopy as a procedure for getting rid of water by subliming ice into water vapour, using
a fixed or unfixed freshly frozen specimen. With fixed specimens, water may
have been replaced with another solvent which is itself frozen and sublimed.
As shown in Fig. 5.2, the term 'freeze-drying' includes a large number of
variations of the basic procedure (which is indicated by the heavy lines). After drying, the specimen may be examined using a variety of methods,
including direct viewing in an SEM, CTEM or STEM; thin sectioning after
embedding; X-ray microanalysis; LAMMA (§5.5.2); autoradiography or
histochemistry.

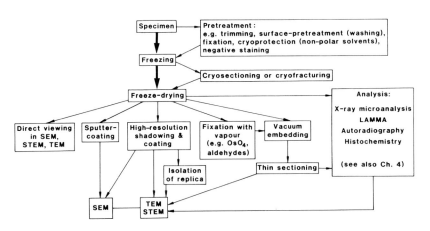

Fig. 5.2.   Flow diagram of freeze-drying procedures.

5.2.1   Theory of drying by sublimation

5.2.1a   Rate of sublimation of ice

Sublimation and condensation phenomena can be understood in terms of
the kinetic gas theory (see also §6.5.1). The absolute rate of sublimation of

TABLE 5.2
Freeze-drying times at different temperature

Temperature		Vapour pressure		Etching rate	Drying time	Drying time per μm					
°C	K	Torr	Pa	(nm s⁻¹)	(s μm⁻¹)	Y	M	D	h	m	s
−60	213	8.08×10^{-3}	1.08×10^{0}	1.48×10^{3}	6.76×10^{-1}						0.68
−70	203	1.94×10^{-3}	2.59×10^{-1}	3.64×10^{2}	2.75×10^{0}						3.0
−80	193	4.02×10^{-4}	5.36×10^{-2}	7.70×10^{1}	1.30×10^{1}						13
−81	192	3.40×10^{-4}	4.53×10^{-2}	6.54×10^{1}	1.53×10^{1}						15
−82	191	2.87×10^{-4}	3.83×10^{-2}	5.54×10^{1}	1.81×10^{1}						18
−83	190	2.42×10^{-4}	3.23×10^{-2}	4.68×10^{1}	2.14×10^{1}						21
−84	189	2.04×10^{-4}	2.72×10^{-2}	3.95×10^{1}	2.53×10^{1}						25
−85	188	1.72×10^{-4}	2.29×10^{-2}	3.33×10^{1}	3.00×10^{1}						30
−86	187	1.44×10^{-4}	1.92×10^{-2}	2.80×10^{1}	3.57×10^{1}						38
−87	186	1.20×10^{-4}	1.60×10^{-2}	2.35×10^{1}	4.25×10^{1}						43
−88	185	1.01×10^{-4}	1.35×10^{-2}	1.97×10^{1}	5.08×10^{1}						51
−89	184	8.40×10^{-5}	1.12×10^{-2}	1.65×10^{1}	6.07×10^{1}					1	1
−90	183	6.99×10^{-5}	9.32×10^{-3}	1.37×10^{1}	7.82×10^{1}					1	13
−91	182	5.81×10^{-5}	7.75×10^{-3}	1.14×10^{1}	8.74×10^{1}					1	27
−92	181	4.81×10^{-5}	6.41×10^{-3}	9.51×10^{0}	1.08×10^{2}					1	45
−93	180	3.98×10^{-5}	5.31×10^{-3}	7.89×10^{0}	1.27×10^{2}					2	7
−94	179	3.29×10^{-5}	4.39×10^{-3}	6.53×10^{0}	1.53×10^{2}					2	33

−95	178	2.71×10^{-5}	3.61×10^{-3}	5.39×10^{0}	1.85×10^{2}					3	5
−96	177	2.23×10^{-5}	2.97×10^{-3}	4.45×10^{0}	2.25×10^{2}					3	45
−97	176	1.83×10^{-5}	2.44×10^{-3}	3.66×10^{0}	2.74×10^{2}					4	34
−98	175	1.49×10^{-5}	1.99×10^{-3}	3.00×10^{0}	3.33×10^{2}					5	33
−99	174	1.22×10^{-5}	1.63×10^{-3}	2.45×10^{0}	4.07×10^{2}					6	47
−100	173	9.93×10^{-6}	1.32×10^{-3}	2.00×10^{0}	4.99×10^{2}					8	19
−101	172	8.07×10^{-6}	1.08×10^{-3}	1.63×10^{0}	6.12×10^{2}					10	12
−102	171	5.54×10^{-6}	8.72×10^{-4}	1.33×10^{0}	7.53×10^{2}					12	33
−103	170	5.29×10^{-6}	7.05×10^{-4}	1.08×10^{0}	9.29×10^{2}					15	29
−104	169	4.26×10^{-6}	5.68×10^{-4}	8.70×10^{-1}	1.15×10^{3}					19	20
−105	168	3.43×10^{-6}	4.57×10^{-4}	7.02×10^{-1}	1.42×10^{3}					23	40
−106	167	2.75×10^{-6}	3.67×10^{-4}	5.65×10^{-1}	1.77×10^{3}					29	30
−107	166	2.20×10^{-6}	2.93×10^{-4}	4.53×10^{-1}	2.21×10^{3}					36	50
−108	165	1.76×10^{-6}	2.35×10^{-4}	3.63×10^{-1}	2.76×10^{3}					46	0
−109	164	1.40×10^{-6}	1.87×10^{-4}	2.89×10^{-1}	3.45×10^{3}					57	30
−110	163	1.11×10^{-6}	1.48×10^{-4}	2.30×10^{-1}	4.34×10^{3}				1	12	20
−115	158	3.34×10^{-7}	4.45×10^{-5}	7.04×10^{-2}	1.42×10^{4}				3	56	40
−120	153	9.31×10^{-8}	1.24×10^{-5}	1.99×10^{-2}	5.02×10^{4}				13	56	40
−130	143	5.54×10^{-9}	7.39×10^{-7}	1.22×10^{-3}	8.17×10^{5}			9	10	56	
−140	133	2.16×10^{-10}	2.88×10^{-8}	4.95×10^{-3}	2.02×10^{7}		7	23	19		
−150	123	5.01×10^{-12}	6.68×10^{-10}	1.19×10^{-4}	8.40×10^{8}	27		2	5		
−160	113	6.02×10^{-14}	8.03×10^{-12}	1.49×10^{-5}	6.71×10^{10}	2157	3	10			

Data taken from Umrath (1983).

ice (J_s) can be calculated according to the following equation (Knudsen equation):

$$J_s = \eta\, P_s \left(\frac{M}{2\pi\varphi\, T} \right)^{0.5}, \quad 0 < \eta < 1 \tag{5.1}$$

where

η = coefficient of evaporation

P_s = saturation vapour pressure of ice

M = molecular weight of water vapour (18.016)

φ = gas constant (= Boltzmann's constant $\times$ Loschmidt's number; note that, although the gas constant is often expressed as R, to ensure consistency throughout this book it is cited here as φ)

T = absolute temperature

J_s = absolute rate of sublimation of ice (in g cm^{-2} s^{-1}).

In deriving this equation two assumptions have been made: (*i*) that there is a dynamic equilibrium between the surface of the ice and its vapour, and (*ii*) that the number of water molecules escaping depends on the temperature of the ice, whereas the number impinging on the ice surface and not escaping depends on the pressure and temperature of the vapour. The actual sublimation rate cannot be calculated but must be measured under vacuum conditions. Several attempts have been made to verify the equation for the absolute rate of evaporation and to determine the exact value of the coefficient of evaporation (η), which ideally should have a value of unity (Strasser et al. 1966; Davy and Branton 1970; Mellor 1978). The theoretical maximum for the sublimation rate and also the observed rates derived from microbalance techniques (Davy 1971; Davy and Branton 1970) are shown in Fig. 6.32 in §6.5.1. As can be seen, at temperatures above about $-85°C$ (188 K), the observed evaporation rate begins to fall below the theoretical rate. Val ues less than unity for η at higher working temperatures have been explained by a rate-limiting step prior to desorption in the evaporation process and a decreasing mean free path (less than the dimensions of the vacuum space). Despite these difficulties, calculations of the maximum theoretical sublimation rates using the Knudsen equation are of great practical value. Based on $\eta = 1$ and using the analytical expression for the saturation vapour pressure of water given by Washburn (1924), Umrath (1983) calculated for the temperature range from -1 to $-160°C$ (272.1–113.1 K), the saturation vapour pressure and the corresponding sublimation rate (g cm^{-2} s^{-1}), the decrease of the ice layer in relation to time of sublimation

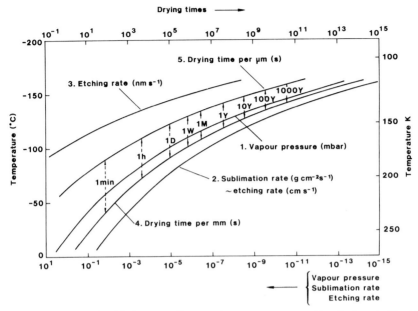

Fig. 5.3. The relationship between: (**1**) saturation vapour pressure (P_s, mbar); and, (**2**) sublimation rate (J_s, g cm^{-2} s^{-1}). The etching rate (J_s/ρ, cm s^{-1}) is obtained by dividing the sublimation rate by the density of water/ice (ρ, g cm^{-3}); it is approximately equal, numerically, to the sublimation rate and is also shown (**3**) expressed as nm s^{-1}. The derived drying times in s mm^{-1} (**4**) and s μm^{-1} (**5**) are indicated with vertical bars to show longer periods of time: minutes (**min**), hours (**h**), days (**D**), weeks (**W**), months (**M**) and years (**Y**). (Redrawn from Umrath 1983.)

(nm s^{-1} = 10^{-7} cm s^{-1}), and the time required for subliming a layer 1 mm thick (see Table 5.2). These results are summarised in Fig. 5.3 and Table 5.2.

5.2.1b   Rate of freeze-drying of biological specimens

In practice, drying times close to the values calculated for maximum sublimation rates of pure water can only be expected for certain kinds of biological specimens such as dilute suspensions of macromolecules (see §5.5.1). For most biological specimens, such as isolated cells or bulk samples (e.g. tissues), the actual rate of freeze-drying can show dramatic deviations from the theoretical calculations for pure ice. In frozen biological specimens, the process of sublimation leads to an increasing surface layer of dry material which subsequently inhibits further free sublimation.

It must be remembered that, for the water being sublimed at the ice surface (and transported to the specimen surface), an amount of energy quantitatively equivalent to its latent heat of evaporation will be required. Thus, freeze-drying is an operation involving both mass transfer (of H_2O molecules) and heat transfer. The rate of drying depends on the magnitude of resistance to these transfers. A schematic view of these processes, including resistance to mass and heat transfer, is given in Fig. 5.4. There are two main combinations of mass and heat transfer during freeze-drying. Heat transfer and mass transfer may pass through the same path (dry layer), but in opposite directions, or heat transfer may take place through the frozen layer and mass transfer through the dry layer. In practice, all possible combinations of mass and heat transfer can occur, and the actual transfers of heat and mass depend on the specimen size, structure and shape; the nature of the contact of the specimen with the support (temperature-controlled plate); the

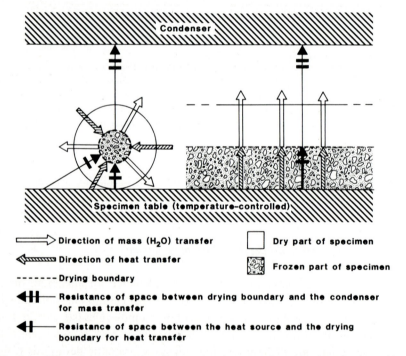

Fig. 5.4.   Schematic representation of the two major types of combination of mass and heat transfer during freeze-drying. The heat and mass transfer resistances are shown for (left) a specimen with poor thermal contact, and (right) a specimen in good contact with the temperature-controlled support. (Redrawn in part from Goldblith et al. 1975.)

vacuum conditions; and the presence, geometry and orientation of condenser(s) in the system. With the small specimens used for electron microscopy, the net rate of heat flux(es) generally requires that the specimen is cooled during the freeze-drying process to avoid a (critical) temperature increase resulting from radiant heat. With large specimens, or specimens which are protected by cold shrouds (where the radiant heat load is very small), the heat transfer itself is the sublimation rate-controlling step when it takes place through a layer of dry material external to the frozen zone.

The large differences in size, shape and structure of biological specimens do not allow the provision of general rules for the selection of freeze-drying procedures. It has been shown, however, that the drying time for a given specimen under optimal vacuum conditions is temperature-dependent (see Table 5.2), and so the shortest drying time is achieved if the specimen is kept just below the temperature at which disturbing recrystallisation of ice may occur.

As discussed in Chapter 2, for most biological specimens with an average water content, recrystallisation phenomena can be expected at temperatures above $-80°C$ (193 K), but since recrystallisation is also a time-dependent phenomenon, no detectable damage may occur during short drying periods at higher temperatures. Frequently, only the outer few tens of micrometres of the well-frozen surface layer of a bulk specimen will be examined (see §5.5.2a). Consequently, it saves time to start by drying the surface layer at a temperature which gives no risk of recrystallisation (e.g. $-90°C$; 183 K) and then to finish by drying the inadequately frozen central part of the specimen at a much higher temperature (e.g. $-60°C$; 213 K). As can be seen from Table 5.2, specimens frozen as a thin layer can be dried within a few minutes (e.g. Table 5.2 shows that, at $-95°C$, 10 μm of pure ice will sublime within approximately 30 min) at a 'recrystallisation-safe' temperature (e.g. $-95°C$; 178 K). (For a discussion of the theoretical advantages of freeze-drying exclusively at low temperature, see also MacKenzie (1965). It is considered important that some bound water remaining in the tissue has a stabilising effect on protein structure. Furthermore, since gases are dissolved in cellular water, some freeze-drying artefacts may be generated by temperature- and pressure-dependent expansions of these gases.) Umrath (1983) gives a survey of published results for a great variety of specimens, including data on drying temperatures, specimen dimensions and vacuum conditions. Many of these results are difficult to compare since in most experiments no accurate temperature measurement of the drying specimen was made. Thus, the possibility of considerable differences between the temperatures of the specimen and of the specimen stage, and/or specimen support has to be kept in mind.

Knowing the maximum sublimation rates for pure water, it becomes obvious that a 1 to 2 mm diameter tissue block cannot be frozen-dried within realistic times at a temperature below $-100°C$ (173 K). Nevertheless such conditions have been described in the literature. This indicates that there was either a poor contact between the specimen stage and the specimen or that the actual drying process took place during the, often slow, warming up period at the end of the freeze-drying schedule.

From this discussion we can conclude that both the recrystallisation temperature for the specimen ice (see Chapter 2) and the maximum sublimation rate for pure water (at the specimen temperature) are the basis for planning freeze-drying experiments. The ratio between the time required for the complete freeze-drying of a defined specimen layer and the sublimation of an identical layer of ice can best be characterised by the so called 'prolongation factor' (f), where:

$$f = \frac{\text{Drying time for } x \text{ } \mu\text{m specimen layer}}{\text{Sublimation time for } x \text{ } \mu\text{m ice layer}}, \quad f \geq 1$$

By comparing different results from the literature, one encounters enormous differences in recommended schedules for optimum drying (Umrath 1983). The course of freeze-drying under high vacuum conditions can be followed if the water is exchanged against deuterium oxide (D_2O) by mass spectroscopic analysis (Wildhaber et al. 1982). From accurate measurements on model systems (Nietsch and Jochem 1973) and tissue samples (Stumpf and Roth 1967) it appears that the sublimation rate decreases exponentially with the thickness of the dry layer on the specimen surface. Thus, for practical reasons, bulk specimens must be trimmed as small as possible to keep the resistances to heat and mass transfers low. Realistic prolongation factors for the drying of 1 mm tissue cubes can be expected to be in the range 20–100 (Umrath 1983). Since heat and mass transfer resistances are dependent on the specimen structure, it is understandable that, for different but equally sized tissue samples, different drying times are required (Stephenson 1960; Pearse 1964; Stumpf and Roth 1967; Schunzel 1968). As an example, Stumpf and Roth (1967) found the relationships of drying times of liver:kidney:brain, to be 1:1.7:3.3.

The sublimation rates during freeze-drying of bulk specimens generally follow the curves shown in Fig. 5.5. During the first period (**A**) when the

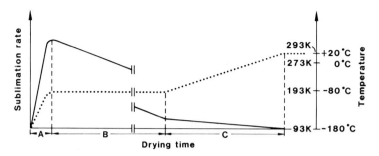

Fig. 5.5. Sublimation rate of freeze-drying (bulk specimens) as a function of time. The solid line shows sublimation rate and the dotted line shows specimen temperature. **A** indicates the zone within which the specimen initially warms to the freeze-drying temperature; in zone **B** drying occurs by the sublimation of ice; while in zone **C** drying is by desorption. (After Goldblith et al. 1975.)

frozen specimen (which has been stored at a low temperature; e.g. under liquid nitrogen) is warmed up to the freeze-drying temperature, the sublimation rate rises rapidly to a maximum value. In circumstances where the frozen specimen is surrounded by a layer of ice, this value may be close to the maximum theoretical sublimation rate given in Table 5.2. As the drying front recedes from the surface to the interior of the specimen (period **B**), the drying rate generally decreases rapidly. This is due to resistance to both water vapour and heat flow in the increasing area between the site of sublimation and the free specimen surface. Finally, with the disappearance of the ice crystals during the third period (**C**), the evaporation rate decreases still more rapidly and finally approaches zero. During this so-called 'secondary' drying phase, the specimen is gradually warmed up to approximately ambient temperature and the residual moisture is desorbed (MacKenzie 1972).

At this point it is necessary to distinguish between the freezeable water, which forms on quenching the water in the specimen matrix, and the so-called (chemically or physically) 'bound' water. No sharp distinction can be made between these two forms and the values given for bound water vary considerably depending on the method of determination (Meryman 1966; Podolsky and Konstantinov 1980). In an average biological tissue, bound water comprises about 5–10% of the total water content and is characterised in terms of g water per 100 g specimen dry weight. Müller (1957) has shown that the amount of water remaining in the specimen after completion of freeze-drying depends on the freeze-drying temperature (Fig. 5.6) and the

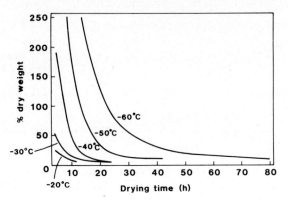

Fig. 5.6.   Residual water in plant tissue (% dry weight) after freeze-drying at different temperatures. The lower the drying temperature, the higher the residual water. (Redrawn from Müller 1957.)

temperature at which the secondary drying process is completed. To avoid contamination and to achieve maximum drying, the specimen is best warmed up to a few degrees above ambient temperature before the vacuum is broken by introducing dry inert gas (e.g. N_2). Alternatively, a variety of additional preparation steps, such as infiltration with resins (with or without previous fixation), sputter-coating, and shadowing and/or replication, can be performed immediately after drying while the specimen is still under vacuum (see §5.5.2). With some specimens exhaustive secondary drying is accompanied by collapse phenomena (see §5.6).

5.3   Practical aspects of freeze-drying

5.3.1   Basic principles

The water vapour released from the specimen during the freeze-drying process must be continuously removed from the system. This can be effected by vacuum pumping systems (see §6.3.2), by adsorption to chemical or mechanical desiccants, by condensation on a surface at a temperature considerably lower than that of the specimen, or by a combination of all these methods.

In a freeze-drying unit the maximum sublimation rate for the specimen water can only be obtained when the partial pressures of all the condensable gases (especially water vapour) in the specimen area are kept at least a factor of 100 (and preferably 1000) lower than the saturation vapour pressure for ice at the specimen temperature (see Table 5.2 and Fig. 5.3). With a higher water vapour pressure in the specimen area, the actual sublimation rate (J_a) is determined by the equation:

$$J_a = \eta(P_s - P_c) \times \left(\frac{M}{2\pi\varphi T}\right)^{0.5} \qquad (5.2)$$

where

J_a = actual sublimation rate
η = coefficient of evaporation
P_s = saturation vapour pressure of ice
P_c = partial pressure of water in the specimen area
M = molecular weight of the water vapour
φ = gas constant (see Eq. 5.1 in §5.2.1a)
T = absolute temperature.

This equation shows that the maximum sublimation rate considered earlier (§5.2.1a) resembles the special case where P_c is much smaller than P_s. The relationship between P_c and P_s and the consequences for the freeze-drying procedure are summarized in Fig. 5.7 (Umrath 1983).

Depending on the specimen temperature, other condensable gases in the system (e.g. CO_2, hydrocarbons) could theoretically mask the specimen surface and reduce the sublimation rate of water. However, with normal pumping systems, such as oil diffusion pumps or turbo-molecular pumps in combination with rotary pumps (see §6.3.2), even without cold traps, water vapour is the main constituent in the residual gas (Table 5.3). Today most commercial freeze-drying equipment includes cold baffles or shrouds which can be cooled either by liquid nitrogen or by refrigerator cryopumps. Such cold traps are kept at considerably lower temperatures than the drying specimen (usually at $-196°C$; 77 K) and, consequently, they dramatically reduce the partial pressure of condensable gases (see also §6.5.2). In such systems, the required factor of 10^2 to 10^3 difference (see Fig. 5.7) between the values for the water saturation vapour pressure at the specimen temperature and the partial pressure of water vapour (and other gases) in the specimen area is easily maintained. This is illustrated by reference to Table 5.2. For example, the figures show that the water saturation vapour pressure of ice at $-80°C$ (193 K) (which is a practical drying temperature for a specimen) is 4.02×10^{-4} Torr (5.36×10^{-2} Pa). To provide the required factor of

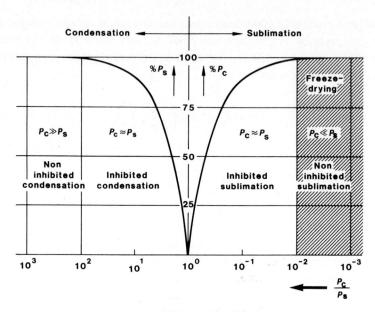

Fig. 5.7.   Relationship between the partial pressure of water in the specimen area (P_c) and the saturation vapour pressure of ice (P_s) and the consequences for the freeze-drying procedure. (Redrawn from Umrath 1983.)

TABLE 5.3

Composition of residual gases at 1×10^{-5} Torr (1.3×10^{-3} Pa)

Gas	Proportion of total (%)	Partial pressure (Torr)	Components (Molecular weights)
H_2O	70	7.0×10^{-6}	H_2O^+ (18),OH^+ (17),O_2(16),H_2(2)
CO, N_2	12	1.2×10^{-6}	CO^+ (28),N_2^+ (28)
CO_2	12	1.2×10^{-6}	CO_2^+ (44)
H_2 + others	6	0.2×10^{-6}	H_2^+(2), CH-groups, noble gases
Total	100	1×10^{-5}	

The partial pressures of the residual gases in a vacuum unit at a total pressure of 1×10^{-5} Torr (1.33×10^{-3} Pa) after pumping with a rotary pump in combination with an oil diffusion pump or turbomolecular pump (without a cold trap) from room atmosphere (ions measured by mass spectrometry). (From Umrath 1977.)

10^3 difference the condenser must have a temperature of about $-115°C$ (158 K), which corresponds to a saturation vapour pressure for ice of 3.34×10^{-7} Torr (4.45×10^{-5} Pa). This clearly shows that cold traps do not necessarily have to be maintained at a much lower temperature than the specimen to ensure an efficient freeze-drying process. For practical reasons, however, cold traps are usually cooled down to 77 K with liquid nitrogen.

To allow the maximum access of vapour molecules to the cold trap (or other pumping system), the pressure in the freeze-drying apparatus should be such that the mean free path of the residual gases is greater than the distance from the specimen to the cold trap (see Fig. 5.8). In practice, however, the actual freeze-drying rate is more likely to be determined by the diffusion resistance for water molecules within the increasing dry layer of the specimen than by the rate of water vapour transport from the specimen surface to the cold trap or vacuum pump (Stephenson 1960). Nevertheless, special

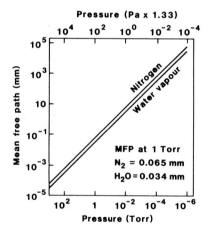

Fig. 5.8. Mean free path (MFP) versus vapour pressure for nitrogen and water vapour (at $0°C$, 273 K). The mean free path (λ) is calculated using the equation:

$$\lambda = \frac{1}{\sqrt{(2\pi n d^2)}}$$

where:

λ = mean free path (the average distance that a molecule will travel before striking another molecule

d = diameter of molecule

n = number of molecules per cm^3.

(Redrawn from Meryman 1966b.)

attention must be paid to the function and design of cold (condenser) surfaces in freeze-drying units. It is best to place the condenser system as close as possible to the specimen, or even to have the specimen surrounded by an 'optically tight' cold shroud, as provided in some freeze-fracture replication units (see §6.3.3b). As discussed in more detail in §6.3.3b, this type of design has the distinct advantage that, in addition to their function as a cryopump, the cold surfaces act as a very efficient anticontaminator and as a total shield for the specimen against radiation heating from warmer surfaces in the system.

Instead of (or in addition to) cryopumps, chemical or mechanical desiccants can be used to remove the water vapour released by the specimen. With chemical desiccants the water molecules undergo a chemical reaction. The most common ones are phosphorus pentoxide, calcium chloride, copper sulphate and sulphuric acid. They have the great advantage that within the gas in contact with them, even at room temperature, the partial pressure of water vapour generally approaches very low values (e.g. P_2O_5, 1×10^{-5} Torr; 1.33×10^{-3} Pa). For this reason, they may be used for long secondary drying procedures to remove the last traces of water from the specimens.

Mechanical desiccants such as zeolite or the synthetic zeolites known as 'molecular sieves' can be regenerated by heating. They are frequently used where a higher residual water vapour pressure is tolerable (e.g. drying of an air flow carrying water vapour with it). At very low temperatures (e.g. 96°C; 77 K), molecular sieves act both as a water vapour trap and a vacuum pump and can produce total pressures as low as 5×10^{-11} Torr (6.67×10^{-9} Pa) (Read 1963). To reach this value, the volume of the molecular sieve and the volume of the vacuum system must be in the correct proportions required by the specification. Freeze-drying without vacuum pumps at all has been performed using desiccants by Stumpf and Roth (1964, 1965).

Irrespective of the freeze-drying system used, it is important to have the possibility of accurate temperature control of the specimen stage and, ultimately, of the specimen itself. As mentioned before, poor thermal contact between the specimen stage, the specimen support and the specimen itself may lead to a considerable temperature gradient. Thus, it is strongly recommended that actual temperatures are checked within a given system to ensure that the specimen temperature is the same as the stage temperature or, at least, that there is a constant and predictable relationship between the two.

As can be seen from Fig. 5.9, a temperature profile can be expected across the drying specimen. It is obvious that the temperatures of greatest interest

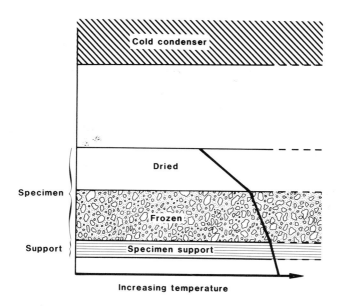

Fig. 5.9. Heat input by conduction from specimen support. This shows the temperature profile that can be expected across the drying specimen when there is a cold condensing surface in the close vicinity of the specimen surface. As heat input occurs by conduction from the specimen support, the surface will be at the lowest temperature. (Redrawn from Meryman 1966b.)

are those of the drying boundary (which determines the sublimation rate) and of the still frozen part of the specimen (which makes it possible to ensure that the temperature is maintained below the recrystallisation and eutectic temperatures). In practice, because of the small size of the specimen, only the temperature of the frozen, and subsequently of the dry, part of the specimen can be measured.

5.3.2 Freeze-drying apparatus designs

Historically, the vacuum for freeze-drying was first generated by water jets, then by mechanical rotary pumps and subsequently by combining rotary pumps with either oil diffusion or turbo-molecular pumps. To remove water vapour more efficiently, firstly chemical desiccants (e.g. phosphorus pentoxide, silica gel, molecular sieves) and then, later, low temperature condenser systems (e.g. liquid nitrogen-cooled traps) were used. Nowadays it is common to use both cold traps and molecular sieves (at low temperature) for maximum water vapour trapping. The first stage-cooling methods used refrigerator systems or thermoelectric devices (Bendet and Rizk 1974), whereas current systems often use liquid nitrogen or cold nitrogen gas that can be further utilised in feeding the cold traps in the apparatus. In future, refrigerator cryopumps will undoubtedly replace liquid nitrogen-cooled systems. With three-stage refrigerators, temperatures down to 163 K ($-110°$C) can be achieved.

A schematic representation of the two main types of freeze-drying system is provided in Fig. 5.10.

5.3.2a   Simple vacuum systems for freeze-drying

The simplest and generally older freeze-drying systems (Fig. 5.10a) use only a rotary pump for evacuating the chamber during the drying process. The specimens are placed onto a pre-cooled, temperature-controlled specimen stage in the drying chamber. After evacuation, the chamber is sealed off and the subliming water vapour is either trapped onto a condenser (not illustrated), which is kept at a temperature considerably lower than that of the specimen, or by using a chemical desiccant (e.g. P_2O_5) positioned between the specimen and the vacuum pump. Using such a system, the subliming water vapour is often not very efficiently trapped since both the condenser and the chemical desiccant tend to be positioned at distances from the specimen much greater than the mean free path of the water molecules attainable under the particular conditions of vacuum. A further disadvantage of such early designs is the possibility of specimen contamination from backstreaming vacuum pump oils (if no adsorption trap is present).

5.3.2b   Modern vacuum systems for freeze-drying

Most modern designs of freeze-drying unit have more efficient vacuum pumping systems, usually incorporating, in addition to a rotary pump, an oil diffusion pump as shown in Fig. 5.10b or a turbo-molecular pump as illustrated for the freeze-fracture replication unit in Fig. 6.10 (§6.3.2) Type II. Again, as with freeze-fracture replication units (see Fig. 6.10, Type III), the technology of refrigerated cryopumps can be expected to replace the other systems because (working as 'closed systems') they provide the ultimate convenience. In the most recent designs of freeze-dryer, the condenser is placed as close as possible to the specimen, thus ensuring optimal trapping efficiency for water vapour together with additional protection of the specimen against condensation contamination.

The most controlled freeze-drying procedures can be best performed in modern freeze-etching units (see §6.3.3b). Within such apparatus, an absolutely contamination-free enviroment is attained by using 'optically opaque' cold shrouds around the specimen. With such shrouds, it is impossible for contaminating molecules to move over long distances in a straight line towards the specimen without being intercepted by a cold surface (see, for

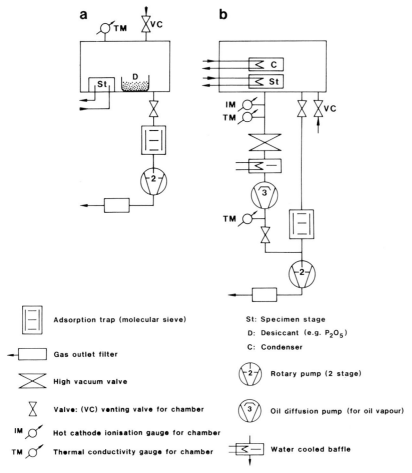

Fig. 5.10. Schematic representation of the two main types of freeze-drying apparatus. (a) The simplest type of freeze-drying system. The specimen chamber is pumped only by a rotary pump and subliming water is generally trapped using a desiccant. (b) Modern freeze-drying units use an efficient pumping system, including both rotary pump and oil diffusion pump, and a cold condensing surface (**C**) positioned closely above the temperature controlled specimen stage (**St**).

example, Figs. 6.15–6.17). Furthermore, effective specimen transfer and air lock systems have been developed which guarantee contamination-free transport of specimens from under liquid nitrogen onto the pre-cooled specimen stage (§6.3.3b). (The phenomenon of condensation contamination is explained in detail in §6.8.3.) In addition, special specimen supports and transfer gadgets for freeze-drying in freeze-etching units have been described (e.g. Iwata and Aita 1976; Smith 1980). Since freeze-etching procedures involve the controlled sublimation of specimen ice, the designs of most specimen supports and stages ensure good thermal contact and enable accurate monitoring of stage and specimen temperatures. Radiation heating can generally be neglected, particularly in models where cold shrouds surround the specimen. Finally, if required, the frozen-dried specimens can be shadowed and/or coated within freeze-etching units without breaking the vacuum.

5.3.2c   Simple freeze-drying devices for use in vacuum systems

Some versions of freeze-drying devices are very simple indeed and can be fitted to almost any vacuum coating unit used for electron microscopical preparation (e.g. Heckly and Skilling 1979; Morioka et al. 1980; Osatake et al. 1980; Roberts and Duncan 1981). Such devices are inexpensive and easily made using routine workshop facilities. Furthermore, they can be used independently of permanent cooling systems because they have pre-cooled metal blocks as heat sinks to keep the specimen cold during the critical period of drying.

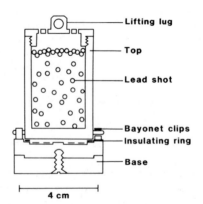

Fig. 5.11.   Diagram of a simple freeze-drying device. This module is commercially available (Appendix 2). (Redrawn from Roberts and Duncan 1981.)

Fig. 5.11 illustrates the simple module developed by Roberts and Duncan (1981) for freeze-drying specimens mounted on electron microscope grids. It consists of a circular base, which is divided into two parts so that its mass can be halved if necessary, and a cylindrical top. The top is recessed a few millimetres into the base, to which it can be locked. It can be filled with lead shot and has several holes which allow the free passage of gaseous or liquid nitrogen, and there is a lifting mechanism within the evaporation unit. Heat transfer between the top and bottom rings is reduced by an insulating ring. In use, the cleaned top and bottom parts are separately cooled in liquid nitrogen. The cooled grids with the mounted specimens are then placed in the wells of the base unit before the cold, top section is attached by means of the bayonet fitting. The loaded and assembled freeze-drying module is subsequently transferred from under liquid nitrogen to the vacuum coating unit and installed at a set distance from the evaporation source. The bayonet clips are disengaged by turning the top section, and the lifting mechanism is engaged with the lug on the lid. During pumping, the unit is allowed to warm up slowly. The insulation ring between the base and top section causes the lower part, containing the specimens, to warm up more quickly than the top. This maintains the required temperature gradient between the freeze-drying specimens and the top plate, which acts as a very efficient condenser. From thermocouple measurements, the authors found freeze-drying to be finished when the bottom part is completely defrosted. This process normally takes about two hours but the time can be considerably varied by changing the amount of lead shot put into the top and/or by using either a double or single base plate. Finally, the upper section is lifted away using a simple crane mechanism to allow shadowing of the dried specimens. The assembly is then left under vacuum overnight while it warms up to room temperature so that contamination of the specimens by condensation of water vapour is avoided when the vacuum is broken. To ensure that the correct physical conditions are fulfilled, is seems advisable to use thermocouples to check the temperature of the upper and lower halves during freeze-drying.

In view of the success of the above approach, it is to be expected that slight modifications to the standard Bullivant type II freeze-fracture replication device (§6.3.3c) – for example, by placing an insulating ring between the base plate and the tunnel cylinder and by closing the tunnel holes – could easily provide an effective freeze-drying module similar to that suggested by Roberts and Duncan (1981). In fact, it has already been shown by Severs and collaborators that, even without such modifications, excellent deep etching (i.e. surface freeze-drying) can be produced using the Bullivant device (Severs and Hicks 1977; Severs and Warren 1978). Nevertheless, all freeze-drying gadgets working without continuous refrigeration only give satisfactory results when the specimen is thin enough to dry before the base plate reaches the temperature at which recrystallisation phenomena in the specimen can be expected, or when only the surface layer of a bulk specimen is to be frozen-dried below the recrystallisation temperature. The latter criterion is adequate when specimen *surface* structures are to be examined in the SEM. Finally, these simple freeze-drying modules can be successfully used to dry specimens in which water has been replaced by an organic solvent prior to freezing (see §5.4; Osatake et al. 1980).

5.3.2d   Simple freeze-drying device for immersion in liquid nitrogen

An elegant but simple technique for freeze-drying specimens exclusively by cryosorption pumping has been described by Edelmann (1978b, 1979). As shown in Fig. 5.12a and b, the drying process takes place within a vacuum- and cryogen-tight sealed container which is immersed in liquid nitrogen. The actual freeze-drying apparatus (Fig. 5.12b) consists of a metal vessel half filled with molecular sieve (in this instance, Zeolite 13X from Leybold-Heraeus, Cologne, but other sources would undoubtedly also be suitable; see Appendix 2). The vessel is sealed by a metal screw cover to which a plug with five pins is fixed. The socket for this plug is fixed to a metal specimen support with small wells for the frozen specimens. The cover is connected to a metal tube, approximately 600 mm long, which houses the cables for the electrical supply to the heating element (power transistor) and the temperature monitoring of the specimen support. The upper end of the tube is closed by a ventilation valve (Fig. 5.12a).

The method used by Edelmann (1978b, 1979) for freeze-drying and embedding chemically unfixed biological specimens is as follows:

(1)   The specimen support (pre-cooled in liquid nitrogen) is loaded with frozen specimens and small drops of evacuated Spurr epoxy resin (Spurr 1968).

(2)   The specimen support is picked up with the plug fixed to its cover.

(3)   The vessel is immediately sealed and transferred into a 30 l container of liquid nitrogen.

(4)   The temperature of the specimen support – which rises during sealing and cooling of the vessel to about 133 K ($-140°C$) – is raised to the

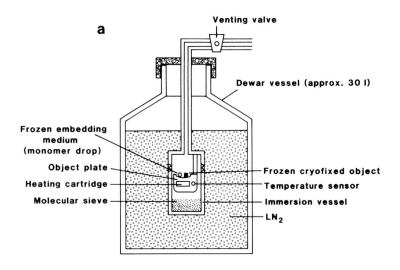

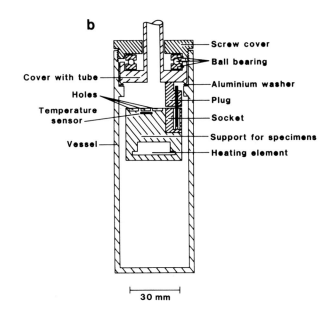

Fig 5.12. (a) General view of a simple freeze-drying device for immersion in liquid nitrogen so that freeze-drying takes place exclusively by cryosorption. (b) Detailed view of the freeze-drying apparatus shown in (a). (Electrical connections not shown.) (Fig. 5.12a redrawn from a diagram kindly provided by H.Sitte. Fig. 5.12b redrawn from Edelmann 1978b.)

chosen freeze-drying temperature. For drying 0.1 mm^3 pieces of muscle tissue, the support was maintained for three days at 193 K (–80°C), followed by six days at 213 K (–60°C). The final pressure obtained with the cooled molecular sieve, together with the cold walls of the container, has been estimated to be about 130 mPa. Under such conditions, the actual partial pressure of water and other gases that are condensable at the specimen temperature must be extremely low.

(5)   At the end of the freeze-drying process, the vessel is vented with dry nitrogen by opening the valve at the upper end of the tube and the support is warmed at a rate of about 10 K h^{-1} to 258 K ($-$15°C), at which temperature it is maintained overnight. At this temperature, the Spurr resin liquefies and infiltrates the dried tissue.

(6)   The support is then warmed further at 10 K h^{-1} to room temperature and the freeze-drying apparatus is removed from the liquid nitrogen container.

(7)   After opening the assembly, the tissue samples are transferred into fresh embedding medium prior to polymerisation.

This technique, which can be used with or without the embedding procedure, has the great advantages of being very simple and of making it possible to freeze-dry specimens with low cryogen consumption and at very low reproducible temperatures.

5.4   Freeze-drying from non-aqueous solvents

There are three methods of drying electron microscopical specimens which involve a step in which the constituent water is replaced by an organic solvent with the aim of avoiding or reducing the damaging effects of surface tension forces (Fig. 5.13, routes 1, 2 and 3). As an additional benefit, ice crystal damage during freezing can be greatly reduced if not avoided in route 3. The freeze-substitution procedure shown as a fourth route in this diagram facilitates the combination of rapid freezing of specimens with resin embedding. This is accomplished by using organic solvents to remove ice from the specimen at temperatures below that at which specimen ice recrystallises. This procedure is described in detail in Chapter 7. As with air-drying from aqueous systems, all solvents possess a finite surface tension which may cause deformation of the sample during fluid loss (Revel and Wolken 1973; Albrecht et al. 1976).

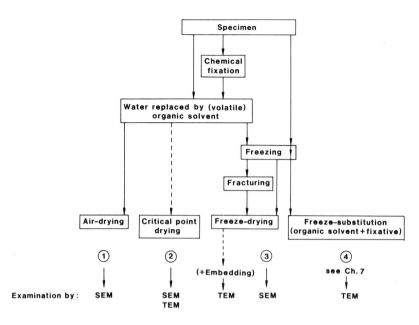

Fig. 5.13. Schematic diagram outlining different specimen preparation routes that involve the replacement of water with an organic solvent at some stage during processing. Route 1 shows air-drying from an organic solvent; Route 2 represents the normal critical point-drying pathway; Route 3 uses freeze-drying (possibly after freeze-fracturing) from an organic solvent rather than water; and Route 4 entails use of the freeze-substitution process (Chapter 7).

Critical point-drying (Fig. 5.13, route 2) provides a much better means of avoiding the damaging effects of surface tension forces than does air-drying. In the first step, water in a fixed or unfixed specimen is exchanged against an organic solvent (e.g. ethanol, acetone, 2,2-dimethoxypropane (DMP)). Subsequently, the specimen is transferred into a closed pressure vessel where it is first infiltrated and surrounded by liquid CO_2 (Anderson 1966) or Freon (Cohen et al. 1968). The temperature and pressure within the vessel are then raised above the critical values at which point (the critical point) the liquid and gas phases have the same density. After passing the critical point (for CO_2, 72 atm (7.29×10^6 Pa) and 31.3°C (304.3 K); for $CClF_3$ (Freon 13), 38 atm (3.85×10^6 Pa) and 28.9°C (301.9 K)), the pressure is slowly reduced so that recondensation of the fluid through adiabatic cooling is avoided. Critical point-drying is one of the most commonly used routes for processing specimens (particularly tisssues) for the SEM. Nevertheless, it has artefact problems of its own (Boyde and Franc 1981). A detailed description of critical point-drying is outside the scope of this book: the reader is referred to Hayat and Zirkin (1973), Willison and Rowe (1980) and Boyde (1980) for further information.

Freeze-drying from volatile organic solvents (Fig. 5.13, route 3) provides the possibility of avoiding both the damaging effects caused by ice crystal growth during freezing and the surface tension forces during air-drying.

5.4.1   Procedure for freeze-drying

Specimens must be fixed before replacing water with non-physiological organic solvents. Usually the dynamic processes of the living specimen are halted by glutaraldehyde fixation. To keep structural changes to a minimum, optimal conditions (temperature, pH, osmolarity, ionic composition) of fixation must be chosen (Glauert 1974; Boyde 1978a). To reduce the solubility of lipids in the substitution fluids, post-fixation with OsO_4, and possibly also uranyl acetate, is necessary. Generally, fixation schedules similar to those used prior to thin sectioning (Glauert 1974; Hayat 1981) or critical point-drying (Boyde 1980) are used (see also §2.3.3).

Subsequently the water of the chemically fixed specimen is replaced with a solvent such as ethanol, iso-amyl acetate or chloroform (Boyde and Wood 1969); camphene (Watters and Buck 1971); or tertiary butyl alcohol (Wheeler et al. 1975). These substitution liquids at 100% concentration solidify on quenching (for techniques see §2.4) to form a glass, or crystals of such small size that they can be neglected. Consequently, structural detail within bulk specimens can be preserved which would otherwise be destroyed if the specimen were to be frozen from the hydrated state.

Due to the high vapour pressure of liquids used just below their melting points (see Fig. 5.14 and Table 5.1a in §5.1), freeze-drying can be extremely rapid. As an example, freeze-drying of a bulk specimen from amyl acetate (melting point, $-71°C$; 202 K) at a temperature of $-75°C$ (198 K) and pressure of 5×10^{-3} Torr (6.65×10^{-1} Pa) may take only about half an hour (Boyde and Wood 1969). However, to have a safe margin to prevent evaporation of liquid rather than sublimation of solid, Boyde (1978a) recommends that specimens infiltrated with organic solvents should be frozen-dried at temperatures at least 10 K below their freezing points.

Since no ice crystal damage occurs on freezing, internal structures of bulk specimens can be exposed by fracturing before drying. This is best done under liquid nitrogen with samples infiltrated with ethanol or halocarbon 13 (Boyde and Maconnachie 1979b). If freeze-fracturing is carried out on a specimen infiltrated with ethanol, then either critical point-drying or freeze-drying may be used (Fig. 5.13) if the specimen is then to be observed in an SEM (Boyde 1980; Humphreys et al. 1974). As discussed in detail in §6.3, specimens frozen in the hydrated state predominatly fracture along membranes. This cannot necessarily be expected after fixation in OsO_4 and infiltration with organic solvents. Thus, relatively little cytoplasmic detail is found in such preparations because the fracturing occurs predominantly

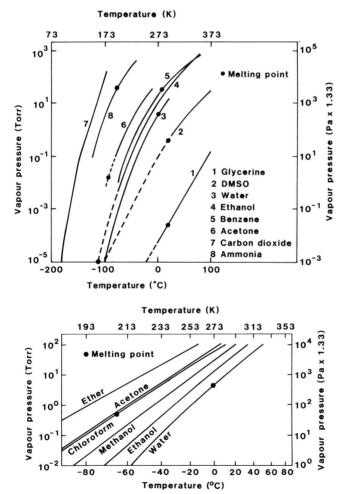

Fig. 5.14.   Vapour pressure of solvents at low temperatures. Two groups are illustrated separately to cover the vapour pressure/temperature range for most of the common solvents. (Redrawn from Willemer 1975 and incorporating additional data.)

independently of membranes. In addition, tissues become very brittle after treatment with organic solvents and may easily fracture during subsequent processing steps or during examination in the SEM. Despite such difficulties, for many routine purposes the advantages of the method (no ice crystal artefacts, rapid preparation) offset the disadvantages.

5.5   Freeze-drying procedures for different types of specimen

5.5.1   Freeze-drying of small biological specimens

The freeze-drying technique can be used in a variety of ways for structural studies of small specimens (e.g. subcellular fractions, macromolecules, viruses, membrane structures, single pro- and eucaryotic cells and monolayer cultures; Fig. 5.15). Most small biological specimens can be frozen-dried without any pretreatment, such as chemical fixation or cryoprotection, provided that the specimen layer is thin enough and the cryofixation technique is appropriate (see §2.4.1 and §6.2.2a). Once the specimen has been liberated from the surrounding (and internal) ice, it can be shadowed, coated or replicated. Some specimens (e.g. viruses), when suspended in a negative-staining solution before freezing, may provide enough contrast in the TEM even without any additional metal deposition.

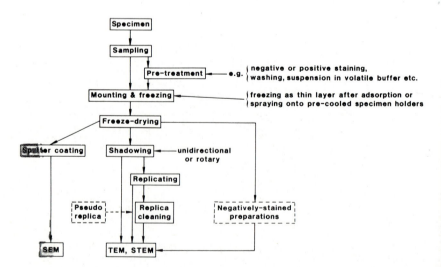

Fig. 5.15.   Different possible stages in the preparation of small specimens by freeze-drying.

5.5.1a   Specimen mounting and freezing

For most mounting procedures it is advantageous to make the surface of the support hydrophilic. For grids with carbon-coated support films this is best done by glow discharge in air at low pressure or by radiation with ultraviolet light. Specimen adsorption may also be improved by the addition of wetting agents (e.g. bacitracin) to the suspension (Gregory and Pirie 1972). No 'wetting' problems should occur if specimens are applied to the surface of a freshly cleaved piece of mica or when the specimens are sprayed onto a pre-cooled support. Supports can be positively charged with poly-L-lysine (see §6.2.2d) (Fischer 1975; Nermut and Williams 1977) or 1% alcian blue 8GX (Sommer 1977; Nermut and Williams 1977, 1980) to achieve a better adsorption of specimens with a net negative charge. Selective adsorption of particles (e.g. viruses) may be obtained by coating the support with specific antisera before use. This is simply done by floating carbon-coated grids on a dilute serum for several minutes (Derrick 1973). Except for cells (e.g. monolayer cultures) *directly* grown on the specimen support, the majority of small specimens are mounted by one of two methods: spray-freezing (Williams 1952, 1953); or adsorption (Nanninga 1968; Nermut et al. 1972; Nermut 1977, 1981a).

In the *spraying technique*, the specimen is directly sprayed with an atomizer (e.g. Effa Spray Mounter, Ernest Fullam; Appendix 2) onto grids or pieces of mica firmly attached to a cold block. This is very conveniently achieved by mounting the specimen support directly onto the temperature-controlled specimen table of a freeze-etching unit (Steere 1973; Nermut 1977) or freeze-drying unit. For spraying, particles are best suspended in distilled water or volatile buffer (e.g. 0.1–0.3 M ammonium acetate, ammonium carbonate or ammonium succinate) (Backus and Williams 1950; Williams 1953). Since the melting point of the eutectic concentration of ammonium acetate buffer is below $-70°C$ (203 K), freeze-drying temperatures must be $\leqslant -80°C$ (193 K) (see Williams 1953). The schedule for the spray-freezing, freeze-drying procedure is then as follows:

(1)   Mount specimen support on the specimen table.
(2)   Evacuate chamber to high vacuum and cool specimen support to about 123 K.
(3)   Vent chamber with dry nitrogen and maintain cold surfaces in vacuum chamber as free as possible from hoar frost by continuing to purge with dry gas.

(4)   Spray specimen onto cooled support.

(5)   Stop dry nitrogen flow and evacuate chamber.

(6)   Warm specimen table to selected freeze-drying temperature ($-85°C$; 188 K) while protecting the specimen from contamination by using a cooled condenser (either a shroud (see §6.3.3b) or a cooled microtome arm (see §6.3.3a)).

If a freeze-fracture replication unit is used which provides for contamination-free transfer of specimens from liquid nitrogen into the vacuum unit (see §6.3.3c), spray-freezing can be performed on a separate cold stage. There are problems when pathogenic organisms have to be examined using the spray-freezing method and there is also the disadvantage that considerably more material is required than for the adsorption technique. Furthermore, special attention has to be paid to the spraying technique itself: the droplets freezing on the support should be as fine as possible and deposited as a monolayer to prevent extensive overlapping of particles after the ice has been sublimed away (see §2.4.1c). As an alternative to spraying small specimens onto cooled supports, they may be sprayed onto supports at room temperature. After repeated washing with distilled water, as with the adsorption technique, and rapid freezing, specimens are transferred to the cold stage of the freeze-drying unit (Walzthöny et al. 1983).

The *adsorption technique* is a relatively more convenient method for applying small specimens to supports. In comparison to the spraying technique, much less material is required and monolayers of particles are more easily obtainable. As illustrated in Fig. 5.16, adsorption of the specimen to the support can be achieved in two ways: by touching the support to the surface of a suitable liquid (distilled water or buffer) on which a monolayer of the specimen has previously been spread (Fig. 5.16a), or by floating the support on a droplet of the particle suspension (Fig. 5.16b). With the floating procedure the required adsorption time depends on the concentration of the particles in the suspension. Thus, even low concentrations can give a sufficient yield after prolonged contact (e.g. 15 min). Fowler and Aebi (1983) obtained excellent results when a diluted solution of protein molecules (typically 10–30 nM) or supramolecular structures (typically 0.5 mg ml^{-1}) was allowed to adsorb to a glow-discharged, carbon-coated copper grid or freshly cleaved mica for 1 min. When coated grids are used as supports, negative-staining techniques can be used to check the success of this preparation step.

The *spreading technique* (Kleinschmidt 1968; Willison and Rowe 1980) requires more concentrated suspensions but monolayers can be compressed with a bar. With virus particles, spreading of 20 μl (10–20 mg protein ml^{-1}) mixed with 10 μl of 0.1% cytochrome *c* (in 0.1 M ammonium acetate) provides a good yield (Nermut 1973a).

Irrespective of which mounting procedure is used before freezing, the best results are obtained if the liquid surrounding the specimen is free of non-volatile (non-sublimable) constituents. Specimen supports loaded by spreading or adsorption techniques must be transferred through several (at least three) washes in distilled water or volatile buffer (Fig. 5.16) to ensure

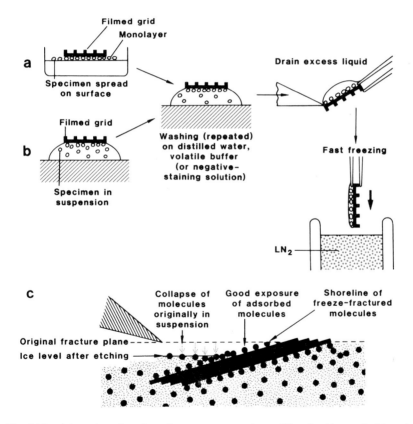

Fig. 5.16. Adsorption of small specimens to supports (e.g. a 'filmed' grid coated with a plastic/carbon film or a piece of mica) for freeze-drying. (a) Adsorption of particles spread as monolayers on the surface of a liquid. (b) Adsorption of particles from suspensions. (c) Diagram illustrating the domains exposed by freeze-fracture and freeze-drying (deep etching) of frozen suspension of mica flakes plus adsorbed macromolecules. (Fig. 5.16c redrawn from Heuser 1983.)

that both non-adsorbed particles and masking material are fully removed. Before freezing, the excess liquid is drained off with filter paper (Fig. 5.16). To avoid complete drying of the thin film, this is best done just above the surface of liquid nitrogen. With hydrophilic supports, freezing by immersion (§2.4.1a) must be done within two to three seconds to avoid air-drying. An even faster transfer into the cooling bath is required if more hydrophobic supports are used. In order to produce a thin film of water prior to freezing, grids can be placed for a few seconds with the specimen side down on a small stack of wet filter paper in a Petri dish. With hydrophilic supports, the filter paper only needs to be kept slightly wet, whereas for hydrophobic specimen carriers the paper must be soaked (Würtz et al. 1976; Kistler et al. 1977).

With some specimens (e.g. suspensions of cultured cells), the formation of usable monolayers by adsorption may be inhibited by surface projections. If the preservation of the gross morphology of the cells is less important, the area of adhesion can be increased by centrifuging the cells onto the support (e.g. mica).

To examine the inner surfaces of membranes, tissue culture cells, firmly attached to a support, are lysed by contact with distilled water. After the unattached membranes and cell contents have been removed with a stream of buffer (e.g. hypotonic phosphate buffer, pH 7.4, applied from a syringe fitted with a hypodermic needle), the inner surface of the plasma membrane can be examined (Mazia et al. 1975; Nermut 1981a, 1981b; see also §6.9.1c).

As an alternative to the common adsorption technique, Heuser (1983) suggests mixing the molecules with an aqueous suspension of tiny flakes of mica and then freezing and fracturing this suspension as with freeze-etching techniques (§6.3.3a). During the subsequent freeze-drying process, which resembles deep etching (§6.5), areas of unfractured mica and thus intact macromolecules are exposed (Fig. 5.16c). When specimens are frozen on cold metal surfaces (§2.4.2) the adsorbed macromolecules are not obscured by salt deposits even if they are suspended in hypertonic solutions. An additional advantage is the fact that the exposed macromolecules are only minimally frozen-dried and consequently retain their three-dimensional topography very well. Using freeze-etching apparatus for this procedure, freeze-drying is best done at $-102°C$ (171 K) for 3 min. After this superficial drying, the specimen is replicated by rotary shadowing with Pt/C and C (§6.4.1). After thawing, the replica is cleaned by treatment with full strength chrome-sulphuric acid and full strength hydrofluoric acid (**Caution:** hydrofluoric acid is extremely hazardous).

The major advantage of this procedure is that macromolecules and small particles, such as viruses, can be viewed unfixed in conditions close to those of their natural ionic environment.

5.5.1b   Freeze-drying and subsequent handling

Frozen specimens mounted by the spreading or absorption technique will only be embedded in ice a few tens of micrometres thick. As can be seen from Table 5.2 in §5.2.1a, under optimal conditions a 50 μm thick layer of ice sublimes at $-85°C$ (188 K) within approximately 25 min. For comparison, at $-90°C$ (183 K) the same layer would take an hour to sublime. These calculations show that, for this type of specimen, relatively short drying times are sufficient. Nevertheless, it will take longer to remove all the ice from the specimen matrix (see §5.3). With some specimens (e.g. procaryotic cells, monolayers of eucaryotic cells, membrane structures), it has been shown that prolonged freeze-drying can lead to certain shrinkage or collapse phenomena (Nei 1962, 1973; Nermut 1981a). Consequently it is best to check whether better structural preservation is achieved if the subsequent preparation step (e.g. replication) is effected *immediately* after the specimen has been liberated from the surrounding ice (Heuser 1983).

As summarised in Fig. 5.15, the freeze-drying procedure for small specimens can be combined with the following preparation steps:

Shadowing (unidirectional or rotary; see §6.4.1) for TEM. To reduce or avoid secondary structural changes caused by the uptake of atmospheric moisture, the shadowed specimens mounted on support films on grids can be stabilised by evaporating a thin (< 0.5 nm) layer of carbon from above. If shadowing or replication of the frozen-dried specimen is performed below ambient temperature then the specimen should be warmed up before venting the coating unit to avoid condensation of atmospheric moisture.

Replication (shadowing, coating (§6.4.1) and cleaning (§6.6)) for TEM. With small specimens (e.g. viruses, subcellular fractions) satisfactory replicas can easily be obtained. Freeze-drying of larger specimens (e.g. isolated cells) generates more pronounced topographical profiles. Specimen swelling during cleaning (see §6.6) frequently causes fragmentation of the replica. This problem can be reduced if the cells are frozen as a densely packed layer. As an example, bacteria are best mounted by smearing a thin layer of a dense pellet over a piece of mica or other suitable support. Since cells mounted in this way cannot be washed by floating the support (see Fig. 5.16), the pellet must be suspended either in distilled water or in volatile buffer to avoid the masking of structural details by the deposition of non-volatile components (culture medium, salts, etc.).

Pseudoreplication (shadowing and coating) for TEM. Small structures (e.g. subcellular fractions) mounted on mica can be examined using the pseudoreplica technique. With this type of preparation the specimen is not removed but remains attached to, or enclosed within, the deposited heavy metal and carbon layers. Pseudoreplicas reveal the same structural details as shadowed specimens mounted on coated grids.

Freeze-drying and negative staining for TEM. A procedure combining negative staining with freeze-drying was developed by Nermut and Frank (1971) for structural studies on viruses. Negative staining is carried out after the last washing step and immediately before freezing (Fig. 5.16a and b). Best results have been obtained using 2% phosphotungstic acid (PTA) and 3% ammonium molybdate but 1% uranyl acetate may also be used (Nermut 1973b). Following freeze-drying, the specimen is warmed up to 30°C (303 K). Immediately after breaking the vacuum, the excess dried negative stain which remains as a powder on top of the specimen has to be removed using a gentle flow of dry nitrogen gas or with a fine brush. Since the dried negative stain is very hygroscopic, the specimens should be examined as soon as possible after preparation and/or stored under vacuum (Nermut 1977).

Sputter-coating for SEM. Examination of frozen-dried single cells and monolayer cultures by SEM is described in §5.5.2a, together with the preparation and examination procedures for bulk specimens.

Embedding. Frozen-dried cells may be vacuum-embedded for thin sectioning with or without previous exposure to fixatives in the vapour phase (see §5.5.2c).

5.5.2   Freeze-drying of bulk specimens

Freeze-drying can be used to prepare bulk specimens for the study of both surface and internal structures (Fig. 5.17). In most unfixed bulk specimens, only a very superficial layer can be well frozen (see §2.3.1). To avoid ice crystal damage at a deeper level, the water must be replaced by an organic solvent before freezing (see Fig. 5.13 in §5.4) unless high pressure freezing techniques are available (§2.4.1d). The surface structure of the frozen-dried specimen can be studied either in the SEM or in the TEM using the high resolution heavy metal/carbon replication technique. This technique resembles that for deep etching specimens during conventional freeze-etching methods (see §6.5). Finally, the frozen-dried bulk specimens can be embedded in res-

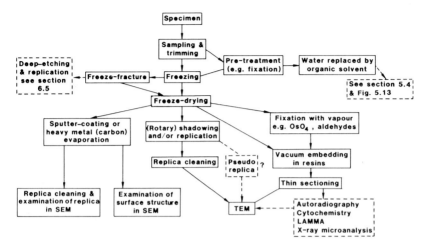

Fig. 5.17. Possible steps in the preparation of bulk specimens by freeze-drying.

in, with or without previous chemical fixation, under vacuum. The subsequent thin sections are suitable for morphological studies, cytochemical procedures, autoradiography (e.g. for the detection of diffusible substances) or LAMMA (laser microprobe analysis) applications.

5.5.2a   Examination of bulk specimens in the SEM

The effects of specimen pretreatment and freezing. For most specimens, the degree of similarity between the native structure and the structure observed in the SEM depends strongly on the treatments preceding the freeze-drying process – in particular the washing and fixation stages. The effects of different pretreatments as well as of different dehydration processes (including freeze-drying) on specimen structure have been discussed in detail by Boyde and his colleagues (Boyde 1976, 1978a, 1978b; Boyde and Franc 1981).

An optimal fixation procedure would instantaneously stabilise the dynamic processes within those areas of the specimen that are subsequently to be studied in the SEM. Usually, this is the surface of the specimen and, in particular, the plasma membrane. Glutaraldehyde-fixed specimens (§2.3.3) usually show less ice crystal damage in frozen-dried images of the plasma membrane than do those frozen from the native state (Boyde 1972, 1980). Fixation may also enhance the resistance to radiation damage and electrical conductivity of a specimen. The latter is of some importance if uncoated frozen-dried specimens are to be studied.

For many SEM studies on biological systems it is important to wash off, or digest, 'masking material' from the specimen surface. This *can* be done before the primary fixation to avoid stabilisation of material that coats the structure of interest. Special attention to this step is necessary. The washing liquid should be isotonic, buffered and physiologically appropriate to the specimen. Sheer forces generated by small jets of water from a syringe fitted with a hypodermic needle or a series of mild centrifugations (e.g. for protozoa) may help to remove the masking material. Since such treatments may cause structural changes they should, if possible, be performed *after* a prefixation stage. For example, mucus on wet epithelial surfaces can be removed by specific mucolytic agents after primary fixation with OsO_4, omitting the use of the cross-linking glutaraldehyde (Boyde 1980). Nevertheless, for most specimens fixation will be performed with glutaraldehyde, where necessary followed by one or more other fixatives known from TEM processing schedules (Glauert 1974), to render the specimen more able to withstand the dehydration process. As an example, little loss of structural material occurs when glutaraldehyde-fixed tissue samples are washed in water to remove the vehicle (buffer system) before freezing. On the other hand, postfixation with OsO_4 is required to render lipids insoluble in organic solvents which are used for replacing the specimen water when freeze-drying is performed from non-aqueous systems (see §5.4) (Boyde 1980).

Although glutaraldehyde will be the first and only fixative used for many applications (e.g. 2–3% glutaraldehyde in 0.1 M sodium cacodylate buffer, pH 7.2, for animal tissues), we should be aware that it can generate changes in membrane structure. Such artefacts have been studied in detail by freeze-fracture replication methods (see §6.8.3) but most of the changes (e.g. displacement of membrane-intercalated particles) occur at a morphological level beyond (below) the resolution limit of the SEM (§5.6). Special attention has to be paid to the composition of the fixative vehicle. As with fixation procedures used for the TEM, severe changes in the surface topography can occur if the fixative is not applied at a physiological temperature and pH and is not isotonic with the ambient fluid in the tissue or cell environment. Furthermore, buffer systems producing precipitates on the specimen surface during fixation and/or dehydration (e.g. phosphate buffers) should be avoided (Boyde 1980). Generally it is recommended that, whenever feasible, a long fixation period (several days) in glutaraldehyde is allowed. This has been shown to make the tissue harder and mechanically more stable – which is of particular importance for SEM (Boyde 1972; Humphreys 1975). These long fixation times apparently do not cause a degradation in the appearance of the tissue fine structure (Humphreys 1975).

Irrespective of any pretreatments, washing in distilled water or in a volatile buffer before freezing is an essential step if 'clean' specimen surfaces are to be exposed by the freeze-drying process.

As discussed in §2.4.1, ultra-rapid quenching allows the superficial layer of a bulk specimen to be frozen without any ice crystal damage that will be visible at the resolution of most SEMs. Nevertheless, special care must be taken that the film of overlying liquid is thin enough to allow high cooling rates to be obtained within the superficial layer while, at the same time, air-drying of the specimen prior to freezing is avoided. If samples are too large, cracking artefacts may be produced by thermal stresses during freezing.

Finally, fixed or unfixed fully hydrated or substituted (see §5.4) frozen specimens may be fractured under liquid nitrogen (see §6.3.3c) before freeze-drying. If the substitution was with ethanol or halocarbon (e.g. Freon 13), critical point-drying may be considered as an alternative dehydration procedure for freeze-fractured specimens (Figs. 5.13 and 5.17) (Boyde 1974a; Boyde and Maconnachie 1979b).

Freeze-drying and examination. Frozen specimens are stored under liquid nitrogen until required for use. Transfer to the temperature-controlled stage of a freeze-drying unit (§5.3) is best carried out under the protection of liquid nitrogen to avoid the risk of ice recrystallisation due to a critical temperature rise in the specimen (Smith 1980; Heuser 1983; see also §6.3.3c).

The individual parameters which determine the freeze-drying procedure for bulk specimens have been discussed in detail in §5.3. Since unmounted specimens (most bulk specimens) have only a poor mechanical fit and thermal contact with the flat, temperature-controlled stage within the freeze-drying unit, special care is required to ensure that the chosen temperature is maintained during the drying process. Boyde and Echlin (1973) and Boyde (1975) suggest that the specimen should be enclosed during drying within a gadget such as that illustrated in Fig. 5.18a. This consists of a cylindrical metal dish with a lid with a diameter larger than that of the dish. The upper part of the sides of the dish has several 'cut-outs' to facilitate the escape of subliming water vapour. The dish is deep enough to carry over some liquid nitrogen from the Dewar flask in which it is loaded with the specimen. This design has the advantages that it protects the specimen from warming above the desired freeze-drying temperature during transfer and that, so long as good thermal contact of the dish with the temperature-controlled platform of the freeze-drying unit is maintained, the specimen will be well protected from radiation heating. In other words, specimens will

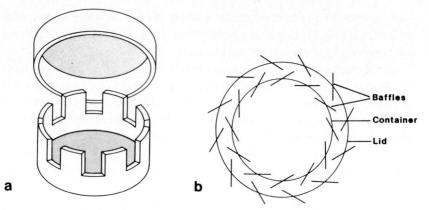

a   **b**

> Baffles
> Container
> Lid

Fig. 5.18.   Specimen containers for freeze-drying bulk specimens. (a) A gadget described by Boyde and Echlin (1973) consisting of a cylindrical metal dish with a lid. The castle-like shape of the dish allows the escape of subliming water vapour. The relatively geometry of the lid/base ensures that the specimen is protected from radiant heating during the drying process. (b) Plan of a similar design principle to (a) but with an improved pathway for the escape of water vapour. The baffles on the dish and lid are arranged so as to allow easy escape of water vapour while totally protecting the specimen from heating by radiation from external sources. (Fig. 5.18a redrawn after Boyde 1975.)

only 'see' surfaces maintained at the desired temperature during the whole freeze-drying procedure. An alternative approach, based on the same principle but with an improved pathway for the escape of water vapour, is illustrated in Fig. 5.18b.

Freeze-drying schedules for bulk specimens depend on the type of specimen and on the scientific information required. As discussed in §5.2.1b, the critical recrystallisation temperature determines the upper temperature limit at which the area of interest can be dried. For SEM purposes, samples are usually dried in the range 173–223 K (-100 to $-50°C$) but even higher temperatures may give satisfactory results (Boyde 1980; Boyde and Franc 1981). It is also difficult to give a rigid rule for freeze-drying procedures, particularly the drying time for bulk specimens that are to be observed in the SEM. As pointed out earlier, since only a superficial layer in most hydrated specimens can be well frozen, it makes little sense to try to dry the more central regions of bulk specimens (where the ice crystals will, in any case, be large) at a temperature chosen to avoid recrystallisation (Boyde 1980) provided that only a thin (5–10 μm) superficial layer of water is present during specimen freezing. At a specimen temperature of $-85°C$ (188 K), for example, the 'drying front' proceeds in this layer of pure water at

2 μm min^{-1} (see Table 5.2 in §5.2.1a). Thus the thin layer of ice covering the specimen surface will have sublimed away after 5–10 min. Considering a much slower drying process (lower sublimation rate) below the cell boundary (cell membrane, specimen surface) freeze-drying for 1–2 h at a 'recrystallisation-safe' temperature is sufficient for good structural preservation of the surface topography.

Since most frozen-dried specimens are very fragile, freeze-drying should ideally be performed with the specimen already mounted on a stub of the type used in the SEM or on a support that can be attached to an SEM stub without risk of specimen damage. The latter approach is possible with cultured cells that have been grown, or mounted by adsorption, on coverslips or other appropriate substrates (see §5.5.1a). Supports carrying the frozen-dried specimens are best attached to stubs using electrically conducting glues, such as silver or graphite paste. For direct mounting of small specimens (e.g. bacteria or tissue culture cells) Boyde (1975) recommends the use of hemispherical specimen supports, such as round-headed aluminium rivets, which have been micro-roughened by treatment with an abrasive unit (see Appendix 2) to render them clean and wettable.

Most frozen-dried specimens are very hygroscopic. Thus, if the vacuum must be broken for mounting, special care is required that all manipulations take place in an atmosphere of low relative humidity for the shortest possible time. Isolated specimens (e.g. organisms, tissue samples, larger cells) can be mounted in the frozen-dried state by carefully attaching them to double-sided sticky tape which has previously been firmly attached to the SEM stub. With large specimens the area of contact can be increased by mounting them on silver or graphite paste. Special care must be taken to prevent the solvent of such pastes from being soaked (by capillary action) into the specimen area of interest. The evaporation of the solvents from adhesive pastes can make the specimen damp and may even cause secondary shrinkage artefacts (see §5.6). To avoid these potential hazards, Boyde (1980) suggests the use of rapid-setting epoxies as adhesives. The viscosity of these resins can be adjusted by chosing the time at which the material is used and, as a further advantage, the vapour pressure of the polymerised resin is so low that it can be neglected.

Frozen-dried biological specimens are primarily composed of constituents with low atomic number which behave as electrical insulators. To prevent charging under the electron bean in the SEM and to enhance secondary electron emission, the following measures can be considered:

(*i*)   coating the dried specimen with a layer of electrically conducting material;

(*ii*)   depositing metal within the specimen;

(*iii*)   using a very low beam accelerating voltage and/or current; and

(*iv*)   replicating the specimen surface and studying the replica.

Coating the specimen with a thin layer of an electrically conducting material is still the most commonly used procedure. For high resolution work this layer should be as thin as possible, have a very small crystallite size and give a sufficient yield of secondary and/or backscattered electrons. As with replica techniques used for the TEM (see §6.4.1a), the actual resolution limit is primarily determined by the grain size of the evaporated layer (Peters 1977, 1979; Blaschke and Boyde 1978; Jakopic et al. 1978). For morphological studies the most common coating materials are gold, alloys of gold and palladium, platinum and silver (Echlin et al. 1982; Kemmenoe and Bullock 1983). For X-ray microanalysis, specimens are coated with carbon, aluminium or chromium (Marshall 1980). The technology for evaporating coating materials by resistance heating and electron-beam evaporation is discussed in §6.4.1, and that for sputter-coating in §3.2.3, §3.3.6 and §3.4.6.

As an alternative to specimen coating, procedures have been developed for depositing osmium within the specimen before dehydration (OTO and OTOTO methods – see Kelley et al. 1973; Munger 1977; Sweeny and Shapiro 1977; Murakami 1978; Murphy 1980). These techniques involve the risk that the benefit of improving specimen conductivity is accompanied by a limitation in resolution arising from metal precipitation on the specimen surface which may thus produce results that are not essentially different from surface-coated specimens.

Examining the specimen at very low accelerating voltage (e.g. 1–2 kV) in the SEM may be a further alternative to the deposition of an electrically conducting material, although high resolution images are then difficult to obtain. Reducing the beam voltage limits the penetration of the electrons but provides the possibility of working in the range where the secondary electron yield is close to unity and, as a consequence, no charging occurs (Reimer 1978). With frozen-dried specimens, charging is also not a problem when a beam voltage of 3–5 kV is used at TV scanning speed. This is of particular interest if specimens are to be micro-dissected in the SEM in which case it would not be sensible to deposit a surface conducting layer (Boyde 1974b). (See further discussion of charging problems related to low temperature observations in the SEM in §3.2.1.)

Finally, particularly for the study of frozen-dried specimens with less irregular topography, even replication techniques can be considered. Firstly, a thick evaporated layer is deposited (e.g. of carbon and gold). After digesting the specimen (see §6.6), it is possible to obtain high resolution SEM images of the replica side originally in contact with the specimen surface (Revel 1978).

5.5.2b   Specimen replication for the TEM

Frozen-dried bulk specimens can be examined in the TEM using high resolution replication techniques (Fig. 5.17). This reveals much of the overall structural information that can be obtained by SEM (§5.5.2a) but, in addition, allows the resolution of structural details down to the limit imposed by the high resolution shadowing technique (§6.4.1b). In particular, the analysis of stereoscopic images makes it possible to obtain excellent three-dimensional information on both the gross morphology and structural details of the frozen-dried specimen (Steere 1973; Heuser 1980, 1981).

High resolution replication of frozen-dried specimens is almost exclusively carried out in freeze-fracture replication units which have temperature controlled stages (§6.3.3). The schedule for freeze-drying and replicating specimens largely resembles the freeze-fracture replication technique for deep etched specimens (§6.5) with the major difference that freeze-fractured or non-fractured specimens are maintained at slightly higher temperatures (e.g. -90 to $-95°C$; 183 to 178 K) to achieve complete sublimation of the specimen ice within a practical period of time (Table 5.2 in §5.2.1a). Specimens that are not fractured under vacuum before being dried need to be protected from hoar frost deposition during transfer onto the pre-cooled specimen stage. If no protective transfer or air lock system is available (§6.3.3b and c), a simple metal lid can be placed over the specimen before it is lifted out of the liquid nitrogen container (Heuser and Kirschner 1980). Once a suitable vacuum (e.g. 10^{-3} to 10^{-4} Torr; 1.33×10^{-1} to 1.33×10^{-2} Pa) has been reached, the protective lid can be removed (using a lever manipulated via a vacuum lead-through) and the specimen warmed up to the drying temperature. If highly sculptured surfaces, and particularly fibrous structures, are exposed by the drying process, it can be advantageous to use rotary shadowing and replication procedures (§6.4.1f) to reveal the specimen structure. With some specimens (e.g. frozen-dried cytoskeletons; see Heuser and Kirschner 1980), it may even be possible to examine the frozen-dried specimen as a pseudoreplica (without digesting the specimen away).

In this case, the replicated, frozen-dried specimen must be warmed up to ambient temperature before breaking the vacuum to avoid water vapour condensation and specimen rehydration. Replicas of rough and heterogeneous frozen-dried specimens have a tendency to disintegrate during the thawing and cleaning procedure. Heuser and Salpeter (1979) suggest that the replicated specimens should be transferred, while still cold, to a small receptacle (e.g. a scintillation vial) of frozen methanol which is then allowed to thaw and to infiltrate (substitute) the material at the same time. After bringing the specimens to ambient temperature, they are then rinsed in water and transferred to the usual cleaning solutions (e.g. sodium hypochlorite; §6.6).

High resolution replicas, particularly of frozen-dried tissue culture cells and purified cytoplasmic proteins, have revealed remarkable new information on the structure and organisation of the cytoskeleton and membrane-associated structures (for reviews, see Heuser 1981, 1983). For this type of

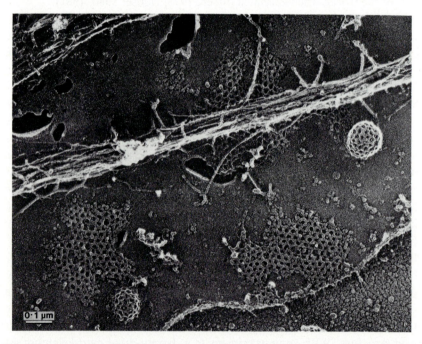

Fig. 5.19.   Micrograph of a high resolution replica of rapidly frozen, deep-etched, rotary shadowed cells showing coated pits and cytoskeletal elements prepared by the method of Heuser (1981). (Micrograph kindly provided by J. Heuser and reproduced with his permission.)

study it has been shown to be essential to use ultra-rapid freezing techniques (e.g. freezing on cold metal surfaces; §2.4.2) to ensure that the ice crystal size in the area of interest is at an absolute minimum. The potency of this technique is demonstrated in Fig. 5.19, which shows coated pits and cytoskeletal elements. Although the study of these replicas does not necessarily require stereoscopic techniques, they are nevertheless highly recommended as they provide a better appreciation of the three-dimensional structures which often closely resemble the images obtained (certainly at considerably lower resolution) in the SEM (Heuser 1981).

5.5.2c   Freeze-drying and embedding for thin sectioning

Methods which combine freeze-drying with embedding have been developed as alternatives to the standard chemical fixation and organic solvent dehydration techniques preparatory to ultramicrotomy. The aims are:

(i)   to achieve a more 'natural' preservation of the native specimen structure by avoiding or reducing artefacts caused by chemical fixation and solvent dehydration; and

(ii)   to avoid or reduce the extraction or displacement of labile or diffusible substances (e.g. during specimen fixation or chemical dehydration).

It was soon realised that this alternative procedure involves its own set of artefacts (for reviews of early developments see Sjöstrand 1967; Rebhun 1972). However, freeze-drying in combination with plastic embedding offers unique possibilities for special electron microscopical applications such as autoradiography of water-soluble substances (Stirling and Kinter 1967; Williams 1977; Pfaller 1979; Edelmann 1980b); cytochemical analysis (Gersh 1973); ion localisation by X-ray microanalysis (Chandler 1973, 1977; Ingram et al. 1974; Edelmann 1978a, 1980a, 1981; Dudek et al. 1981); laser microprobe mass analysis (Edelmann 1980a, 1981); and investigations of the molecular structure of cell compartments (Sjöstrand and Kretzer 1975). As with most other low temperature techniques, the success of these preparation methods depends primarily on optimal cryofixation of the specimen and many past failures can be attributed to this problem. Again this will be more of a limitation for high resolution studies, where the aim is optimal preservation of ultrastructure, than for cytochemical or microanalytical investigations. Finally, infiltration of the frozen-dried specimen with the embedding medium can cause considerable structural damage, some of which may be attributable to surface tension forces created by the advancing plastic (Hanzon and Hermodsson 1960). For these reasons, in many studies the dried specimens are fixed before embedding. This can be done by exposing the specimen to OsO_4 or aldehyde (e.g. glutaraldehyde or formaldehyde) vapour. It is obvious that vapour fixation is contrary to the original concept of using cryofixation as an alternative to chemical fixation but it may be tolerable for some experimental approaches (e.g. localisation of diffusible substances).

A reliable procedure involving a simple apparatus for freeze-drying and embedding small pieces of unfixed biological tissue was described in §5.3.2d. Using pieces (<0.1 mm^3) of frog sartorius muscle or human blood platelets as model systems, it was convincingly demonstrated that, with optimised freeze-drying and embedding methods, the quality of structural preservation is directly related to the effectiveness of the freezing stage (Edelmann 1978a,b, 1979). As summarised by the author, the following steps seem to be particularly important in ensuring good structural preservation of chemically unfixed biological specimens:

(*i*)   'Pseudovitrification' of the specimen area of interest.

(*ii*)   Freeze-drying *below* the recrystallisation temperature.

(*iii*) The embedding procedure should be performed at low temperature (see also §7.5). Using Spurr's resin (Spurr 1968) as an embedding medium, a temperature rise from 258 to 293 K leads to a much poorer preservation of muscle tissue structure. Similarly, Edelmann has shown that better structural preservation of the unfixed specimen was obtained when the polymerisation was carried out at 40°C (313 K) in comparison to 70°C (343 K). The recently introduced resins, specifically designed for low temperature infiltration and polymerisation, should prove ideal for such applications (§7.5).

Although it might be expected that embedding at reduced pressure would lead to better infiltration of the frozen-dried specimen, Edelmann (1979) obtained improved structural preservation when infiltrating at atmospheric pressure. The schedule for embedding pieces of frozen-dried tissue in Spurr's resin is given in §5.5.2c and for the low temperature resins in §7.5. At present this technique represents the simplest approach for achieving optimal structural preservation of unfixed, frozen-dried specimens. An example of the results is provided in Fig. 5.20.

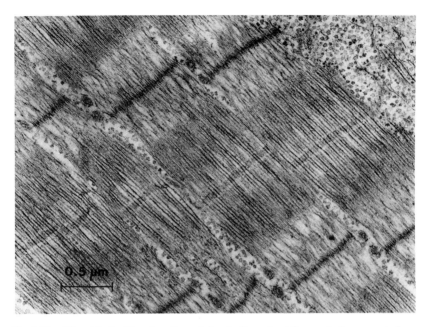

Fig. 5.20. Ultrathin (<40 nm) section of chemically unfixed frog sartorious muscle freeze-dried, embedded in Spurr's resin, and then stained with 5% uranyl acetate (5 min) and lead citrate (5 min). (Micrograph kindly provided by L. Edelmann and reproduced with his permission.)

The technique of directly infiltrating the frozen-dried specimen with resin has the advanatage that, once the frozen specimen and frozen resin have been put into the drying chamber, very few subsequent manipulations are required. To avoid bubbling of the resin on warming up (especially if using vacuum infiltration), it must be degassed before freezing. This can be done directly in the freeze-drying unit before cooling the specimen stage and introducing the specimen or in a separate vacuum apparatus. The latter procedure has the advantage that the frozen specimen and the resin beads can be put into the freeze-drying chamber simultaneously. If the frozen-dried specimen is to be vapour fixed then this must be done at a temperature below that of the melting point of the resin. Alternatively, the embedding medium can first be degassed under vacuum and then poured over the frozen specimen from a side-arm of the main vacuum chamber. Using this system, the dried specimen can be vapour fixed at any appropriate temperature (Sjöstrand 1967; Sjöstrand and Kretzer 1975; Pfaller 1979).

For *in situ* vapour fixation, the frozen-dried specimen has to be sealed from the vacuum of the main chamber. Different techniques have been used and Fig. 5.21 shows one such design (Pfaller 1979) which is commercially available (see Appendix 2) and which may also be suitable for adaptation to other types of freeze-drying unit. It consists of receptacles for both the resin and the fixative (e.g. OsO_4) and a flexible (vacuum lead-through) rod with a glass flange (at its lower end) which can be connected to the specimen table. Through the connecting tubes, the specimen can either be exposed to the fixative vapour or infiltrated with the resin which has been previously degassed with a rotary pump. Fixation with OsO_4 at reduced pressure is usually carried out for one to several hours. At atmospheric pressure, longer

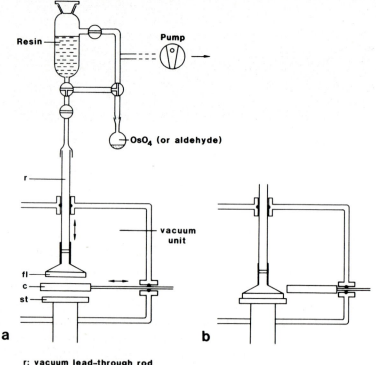

r: vacuum lead-through rod
fl: glass flange
c: condenser (LN$_2$-cooled on vacuum lead-through rod)
st: specimen stage

Fig. 5.21.   Apparatus (Balzers BUO 20 004-T; see Appendix 2) for vacuum embedding and vapour fixation of frozen dried specimens. (a) Position of the components during freeze-drying. (b) Position during exposure to the fixative vapour and infiltration with resin. (Diagram redrawn and reproduced by courtesy of Balzers AG.)

fixation times would be required (e.g. 24 h; Rebhun 1972). Postfixation can be performed (e.g. for one hour) with formaldehyde vapour, which is generated by heating paraformaldehyde crystals to 80°C (353 K) (Bondareff 1967).

For freeze-drying and embedding, special specimen tables have been designed (Hanzon and Hermodsson 1960; Sjöstrand and Kretzer 1975). It is particularly convenient for the embedding procedure if the frozen specimens can be placed (under liquid nitrogen) directly in gelatin capsules (e.g. 5.8 mm in diameter) which fit into indentations in the holder (see Fig. 5.22). To avoid a critical temperature rise during the transfer of the table onto the cold stage, specimens are best covered with liquid nitrogen. After the specimens have been infiltrated with the resin, they are removed from the freeze-drying unit for the final polymerisation step. Embedding of frozen-dried specimens can also be carried out outside the drying chamber but, especially with the more hygroscopic unfixed specimens, care will be required to avoid rehydration.

For the analysis of membrane structures, attempts have been made to combine freeze-drying with embedding at low temperatures (Sjöstrand and Kretzer 1975). Using this procedure, the material is first stabilised by means of brief cross-linking with glutaraldehyde and is subsequently infiltrated with 30% ethylene glycol to prevent ice crystallisation during freezing. After freeze-drying at −80°C (193 K), specimens are infiltrated with pre-evacuated hydroxypropyl methacrylate at −30°C (243 K). Polymerisation by means of ultraviolet radiation is performed at −27°C (246 K). It remains to be seen whether the polar and non-polar methacrylate-based low temper-

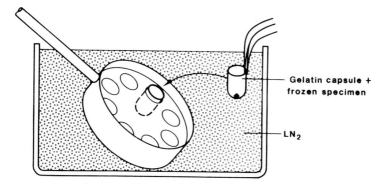

Gelatin capsule + frozen specimen

LN$_2$

Fig. 5.22.   A typical specimen holder for freeze-drying and vacuum embedding. The holder can conveniently be made from aluminium.

ature resins developed by Carlemalm and collaborators (§7.5) will provide similar or improved membrane images.

Sections of frozen-dried, unfixed embedded specimens can be stained by floating on uranyl acetate (e.g. 5% uranyl acetate, 5 min) and lead citrate (Reynolds 1963) (see Fig. 5.20). For microanalytical studies, it is important to ensure that the benefit of trapping diffusible specimen components by freeze-drying is not lost during the infiltration process (Ingram et al. 1974). In this respect, structural preservation may even have to be sacrificed in favour of reduced translocation of diffusible specimen constituents (e.g. electrolytes). This is certainly a field in which only systematic work on defined model systems will help to resolve the numerous questions.

5.5.2d   Freeze-drying of cryosections

Freeze-drying of cryosections (§4.3.2) may be performed either at atmospheric pressure or under vacuum (Christensen 1971; Appleton 1974; Barnard and Sevéus 1978). In either procedure, the dry sections (cut without a trough liquid) are first picked up on coated grids. Drying at atmospheric presure is most conveniently done directly in the cold nitrogen atmosphere of the cryochamber with the specimen grid attached to a temperature-controlled cold block, or within a holder containing molecular sieve. Drying under vacuum involves a contamination-free transfer of the cryosection from the cryomicrotome to the pre-cooled stage of the freeze-drying unit (Spriggs and Wynne-Evans 1976; Geymayer et al. 1977). During this transfer, special care is required to maintain the specimen below the critical recrystallisation temperature. After freeze-drying (for drying times see Table 5.2 in §5.2.1a), the sections can be carbon-coated within the drying unit to reduce rehydration from atmospheric moisture during subsequent transfer into the electron microscope (Zingsheim 1984). Zingsheim (1984) also found that the sublimation rates of ice from cryosections in a cryoultramicrotome were about four to five orders of magnitude slower that they would be under vacuum. This means that there is less worry that such cryosections will inadvertently freeze-dry.

Frozen-dried cryosections of most unfixed biological specimens are not suitable for high resolution morphological studies, but they are of unique value for depicting the distribution of diffusible substances (such as electrolytes or specimen components which are lost during chemical fixation and dehydration with organic solvents as used for conventional thin sectioning procedures) by X-ray microanalysis (Baker and Appleton 1976; Chandler

1977; Barnard 1982), autoradiography (Frederik and Klepper 1976; Baker and Appleton 1976) and ion microscopy (Stika et al. 1980).

5.6 Freeze-drying artefacts and problems of specimen interpretation

As already indicated in the previous sections, specimens can show great differences in their 'sensitivity' to the individual steps of freeze-drying procedures. The most common reasons for artefact formation are:

5.6.1 Fixation artefacts

Specimen fixation may be required as a preliminary step for freeze-drying of bulk specimens. In particular, glutaraldehyde fixation is used to make specimens more resistant to damage by ice crystals during freezing and to shrinkage following ice sublimation during drying (§5.5.2). The structural changes produced by fixation are listed in Table 2.12 in §2.3.2, but several of these artefacts will not be revealed after freeze-drying.

5.6.2 Adsorption artefacts

Strong electrostatic interactions between very flexible specimens (e.g. membrane vesicles) and the support (e.g. positively charged, coated grids or mica; §5.5.1a) may induce collapse or distortion during adsorption (Nermut and Williams 1977; Nermut 1981a).

In contrast, collapse may aid better structural preservation during the subsequent preparation steps (Trinick and Elliott 1982).

5.6.3 Washing artefacts

Since (pseudo)eutectic layers generated around ice crystals during freezing sublime at a lower rate than the ice, the remaining eutectic network, together with the non-volatile debris, can mask the frozen-dried specimen (Miller et al. 1983). To reveal 'non-contaminated' surfaces during freeze-drying, specimens must be rinsed in distilled water or a volatile buffer before freezing. Especially if unfixed specimens are used, this technique can lead to artefacts (lysis, structural alterations, extraction of constituents, etc.) through change in pH, temperature differences, osmotic shock and differences in ion distribution. Therefore, washing should be for as short a time as possible. If specimens were originally suspended in solutions of high salt or sugar content (e.g. from a sucrose gradient), repeated washing in (double) distilled water will be required to ensure the complete removal of non-volatile components. On the other hand, suspended specimens mounted by adsorption (§5.5.1a) may become detached during washing with distilled water due to changes of ionic strength. This can be particularly important if statistical data are to be accumulated. A parallel preparation using negative-staining methods after each preparation (washing) step will detect this 'desorption' artefact.

5.6.4 Air-drying artefacts

With most specimens, the surface tension forces created at the vapour/liquid (water) interface during air-drying cause severe structural changes (e.g. collapse, disruption, aggregation, dimensional changes). Specimens dried by solvent evaporation generally appear smaller than the equivalent frozen-dried specimens (Fig. 5.23). Air-drying can occur in the interval between specimen mounting and freezing or during transfer of the frozen specimen to the cold stage of the freeze-drying unit, where drying is preceded by specimen melting. Special care is required with specimens that are frozen as a thin layer (§5.5.1a). With bulk specimens (§5.5.2a) it must be ensured that a thin film of water (or other volatile solvent) covers the specimen surface at the moment of freezing.

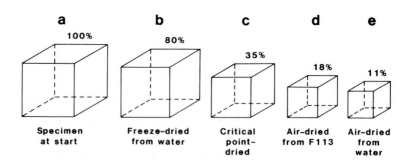

Fig. 5.23. Diagram, drawn to scale, showing cubes of tissue (a) frozen-dried from water (b), critical point-dried (c), air-dried from a volatile non-polar solvent (trichlorotrifluoroethane, Freon 113) (d), and from water (e). Data were obtained from the results of Boyde and Franc (1981) and from Boyde and Maconnachie (1979a,b). (Redrawn from Boyde and Franc 1981.)

5.6.5 Freezing and recrystallisation artefacts

As discussed in detail in Chapter 2, the formation of ice crystals during freezing may cause severe morphological changes. Similar artefacts can occur if initially well-frozen specimens exceed the recrystallisation temperature during the transfer to the cold stage of the freeze-drying unit or during the period of drying (e.g. due to radiation heating and/or a poor thermal contact betwen the specimen and the cold stage). Ice crystal formation in bulk specimens will only be disturbing if crystals grow through the specimen surface (ice crystal puncture artefacts) or if fracture faces are examined by freeze-drying techniques.

Depending on the specimen texture, shape and size, thermal stress cracking may occur during freezing. This artefact, which can be expected particularly with large specimens, is caused by the differential dimensional changes in ice and ice-encapsulated water at various temperatures through the thickness of the specimen (Boyde 1978a). Thermal stress cracking may not only happen with hydrated specimens but also on freezing specimens that have had their water replaced with organic solvents. The only solution is to reduce the specimen size.

At too high freeze-drying temperatures, melting of the eutectic surrounding the ice crystals may occur. For most biological specimens the critical temperature for eutectic melting is close to $-35°C$ (238 K) and thus far above the critical recrystallisation temperature. Eutectic melting occurs throughout the frozen portion of the specimen and liberates large volumes of liquid. This means that evaporation of water occurs simultaneously with the sublimation of ice. In common with air-drying of non-frozen specimens, the surface tension forces created at the interface betwen the vapour phase and the molten eutectic can cause severe collapse phenomena (MacKenzie 1972, 1975a; Boyde 1978b).

5.6.6 Dehydration artefacts

Several kinds of artefact can be attributed to the freeze-dehydration stage. Collapse phenomena (frequently associated with the shrinkage of bulk specimens) represent the main structural change that can be expected. Valuable information about the mechanisms involved has been obtained from model systems and, particularly, by following the freeze-drying process by observation with specially designed light microscopes or in the SEM (MacKenzie 1965, 1975a, 1976; Nei and Fujikawa 1977a,b; Boyde 1978b; Boyde and

Franc 1981). It was shown that the direct vacuum sublimation of ice and the disappearance of water vapour through condensed amorphous barriers, leaves the specimen structure essentially intact. Cracking, collapse and shrinkage have been shown to occur predominantly *after* the passage of the freeze-drying front during the desorption of water (e.g. during the release of unfrozen water from protein). At this point it should be remembered that desorption behaviour is dependent on specimen temperature and water vapour pressure in the specimen area (relative humidity). Thus, shrinkage artefacts due to desorption should be avoidable if the partly frozen-dried specimen can be held at the temperature at which freeze-drying is conducted and at 100% relative humidity with respect to that temperature (Boyde 1978a). Artefact problems (shrinkage, wrinkling, folding and collapse) related to the freeze-drying of viruses and other small biological structures have been reviewed by Nermut (1977, 1981a). In practice, these small specimens exhibit less shrinkage when frozen-dried at $-90°C$ (183 K) than at $-70°C$ (203 K). Similarly, cytoskeleton structures do not appear to collapse

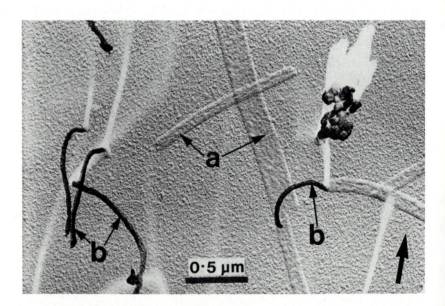

Fig. 5.24.   Frozen-dried, platinum/carbon shadowed self-assembly products (obtained by *in vitro* reaggregation of bacterial cell wall glycoproteins). Only the hollow protein tubes which completely collapsed (**a**) onto the support are well preserved. The non-attached parts (**b**) are extensively shrunken. Arrow indicates the direction of shadowing. (Micrograph by P. Messner and U.B. Sleytr, unpublished.)

very much when frozen-dried at −90°C but became distorted if warmed much above −80°C (Heuser and Kirschner 1980). Furthermore, shadowing and replication of frozen-dried specimens is best done with the specimen still at this, or even lower, temperature. Less rigid structures, such as bacteriophages (polyheads) or cylindrical protein self-assembly structures (Kistler and Kellenberger 1977; Kistler et al. 1977; Sleytr 1978; Kellengberger and Kistler 1979) may collapse during adsorption, during freeze-drying, or during shadowing and replication. A pseudoreplica of a frozen-dried self-assembly cylinder composed of bacterial cell wall surface proteins (Sleytr 1978; Sleytr et al. 1982) partially collapsed onto the support is illustrated in Fig. 5.24. The non-attached part of the cylinder is extensively shrunken and shows less structural detail than the part that has collapsed onto the support film. Such an elevated, dehydrated, protein framework can be expected to have poor thermal conductivity and frequently it may collapse or even disintegrate from the heat load imposed during shadowing and replication. Similar effects are known from freeze-fracture replication techniques and should be considered in this context (§6.8.3). Even without this energy input, extended, protruding areas of frozen-dried specimens (e.g. bacterial flagella) are likely to vibrate under vacuum due to thermally

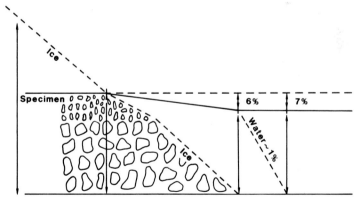

Fig. 5.25. Schematic diagram showing the changes in linear dimensions of specimens during freeze-drying. At the beginning, at the left, ice is outside the specimen. The specimen reaches the original true specimen dimensions at the second vertical arrow from the left. These dimensions are reduced by 6% linear when all the ice has sublimed out of the specimen and by 7% when the remaining firmly bound water has left the specimen. The diagram indicates that the rate of shrinkage may be related to the pore volume within the tissue created by the ice crystals which are larger at deeper levels within the specimen. However, this different rate of shrinkage may also be related to the change in the mass of the ice within the specimen. A smaller mass of ice would cause less self-cooling by sublimation. (Redrawn from Boyde and Franc 1981.)

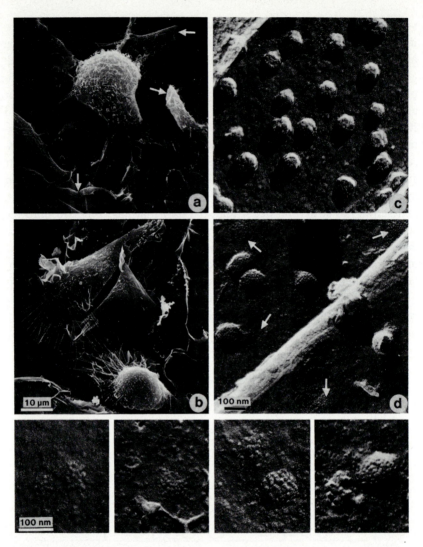

Fig. 5.26. Structural features of cells infected with mouse leukemia virus as visualised by freeze-drying (a, d) and critical point-drying (b, c). (a) Cells freeze-dried for SEM are reasonably well preserved. Some fractures and cracking occur (arrows) and the protrusions and fibres seen in (b) are largely destroyed. (b) Cells critical point-dried for SEM show well-preserved protrusions and fibres. Again, the viruses are clearly visible. (c) After critical-point drying for TEM, the viral surface projections (knobs) are still faintly visible but early budding stages (knobs assembly) are rarely recognisable. Resolution is hampered by the dehydration process involving organic solvents. (d) In comparison with critical point-drying (c), freeze-drying clearly gives better structural resolution for TEM work. Knobs, even in early budding stages (arrows), are easily observable (shown more clearly at higher magnification in the lower micrographs). In conclusion, for this specimen, freeze-drying is the method of choice for TEM while critical point-drying is better for SEM and overall preservation. (Micrographs kindly provided by H. Frank and reproduced with his permission.)

induced movement (Anderson 1954). This may either lead to an irreversible collapse onto the support or may cause the formation of blurred 'shadows' (Kellenberger and Kistler 1979). However, collapse (flattening of structures) may be advantageous if regular arrays of the macromolecules which make up hollow structures are to be evaluated by image analysis methods (Kistler et al. 1977). Chemical fixation (e.g. with glutaraldehyde) can render flexible structures more resistant to collapse (Nermut 1981a).

Especially with bulk specimens, an overall dimensional change can be expected during freeze-drying. Volume changes of biological samples during different preparation procedures for the SEM, including air-drying from water, air-drying from organic solvents, critical point-drying and freeze-drying have been carefully analysed by Boyde and his colleagues using direct, quantitative on-line measurements (Boyde 1978b, 1980; Boyde and Maconnachie 1979a,b, 1980; Boyde and Franc 1981). A variety of biological samples (mostly glutaraldehyde-fixed) have been shown to shrink linearly by about 7% (20% by volume) during freeze-drying (Campbell and Roach 1983). As can be seen from Fig. 5.23, this is still considerably less than can be expected from the alternative dehydration procedures. Using 1.0 mm cubes of glutaraldehyde-fixed mouse liver as a model system, it was shown that most of this shrinkage ($\simeq 6\%$ linear) is associated with 'secondary' drying when the firmly bound water, which can be up to about 10% of the specimen dry weight, is removed (Fig. 5.25).

Freeze-drying artefacts may also arise from the pressure- and temperature-dependent expansion of gases dissolved in the specimen water prior to freezing. Finally, at too high freeze-drying temperatures, macromolecular particles may be lost from the specimen surface or from the withdrawing ice front along with the dynamic stream of subliming water molecules (water vapour wind) (Anderson 1954, 1956) and isolated particles that have been liberated from the ice may diffuse laterally on the frozen surface and re-arrange into aggregates (Lepault and Dubochet 1980).

5.6.7 Rehydration artefacts

Frozen-dried specimens must be warmed up to ambient temperature before breaking the vacuum of the freeze-drying unit so that water condensation is avoided. Most frozen-dried specimens are very hygroscopic. A potential source of specimen rehydration is the relatively high water vapour pressure that is inevitably near to the surface of any manipulator (forceps or other handling devices). Specimens are therefore either transferred immediately into the microscope or are stored in a good desiccator until use. Rehydration can be slowed down by coating the specimen with a thin layer of carbon.

5.6.8   Shadowing and/or replication artefacts

Structural alterations caused by thermal load during shadowing and replication are discussed in §6.4 and §6.8.3.

5.6.9   Freeze-drying: structural comparisons with other dehydration procedures

Comparative studies on bulk specimens have shown that specimen shrinkage during critical point-drying is always greater than that from freeze-drying (Boyde 1978a,b, 1980; Boyde and Maconnachie 1979a) (see Fig. 5.23). Nevertheless, for many specimens it appears useful to compare the structural preservation achieved by both techniques. Fig. 5.26 shows an example of structural differences in leukaemic cells after freeze-drying or critical point-drying. Whereas protrusions and fibres from the virus-producing cells appear to be better preserved after critical point-drying, the different budding stages are more clearly depicted in replicas of frozen-dried cells (Frank, personal communication).

References

Albrecht, R.M., D.H. Rasmussen, C.S. Keller and R.D. Hindsill (1976), Preparation of cultured cells for electron microscopy: air drying from organic solvents, J. Microscopy *108*, 21.

Anderson, T.F. (1951), Techniques for the preservation of three-dimensional structure in preparing specimens for the electron microscope, Trans. N.Y. Acad. Sci. *13*, 130.

Anderson, T.F. (1954), Some fundamental limitations to the preparation of three-dimensional specimens for the electron microscope, Trans. N.Y. Acad. Sci. *16*, 242.

Anderson, T.F. (1956), Electron microscopy of micro-organisms, in: Physical technique and biological research, Vol. 3, G. Oster and A.W. Polister, eds. (Academic Press, New York), p. 177.

Anderson, T.F. (1966), Electron microscopy of micro-organisms, in: Physical technique and biological research, Vol. 3, A.W. Polister, ed. (Academic Press, New York), p. 319.

Appleton, T.C. (1973), Cryomicrotomy; possible applications in cytochemistry, in: Electron microscopy and cytochemistry, E. Wisse, W.Th. Deams, I. Molenaar and P. van Duijn, eds. (North-Holland Publishing Company, Amsterdam), p. 299.

Appleton, T.C. (1974), A cryostat approach to ultrathin "dry" frozen sections for electron microscopy: a morphological and X-ray analytical study, J. Microscopy *100, 49*.

Backus, R.C. and R.C. Williams (1950), The use of spraying methods and of volatile suspending media in the preparation of specimens for electron microscopy, J. appl. Phys. *21*, 11.

Baker, J.R.J. and T.C. Appleton (1976), A technique for electron microscope autoradiography (and X-ray microanalysis) of diffusible substances using freeze-dried fresh frozen sections, J. Microscopy *108*, 307.

Barnard, T. (1982), Thin frozen-dried cryosections and biological X-ray analysis, J. Microscopy *126*, 317.

Barnard, T. and L. Sevéus (1978), Preparation of biological material for X-ray microanalysis of diffusible elements, J. Microscopy *112*, 281.

Bendet, I.J. and N.I. Rizk (1974), Thermoelectric freeze-drying, J. Microscopy *101*, 311.

Blaschke, R. and A. Boyde (1978), Particle size in conductive coating and resolution in the SEM, Scanning *1*, 64

Bondareff, W. (1967), Demonstration of an intercellular substance in mouse cerebral cortex, Z. Zellforsch. *81*, 366.

Boyde, A. (1972), Biological specimen preparation for the scanning electron microscope: an overview, Scanning Electron Microscopy 1972, 257.

Boyde, A. (1974a), Freezing, freeze-fracturing and freeze-drying in biological specimen preparation for the SEM, Scanning Electron Microscopy 1974, 1063.

Boyde, A. (1974b), Volumenänderung biologischer Gewebe bei Entwässerung und Trocknung, Beitr. elektronenmikr. Direktabb. Oberfl. *7*, 231.

Boyde, A. (1975), Freeze-drying in the preparation of animal tissue and cell specimens for SEM, in: Publication no. P1 (Edwards High Vacuum Ltd., Crawley).

Boyde, A. (1976), Do's and don't's in biological specimen preparation for the SEM, Scanning Electron Microscopy 1976, *1*, 683.

Boyde, A. (1978a), Pros and cons of critical point drying and freeze-drying for SEM, Scanning Electron Microscopy 1978, *2*, 303.

Boyde, A. (1978b), Volumenandesungen biologischer Garebe bei Eutwasserung und Trocknung, Beitr. elektronenmikr. Direktabb. Oberfl. *11*, 231.

Boyde, A. (1980), Review of basic preparation techniques for biological scanning electron microscopy, Proc. 7th. Eur. Reg. Conf. Electron Microscopy, The Hague *2*, 768.

Boyde, A. and P. Echlin (1973), Freezing and freeze-drying: a preparative technique for SEM, Scanning Electron Microscopy 1973, 759.

Boyde, A. and F. Franc (1981), Freeze-drying shrinkage of glutaraldehyde fixed liver, J. Microscopy *122*, 75.

Boyde, A. and E. Maconnachie (1979a), Volume changes during preparation of mouse embryonic tissue for scanning electron microscopy, Scanning *2*, 149.

Boyde, A. and E. Maconnachie (1979b), Freon 113 freeze-drying for scanning electron microscopy, Scanning *2*, 164.

Boyde, A. and E. Maconnachie (1980), Treatment with lithium salts reduced ethanol dehydration shrinkage of glutaraldehyde fixed tissue, Histochemistry *66*, 181.

Boyde, A. and C. Wood (1969), Preparation of animal tissues for surface SEM, J. Microscopy *90*, 221.

Butler, E.P. and K.F. Hale (1981), Dynamic experiments in the electron microscope, in: Practical methods in electron microscopy, Vol. 9, A.M. Glauert, ed. (Elsevier/North-Holland, Amsterdam).

Campbell, G.J. and M.R. Roach (1983), Dimensional changes associated with freeze-drying of the internal elastic lamina from cerebral arteries, Scanning *5*, 137.

Chandler, J.A. (1973), The use of wavelength dispersive X-ray microanalysis in cytochemistry, in: Electron microscopy and cytochemistry, E. Wisse, W.Th. Deams, I. Molinaar and P. van Duijn, eds. (North-Holland, Amsterdam), p. 203.

Chandler, J.A. (1977), X-ray microanalysis in the electron microscope, in: Practical methods in electron microscopy, Vol. 5, A.M. Glauert, ed. (Elsevier/North-Holland, Amsterdam).

Chandler, J.A. and S. Battersby (1976), X-ray microanalysis of ultrathin frozen and freeze-dried sections of human sperm cells, J. Microscopy *107*, 55.

Christensen, A.K. (1971), Frozen thin sections of fresh tissue for electron microscopy, with a description of pancreas and liver, J. Cell Biol. *51*, 772.

Cohen, A.L., D.P. Marlow and G.E. Garner (1968), A rapid critical point method using fluor-ocarbons (Freons) as intermediate and transitional fluids, J. Microscopie *7*, 331.

Davy, J.G. (1971), Vacuum microbalance apparatus for rapid determination of low-temperature vapourization rates, Vac. Microbalance Tech. *8*, 155.

Davy, J.G. and D. Branton (1970), Subliming ice surfaces: freeze-etch electron microscopy, Science *168*, 1216.

Derrick, K.S. (1973), Quantitative assay for plant viruses using serologically specific electron microscopy, Virology *56*, 652.

Dudek, R.W., A.F. Boyne and N. Freinkel (1981), Quick-freeze fixation and freeze-drying of isolated rat pancreatic islets, J. Histochem. Cytochem. *29*, 321.

Echlin, P., B. Chapman, L. Stoter, W. Gee and A. Burgess (1982), Low voltage sputter coating, Scanning Electron Microscopy 1982, *1*, 29.

Edelmann, L. (1978a), Visualization and X-ray microanalysis of potassium tracers in freeze-dried and plastic embedded frog muscle, Microscopica Acta Suppl. *2*, 166.

Edelmann, L. (1978b), A simple freeze-drying technique for preparing biological tissue without chemical fixation for electron microscopy, J. Microscopy *112*, 243.

Edelmann, L. (1979), Freeze-drying of chemically unfixed biological material for electron microscopy, Mikroskopie *35*, 31.

Edelmann, L. (1980a), Preferential localized uptake of K^+ and Cs^+ over Na^+ in the A-band of freeze-dried embedded muscle section: detection by X-ray microanalysis and laser microprobe mass analysis, Physiol. Chem. Phys. *12*, 509.

Edelmann, L. (1980b), Potassium binding sites in muscle: electron microscopic visualization of K, Rb and Cs in freeze-dried preparations and autoradiography at liquid nitrogen temperature using ^{86}Rb and ^{134}Cs, Histochemistry *67*, 233.

Edelmann, L. (1981), Selective accumulation of Li^+, Na^+, K^+, Rb^+ and Cs^+ at protein sites of freeze-dried and embedded muscle detected by LAMMA, Fresenius Z. Anal. Chem. *308*, 218.

Fischer, K.A. (1975), "Half" membrane enrichment: verification by electron microscopy, Science *190*, 983.

Fowler, W.E. and U. Aebi (1983), Preparation of single molecules and supramolecular complexes for high resolution shadowing, J. Ultrastruct. Res. *83*, 319.

Frederik, P.M. and D. Klepper (1976), The possibility of electron microscopic autoradiography of steroids after freeze-drying of unfixed testes, J. Microscopy *106*, 209.

Gersh, I. (ed.) (1973), Submicroscopic cytochemistry. I. Proteins and nucleic acids, (Academic Press, London), p. 1.

Geymayer, W., F. Grasenick and Y. Hodl (1977), Stabilising ultrathin cryosections by freeze-drying, J. Microscopy *112*, 39.

Glauert, A.M. (1974), Fixation, dehydration and embedding of biological specimens, in: Practical methods in electron microscopy, Vol. 3, A.M. Glauert, ed. (North-Holland, Amsterdam).

Goldblith, S.A., L. Rey and W.W. Rothmayr (eds.) (1975), Freeze-drying and advanced food technology (Academic Press, London).

Gregory, D.W. and B.J.S. Pirie (1972), Wetting agents for electron microscopy of biological specimens, Proc. 5th Eur. Reg. Conf. Electron Microscopy, Manchester, p. 234.

Hanzon, V. and L.H. Hermodssen (1960), Freeze-drying of tissue for light and electron microscopy, J. Ultrastruct. Res. *4*, 332.

Hayat, M.A. (1981), Fixation for electron microscopy (Academic Press, New York).

Hayat, M.A. and B.R. Zirkin (1973), Critical point drying method, in: Principles and techniques of electron microscopy, Vol. 3, M.A. Hayat, ed. (Van Nostrand Reinhold, New York), p. 297.

Heckly, R.J. and D. Skilling (1979), Device to facilitate freeze-drying specimens for scanning electron microscopy, Cryobiology *16*, 196.

Heuser, J.E. (1980), 3-D visualization of coated vesicle form, J. Cell Biol. *84*, 560.

Heuser, J.E. (1981), Preparing biological samples for stereomicroscopy by quick-freeze, deep-etch, rotary replication technique, in: Methods in cell biology, Vol. 3, J.N. Turner, ed. (Academic Press, New York), p. 97.

Heuser, J.E. (1983), Procedure for freeze-drying molecules adsorbed to mica flakes, J. mol. Biol. *169*, 155.

Heuser, J.E. and M.W. Kirschner (1980), Filament organisation revealed in platinum replicas of freeze-dried cytoskeletons, J. Cell Biol. *86*, 212.

Heuser, J.E. and S.R. Salpeter (1979), Organization of acetylcholine in quick-frozen, deep-etched, and rotary replicated *Torpedo* post-synaptic membrane, J. Cell Biol. *82*, 150.

Humphreys, W.J. (1975), Drying soft biological tissue for scanning electron microscopy, Scanning Electron Microscopy 1975, 707.

Humphreys, W.J., B.O. Spurlock and J.S. Johnson (1974), Critical point drying of ethanol infiltrated cryofractured biological specimens for scanning electron microscopy, Scanning Electron Microscopy 1974, *1*, 275.

Ingram, F.D., M.J. Ingram and C.A.M. Hogben (1974), An analysis of the freeze-dried, plastic embedded electron probe specimen preparation, in: Microprobe analysis as applied to cells, T. Hall, P. Echlin and R. Kaufmann, eds. (Academic Press, London, New York), p. 119.

Iwata, H. and S. Aita (1976), Freeze-drying technique for small biological materials using a modified freeze-etching device, J. Electron Microscopy *25*, 305.

Jakopić, E., A. Brunegger, R. Essl and G. Windisch (1978), A sputter source for electron microscopic preparation, Proc. 9th Int. Congr. Electron Microscopy, Toronto *1*, 150.

Kellenberger, E. and J. Kistler (1979), The physics of specimen preparation, in: Unconventional electron microscopy for molecular structure determination, W. Hoppe and R. Mason, eds. (Friedr. Vieweg and Sohn, Wiesbaden), p. 49.

Kelley, R.O., R.A.F. Dekker and J.G. Bluemink (1973), Ligand-mediated osmium binding: its application in coating biological specimens for scanning electron microscopy, J. Ultrastruct. Res. *45*, 254.

Kemmenoe, B.H. and G.R. Bullock (1983), Structure analysis of sputter-coated and ion-beam sputter-coated films: a comparative study, J. Microscopy *132*, 153.

Kistler, J. and E. Kellenberger (1977), Collapse phenomena in freeze-drying, J. Ultrastruct. Res. *59*, 70.

Kistler, J., Y. Aeb and E. Kellenberger (1977), Freeze-drying and shadowing a two-dimensional periodic specimen, J. Ultrastruct. Res. *59*, 76.

Klein, H.P. and W. Stockem (1976), Präparation biologischer Objekte für das Rasterelektronenmikroskop, Microscopica Acta *78*, 388.

Lepault J. and J. Dubochet (1980), Freezing, fracturing and etching artefacts in particulate suspensions, J. Ultrastruct. Res. *72*, 223.

MacKenzie, A.P. (1964), Apparatus for microscopic observations during freezing, Biodynamica *9*, 213.

MacKenzie, A.P. (1965), Factors affecting the mechanism of transformation of ice into water vapour in the freeze-drying process, Ann. N.Y. Acad. Sci. *125*, 522.

MacKenzie, A.P. (1967), The collapse phenomena in the freeze-drying process, Cryobiology *3*, 387.

MacKenzie, A.P. (1972), Freezing, freeze-drying and freeze-substitution, Scanning Electron Microscopy 1972, *2*, 273.

MacKenzie, A.P. (1975), Collapse during freeze-drying – qualitative and quantitative aspects, in: Freeze-drying and advanced food technology, S.A. Goldblith, L. Rey and W.W. Rothmayr, eds. (Academic Press, New York), p. 277.

MacKenzie, A.P. (1976), Principles of freeze-drying, Transplantation Proc. *8*, 181.

Marshall, A.T. (1980), Freeze-substitution as a preparation technique for biological X-ray microanalysis, Scanning Electron Microscopy 1980, *2*, 395.

Mazia, D., G. Schatten and W. Sale (1975), Adhesion of cells to surfaces coated with polylysine. Applications to electron microscopy. J. Cell Biol. *66*, 198.

Mellor, J.D. (1978), Fundamentals of freeze-drying, (Academic Press, London), p. 3.

Merymann, H.T. (1959), Sublimation freeze-drying without vacuum, Science *130*, 628.

Merymann, H.T. (1966), Freeze-drying, in: Cryobiology, H.T. Meryman, ed. (Academic Press, New York), p. 609.

Miller, K.R., C.S. Prescott, T.L. Jacobs and N.L. Lassignal (1983), Artefacts associated with quick-freezing and freeze-drying, J. Ultrastruct. Res. *82*, 123.

Moroika, H., H. Ozasa, T. Kishida and A. Suganuma (1980), A simple freeze-drying method for electron microscopy of bacteriophages, J. Electron Microscopy *29*, 64.

Müller, H.R. (1957), Gefriertrocknung als Fixierungsmethode an Pflanzenzellen, J. Ultrastruct. Res. *1*, 109.

Munger, B.L. (1977), The problem of specimen conductivity in electron microscopy, Scanning Electron Microscopy 1977, *1*, 481.

Murakami, T. (1978) Tannin-osmium conductive staining of biological specimens for noncoated scanning electron microscopy, Scanning *1*, 127.

Murphy, J.A. (1980), Non-coating techniques to render biological specimens conductive: 1980 update, Scanning Electron Microscopy 1980, *1*, 209.

Nanninga, N. (1968), The conformation of the 50S ribosomal subunit of *Bacillus subtilis*, Proc. natl. Acad. Sci. U.S.A. *61*, 614.

Nei, T. (1962), Freeze-drying in the electron microscopy of microorganisms, J. Electron Microscopy *11*, 185.

Nei, T. (1973), Some aspects of freezing and drying of microorganisms on the basis of cellular water, Cryobiology *10*, 403.

Nei, T. and S. Fujikawa (1977a), Observation of the freeze-drying process of biological materials with a scanning electron microscope, Dev. Biol. Stand. *36*, 243.

Nei, T. and S. Fujikawa (1977b), Freeze-drying process of biological specimens observed with a scanning electron microscope, J. Microscopy *111*, 137.

Nermut, M.V. (1973a), Freeze-drying for electron microscopy, Balzers High Vacuum Report, VB2.

Nermut, M.V. (1973b), Freeze-drying and freeze-etching of viruses, in: Freeze-etching, techniques and applications, E.L. Benedetti and P. Favard, eds. (Soc. Française de Microscopie Electronique, Paris), p. 135.

Nermut, M.V. (1977), Freeze-drying for electron microscopy, in: Principles and techniques of electron microscopy, Vol. 7, M.A. Hayat, ed. (Van Nostrand Reinhold Company, New York), p. 79.

Nermut, M.V. (1981a), Advanced methods in electron microscopy of viruses, in: Newer techniques in practical virology, C. Howard, ed. (A. Liss, New York).

Nermut, M.V. (1981b), Visualization of the "membrane skeleton" in human erythrocytes by freeze-etching, Eur. J. Cell Biol. *25*, 265.

Nermut, M.V. and H. Frank (1971), Fine structure of influenza A2 (Singapore) as revealed by negative staining, freeze-drying and freeze-etching, J. gen. Virol. *10*, 37.

Nermut, M.V. and L.D. Williams (1977), Freeze-fracturing of monolayers (capillary layers) of cells, membranes and viruses: some technical considerations, J. Microscopy *110*, 121.

Nermut, M.V. and L.D. Williams, (1980), Freeze-fracture autoradiography of the red blood cell plasma membrane, J. Microscopy *118*, 453.

Nermut, M.V., H. Frank and W. Schafer (1972), Properties of mouse leukemia viruses. III. Electron microscopic appearance as revealed after conventional preparation techniques as well as freeze-drying and freeze-etching, Virology *49*, 345.

Nietsch, W. and E. Jochem (1973), Gerüst- und Grenzflächenwiderstand bei der Gefriertrocknung dünner Schichten, Vakuum-Technik *22*, 73.

Osatake, H., K. Atoh, A. Mitsushima, K. Tanaka and H. Takahashi (1980), A simple method of freeze-drying for scanning electron microscopy, J. Electron Microscopy *29*, 72.

Pearse, A.G.E. (1964), Ein thermoelektrisches Gerät zum raschen Gefriertrocknen biologischer Gewebe, Die Kalte *17*, 75.

Peters, K.R. (1977), Stereo surface replicas of cultured cells for high resolution electron microscopy, J. Ultrastruct. Res. *61*, 115.

Peters, K.R. (1979), Scanning electron microscopy at macromolecular resolution in low energy mode on biological specimens coated with ultrathin metal films, Scanning Electron Microscopy 1979, *2*, 113.

Pfaller, W. (1979), Freeze-drying and vacuum embedding of spray fozen unicellular organisms, Mikroskopie *35*, 37.

Podolsky, M.V. and J.A. Konstantinov (1980), A study of the final period of freeze-drying and determination of residual moisture of dry biological materials, Cryobiology *17*, 585.

Read, P.L. (1963), Sorption pumping at high and ultra-high vacuum, Vacuum *13*, 271.

Rebhun, L.I. (1972), Freeze-substitution and freeze-drying, in: Principles and techniques of electron microscopy, Vol. 2, M. A. Hayat, ed. (Van Nostrand Reinhold Company, New York), p. 3.

Reimer, L. (1978), Scanning electron microscopy: present state and trends, Scanning *1*, 3.

Revel, J.P. (1978), Biological scanning electron microscopy for physicists and engineers, Scanning Electron Microscopy 1978, *1*, 829.

Revel, J.P. and K. Wolken (1973), EM investigations of the underside of cells in culture, Exp. Cell Res. *78*, 1.

Reynolds, E.S. (1963), The use of lead citrate at high pH as an electron-opaque stain in electron microscopy, J. Cell Biol. *17*, 208.

Rick, R., A. Dorge and K. Thurau (1982), Quantitative analysis of electrolytes in frozen dried sections, J. Microscopy *125*, 239.

Roberts, I.M. and G.M. Duncan (1981), A simple device for freeze-drying electron microscope specimens, J. Microscopy *124*, 295.

Schunzel, G. (1968), Uber den Einsatz der Peltier-Kaskade fur die biologische Gefriertrocknung, Acta Histochemica *30*, 201.

Severs, N.J. and R.M. Hicks (1977), Frozen-surface replicas of rat bladder luminal membrane, J. Microscopy *111*, 125.

Severs, N.J. and R.C. Warren (1978), Analysis of membrane structure in the transitional epithelium of rat urinary bladder. I. The luminal membrane, J. Ultrastruct. Res. *64*, 124.

Sjöstrand, F.S. (1967), Electron microscopy of cells and tissues, Vol. 1, Instrumentation and techniques, (Academic Press, New York, London), p. 188.

Sjöstrand, F.S. and F. Kretzer (1975), A new freeze-drying technique applied to the analysis of the molecular structure of mitochondria and chloroplast membranes, J. Ultrastruct. Res. *53*, 1.

Sleytr, U.B. (1978), Regular arrays of macromolecules on bacterial cell walls: structure, chemistry and function, Int. Rev. Cytol. *53*, 1.

Sleytr, U.B., P. Messner, P. Schiske and D. Pum (1982), Periodic surface structures on procaryotic cells, Proc. 10th. Int. Congr. Electron Microscopy, Hamburg *3*, 1.

Smith, P. R. (1980), Freeze-drying specimens for electron microscopy, J. Ultrastruct. Res. *72*, 380.

Sommer, J.R. (1977), To cationise glass, J. Cell Biol. *75*, 245a.

Spriggs, T.L.B. and D. Wynne-Evans (1976), Observations on the production of frozen-dried thin sections for electron microscopy using unfixed fresh liver, fast frozen without cryoprotectants, J. Microscopy *107*, 35.

Spurr, A.R. (1968), A low viscosity epoxy resin embedding medium for electron microscopy, J. Ultrastruct. Res. *26*, 31.

Steere, R.L. (1973), Preparation of high resolution freeze-etch, freeze-fracture, frozen surface and freeze-dried replicas in a single freeze-etch module, and the use of stereo electron microscopy to obtain maximum information from them, in: Freeze-etching, techniques and applications, E.L. Benedetti and P. Favard, eds. (Societe Française de Microscopie Electronique, Paris), p. 224.

Stephenson, J.L. (1960), Fundamental physical problems in the freezing and drying of biological materials, in: Recent research in freezing and drying, A.S. Parkes and A.U. Smith, eds. (Blackwell Scientific Publications, Oxford), p. 122.

Stika, K.M., L.K. Bielat and G.H. Morrison (1980), Diffusible ion localization by ion microscopy: a comparison of chemically prepared and fast-frozen, freeze-dried, unfixed liver sections, J. Microscopy *118*, 409.

Stirling, C.E. and W.B. Kinter (1967), High resolution radio autography of galactose-^{3}H accumulation in the hamster intestine, J. Cell Biol. *35*, 585.

Strasser, J., R. Heiss and P. Gorling (1966), The water vapour and heat transport during vacuum freeze-drying with reference to optimal process conditions, Kaltetechnik *18*, 286.

Stumpf, W.E. and L.J. Roth (1967), Freeze-drying of small tissue samples and thin frozen sections below −60°C, J. Histochem. Cytochem. *15*, 243.

Sweeney, L.R. and B.L. Shapiro (1977), Rapid preparation of uncoated biological specimens for scanning electron microscopy, Stain Technol. *52*, 221.

Trinick, J. and A. Elliott (1982), Effect of substrate on freeze dried and shadowed protein structure, J. Microscopy *126*, 151.

Umrath, W. (1977), Prinzipien der Kryopräparationsmethoden: Gefriertrocknung und Gefrierätzung, Mikroskopie *33*, 11.

Umrath, W. (1983), Berechnung von Gefriertrocknungszeiten für elektronenmikroskopische Praparation, Mikroskopie *40*, 9.

Walzthöny, D., M. Bähler, Th. Wallimann, H.M. Eppenberger and H. Moor (1983), Visualization of freeze-dried and shadowed myosin molecules immobilized on electron microscopic films, Eur. J. Cell Biol. *30*, 177.

Washburn, E.W. (1924), The vapour pressure of ice and of water below the freezing point, Monthly Weather Rev. *52*, 488.

Watters, W.B. and R.C. Buck (1971), An improved simple method of specimen preparation for replicas or SEM, J. Microscopy *94*, 185.

Wheeler, E.E., J.B. Gavin and R.N. Seelye (1975), Freeze-drying from tertiary butanol in preparation of endocardium for SEM, Stain Technol. *50*, 331.

Wildhaber, I., H. Gross and H. Moor (1982), The control of freeze-drying with deuterium oxide (D_2O), J. Ultrastruct. Res. *80*, 367.

Willemer, H. (1975), Low temperature freeze-drying of aqueous and non-aqueous solutions using liquid nitrogen, in: Freeze-drying and advanced food technology, S.A. Goldblith, L. Rey and W. Rothmayr, eds. (Academic Press, London), p. 461.

Williams, R.C. (1952), Electron microscopy of sodium deoxyribonucleate by use of a new freeze-drying method, Biochim. Biophys. Acta *9*, 237.

Williams, R.C. (1953), A method of freeze-drying for electron microscopy, Exp. Cell Res. *4*, 188.

Williams, M.A. (1977), Autoradiography and immunocy$«chemistry. in: Practical methods in electron microscopy, Vol. 6, A. M. Glauert, ed. (Elsevier/North-Holland, Amsterdam).

Willison, J.H.M. and A.J. Rowe (1980), Replica, shadowing and freeze-etching techniques, in: Practical methods in electron microscopy, Vol. 8, A.M. Glauert, ed. (Elsevier/North-Holland, Amsterdam).

Würtz, M., J. Kistler and T. Hohn (1976), Surface structure of *in vitro* assembled bacteriophage lambda polyheads, J. mol. Biol. *101*, 39.

Zingsheim, H.P. (1984), Sublimation rates of ice in a cryo-ultramicrotome, J. Microscopy *133*, 307.

Chapter 6

Freeze-fracture replication

6.1   Principles of freeze-fracture replication and freeze-etching techniques

Freeze-fracture replication and freeze-etching techniques both involve the production of a vacuum-deposited replica from a fracture plane in a frozen specimen. Generally, the cleaned replica is examined in a conventional transmission electron microscope, although methods are also available for studying the topography of the coated fracture face in a scanning electron microscope (§3.3) if a resolution of less than about 10 nm is sufficient. The more common term for 'freeze-fracture replication' is still 'freeze-etching' but because not all methods developed involve, or allow, sublimation of ice (etching) from the fracture face before replication, one should distinguish between 'freeze-fracture replication' or 'freeze-fracturing' and 'freeze-fracturing and etching' or 'freeze-etching' methods. In this Chapter we will use *freeze-fracture replication* as a term which covers *both* methods and *freeze-etching* for methods where controlled freeze-drying (etching) of the fracture face is an essential part of the schedule.

Nowadays, freeze-fracture replication is an established method in electron microscopy, especially in cell biology and membrane biology but also in some areas of materials science, such as colloid research. The main reasons for the remarkable expansions and success of this preparation technique since the early 1960s are that:

(*i*)   In freeze-fracture replication, chemical fixation is replaced by freeze-fixation, thus the molecular architecture of hydrated and in some instances even potentially living, cells can be examined.

(*ii*)   Although the precise fracture path within a cell or tissue is generally unpredictable, it preferentially follows membranes or other structurally weak planes, such as phase boundaries. In consequence, the three-dimensional information derived from freeze-fracture replicas has led to an understanding of the molecular architecture of cells, and particularly of membranes, which is not obtainable with any other method of specimen preparation.

Nevertheless, as with all other methods in electron microscopy, there are very few instances where the morphological details in a freeze-fracture replica represent the actual 'true-to-life' structure. Specific artefacts can occur at each preparation step and great efforts are still being made to learn more about their generation. However, since it is only a relatively beam-stable *replica*, containing information about the topography of the fracture face, that is examined in a transmission electron microscope, there are few problems arising from possible specimen beam damage.

The present-day freeze-fracture replication techniques are based on the pioneering work of Hall (1950), Meryman (1950) and Meryman and Kafig (1955), who were concerned with the production of vacuum-deposited replicas from the surfaces of frozen specimens. Subsequently, Steere (1957) modified and improved these methods. His approach was to freeze pieces of tissue in a cold chamber on a brass block. In the isolated chamber the specimen was cleaved, at atmospheric pressure, with a cold scalpel under a dissecting microscope and subsequently transferred to a high vacuum unit equipped with a cold stage and cold trap. After superficially freeze-drying the fracture face, a replica was produced by evaporating chromium and carbon onto it. Although, compared with present-day technology, Steere's method has its limitations, he was the first to apply the freeze-fracture replication and etching method successfully to biological specimens by demonstrating the structure of virus crystals within frozen cells.

The actual breakthrough in the application of freeze-fracture replication techniques, both for biological and non-biological material, came with the development of the rather sophisticated 'freeze-etching' apparatus described by Moor et al. (1961). The basic design consisted of an evaporation unit incorporating a liquid nitrogen-cooled microtome for specimen fracturing and a specimen cold stage which could be maintained at a precise temperature. The microtome, which was originally designed to produce frozen sections, failed in this purpose (Mühlethaler 1973) but, since it allowed controlled fracturing of a frozen specimen under vacuum for the first time, it proved to be superior to Steere's freeze-fracture process. The Moor and Mühlethaler equipment, the modern version of which, with a simpler microtome, is still commercially supplied by Balzers AG (see Appendix 2), allowed reliable, routine preparations to be made in many laboratories for the first time and contributed significantly to the widespread use and application of freeze-fracture techniques.

The complexity and price of the Moor device subsequently stimulated a number of electron microscopists to develop simpler instruments and apparatus to fulfil essentially similar functions. The most important simplifications and stimuli for further developments came from:

(*i*) applying simpler fracturing methods, not involving a complex microtome device, and consequently making possible the study of both faces of the fracture, i.e. the complementary fracture faces (Steere and Moseley 1969), and

(*ii*) showing that it is possible to look at fracture faces that have been obtained by freeze-fracturing under liquid nitrogen outside an evaporation unit (Bullivant and Ames 1966; Geymayer 1966; Bullivant et al. 1968).

More recent technical developments, which have again led to more complex equipment, have been stimulated by attempts to work with maximum automation, flexibility, and reproducibility as well as by efforts to reduce condensation of contaminants on specimen fracture faces before replication, particularly at very low temperatures. Contamination can be reduced, or even eliminated, by using either highly sophisticated ultra-high vacuum technology (Gross et al. 1978a) or by using very efficient cold traps (cold shrouds) around the specimen area (Steere 1969; Sleytr and Umrath 1976; Umrath 1978). (Cold shrouds are sometimes termed 'optically tight'. This implies that there is no direct line of sight through them and, therefore there can be no direct access by a molecule or particle moving in a straight line.)

Despite the wide variety of methods and equipment developed so far, all freeze-fracture replication techniques normally involve the following major preparation steps:

(*i*) pretreatment of the specimen;

(*ii*) freezing;

(*iii*) fracturing;

(*iv*) (optional) sublimation of ice from the fracture face (etching);

(*v*)   shadowing and replication; and
(*vi*)   cleaning the replica free of specimen residues.

A schematic diagram showing the individual steps of freeze-fracture replication procedures is provided in Fig. 6.1. The heavy lines indicate the sequence of essential steps. As can be seen, the greatest variation is in the fracturing process itself. Freeze-fracturing can either be performed under vacuum in the evaporation unit or outside at atmospheric pressure under liquid nitrogen (or helium), or dry nitrogen gas to protect the fracture faces from contamination. Both fracturing techniques allow the use of either simple fracturing methods or complex cryoultramicrotome assemblies.

This chapter deals particularly with the practical aspects of the individual preparative steps of freeze-fracture replication. Nevertheless it must be emphasised that, for the correct interpretation of information provided in

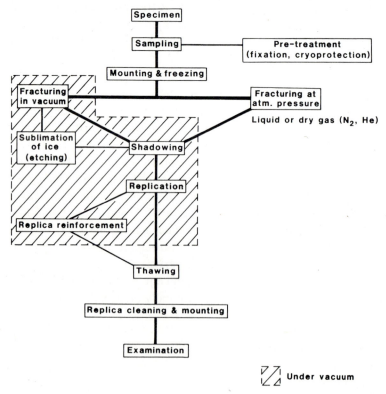

Fig. 6.1.   Freeze-fracture flow diagram.

the final two-dimensional image of the final replica, it is essential to understand fully the principles involved in the various preparation steps (see reviews by Benedetti and Favard 1973; Sleytr and Robards 1977b; Rash and Hudson 1979; Willison and Rowe 1980). An annual survey of the freeze-fracture replication literature was published for some years by Balzers AG, Liechtenstein.

6.2 Specimen preparation

6.2.1 Sampling and pretreatment

When using freeze-fracture replication techniques, the main purpose of all sampling, pretreatment and cryofixation methods is preservation of the detailed molecular organisation of the specimen, as close as possible to that *in vivo*. These requirements are common to all cryotechniques and have already been discussed in §2.3.

6.2.2 Specimen supports and specimen mounting

There are many different designs of holder for mounting specimens for freeze-fracturing. The best design can be judged from the specimen cooling rates that can be obtained and also from the ability to facilitate the specific mechanical operations that are essential during processing of freeze-fracture specimens. As shown in a detailed study by Costello and Corless (1978), specimen cooling rates can be increased by decreasing the specimen mass, decreasing the sample holder mass and increasing the surface area of the specimen in contact with metals of high heat conductivity. A further improvement in cooling rates of specimens frozen in liquid coolants at high immersion velocities can be expected by designing the specimen and holder shapes to optimize heat transfer (§2.2.3). Consequently, specimen holders are made as small as possible and are constructed from metals with high thermal conductivity, such as gold, silver, copper and brass. As discussed in §2.2.6, heat can only be extracted from the specimen surface during cooling and, due to the low thermal conductivity of water, adequately high freezing rates can only be achieved in a very thin superficial layer in non-cryoprotected specimens (§2.3.1). Therefore, the specimen holder of choice depends not only on the type of specimen (e.g. bulk specimen or suspension) but also on whether a cryoprotectant is used. A selection of commonly used specimen holders, in plan and side elevation view, together with the way that they are loaded, is illustrated in Fig. 6.2. Some of these designs can be used directly for fracturing, while others are easily adapted to fit the specimen tables of different types of freeze-fracture replication apparatus. Fig. 6.2 also shows

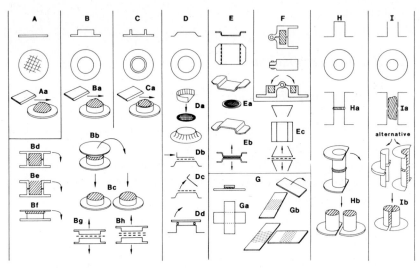

Fig. 6.2.   A selection of commonly used specimen holders, in plan and side elevation, showing loading methods. (A, Aa, B, Ba, C and Ca) Supports for freezing, and microtome fracturing, of suspensions (Moor et al. 1961 and commercial products of Balzers AG; Appendix 2). The discs in A are punched out of thin metal (e.g. copper) foils. Holders B and C are manufactured from gold. (Bb to Bh) Assemblies for sandwich-freezing of suspensions suitable for single fracture and complementary replica techniques (Mühlethaler et al. 1970, 1973; Nermut 1973; Nickel et al. 1978). One or two EM grids or a ring can be used as spacers between the holders. (Da to Dd) Assembly for sandwich-freezing of suspensions suitable for single replica techniques. The discs (Da and Db) are punched out of thin metal foil (e.g. copper) (Gulik-Krzwicki and Costello 1978). (Dd) Sandwich assembly for 'unidirectional' propane jet-freezing consisting of a metal disc, a spacer ring (approximately 20 μm thick) and a flat Thermanox plastic sheet (Pscheid and Plattner 1980; Pscheid et al. 1981). Espevik and Elgsaeter (1981) suggest the same assembly for bi-directional propane jet-freezing. (Ea to Ec) Sandwich assemblies for bi-directional propane jet-freezing (Ea and Eb, holders manufactured by Balzers AG; Ec, holders described by Müller et al. 1980b). (F) Hinged complementary replica holders described by Steere and Moseley (1970). (Ga and Gb) Assembly of thin mica or metal foils for sandwich-freezing and complementary replica techniques (Nermut and Williams 1977; Robards et al. 1980). (Ha and Hb) Assembly of hollow rivets for freezing suspensions and complementary replica techniques (Sleytr 1970a; Sleytr and Umrath 1974a,b). (Ia and Ib) Assembly of hollow rivets for freezing tissue samples for complementary replica techniques (Chalcroft and Bullivant 1970; Sleytr and Umrath 1974a). (Cross-hatched or stippled areas indicate specimens; dashed lines between holders indicate spacers.)

how combinations of supports can be used for special methods such as sandwich-freezing, or complementary replica techniques, which allow the study of both fracture faces obtained from a single fracture (see §6.3.3a). For bulk specimens, hat-shaped holders are used (Fig. 6.2; types C, F and I). This form prevents the tissue from being lost or pulled out during the fracturing process. Sand-polishing, or scratching the surface of the specimen holder with coarse emery paper, reduces the likelihood that the fracture plane will follow the metal surface of the holder.

6.2.2a   Single cells and cell components

Suspensions of single cells or cell components are best applied to the holder from a modified Pasteur pipette. The original fine tip of a pipette is pulled out by hand, over a small burner, so that the diameter is reduced to about half its original size. With this reduced bore, small volumes (about 0.3–0.7 mm^3) can be accurately placed as a hemisphere or less on type A–D specimen holders. To be pipettable, the sample must not be too viscous. If denser pellets are used, they are best applied from the top of a needle which is loaded by touching the surface of the pellet. As with all specimens, mounting and subsequent freezing must take place as quickly as possible to avoid air-drying of the sample. Compared with this drop-freezing method, much less material is necessary using the sandwich method (Fig. 6.2; holders B–E, G and H) and this can be a great advantage if valuable and/or scarce material is under study. In addition, even if it is not necessary to obtain *complementary* replicas, the amount of replica obtained from a small drop is doubled using this technique. A small amount is placed between the two holders and the excess is drained off with a filter paper immediately before freezing. With some sandwich methods and mounting procedures for preparing complementary fracture faces (§6.3.3a) a 'spacer' in the form of an electron microscope grid or small metal ring is inserted (Fig. 6.2; methods B, D, E and H). For propane jet-freezing (§2.4.1b), a gold 400 mesh grid, 2.3 mm in diameter and 12 μm in thickness, can be used (Müller et al. 1980b) or, alternatively, an approximately 20 μm thick spacer-ring (Pscheid et al. 1981).

A simple method of mounting spherical cells so that selective fracture planes are obtained has been described by Peng and Jaffe (1976) (Fig. 6.3). This method is suitable for cells that are larger than 70 μm. Cells are loaded into the holes of a thin nickel screen (Perforated Products; see Appendix 2). These screens have many closely spaced, round, funnel-shaped holes into which the eggs fit. To prevent the eggs from being lost during subsequent

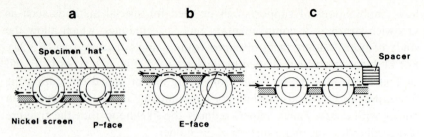

Fig. 6.3.   Mounting method for producing selective fracture planes through spherical cells. P-faces (a), E-faces (b) and cross-fractured cells (c) can be produced using this technique. (Redrawn after Peng and Jaffe 1976.)

manipulation, the surface of the screen is positively charged by coating with a layer of poly-L-lysine. (For poly-L-lysine coating procedures see §6.2.2d). A specimen mounting 'hat' is positioned against either the filling side (Fig. 6.3a and c), or the opposite side (Fig. 6.3b), of the screen and the assembly is frozen by immersion in a liquid cryogen or by propane jet-freezing. A suitably sized ring-spacer, as shown in Fig. 6.3c, can be placed between the screen and the 'hat'. The specimens are fractured by pulling the hat away from the screen in a freeze-fracture apparatus. By using two different kinds of screen, together with different methods of assembly, E-faces, P-faces or cross-fractured cells (for nomenclature see §6.8.2) can be obtained using this technique.

6.2.2b   Bulk specimens

With bulk material, such as animal or plant tissues, the chemical fixation and/or cryoprotection steps are usually carried out on pieces that are larger than will be required for mounting and freezing. Final trimming is best done with a sharp razor blade on a sheet of dental wax in a few drops of the solution with which the tissue was last treated. Whenever possible, samples from bulk specimens should have dimensions of less than 1 mm and preferably *much* less, when cut as a cube or as a cylinder. A stereo microscope can be a special help for trimming the tissue. Particular care has to be taken that the liquid on the wax plate does not evaporate during trimming, since this would lead to an undesirable increase in the concentration of cryoprotectant or other solutes. Transfer of the trimmed tissue to the support is usually done with fine tweezers but some people are sufficiently skilled to use a needle or a pointed wooden applicator stick for this delicate procedure. The transfer should be as fast as possible and it is important that the specimen is not unduly squeezed or squashed. The trimmed tissue can be mounted directly, using the adherent liquid to fill any remaining gaps between it and the holder. When necessary, gum arabic or a suspension of yeast can be used as a glue to stick the tissue to the holder. The specimen is usually gently coated with the suspending medium or placed directly into a droplet of this glue on a specimen mount. To avoid osmotic and other effects, such additional media must be adjusted to the pH and osmolarity of the buffer and/or cryoprotectant in which the tissue has previously been immersed.

For some fragile specimens, which cannot be trimmed to a suitable size without excessive damage before freezing, special mounting techniques have been developed (Johnson 1968; Johnson et al. 1976; Willison and Brown 1977). Such specimens are first frozen by pouring a liquid cryogen over them or by plunging them into cold liquid propane or Freon (§2.4.1a). Subsequently, the frozen samples are quickly transferred to liquid nitrogen using pre-cooled tweezers (Fig. 6.4). Pieces that are small enough for mounting on conventional specimen holders (Fig. 6.2; types C, F or I) are then broken from the rapidly frozen specimen (Fig. 6.4a). Meanwhile, specimen mounts carrying a droplet of cold, but still just liquid, butyl benzene (Willison and Brown 1977) or trichlorethylene, kept at a temperature of 188 or 193 K (-85 or $-80°C$), respectively, or a low temperature 'glue' (Table 4.1

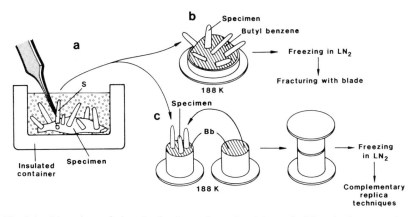

Fig. 6.4. Mounting technique for fragile specimens involving embedding the frozen specimen pieces (S) in butyl benzene (**Bb**) prior to fracturing. Fragments are removed from a rapidly frozen specimen while immersed in liquid nitrogen (a) and are pushed into viscous butyl benzene adhering either to a disc (b) or to a rivet-shaped specimen support (c) maintained at 188 K. The whole assembly is then transferred back to liquid nitrogen before continuing with the fracturing process. (Redrawn and modified after Willison and Brown 1977.)

in §4.2.1; §4.3.1c; §4.3.2c) are prepared (Fig. 6.4b). The supports are maintained at this temperature with simple cold blocks as used in the spray-freezing method (§2.4.1c; Fig. 2.54). The frozen specimen pieces (**S**) are quickly transferred, with cooled tweezers, from liquid nitrogen to the droplet of butyl benzene (**Bb**) or trichlorethylene, which acts as a glue, and the whole assembly is frozen in liquid nitrogen. This method allows the specimen to be fractured with either a cryomicrotome assembly (Fig. 6.4b) or a complementary replica device (Fig. 6.4c). For complementary fracturing, the glue droplet containing the frozen sample is frozen as a sandwich between two suitable specimen holders.

6.2.2c   Monolayer cultures

A variety of special mounting and pretreatment procedures has been developed for cell and tissue cultures, which allow the manipulation of the cells in their *in vivo* positions while still attached to the support on which they were grown (Table 6.1).

Collins et al. (1975) describe a simple technique for the *in situ* freeze-fracture of cultures grown on 3 mm diameter gold carriers of the type B shown in Fig. 6.2, coated with a thin layer of vacuum-deposited silicon monoxide. A silicon layer is a suitable substrate for the growth of many different cells. After fixation (the authors used a two-step procedure involving paraformaldehyde and glutaraldehyde), incubate the cells in cold 30% glycerol, remove the carriers from the dish and carefully drain-off the excess liquid. Then spray the carrier (Fig. 6.5) with a 1% suspension of 9.7 μm polystyrene latex beads (see Appendix 2). Place a small copper hat on top of the layer and freeze the whole assembly. The polystyrene latex spheres function as spacers between the cell monolayer and the copper hat and compensate for the uneven surface of the cell culture. To increase the probability of fracture

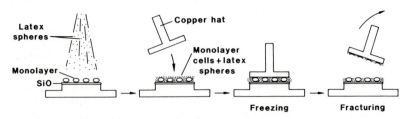

Fig. 6.5.   Mounting method for *in situ* freeze-fracture of cell monolayers using polystyrene latex spheres as spacers. (Redrawn after Collins et al. 1975.)

TABLE 6.1

Mounting methods for cell monolayers

Reference	Substratum	Features of method
Collins et al. (1975)	Silicon-coated gold disc	Latex beads as spacers
Pfenninger and Rinderer (1975)	Collagen-coated gold grid	
Pauli et al. (1977)	Plastic coverslip	Mounted in medium containing PVA[a]
Bendicht et al. (1979)	Plastic coverslip	Mounted in medium containing PVA
Prescott and Brightman (1978)	Collodion and collagen-coated glass coverslip	Cells stained with ruthenium red
Griepp et al. (1978); Griepp and Bernfield (1978)	Tissue culture foil	Cell layer mounted in gelatin
Umrath and Isenberg (1980)	Fluorethylene-propylene foil membrane	Holders attached to foil membrane with cyanoacrylate glue
Pscheid and Plattner (1980)	Thermanox sheets	Suitable for cryojet-freezing

[a] PVA, polyvinyl alcohol

planes passing through the monolayer of cells, beads with a smaller diameter (4.7 μm) can be used and both the surface of the specimen carrier and
of the hat can be polished. This technique has the advantage that silicon
monoxide can be sterilized at 132°C without inducing any structural changes, something that is not possible with the method of Pfenninger and Rinderer (1975) who used a collagen layer attached to a gold grid as the substrate. (However, this method has the advantage that the collagen layer is
transparent and consequently the cells can be examined in the light microscope).

Another versatile method for *in situ* freeze-fracturing of monolayer cultures, which can be used for cell and tissue cultures grown on standard plastic coverslips, has been described by Pauli et al. (1977) (Fig. 6.6). This method has the advantages that the same culture can be used for both light and
electron microscopy and, also, that it is possible to preselect areas to be
freeze-fractured. Cells grown on plastic coverslips are fixed *in situ* (the authors used 0.1 M cacodylate-buffered (pH 7.3) 2% glutaraldehyde at 4°C for
30 min) and subsequently infiltrated with 25% glycerol by immersion for
5 h at 4°C. For freeze-fracturing, 1.5 mm squares (selected by light microscopy) are cut out and the excess glycerol solution is drained off with filter
paper, so leaving only a thin layer of liquid over the culture. The squares
are then placed, with the cells facing down, on a drop of polyvinyl alcohol
(PVA)-containing mounting medium on the specimen holder. After freezing, this PVA layer encourages the propagation of fractures through the frozen cell monolayer rather than through the mounting medium or along the
specimen carrier. In the original study, Elvanol 51-05 (DuPont de Nemour;
see Appendix 2) was used in the mounting medium, but this is no longer
commercially available. Bendicht et al. (1979) suggest the use of a chemically identical polyvinyl alcohol, Vinol 205 S (Air Products and Chemicals,

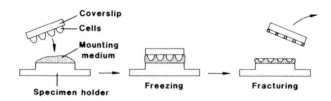

Fig. 6.6. Technique for the *in situ* freeze-fracture of cell monolayers using polyvinyl alcohol
in the mounting medium to encourage fractures through the cell layer. This method allows the
preselection of the area to be freeze-fractured. (Redrawn after Pauli et al. 1977.)

Inc.; Appendix 2). The following procedure is recommended for preparing the mounting medium:

(1)   20 ml of 100% glycerol is added to 80 ml of 0.1 M phosphate buffer, pH 7.3, containing 0.14 M NaCl (phosphate buffered saline; PBS). The mixture is heated to 368 K (95°C) and stirred.

(2)   20 g of Vinol 205 S powder is added slowly, with constant stirring, until the powder is completely dissolved in the hot PBS/glycerol mixture.

(3)   The residual white foam on the Vinol 205 S mixture is removed from the surface with a pipette attached to a water-jet vacuum pump.

(4)   The mixture is cooled to room temperature while gently agitating with a magnetic stirrer to prevent 'surface skin' formation.

(5)   A crystal of thymol is added to prevent mould growth and the mixture is stored at room temperature.

Prescott and Brightman (1978) developed a method for controlled fracturing of monolayer cultures in which the cells are stained with ruthenium red. Cultures are grown on glass coverslips which have first been coated with Collodion and then with collagen. Fixation is in glutaraldehyde to which ruthenium red has been added. After fixation the composite three layers are glycerinated, cut into pieces less than 2 mm in diameter, mounted on a specimen holder and frozen. During the fracturing process, the red-stained cells can be distinguished from the underlying Collodion and overlying frost using a light microscope (magnification about 200–300 ×). The appearance of pink flakes on the edge of the microtome blade serves as an indication that the fracture plane has passed through the monolayer culture.

The method of mounting (Fig. 6.7) described by Griepp et al. (1978) (see also Griepp and Bernfield 1978) was developed for producing large areas of orientated, frozen-fractured, embryonic heart cell aggregates and layers and has the potential for successful application to a variety of specimens. Aggregate layers are first fixed *in situ* with glutaraldehyde and are subsequently impregnated with 25% glycerol, both chemicals being made up in Earle's basic salt solution (EBSS). An area of approximately 1 mm² is scratched around each aggregate and each aggregate layer is then removed from the substratum by aspiration and placed in a dish containing a warm solution of 6% gelatin in the glycerol/EBSS mixture (Fig. 6.7a). After solidification, the gelatin is fixed at 4°C in a solution of 2.5% glutaraldehyde in glycerol/EBSS. After 1–5 h, small cubes of the hardened gelatin, containing the aggregate layers, are cut out (Fig. 6.7b and c) and mounted in the correct

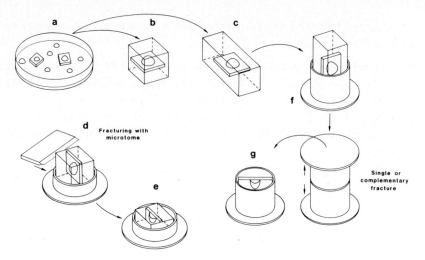

Fig. 6.7.   Method for the orientated mounting of cell aggregates and layers. Pieces of cell aggregates or monolayers that have been fixed and cryoprotected *in situ* are embedded in gelatin (a). After fixation and solidification of the gelatin by fixation with glutaraldehyde, cubes (b) or blocks (c) of the hardened gelatin containing the aggregate layers are cut out. The blocks can be mounted for single or complementary fracture (f, g) or for freeze-fracturing using a microtome (d, e). (Redrawn and modified from Griepp et al. 1978.)

orientation, in relation to the predicted fracture planes, on specimen holders (Fig. 6.7d and f). The samples are frozen by immersion in liquid Freon or propane. For fracturing, both single (Fig. 6.7d and e) and complementary (Fig. 6.7g) fracture methods are possible.

The method of Umrath and Isenberg (1980) for the observation of specific areas of monolayer cultures is illustrated in Fig. 6.8. Cells are grown in Petriperm (registered trade mark of W.C. Heraeus GmbH.; see Appendix 2) tissue culture dishes (Fig. 6.8a). The bottom of each dish consists of a 25 μm thick fluorethylene-propylene foil membrane, which is permeable to gases but not to liquids. Before use, both sides of this membrane are made hydrophilic by a 12 s treatment in a glow discharge, at 10 Pa pressure and 650 V, 4 A m^{-2} at a distance of 14 mm from the electrodes. The dishes are then sterilised by ultra-violet or electron irradiation in the normal way. Monolayers are either chemically fixed and glycerinated, or frozen without pretreatment. For specimen mounting, hat-shaped specimen holders (e.g. Fig. 6.2; types D, E or H) are roughened and carefully degreased by sonication in ethanol or acetone. With greatest care, the specimen carriers are glued to selected areas of the foil membrane (Fig. 6.8b). Two kinds of glue

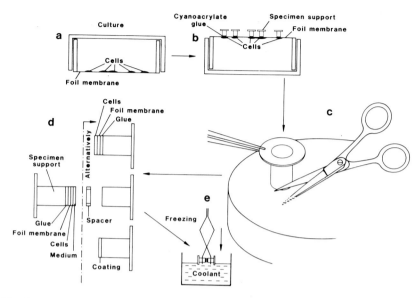

Fig. 6.8. Mounting method for cells grown on a foil membrane (a). Specimen supports are glued to selected areas of the foil membrane (b) using 'super-glue' (cyanoacrylate). After cutting out the specimen/foil membrane/support assembly (c), various sandwich mounting methods (including complementary replica techniques) can be used (d) prior to rapid freezing by plunging (e). (Redrawn from Umrath and Isenberg 1980.)

have been used with equal success: *Hystoacryl blue* (Braun-Dexon GmbH.; see Appendix 2); and *CA-15* (Delo GmbH.; see Appendix 2). Both types polymerise after a few seconds and tolerate high cooling rates (Heinzmann 1974). The specimen holder, with the attached foil membrane and the adhering cells underneath, is dissected out using small scissors (Fig. 6.8c) and surplus areas are removed. Subsequently, two holders are mounted, with the monolayers facing each other, in a special pair of forceps (Figs. 6.8e and 6.9) or other suitable tool for sandwich-mounting. After the opposing layers have been brought into contact, the assembly is frozen by immersion in the coolant (§2.4.1a).

Different methods of mounting can be used, as illustrated in Fig. 6.8d. As an alternative to the assembly of two monolayers, a single monolayer may be frozen against a blank holder which can be covered with latex beads (as in the method of Collins et al. (1975)) or other spacers, and glycerol or tissue culture medium may be replaced by other 'connection liquids', such as polyvinyl alcohol (Pauli et al. 1977), if variations in the fracture plane

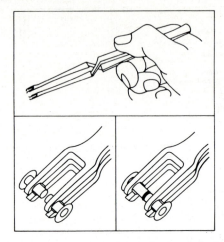

Fig. 6.9.   Special forceps designed to ensure correct alignment of paired rivet-type specimen holders during specimen mounting and freezing. (Redrawn from Sleytr and Umrath 1974a.)

are desirable. Finally, to avoid excessive and/or uncontrolled squeezing of the cells, a spacer can be placed between the specimen holders.

A mounting procedure particularly developed for uni-directional cryojet-freezing (§2.4.1b) of monolayer cell cultures has been described by Pscheid and Plattner (1980) and Pscheid et al. (1981). Cells are grown directly on thin Thermanox plastic sheets, 0.2 mm thick (Lux Scientific; see Appendix 2). Small discs with cell layers are punched out from the plastic sheet and mounted on a U-shaped, low mass specimen holder (Fig. 6.2; type D or E) made from 0.1 mm thick copper or 0.3 mm gold sheet. Melinex (ICI, Appendix 2) can be used as an alternative to Thermanox. To avoid mono-layers becoming damaged or detached during the cutting or punching pro-cedures, cells can also be grown on accurately shaped plastic sheets. To pro-tect the cells from being squeezed, a spacer ring is placed between the speci-men holder and the Thermanox sheet. The sandwich (Fig. 6.2; Dd) is frozen by squirting liquid propane onto the metal side only (§2.4.1b). This techni-que has the advantage that it is suitable not only for the examination of cells grown as monolayers but also for suspensions of macromolecules, subcellu-lar fractions or whole cells in suspension. Knoll et al. (1982) were able to show excellent cryofixation in a variety of specimens without any chemical pretreatment.

6.2.2d   Specimen mounting for complementary replica methods

Various designs of specimen holder can be used to produce complementary fracture faces as illustrated in Fig. 6.2. To ensure the proper alignment of rivet-type supports (Fig. 6.2; types H and I) during specimen mounting and freezing, special forceps (Fig. 6.9) can be used (Sleytr and Umrath 1974a). For some complementary replica methods, cell monolayers or membrane fractions are sandwiched between two thin sheets of metal (Müller et al. 1980a) (Fig. 6.2; E), mica (Nermut and Williams 1977), or a combination of mica, metal plates and glass (Nermut and Williams 1977; Fisher 1978) (Fig. 6.2; G). Each of these techniques has the advantage that the specimen can be rapidly frozen and that only very small amounts of material are needed (1–10 μl). Furthermore, if one side of the sandwich is coated with a polycationic material, it is often possible to achieve uniform fracturing of monolayers.

Positively charged, smooth surfaces of glass, freshly cleaved mica, or carbon-coated metal sheets are produced either:

(*i*)   by a glow discharge in amylamine vapour (Dubochet et al. 1971);

(*ii*)   by coating with poly-L-lysine (Mazia et al. 1974; Fisher 1975, 1978); or

(*iii*)   by coating with Alcian blue (Sommer 1977; Nermut 1982a).

Poly-lysine- or Alcian blue-coated supports have been used for a variety of studies, including controlled fracturing of erythrocytes and bacterial purple membranes (Nermut and Williams 1977; Fisher 1978; Fisher et al. 1978; Nermut 1982a).

Nermut and Williams (1977) and Nermut (1982a) suggested the following procedure for coating glass coverslips and either freshly cleaved or carbon-coated mica with poly-L-lysine. Pieces of freshly cleaved mica or glass (e.g. with the shape shown in Gb in Fig. 6.2) are stored in chromosulphuric acid (10% sulphuric acid in saturated potassium dichromate, diluted 1:2 before use), rinsed thoroughly with glass-distilled water and dried by touching one edge with filter paper. A drop of poly-L-lysine (2.5 mg ml^{-1}; Sigma Chemicals Mark II, MW about 4000; see Appendix 2) is placed on the mica. After 5 min, excess poly-L-lysine is washed off with a stream of glass-distilled water. The coated mica is either used immediately or is allowed to dry and then used within 20 min. To avoid adsorption to glass surfaces, the poly-L-lysine solution must be stored in plastic bottles and, to ensure uniform coating, all tools used in the process must be scrupulously clean.

Fisher (1982c) recommends the following alternative procedure for coating coverslips with poly-L-lysine.

(1)   Wash coverslips, preferably in porcelain staining racks, in hot tap water plus a few drops of detergent. Sonicate for 60 s (e.g. with 80 kHz, 80 W bath-type sonicator).

(2)   Wash coverslips with running tap water for 30 s.

(3)   Immerse coverslips in chromic-sulphuric acid (4.0 kg concentrated H_2SO_4 + 25 ml Chromerge (Fisher Scientific Co.; see Appendix 2)) and clean at 60–70°C for 1–2 h.

(4)   Rinse the rack of coverslips in running ion-exchanged distilled water and then rinse in glass-distilled water.

(5)   Remove coverslips from rack with clean forceps and quickly dry in a stream of compressed nitrogen gas. The cleaned coverslips should be used within 30 min.

(6)   Apply 5.0 mM poly-L-lysine (poly-L-lysine hydrobromide, MW 1000–4000, Type II (Sigma Chem. Co.) or MW 1500–8000 (Miles Laboratories) see Appendix 2) from a hydrophobic pipette to one side of the clean, dry coverslip (25 μl for 11×22 mm coverslips). Gyrate the coverslip in the horizontal plane for 30 s at 20–22°C so that the coverslip is uniformly coated.

(7)   Wash away excess, unbound poly-L-lysine with glass-distilled water for 30 s and dry the coverslip with a burst of dry nitrogen from an air gun.

(8)   The poly-L-lysine-treated coverslip should be used within 30 min.

The specimen is best applied to the positively charged surface by spreading a drop of specimen suspension (less than 10 μl) over the surface and leaving it for between a few seconds and 5 min depending on the concentration of the suspension. Excess material is removed by washing in either a stream of water or in phosphate-buffered saline (Fisher 1975; Nermut and Williams 1977).

To make specimen supports positively charged using Alcian blue (e.g. Alcian blue 8GX p.a., C.I. 74240, Serva; see Appendix 2), the surfaces are brought into contact with a 5.0 mM solution for 5–10 s and are then rinsed with water (e.g. as a fast stream from a pipette) for approximately 10–15 s. After drying in a stream of clean, dry nitrogen gas, the treated surfaces will preferentially adsorb specimens with a net negative charge (Sommer 1977).

In practice, many different combinations of sheet-like material can be used for sandwich mounting. Nermut and Williams (1977) and Robards et al. (1980) have developed special gadgets for cleaving apart 10×15 mm or 5×10 mm sized sheets of mica which are frozen together in a cross-wise arrangement (Fig. 6.2; Ga). An alternative technique has been developed by Fisher (1975) where the specimen material is sandwiched between disc-shaped glass and copper sheets.

Specimen mounting can be done in different ways for preparation of complementary fracture faces (Fig. 6.2). As an aid to fracturing suspensions in the right plane, one or two gold finder grids (grids with marks so that individual squares can be recognised and recovered; see Appendix 2 for suppliers) can be placed between two adjacent specimen holders (Mühlethaler et al. 1973; Nermut 1973). If one grid is sandwiched (Fig. 6.2; Bg), more even fracture planes may be obtained, with the additional advantage that the replica from one fracture face frequently remains attached to the grid throughout the cleaning procedure. If two finder grids are frozen in register (Fig. 6.2; Bh), replicas of both fracture faces may be easier to locate. Indexed (two grids with their meshes aligned; see Appendix 2) gold grids are best attached to each other using a thin layer of Vaseline which also increases the likelihood that fracturing will occur *between* the grids. Since handling of the two grids must be done under a dissecting microscope, this method is frequently considered to be too laborious. As another aid to identifying complementary areas, morphologically distinct, easily identifiable, marker cells (yeast cells, erythrocytes or pollen grains) can be mixed with the suspension under study before freezing (Sleytr and Umrath 1974a).

Special mounting methods are used for spray-freezing (Bachmann and Schmitt-Fumian 1973; §2.4.1c), or for fast-freezing on cold surfaces (Heuser et al. 1976; §2.4.2). A mixture of spray- or jet-frozen droplets with butyl benzene can be transferred and mounted on almost any type of specimen holder in contact with a cold table. The droplets can be enclosed in holders suitable for both single-fracture methods, such as complementary replica techniques, and also for cryomicrotomy. Dempsey and Bullivant (1976a) used specimens mounted on thin Teflon supports for freeze-substitution and on more complex assemblies for freeze-fracturing (Dempsey and Bullivant 1976b).

6.2.3 Freezing and storage

Freezing methods have been described in detail in §2.4. The storage methods suggested in §2.5 are perfectly suitable for specimens to be used for freeze-fracture replication. No changes in structure have been noted during storage of frozen specimens in liquid nitrogen. Frozen samples are generally stored in 25 litre containers. A special storage container for freeze-etch specimens, which reduces the loss of liquid nitrogen and allows the storage of over 300 different specimens in a 30 litre container, is recommended by Krah et al. (1973) (see Fig. 2.65 in §2.5).

6.2.4   Transfer of frozen specimens

During the various transfer stages, the specimens must not warm up above the recrystallisation temperature which, for most biological specimens, is around 180–190 K (§2.1). Consequently, all transfer devices (tweezers, containers, etc.) must be carefully pre-cooled before use. The various freeze-fracturing methods involve different transfer procedures but, as a general rule, the specimens should only be exposed to conditions where hoar frost will be formed for as short a time as possible. If specimens are not freshly frozen, they are first transferred from their large, long-term, storage container to a small (0.5–1.0 litre) container which should have compartments to keep (at least 3) specimens separately. Depending on the particular freeze-fracture replication method, specimens are then either transferred directly with a pair of cooled tweezers from this interim container to the cold stage of the freeze-fracturing apparatus, or they are first mounted under liquid nitrogen on a special gadget, such as a complementary replica device, and subsequently transferred to a cold stage. The loading or transfer of fracturing devices under liquid nitrogen is best done in small styrofoam (expanded polystyrene) containers which have excellent thermal insulation properties and can be handled with less care than glass Dewar flasks. When specimens are frozen in cryogens with a melting point higher than 77 K (e.g. Freon 12 or 22; see also §2.2.3a, Table 2.6), remnants of the liquid solidify around the specimen when it is plunged into liquid nitrogen for storage. This can create difficulties if specimen handling devices are loaded under liquid nitrogen. As an alternative, loading can take place under Freon (12 or 22) or specimens can be frozen in cryogens with a lower melting point (e.g. propane or melting nitrogen). Before specimens are transferred from the liquid nitrogen storage container to a cold stage, it is a good practice to 'lubricate' the stage with liquid Freon (using a small brush to remove hoar frost and to ensure good thermal contact). This also improves the transitory thermal contact of the specimen with the stage during the transfer/ mounting procedure. On the other hand there is the disadvantage that impurities in the Freon may cause contamination of both the vacuum system and the fractured specimen (§6.5.2).

6.3   Fracturing

6.3.1   General considerations

The fracturing process parts the frozen specimen along the plane which offers least resistance to the applied forces. In practice, fracturing of the specimen is done either by tensile stress, which produces a single cleavage plane, or with a cold scalpel or microtome knife which, due to shear forces, produces conchoidal flakes in front of the knife edge. Both single and multiple fracturing processes can be accomplished either under vacuum or at atmospheric pressure (see Fig. 6.1 in §6.1). For freeze-fracture replication, the following fracturing methods have been developed and are described in §6.3.3 in detail.

(*a*) Freeze-fracturing under vacuum:

   (*i*) Methods for single fractures using devices for producing single, or complementary, fracture faces.
   (*ii*) Freeze-fracturing with a scalpel or a microtome assembly.

(*b*) Freeze-fracturing under improved vacuum conditions using ultra-high vacuum technology or cold shrouds.

(*c*) Freeze-fracturing outside the vacuum unit under a liquid gas:

   (*i*) Freeze-fracturing under liquid nitrogen using single fracture methods or microtome assemblies.
   (*ii*) Freeze-fracturing under liquid helium.

(*d*) Freeze-fracturing under a dry nitrogen atmosphere using a cryoultramicrotome.

To characterise the fracture planes obtained from the different techniques listed above, it is useful to distinguish between gross and fine topography. The former is defined by the extent of coarse undulations ($> 1.0 \ \mu$m) in relation to the mean level of the fracture plane, whereas the latter defines structural details on the replica with dimensions much less than $1.0 \ \mu$m.

6.3.1a   Fine topography of the fracture plane

At sufficiently low temperatures (< 200 K), simple cleavage induced by tensile stress or fracturing with a microtome knife both lead to the production of identical fine topography in the fracture plane. In heterogeneous specimens, the fracture plane is determined by the distribution of chemical bonds and predominantly proceeds along intermolecular paths requiring the least energy to break these bonds.

The evaluation of replicas of complementary fracture faces has been shown to be particularly useful when analysing the fine topography of the fracture plane. As will be discussed in detail in §6.8.3, the fine topography of freeze-fracture replicas of biological specimens indicates that potentially 'brittle' and 'ductile' regions alternate along the cleavage plane. Experiments on model systems have shown that many polymers, biological as well as synthetic, deform plastically during freeze-fracturing at temperatures as low as 4 K and that considerable energy must be dissipated during the fracturing process. Heat dissipation during local deformation processes can cause a rearrangement of molecules in adjoining regions which may have brittle fracture characteristics. Alterations to the fracturing behaviour, and particularly the deformability of biological structures, observed after different pretreatments, such as fixation or infiltration with cryoprotectants, can be interpreted as the result of such interface (surface) phenomena and changes in intermolecular bonds (for reviews see Sleytr and Robards 1977a, 1982).

In biological specimens, the plane of weakness is often predetermined by the disposition of membranes, which have been shown generally to fracture along a central hydrophobic region (Branton 1966). Other structures which may have an influence on the path of the fracture plane include lipid bodies, such as plant spherosomes, which frequently show a concentric arrangement of bilayers. In addition, regular aggregates of macromolecules such as proteins (e.g. protein crystals, microtubules and flagella) can act as preferential cleavage sites. In procaryotic cells the fracture path is frequently deflected, not only by the cytoplasmic membrane, but also by other layers such as the outer membrane of the Gram-negative cell envelope or the coats of endspores. Similarly, the multi-layered cell walls of some eucaryotic cells are cleaved by the fracturing processes.

6.3.1b   Gross topography of the fracture plane

As opposed to the fine topography of the fracture plane, differences in gross topography can frequently be observed between microtomed or simply fractured surfaces. In general, at low temperatures, the chances of producing a less highly sculptured fracture plane are better when a microtome is used.

This has the advantage that larger areas of the surface can be placed at an optimal angle to the shadowing source, so leading to the production of good replicas.

By lowering the microtome blade between consecutive strokes, small specimen chips can be fractured away. Consequently, the final fracture plane does not necessarily represent the product of the last cut, because conchoidal chips from a deeper level than the plane of the microtome advance between two consecutive cuts may have been produced by the preceding movement of the blade. Furthermore, most replicas of specimens smoothed-down with a microtome show areas with knife marks or 'flat areas' generated by localised friction between the knife and the specimen surface. These areas of localised surface melting are devoid of any information. It is remarkable that such knife marks can also be observed when freeze-fracturing is done with a microtome under liquid nitrogen (§6.3.3c). The layers destroyed by frictional heating can be very thin and subsequent etching sometimes reveals structures that show no clear sign of heat damage. In addition, frozen fragments produced during the fracturing process are frequently in poor thermal contact with the microtome knife unless freeze-fracturing is done under the protection of a liquid gas (e.g. liquid nitrogen). Frequently, chips generated by the trimming process with the microtome fall into deeply concave fractured areas of the specimen fracture plane where they are not removed by the final strokes of the knife (Staehelin and Bertaud 1971). Consequently, these small chips and fragments attach to the microtome blade and represent a potential source of subliming water for condensation onto the specimen surface (Staehelin and Bertaud 1971; Dunlop and Robards 1972; Walzthöny et al. 1982). In contrast to microtome fracturing methods, single fracture methods, as needed for the production of complementary replicas, produce no, or very little, contamination of the fracture face from localised surface melting or sublimation of ice from fractured fragments.

Few systematic examinations have been carried out but it is evident that the steel blade edge, or any other microtome knife, is very rapidly damaged by 'cutting' the hard frozen specimen (Parish 1973). With simple specimens, such as suspensions of cell fragments, isolated cells or homogeneous tissues, this has little implication for the topography of the final cleavage plane. However, for other specimens, like plant cells with thick cell walls, it has been shown that a more elaborate microtome assembly can be useful, such as the system described by Robards et al. (1970) and Robards and Parish (1971) which has two razor blades: one for preliminary cutting and the other for the final cut(s). In addition, it has been shown that different knife angle settings can have a significant influence on the final gross topography of the fracture plane. For smoothing down the rough surface of a fracture through a very heterogeneous tissue, it has been recommended that a few fine cuts in approximately 100–500 nm steps should be made and that the knife angle should be reduced to approximately 45–55° (Parish 1973; Robards 1978b).

If no replica reinforcement procedures (§6.4.2), particularly the 'silver technique' (Robards and Umrath, 1978) are available, the excessively rough fracture face of a heterogeneous tissue can make it difficult or impossible to obtain intact replicas. After smoothing down the fracture face by the removal of small conchoidal chips, fragmentation of the replica during thawing and cleaning (§6.6) will generally be less of a problem.

There is little or no difference in the gross topography of the fracture plane, as exposed by either simple fracture methods or microtomes, when homogeneous tissues, suspensions of cells, or cell fragments are examined. For single fracture methods and analogous complementary replica techniques, specimen holders with flat ends can be used (see Fig. 6.2; types B, D, E and H). With these holders, only a thin layer of a suspension or monolayer culture is sandwiched between the holders and consequently a smooth fracture plane, without the potential hazards from microtome fracturing, is obtained.

6.3.1c   Conditions for freeze-fracturing

Most specimens are freeze-fractured under vacuum using single fracture methods or microtomes at temperatures down to 77 K. The first departure from this generalisation was by Branton (1973) who tried to freeze-fracture specimens under vacuum with a microtome at 16 K. Freeze-fracturing at specimen temperatures around 10 K has become less of a problem with the introduction of a new generation of freeze-fracture replication units (Escaig et al. 1980; Niedermeyer 1982a; and the Balzers UHV apparatus BAF 500K; see Appendix 2). The lowest fracturing temperature to have been reached so far has been that of liquid helium (Sleytr 1974; Sleytr and Umrath 1974b) using a fracturing technique to produce complementary fracture planes (§6.3.3c). Within the range tested, from approximately 180 K down to 4 K, it has been shown that, in contrast to the *fine* topography (§6.8.3), no significant differences in the *gross* topography of fracture planes through biological specimens are revealed when freeze-fracturing is done at either higher or lower temperatures, or whether under vacuum or liquid gases.

In summary, most specimens can be freeze-fractured satisfactorily using a single fracture method. The possibilities for studying small samples of homogeneous material or monolayers of cells, for obtaining complementary fracture planes and for reducing the hazards of contamination by condensation of water, all favour single fracture methods. However, for fracturing extremely heterogeneous tissues, a microtome can be advantageous although, even here, replica reinforcement methods can help to overcome the problems associated with the unduly rough surfaces created by a single fracture method. Fracturing with a microtome is also required if replicas are to be prepared from the limited, well-frozen zone of specimens that have been frozen on the surface of a cold metal block (§2.4.2; Heuser 1981).

6.3.2 *Vacuum units for freeze-fracture replication*

Irrespective of which freeze-fracture method is used, deposition of the replica onto the specimen fracture face must be done in a vacuum coating unit. The degree of complexity and performance of different commercially available units vary considerably and examples of different vacuum systems used for freeze-fracture replication are illustrated in Fig. 6.10. The schematic drawings show the *essential parts* of the most common systems and also the additional parts required if specimens or evaporators are to be interchanged through an air-lock system (the area in Fig. 6.10 indicated by a dashed line). The most usual vacuum units consist of a bell-jar, or a similar working chamber, pumped by a two-stage system comprising an oil diffusion pump backed by a rotary pump (Fig. 6.10; system I). The operation of such a system has been described in detail by Willison and Rowe (1980). As an alternative to a diffusion pump, some manufacturers offer systems incorporating turbo-molecular pumps (Fig. 6.10; system II). These have the advantage that they can be operated more easily than diffusion pumps which must always work below the critical backing pressure to avoid the possibility of gas molecules breaking through the oil jet and stopping the pump action. A vacuum with low partial pressure of hydrocarbons is achieved more readily with turbo-molecular pumps than with diffusion pumps which also require a baffle cooled with liquid nitrogen to condense organic vapours. It must, however, be remembered that turbo-molecular pumps have to be vented (preferably with dry nitrogen gas) to about 1 kPa (7.5 Torr) when the system is turned off so as to avoid back-streaming of oil from the rotary pump.

The latest designs of freeze-fracture replication unit use vacuum systems incorporating refrigerator cryopumps (these are cryopumps that are cooled by a closed refrigerator system: see Fig. 6.10; system III), adsorption and

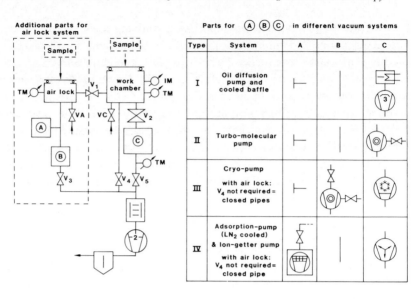

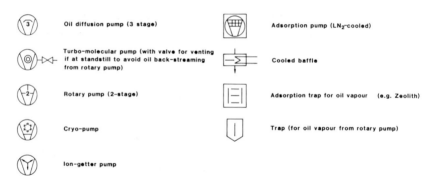

Fig. 6.10.   Schematic diagram illustrating the basic principles of vacuum units used for freeze-fracture replication. The components that are required if specimens and/or evaporators are interchanged via an air-lock system are shown enclosed within the dashed line. Type I is the commonest, two-stage, pumping system comprising a three-stage oil diffusion pump backed by a rotary pump; alternatively (Type II) a turbo-molecular pump may be used in place of a diffusion pump. Ultra-high vacuum pumping systems frequently use either refrigerator cryo-pumps (Type III) or adsorption and sputter-ion pumps (Type IV). V_1 to V_5, V_A and V_C are valves: V_1, valve connecting air-lock system to work chamber; V_2, high-vacuum valve; V_3, valve for evacuation of air-lock system; V_4, valve for pre-evacuating the work chamber; V_5, valve for pre-evacuating the high-vacuum pumps (or work chamber); V_A, valve for venting air-lock; V_C, valve for venting the work chamber. **IM** is a hot cathode ionisation gauge and **TM** indicates thermal conductivity vacuum gauges.

sputter-ion pumps (Fig. 6.10; system IV), or combinations of these.

Besides the inherent limitations of some vacuum systems (e.g. oil back-streaming), frequent causes of contamination are degassing gaskets, grease, oil and finger prints (hence use gloves!) in the equipment, which can create a poor vacuum over long pumping periods.

Different values are given by manufacturers for the final pressure achievable with their systems. The cited 'working pressure' frequently varies between 1×10^{-4} and 1×10^{-7} Pa (7.5×10^{-7} and 7.5×10^{-10} Torr) but, as discussed below and in §5.2.1, this value can be misleading, since it represents the *total pressure* of *all* residual gases in the system. The pressure which is of interest for freeze fracturing is the *partial pressure* of the gas components which condense at temperatures down to the fracturing temperature of the specimen and thus can lead to condensation contamination. There are also considerable variations in the degree of automation and in the number of fail-safe devices with different pumping systems. Complete automation of a vacuum system is generally expensive but gives the great advantage of fast and correct sequential operations. So far the highest degree of automation has been achieved by the Leybold-Heraeus Bioetch 2005 freeze-etching unit (see Appendix 2) where the automatic vacuum system is interlinked with a fully automatic freeze-fracture replication process (§6.3.3b). Most new freeze-fracture units have at least a fully automated pumping system.

The most frequent cause of poor results with freeze-fracture replication is still contamination of specimen fracture faces with, particularly, water vapour condensing from either the vacuum system or parts covered with hoar frost. Thus, excellent vacuum conditions, with a minimum of condensable gases in the neighbourhood of the specimen during exposure of the fracture planes for etching and replica deposition, are essential. There are two different approaches to obtaining an excellent vacuum environment with a minimum of condensable gases in the specimen area. The simplest involves liquid nitrogen-cooled traps which act as very efficient cryopumps for both hydrocarbons and water vapour. Cold shroud (and specimen stage) temperatures as low as about 10 K have become easily obtainable by using closed refrigerator cryopump systems which do not require a continuous supply of liquid helium. The other method of improving the vacuum conditions involves complicated ultra-high vacuum (UHV) technology. For UHV freeze-fracture replication, not only must the vacuum around the specimen area be improved, but the whole system must be evacuated to the same level. The necessary technology requires a system in which every part can be baked to between 200 and 400°C (473 and 673 K) to desorb molecules trapped on metal surfaces and within metals. In such systems the total pressure can be reduced to approximately 1.0×10^{-8} Pa (7.5×10^{-11} Torr) or even lower. Elaborate air-lock and specimen transfer devices are necessary to maintain such a vacuum (Gross 1977; Gross et al. 1978a; Niedermeyer 1982b) and the whole unit must be baked out again after it has been vented (e.g. to reload the evaporation sources if no special air-lock system for the evaporators is available). In such a system, the main sources of gas are surfaces bearing hoar frost (particularly on the specimens, their supports and the transfer devices) and the water vapour released from warming ice fragments during fracturing (Walzthöny et al. 1982). The hoar frost can be removed by sublimation at temperatures from about 173 K ($-100°C$) but the released water vapour will adsorb to metal surfaces in the unit and subsequently lead to a poorer vacuum unless it is trapped by a cold shroud (Escaig et al. 1980). In practice, it has been shown that good shielding with an optically tight cold shroud, that is kept at a temperature at least 20 K below that of the specimen, is the most efficient protection of the fracture surface from contamination (Steere 1973, 1982; Sleytr and Messner 1978; Steere et al. 1980; Sleytr et al. 1981; Niedermeyer 1982a).

Apart from the vacuum conditions, there are other important criteria in the selection of freeze-fracture replication methods and apparatus (e.g. specimen temperature control, evaporation methods) as discussed below.

6.3.3   Instruments for freeze-fracturing

There are several ways of fracturing frozen biological specimens (see Fig. 6.1 in §6.1). Freeze-fracturing can be carried out either under vacuum or at atmospheric pressure, and the fracture process itself is done either by applying tensile stress or with a cooled knife. This diversity of methods is reflected in the many different types of commercially available apparatus or 'workshop prototypes' developed in the last fifteen years. It is not possible to discuss all variations in this Chapter but the major procedures will be described in detail.

6.3.3a   Freeze-fracturing under vacuum

(i)   Single fracture methods
These are the simplest freeze-fracture replication techniques and much of the pioneer work was carried out using them. Single fracture methods are

primarily designed so that replication takes place immediately after fracturing and, therefore, a specimen temperature control for etching is usually either crude or non-existent. Due to their simplicity, the majority of such devices are easy to build and can be fitted into existing coating units or adapted for use in freeze-etching units which already incorporate a microtome assembly.

A simple single fracture method involves freezing the specimen between two specimen holders of the type used for producing complementary fracture faces (Fig. 6.2; types B–I). The frozen sandwich is either mounted directly onto a pre-cooled cold stage in a coating or freeze-fracture replication unit (Nickel et al. 1978), or is first fixed to a transport plate under liquid nitrogen and subsequently transferred to the stage (Steere and Moseley 1970; Sleytr et al. 1972). After the working vacuum has been obtained and the specimen has reached the correct temperature, fracturing takes place by breaking off the free specimen support from the fixed one (Fig. 6.11). Using this method, a replica can only be produced from one of the complementary fracture faces since the other is sacrificed with the broken-off specimen sup-

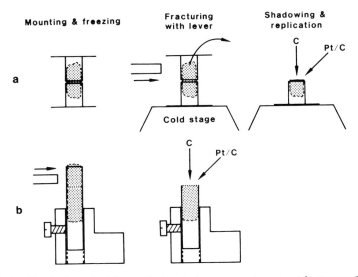

Fig. 6.11. Two basic methods for producing single (non-complementary) fracture surfaces for replication. (a) Specimens, frozen between two rivet-shaped holders, are fractured by breaking off the free specimen support from the fixed one. Alternatively (b) specimens (particularly suspensions) can be frozen in pre-scored capillary tubes which are subsequently broken apart. The arrow labelled **Pt/C** indicates the direction of platinum/carbon shadowing while the arrow labelled **C** shows carbon deposition onto the fracture face.

port. Replication takes place either immediately after fracturing or after etching. Such methods, using tensile stress to separate the two specimen holders, are suitable for both cell suspensions and bulk tissues.

A single fracture method for fracturing frozen suspensions or cells sandwiched between two pieces of aluminium foil has been described by Winkelmann and Meyer (1968) and Winkelmann and Wammetsberger (1968). The frozen sandwich is clamped in a two-part brass block that has been precooled in liquid nitrogen and is then transferred to the evaporation unit. When the fracturing (or etching) temperature is reached the specimen is fractured by tearing away the upper foil.

In other single fracture methods one stroke of a blade or a wedge mechanism is used to fracture the specimen under vacuum, as in the sliding cold block freeze-fracture device originally described by Bullivant and Ames (1966). Based on this principle, Akahori et al. (1972) and Akahori and Nishiura (1972) developed a simple device which provides good anticontamination shielding and can easily be constructed in a laboratory workshop. As shown in Fig. 6.12, the device consists of two matching wedge-like brass blocks. The specimen support holder is inserted into the lower block. The upper block, which contains the knife for fracturing, is held in position by a stop. When the stop is removed, the upper block slides down the inclined plane, fractures the specimen and the two shadowing tunnels come into alignment over the fracture face so that heavy metal and carbon evaporation can take place. As in the Bullivant and Ames type II device (§6.3.3c), the narrow tunnels of the upper cold block provide excellent protection against contamination of the fractured specimen surface by condensation.

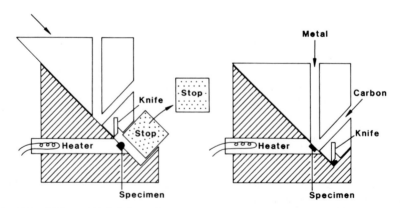

Fig. 6.12.   Sliding cold block freeze-fracture device providing good anticontamination shielding. (Redrawn from Akahori et al. 1972 and Akahori and Nishiura 1972.)

The lower block contains a heater to raise the specimen temperature for etching. The Bullivant and Ames type II single fracture device (§6.3.3c), originally designed for fracturing under liquid nitrogen, was modified by Mc-Alear and Kreutziger (1967) to allow fracturing of the specimen under vacuum by means of a spring-loaded upper block which houses a cutting device. Using the spring release, the upper block rotates, fractures the specimen and the tunnels are brought into alignment with the evaporation source for shadowing and replication.

An apparatus for fracturing either bulk specimens or frozen suspensions under vacuum, developed by Stolinski (1975), can be fitted to conventional coating-units. Fracturing is done using a cleaving mechanism (Fig. 6.13) in the form of a double-sided converging wedge which is operated from outside the vacuum chamber. The location of the fracture plane can be roughly

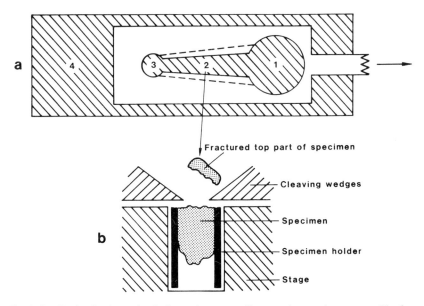

Fig. 6.13. Device for fracturing bulk specimens or cell suspensions under vacuum. The fracture occurs as the two cleaving wedges converge, leaving most of the fracture surface untouched. (a) Top view of the cleaver. The circled numbers show the relative positions of the specimen in relation to the moving cleaver. The specimen is loaded into the specimen holder (b) through an opening in the cleaver (position 1). Once under vacuum, the cleaver is moved in the direction of the arrow so that (position 2) the two wedges induce fracturing as shown in side view (b). Shadowing and carbon deposition take place with the specimen in position 3 before the cleaver is moved out of the way (position 4) so that the fractured specimen can be removed. (Redrawn from Stolinski 1975.)

predetermined by pre-aligning a chosen region of tissue prior to freezing in the specimen holder. This method has the advantage that the cleaving wedges only come into contact with the outer edges of the specimen, so leaving most of the surface untouched (Fig. 6.13b). The cleaver, which is maintained at a lower temperature than the specimen, also functions as an anti-contamination shield for the freshly fractured specimens during etching. The temperature of the specimen stage can be controlled by balancing cooling using evaporating liquid nitrogen with heating using a small block heater. A modified form of this freeze-fracture stage has been incorporated into a Balzers BAF 400D freeze-etching apparatus (Stolinski et al. 1984).

Single fracture methods for use under vacuum have also been adapted for use in more elaborate apparatus for freeze-fracture replication. In the Steere-type freeze-etch module (Steere 1969, 1973), the specimen is frozen onto a metal support, mounted on a temperature-controlled cold stage and fractured under vacuum with a cold steel blade, mounted on a rod (§6.3.3b(i)).

(ii)   Devices to produce complementary fracture faces

The development of freeze-fracture devices which allow the retention of both faces produced by a single fracturing process was stimulated by problems concerning the interpretation of membrane fracture planes, and the nature and mechanism of artefact formation (see reviews by Sleytr and Robards 1977a, 1982). Within one year, four separate groups independently developed methods for obtaining complementary replicas (Steere and Moseley 1969, 1970; Wehrli et al. 1970; Chalcroft and Bullivant 1970; Sleytr 1970a), applying two basic principles.

Using one method, the so-called 'broken-capillary' or 'snapped-tube' device, the specimen is frozen in a small tunnel-topped metal capillary tube which is scored to predetermine the point of fracture. The tube is fixed to the cold stage of a freeze-etching unit and fractured, after obtaining a high vacuum, with the arm of a microtome. The broken-off part falls, upside-down, into a cage alongside the lower part, and both fracture faces are replicated simultaneously (Wehrli et al. 1970). A similar method, adapted for tissues, was later devised by Tonosaki and Yamamoto (1974). Although high quality replicas of complementary fracture faces were obtained with these methods, they had the disadvantages that the specimen holders used were too complicated, that the broken-off part had only poor thermal contact with the cold stage and that both fracture faces were not accurately aligned with respect to one another during shadowing. Using the snapped-tube

principle, the problem of aligning the fracture faces was only properly overcome when methods were perfected for fracturing under liquid nitrogen or
helium (§6.3.3c).

The second and most common method for producing complementary
fracture faces under vacuum involves the use of hinged devices. Steere and
Moseley (1969) were the first to develop a hinged holder to yield complementary fracture faces (Fig. 6.14a). Their complementary fracture device
(Steere and Moseley 1970; Steere 1973) is now available from Denton (see
Appendix 2). Subsequently, other designs of complementary replica device
were described by Mühlethaler et al. (1970) for the Balzers apparatus (Fig.
6.14b) and by Sleytr and Umrath (1974a) for both the Leybold-Heraeus
universal electron microscopical preparation plant EPA 100 and the Bioetch
2005 automatic freeze-fracture replication unit (Fig. 6.14c). Both the Steere
complementary replica device (Dempsey et al. 1974) and the Sleytr and

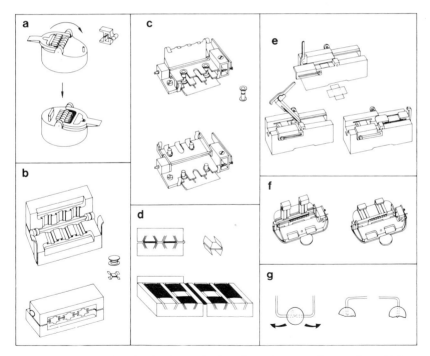

Fig. 6.14.   A variety of devices for the production of complementary fracture faces. (a) Steere
and Moseley (1969); (b) Mühlethaler et al. (1970); (c) Sleytr and Umrath (1974a); (d) Müller
et al. (1980b); (e) Nermut and Williams (1977); (f) Robards et al. (1980); (g) Menold et al.
(1972, 1976). The modes of operation of these devices are described in the text.

Umrath device (Sleytr 1975) can be adapted for use with the Balzers freeze-etching unit.

The complementary replica device for the Denton modular freeze-fracture replication unit (Fig. 6.14a) is loaded under liquid Freon (or liquid nitrogen) which is cooled with liquid nitrogen to a temperature close to its melting point (Steere 1973). Up to six specimens, frozen in the hinged holders illustrated in Fig. 6.2 (type F), can be mounted in the fracturing device. Fracturing takes place under vacuum according to the Steere-type freeze-fracture replication routine (§6.3.3b).

For the hinge mechanism developed by Mühlethaler and collaborators (1970), the specimens are mounted and frozen between a pair of specimen supports of type B, C or E (Fig. 6.2). The device shown in Fig. 6.14b represents a slight modification of the original design which is commercially available from Balzers (see Appendix 2). The mechanism is loaded, under liquid nitrogen, with three frozen specimen 'sandwiches' and subsequently transferred to the pre-cooled cold stage (see Fig. 6.15 below) of the Balzers unit. The device is opened under vacuum by pulling the upper part over using the microtome arm or lever. A hinged device similar to the Mühlethaler design, but especially adapted for type E holders (Fig. 6.2) which are suitable for sandwich- or propane jet-freezing of suspensions, has been built and described by Müller et al. (1980b) (Fig. 6.14d) (see also Müller et al. 1984).

The Sleytr and Umrath (1974a) hinged complementary fracture device designed for the Leybold freeze-fracture replication units consists of a hinged plate with an adjustable cover (Fig. 6.14c). The cover plate has spring-clip holders to ensure good thermal contact between the cold stage and the upper specimen holder after fracturing and, therefore, equal etching of complementary fracture faces. The hinged plate is held together, with clamps, against the force of a spring and can be released either mechanically or with an electromagnet incorporated in the universal specimen cup of the Leybold-Heraeus Bioetch 2005 freeze-etching unit (§6.3.3b). Since hollow or closed rivet-type specimen holders (Fig. 6.2; types H and I) can be used, the device is suitable for the production of complementary fracture faces from either tissues or suspensions. The hinge mechanism is loaded under liquid nitrogen and transferred to the cold stage.

Nermut and Williams (1977) described a hinged, spring-loaded device for freeze-fracturing cell monolayers (Fig. 6.14e). The gadget was designed for the Balzers cold table but can easily be adapted for other freeze-etching units. The specimen is frozen between two pieces of mica or sheets of other material (metal, glass) aligned in a cross-wise fashion (Fig. 6.2; type G). The frozen sandwich is mounted on the pre-cooled device and fracturing takes place under vacuum by releasing the catch. A modification of the Nermut and Williams device, which allows fracturing of two sandwiches at once, has been designed by Robards et al. (1980) (Fig. 6.14f) for the Leybold-Heraeus Bioetch 2005 unit. With this modification, the fracturing device is loaded with specimens under liquid nitrogen and released automatically under vacuum by an electromagnet. This device can thus easily be adapted for standard evaporation units.

Menold and collaborators (1972, 1976) described a 'drop-freeze-fracturing device' suitable for preparing complementary fracture faces from small droplets of suspensions. The basic principles of the method are shown in Fig. 6.14g. A drop (0.5–1.0 mm in diameter) of suspension is frozen between the ends of two wires which are attached to a hinged copper block mechanism (see §2.4.1f). Under vacuum the drop is fractured by pulling the two wires apart. To avoid heat transfer to the drop from the warm parts of the device during freezing, wires with low thermal conductivity (tungsten) are used. After the drops have been rapidly frozen, the copper blocks are cooled down in liquid nitrogen. The device, which is preferably mounted on a supporting flange, is then transferred to any type of coating unit. The slow temperature increase of the specimen during pumping down the vacuum unit is measured with a thermocouple and fracturing is performed when a suitable temperature has been reached (Menold et al. 1976).

(iii) Freeze-fracturing with a microtome

The first successful and still most commonly used, microtome assembly for freeze-fracturing was developed by Moor et al. (1961) using a design similar to that originally described by Porter and Blum (1953). The microtome is mounted in the Balzers type BA 360M freeze-etching unit, or one of its successors: types BAF 300 and BAF 301. The microtome arm and the attached razor blade, which is used as a knife (Fig. 6.15a), can be cooled with liquid nitrogen to about 77 K ($-196°C$). Cutting (i.e. fracturing) of the frozen specimen can be done using either a mechanical or thermal advance mechanism. The specimen is transferred from liquid nitrogen (storage Dewar vessel) to a pre-cooled (approximately 123 K; $-150°C$) specimen stage. The stage incorporates both liquid nitrogen cooling as well as a heating element and can be maintained at temperatures from 77 K ($-196°C$) to above 293 K (20°C) with an accuracy of ±0.01 K. Depending on the number of samples to be fractured in one run, there are different designs of specimen table

Fig. 6.15.   Balzers freeze-etching unit. (a) Microtome in models BAF 300 and 301; (b) new model BAF 400; (c) microtome in model BAF 400; (d) evaporators and anticontamination shielding in model BAF 400. (Photographs reproduced by courtesy of Balzers AG.)

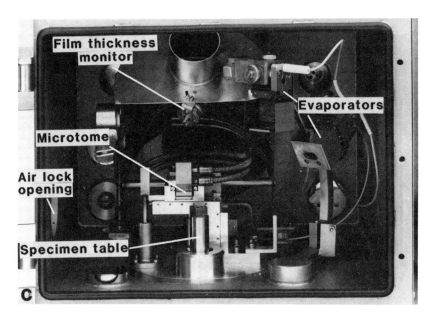

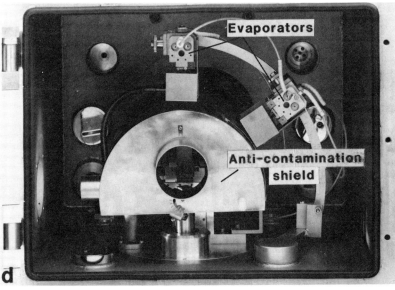

and different ways of mounting the sample (Nickel et al. 1978). The most common specimen supports used in the Balzers unit are types A–E (Fig. 6.2). To ensure good thermal contact, hoar frost from the specimen stage is washed off with a brush soaked in liquid Freon 22 (or 12) before the specimen is transferred to it. In the first model (BA 360 M) cutting is manually controlled but in the later models (BAF 300 and BAF 301) cutting can be both automatically or manually controlled. Experience with many specimens has shown that the mechanical advance alone is usually adequate.

A typical freeze-fracture run would proceed as follows. Cool the specimen stage under vacuum to 123 K ($-150°$C). After venting the unit (preferably with dry nitrogen), transfer the frozen specimen quickly to the specimen stage using cooled forceps and then evacuate the chamber again. During evacuation the microtome is cooled down to liquid nitrogen temperature and the specimen is warmed to the fracturing temperature. Fracturing, etching and replication are then done at a working pressure of about 2×10^{-4} Pa (1.5×10^{-6} Torr). If etching is not required, the specimen temperature is usually maintained below 173 K (in the range of 153 to 163 K; -120 to $-110°$C) and the replica is evaporated onto the surface immediately after the final cut. If etching *is* required (§6.5) fracturing takes place at a specimen temperature of about 173 K. After the final cut, place the microtome arm as rapidly as possible above the specimen to act as a cold trap to reduce any possible contamination. After etching, remove the knife and evaporate the shadowing material and the carbon backing layer (§6.4). Again, place the knife above the specimen before admitting air to reduce hoar frost formation on the replica. This precaution improves the chances of the replica floating on the thawing fluid. After each run, vent the bell-jar and warm up and defrost all cooled parts (microtome and specimen table). The single-edged razor blade, which is used as the microtome knife, is replaced after each fracturing process and must be carefully degreased before use (by washing in acetone or chloroform) to avoid contamination of the fracture surface with outgassing hydrocarbon vapours. The Balzers apparatus is available with both resistive evaporators and electron guns. Deposition rates during evaporation can be checked using a quartz crystal monitor (§6.4.1e). On later Balzers freeze-etching units there is the option of having either an oil diffusion pump or a turbo-molecular pump for the vacuum system.

The latest version of the Balzers freeze-fracture replication unit, model BA 400, incorporates essentially the same features as the earlier models (BAF 300 and BAF 301). This unit (Fig. 6.15b, c and d) has a large, although not completely closed, liquid nitrogen-cooled anticontamination shield over the specimen area. The unit can be equipped with a complementary replica device, a rotary shadowing specimen table (§6.4.1f) or a liquid nitrogen-cooled microtome which, compared with the previous model (Fig. 6.15a), is considerably smaller and consequently more economical in liquid nitrogen consumption. Furthermore, it does not have a thermal advance mechanism. As with the earlier models (BAF 300 and BA 301), the microtome can be controlled either automatically or manually. To avoid hoar frost formation on the liquid nitrogen-cooled cold shrouds and specimen stage, the frozen specimens, attached to the specimen table, can be transferred from liquid nitrogen to the specimen stage against a counter-flow of dry nitrogen gas through a simple door or more complex air-lock system. Thus, when equipped with electron guns, several preparations can be made without warming up the cold parts for defrosting, as was necessary with the earlier models. The pumping system is automatically controlled and can be equipped with an oil diffusion pump, a turbo-molecular pump or a refrigerator cryopump system (see Fig. 6.10 in §6.3.2). The specimen can be observed through a window in the work chamber and an opening in the cold shroud, using a binocular microscope attached to the front door (Fig. 6.15a and b). Due to the large anticontamination shields over the specimen area, the new Balzers freeze-fracture replication unit gives excellent results even when freeze-fracturing is done close to 77 K.

A basically similar microtome design to the early Balzers model (BA 360M) was used in the NGN freeze-etching unit developed by Preston and Barnett (Barnett 1969; Robards and Cooper 1971). The essential difference is that the blade holder of the microtome arm is moved directly across the specimen and not in a circular path as in the Balzers unit. In addition, special knife systems for this unit have been devised by Robards et al. (1970) which use two razor blades on the microtome arm for rough and final cutting and for varying the knife settings. Other instruments which resemble the Balzers unit were developed by Koehler (1966, 1972b) and Elgsaeter (1978). There is now a variety of commercially available apparatus (see Appendix 2) incorporating less complex knife mechanisms for controlled multiple specimen fracturing.

6.3.3b   Freeze-fracturing under improved vacuum conditions

Condensation of material onto the fracture face of the freshly cleaved specimen in the period between fracturing and replication is one of the main

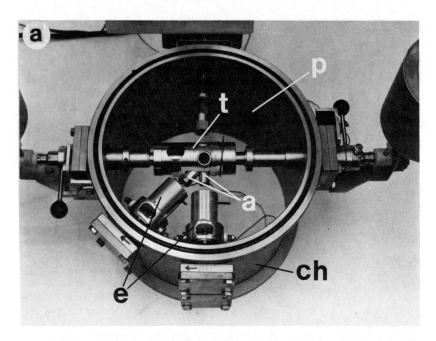

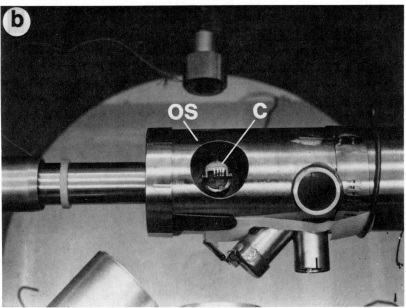

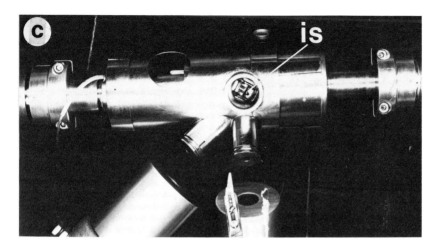

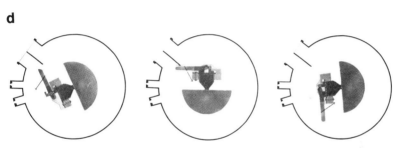

Fig. 6.16. The Steere-Denton freeze-fracture replication unit with liquid nitrogen-cooled double shroud. (a) View of the module consisting of a metal chamber (**ch**) with a glass cover-plate (**p**); the fracturing assembly sits within opposing hollow tubes (**t**), serving as cold shrouds, which have apertures (**a**) through which the evaporation sources (**e**) can deposit the shadowed replica. (b) Double shroud with complementary replica cap (**c**) in place in outer shroud (**os**) compartment. (c) Double shroud with complementary replica cap in inner shroud (**is**) compartment. (d) Diagram of specimen tube and specimen stage, with specimen holder clamped in place and surrounded by inner chamber of shroud. Left to right: units in position for fracturing; locked-down position after fracturing (e.g. during etching); aligned for shadowing and carbon coating. (Reproduced, with permission, from Steere and Moseley 1970.)

sources of artefacts (§6.8.3) since fine structual details can become masked or enhanced by specific decoration. Therefore, it is important that the vacuum is as good as possible. Two different approaches have been made to obtaining an excellent vacuum environment, with a minimum of condensable gases in the specimen area during exposure of the fracture plane. One method consists of completely enclosing the specimen before fracturing within a liquid nitrogen- (or helium-) cooled cold trap or cold shroud (Steere 1973; Sleytr and Umrath 1976; Niedermeyer 1982a). The other method involves more complex ultra-high vacuum technology.

(i)   Cold shroud devices

The first freeze-fracture replication unit with a liquid nitrogen-cooled 'double shroud' around the specimen area was developed by Steere (1969) and Steere and Moseley (1970). The system (for detailed description see Steere 1983) is available from Denton Vacuum Incorporation (see Appendix 2). As illustrated in Fig. 6.16, the freeze-fracture replication module consists of a chamber (**ch**) with a glass cover plate (**p**) and two opposing hollow tubes (**t**) which can be rotated. A 'double chamber shroud' consisting of an inner and an outer chamber is attached to the moveable hollow tubes. The outer chamber (**os**) has one large opening, through which the specimen is mounted on the specimen stage on the inner tube, and two small apertures (**a**) for evaporation. Both the specimen stage and the cold shroud are cooled with liquid nitrogen through tubes from the outside and, after introducing the specimen, they can be rotated with respect to each other so that the specimen is completely enclosed by cold surfaces during fracturing and etching. Thus the fracture planes are only exposed to warm surfaces during replication. This complete protection with cold surfaces is only provided when the complementary fracture specimen cup is used (see Fig. 6.14a). In the Steere freeze-fracture replication unit, fracture faces can be also obtained by 'shaving' the specimen surface with a scalpel blade mounted at the end of a push-pull rotary probe, although the specimen cannot be surrounded by the double shroud while this takes place.

For a typical freeze-fracture replication run using the complementary replica technique, the specimens are frozen in hinged holders (Fig. 6.2; type F). The frozen specimens are then mounted under liquid Freon 22 in the specimen cup illustrated in Fig. 6.14a and transferred with special forceps to the pre-cooled (123 K; $-150°C$) cold stage. The unit is pumped down and the cold shrouds are cooled to liquid nitrogen temperature. After a good vacuum $(1.3 \times 10^{-3}$ Pa, $< 1 \times 10^{-5}$ Torr) is obtained, the specimen is warmed to 173 K ($-100°C$) and held at this temperature until any frost which has formed during the transfer has disappeared from the cup and the specimens. When the shrouds have cooled down to liquid nitrogen temperature and a vacuum of 2.6×10^{-4} Pa (2×10^{-6} Torr) or better is reached, the specimen is cooled to 93 K ($-180°C$) or colder and the shroud is moved inwards until the specimen is enclosed in the inner chamber of the shrouds. Specimens are fractured by rotation of the shroud and the specimen tube in opposite directions, which lifts the bar attached to the specimen cup and turns the upper part of the hinged specimen holder through 180° (Fig. 6.14a and 6.16d). If etching (§6.5) is required, the specimen stage is warmed to about 173 K while the fracture faces remain enclosed by the shroud. For shadowing (§6.4), the specimens are recooled to 123 K ($-150°C$) or colder and the specimen tube and shroud are rotated to the 45° shadowing position. The specimen temperature is then raised to 163 K ($-110°C$) for evaporation of a carbon backing film. Finally, the vacuum system is brought to atmospheric pressure with dry nitrogen gas and the complementary replica device is removed from the vacuum chamber.

A cold trap-type freeze-fracture replication unit with a high degree of automation was developed by Sleytr and Umrath (1976) and Umrath (1978) and manufactured by Leybold-Heraeus as the Bioetch 2005. The apparatus (Fig. 6.17a) consists of a cylindrical vacuum chamber equipped with quick-connection flanges for the specimen cup transport device and the evaporators, and a liquid nitrogen tank, which acts as a vacuum baffle. Two concentric liquid nitrogen-cooled cylinders act as cold shields for the specimens; one of these cylinders is fixed to the nitrogen tank, the other to the moveable specimen stage. A pneumatically driven bellows-sealed cylindrical air-lock can be moved over the outer cold shield to avoid ice condensation on the shields during venting of the air-lock for specimen and evaporator exchange. In the schematic drawing (Fig. 6.17a) the air-lock is shown cut through the central line to demonstrate the opened and closed positions. The temperature of the specimen stage and of the specimen cup (Fig. 6.17b) can be accurately controlled using two independent platinum resistance thermometers. The specimen stage and the inner cold shield attached to it can be rotated to predetermined positions during the preparation cycle using a motor. For routine preparation, specimens are frozen in specimen supports of type H or I (Fig. 6.2) and mounted under liquid nitrogen on the complementary fracture device (Fig. 6.14c) in the universal specimen cup shown in Fig. 6.17b and c. Before use, the unit is evacuated and the liquid nitrogen tank is filled. This automatically cools the two cold shrouds and the specimen stage. The specimen cup, with the specimens, is now transferred from liquid nitrogen to the preparation plant using a bellows-sealed transfer device (Fig. 6.17a). When a vacuum of 10.6 Pa (8×10^{-2} Torr) is attained in the air-lock, the air-lock is opened automatically and the specimen cup can be inserted through the aligned openings of the outer and inner

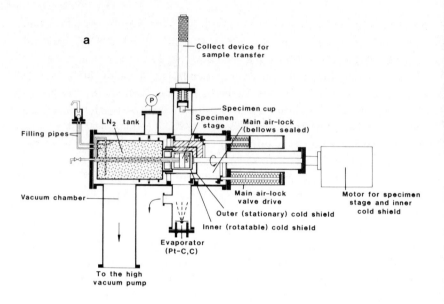

a

Collect device for
sample transfer

Specimen cup

Specimen
stage

Main air-lock
(bellows sealed)

LN$_2$ tank

Filling pipes

Vacuum chamber

Main air-lock
valve drive

Motor for specimen
stage and inner
cold shield

Outer (stationary) cold shield

Inner (rotatable) cold shield

Evaporator
(Pt–C,C)

To the high
vacuum pump

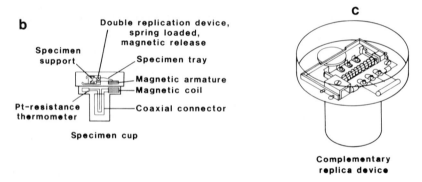

b

Double replication device,
spring loaded,
magnetic release

Specimen
support

Specimen tray

Magnetic armature
Magnetic coil

Pt-resistance
thermometer

Coaxial connector

Specimen cup

c

Complementary
replica device

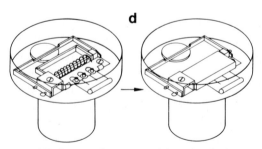

d

LN$_2$-fracture (He-fracture) transfer device

cold shields into the socket in the pre-cooled specimen stage. The transfer device is then locked in the upper position shown in Fig. 6.17a. Completion of the preparation run is now fully automatic. The individual operating parameters, such as specimen temperature and vacuum status during the fracturing process, time of etching and evaporation conditions, can all be preset on an electronic process control unit.

The first automatic action in the cycle is that the specimen stage, together with the inner cold shield, is rotated in relation to the outer cold shield so that the specimen is in a completely enclosed position. The two liquid nitrogen-cooled shields now act as an opaque screen against any contamination from condensable gases. This is the position in which fracturing and etching are performed after the preset criteria for temperature and pressure have been achieved. Only immediately prior to evaporation is the specimen stage turned so that the cold shroud is opened again. To reduce any sudden outgassing during evaporation, both resistive sources (or an alternative electron gun evaporation system) are degassed by automatic pre-heating at the beginning of the cycle while the specimen is still enclosed within the cold shrouds. Finally, after the specimen has been rotated back to its original position, the specimen cup can be removed via the air-lock using the transfer device.

Instead of a double replica device (see Fig. 6.14), other gadgets can be mounted in the universal specimen cup, such as a device for contamination-free transport of specimens from liquid nitrogen into the vacuum chamber (Fig. 6.17d). This is particularly useful when fracturing or handling specimens under liquid gases (§6.3.3c) such as is called for when using freeze-etching in combination with cytochemical techniques (§6.9). When equipped with an oil diffusion pump, the attainable pressure of the cold-trapped chamber is in the range of 2–3×10^{-6} Pa (1.5–2.25×10^{-8} Torr) as measured outside the cold shields. Within the liquid nitrogen-cooled shields the partial pressure of water is considerably better and should reach the order of

Fig. 6.17. Leybold-Heraeus Bioetch 2005 freeze-fracture replication unit with sophisticated automatic system and various devices for the study of specimen fracture faces obtained under vacuum, liquid nitrogen or liquid helium. (a) Cross-section of the apparatus. (b) Detailed view of the specimen cup with complementary replica device. (c) Complementary replica device after fracturing. (d) Device for contamination-free transport of specimens freeze-fractured under liquid nitrogen or liquid helium into the vacuum chamber. The anticontamination hood is placed over the fractured specimens under liquid nitrogen. (Redrawn from Sleytr and Umrath 1976 and Sleytr et al. 1981.)

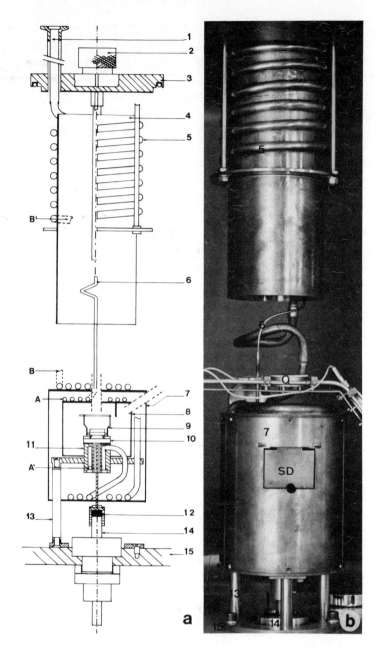

1.2×10^{-11} Pa $(9 \times 10^{-14}$ Torr). Thanks to the air-lock system, the cold shrouds remain continuously under high vacuum and thus maintain their pumping capacity during specimen and evaporator exchanges.

A helium-cooled cold shroud device suitable for installing in a standard Balzers 301 vacuum unit for freeze-fracture replication at about 10 K has been developed by Niedermeyer (1982a). A schematic drawing and illustration of the cooling device are shown in Fig. 6.18. Cooling of the whole device is achieved using a continuous flow cryostat in which the coolant is drawn, by a vacuum pump, from a liquid helium container through the cooling coils. The required temperature is adjusted by changing the helium flow-rate using a valve situated between the cooling coils and the pump. As illustrated in Fig. 6.18a, the vacuum-isolated helium transport pipe (2) is connected to the cooling device which is on top of the vacuum chamber. The pipe inside the vacuum chamber (6) leads directly to the inner shroud (8) where the lowest temperature is achieved (approximately 6.0 K). Subsequently, the cryostat (11), with the specimen stage (10) and specimen holder (9), is cooled. After cooling the cryostat (11), the helium cools the outer shroud (7) which reaches a temperature of approximately 20 K. Finally, the helium passes through a cryopump (4) to improve the vacuum in the chamber before it leaves the apparatus (1).

To allow rotary shadowing (§6.4.1f), the specimen stage can be rotated using an external motor via a rotary lead-through (14). The connection to the cryostat is achieved with a strong permanent magnet (12). To allow the quick, hoar-frost free transfer of the specimen holder through a swing door, a simple air-lock system has been developed which can easily be connected to the chamber with a flange (Niedermeyer 1982b). At the start of the cooling process, the vacuum is in the range of 13×10^{-6} Pa $(9.7 \times 10^{-8}$ Torr). During the cooling process, this improves to 3×10^{-6} Pa $(2.2 \times 10^{-8}$ Torr). It is not possible to measure the actual vacuum within the shroud but the

---

Fig. 6.18. Liquid helium-cooled cold shroud (developed by Niedermeyer 1982a) for the Balzers 301 vacuum unit. This attachment allows freeze-fracture replication at about 10 K. (a) Schematic drawing of the system. **1**, Helium outlet; **2**, helium inlet; **3**, vacuum seal from the normal Balzers apparatus; **4**, cryopump; **5**, tubing around cryopump; **6**, helium pipe; **7**, outer shroud; **8**, inner shroud; **9**, specimen holder; **10**, specimen table; **11**, cryostat; **12**, permanent magnet; **13**, support pillar for the shroud; **14**, rotary lead-through; **15**, base of the vacuum chamber. **A...A′** and **B...B′**, helium pipes leading from **A** to **A′** and **B** to **B′** (dashed lines, not drawn to scale). (b) Photograph of the device showing the position of the swing-door (**SD**) for specimen transfer and the position of the quartz crystal. (Illustration kindly provided by W. Niedermeyer and reproduced with his permission.)

author estimates a vacuum of about 1.3×10^{-8} Pa (1×10^{-10} Torr). Analysis of the residual gases outside the cold shrouds, before and during replication, by mass spectrometry revealed that the contamination-free vacuum conditions were equivalent to those of ultra-high vacuum systems, even down to a specimen temperature of 12 K. The effectiveness of the helium-cooled cryoshield and the cryopump in the standard high vacuum apparatus was confirmed by analysis of the quality of yeast plasmalemma fracture faces as test specimens.

(ii)   Ultra-high vacuum technology

Relatively complex ultra-high vacuum (UHV) technology provides an alternative means of obtaining an excellent vacuum around the specimen. UHV pumping units are designed to pump out all gases in the system and not only those which represent the potential hazard of leading to condensation contamination when the specimen is sufficiently cold. Due to its complexity, UHV technology has rarely been used for freeze-fracture replication, although some relatively early work was done with such systems (Kreutziger 1970). Koehler (1972a) described the first attempt to use a UHV freeze-fracture replication unit. The apparatus incorporated a titanium getter pump and an orb-ion pump. The roughing system consisted of a Gast carbon vane turbine pump and two cryosorption pumps filled with Linde molecular sieve material. Fracturing was done with a simple freeze-fracture block device for single fracturing under vacuum. Since the block was loaded with specimens under liquid nitrogen external to the apparatus, the problem of frost formation during transfer into the unit does not seem to have been overcome in this experimental system.

UHV freeze-fracture units with air-lock systems have been developed by Escaig and Nicolas (1976), Gross et al. (1978a) and Balzers Union. The apparatus from the French group (Escaig and Nicolas 1976), illustrated in Fig. 6.19, is manufactured by Reichert-Jung (see Appendix 2). The vacuum system comprises a rotary vane pump (**6**), a liquid nitrogen-cooled sorption pump (**7**) and an ion-getter pump (**4**). The manual and electric valves have stainless steel bellows and Viton gaskets and the rotary feed-throughs of the work chamber are also bellows-sealed. The system can be baked at temperatures up to 250°C. An ultimate pressure of 2.67×10^{-6} Pa (2×10^{-8} Torr) is reached in 4 h without baking and a pressure of 4×10^{-7} Pa (3×10^{-9} Torr) in 10 h with a baking cycle of 4 h. For the primary vacuum, the chamber is pumped by the rotary pump (**6**) and the liquid nitrogen-cooled sorption pump (**7**). The rotary pump is used alone until a pressure of approximately 13.3 Pa (1×10^{-1} Torr) is obtained; back-streaming of hydrocarbons from the pump (**6**) to the chamber (**1**) is prevented by a molecular sieve trap (**8**). The specimen stage (**11**), which can be cooled down to 80 K

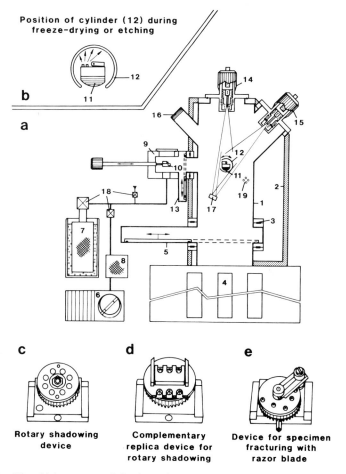

Position of cylinder (12) during freeze-drying or etching

Rotary shadowing device

Complementary replica device for rotary shadowing

Device for specimen fracturing with razor blade

Fig. 6.19. Ultra-high vacuum unit for freeze-fracture replication at temperatures down to 12 K (Escaig and Nicolas 1976). (a) The unit incorporates a rotary pump (**6**), a liquid nitrogen-cooled sorption pump (**7**) and an ion getter pump (**4**). Specimens attached to the holder (**10**) are inserted via an air-lock system (**9**). For explanation of other numbers see text. (b) Detail showing the position of the cooled cylinder (**12**) in relation to the specimen stage (**11**) during etching or freeze-drying. (c–e) Several types of specimen holder are available, including a rotary shadowing device (c), a complementary replica device for rotary shadowing (d), and a device for freeze-fracturing using a razor blade (e). (Redrawn from original illustrations provided by courtesy of Reichert-Jung.)

with cold nitrogen gas or to 12 K with liquid helium, can be tilted, using an external control, to any shadowing angle. The stage is partially encased in a cylinder (12) which can be cooled with liquid nitrogen to trap water vapour liberated from the specimen if etching is done. A modification of the cooling system for the shroud, in which the flexible nitrogen-supply pipes are replaced by flexible copper braid, has been suggested by Allain and Ardonceau (1983). This has the effect of increasing reliability while retaining the high quality of replicas. The electron guns for heavy metal (14) and carbon (15) evaporation are water-cooled. It is possible to adjust an external control so that the electrodes of the electron guns are moved up to 10 mm. This means that several evaporations can be made (depending on the thickness of the evaporated layers) without the need to vent the work chamber to change the evaporants or adjust the guns. A quartz crystal oscillator (17) allows deposition rates to be monitored (§6.4.1e), while a quadrupole mass spectrometer (19) provides information about both total and partial pressures in the system.

The sample holder, having been loaded with specimens under liquid nitrogen, is introduced into the air lock (9) which can subsequently be pumped down to 1.3×10^{-1} Pa (1×10^{-3} Torr) within 2 min. During this time, the temperature of the specimen holder rises by about 20 K. When the valve (13) is opened to allow the specimen to be inserted into the chamber, the pressure within the work chamber only increases to about 1.3×10^{-4} Pa (1×10^{-6} Torr). The pressure falls to 4×10^{-6} Pa (3×10^{-8} Torr) as soon as the valve (13) closes and reaches approximately 1×10^{-6} Pa (7×10^{-9} Torr) after about 5 min. This good vacuum, together with the very limited pressure increase during evaporation, means that fracturing can take place without any problems of contamination down to a temperature of 12 K (Escaig et al. 1980).

A variety of specimen holders has been developed for the Escaig apparatus. In addition to the standard hinged complementary replica device, revolving specimen supports for rotary shadowing (Fig. 6.19c), complementary replica devices (Fig. 6.19d) and devices for fracturing with a razor blade (Fig. 6.19e) are available. During freeze-drying or freeze-etching processes, the cylinder (12) is cooled with liquid nitrogen and maintained in the position shown in Fig. 6.19b to trap on the internal walls the subliming water vapour molecules, which would otherwise destroy the good ultra-high vacuum.

The UHV equipment designed by Gross (1977) and Gross et al. (1978a) is based on a converted Balzers BA 350 U unit. The system is evacuated using five pumps working in series: a rotary vane pump, two oil diffusion pumps, a liquid nitrogen-cooled baffle plus Meissner trap and a titanium sublimation pump. A Viton gasket is only used for the main flange of the chamber, all other flanges being sealed with gold wire. The unit allows fracturing, shadowing and coating of specimens at 77 K in a vacuum of about 1.33×10^{-7} Pa (1×10^{-9} Torr). Specimens are frozen in scored capillary tubes 1 mm in diameter (see Fig. 6.11b in §6.3.3a) and loaded under liquid nitrogen (Fig. 6.20) in a specimen cartridge which is then screwed to a manipulator. Together with the cartridge, a cylinder with a lid is attached to the manipulator which allows transfer of the cartridge into the air-lock under the protection of liquid nitrogen. The air-lock itself consists of a double chamber system. In the first chamber, the liquid nitrogen is poured out by removing the lid from the cylinder and the cartridge is then transferred to the second chamber. When a vacuum of approximately 4 Pa (3×10^{-2} Torr) has been achieved, a gate connecting the main chamber to the UHV chamber is opened and the cartridge is screwed to the pre-cooled specimen stage. The specimen temperature is then raised to $-100°C$ (173 K) for sublimation of both Freon (if it was used for freezing) and ice contaminants, and is then cooled back down to $-196°C$ (77 K). Fracturing is done with a liquid nitrogen-cooled lever which breaks the capillaries and the contained specimens in the region of the score-mark (as indicated in Fig. 6.11b). The

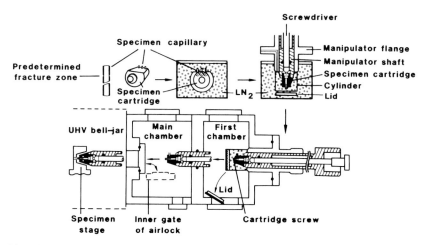

Fig. 6.20. Specimen mounting and two-chamber air-lock system for inserting specimens into an ultra-high vacuum unit (from Gross 1977 and Gross et al. 1978a). Specimens are frozen in scored capillary tubes (see Fig. 6.11b in §6.3.3a) which are loaded into the specimen cartridge while under liquid nitrogen. (Redrawn after Gross 1977 and Gross et al. 1978a.)

broken-off parts are collected in a liquid nitrogen-cooled baffle, thus preventing contaminative outgassing. Heavy metal and carbon are evaporated with outgassed electron guns. With the exception of a very short (<1.0 s) pressure increase during fracturing, the whole process, including evaporation, is performed in a vacuum of approximately 2.6×10^{-7} Pa (1.95×10^{-9} Torr). The deposition rate is about 0.2 nm s^{-1} during platinum/carbon shadowing (§6.4.1).

Balzers Union has produced a commercially available UHV freeze-fracture replication system (Type BAF 500K). The automatic vacuum unit comprises a refrigerator cryopump system (see also Fig. 6.10 (type III) in §6.3.2) which produces an ultimate vacuum of 5×10^{-8} Pa (3.75×10^{-10} Torr), together with a catalyst trap to remove oil vapour and an integrated closed-circuit hot water system for heating the chamber to 80°C. A gate valve is positioned between the cryopump and the UHV vacuum chamber and by-pass pumping is possible using a turbo-molecular pump. An automatic pumping system control monitors the different vacuum systems and valves but the individual elements can also be manually controlled. The specimen table is refrigerator-cooled and, in combination with a heating element, temperatures between 8 and 293 K (-265 and $+20$°C) can be achieved with an accuracy of $\leqslant 0.2$ K. A tilting device on the cooled UHV specimen table allows for continuous adjustment of the evaporation angle between 0 and 90°. Complementary replica devices (Müller et al. 1984) and specimen supports for TEM grids (e.g. for freeze-drying) can also be used. The triggering mechanism on the fracturing device is cooled to $\leqslant 53$ K (-220°C) via the first cooling stage of the refrigerator cold table. Three automatic water-cooled electron-beam evaporators allow up to eight freeze-fracture replication runs per UHV cycle. Since the evaporation is in an upward direction, not only platinum/carbon but also tantalum/tungsten, silver, gold and other substances can easily be evaporated (§6.4.1). A manually controlled, pneumatically operated shutter is cooled to $\leqslant 53$ K (-220°C) and coupled to the quartz crystal film-thickness monitor. The high-vacuum lock for specimen transfer can be pumped with the turbo-molecular pump to an ultimate vacuum of $\leqslant 3 \times 10^{-5}$ Pa (2.25×10^{-7} Torr). If the pressure in the lock is $<5 \times 10^{-5}$ Pa (3.8×10^{-7} Torr), then the pressure in the UHV vacuum chamber rises from $<1 \times 10^{-7}$ Pa to $\leqslant 5 \times 10^{-6}$ Pa (3.8×10^{-8} Torr) during specimen transfer but recovers within 5–10 min to the original value. The high-vacuum lock allows UHV cycles of less than 15 min. A loading system using a counter-flow of dry nitrogen gas allows the specimen table to be immersed in liquid nitrogen and then passed through a fore-vacuum lock into the high-vacuum lock without frost contamination. An unacceptable increase in specimen temperature during transfer through the high-vacuum lock is prevented by cooling the cold table with liquid nitrogen in the high-vacuum lock. The temperature of this table can be adjusted to between 293 and 123 K ($+20$ and $-150°C$), enabling freeze-drying to be done under good vacuum conditions. The valve between the high-vacuum lock and the UHV vacuum chamber is operated automatically by the specimen insertion device.

6.3.3c *Freeze-fracturing outside the coating unit under a liquid gas*

(i) *Freeze-fracturing under liquid nitrogen*
Methods for the evaluation of fracture faces obtained under liquid nitrogen were originally developed independently by Bullivant and Ames (1966) and Geymayer (1966).

The routine preparation procedure using the Bullivant-Ames device is illustrated in Fig. 6.21. The apparatus consists of three cylindrical brass tiers, each of 70 mm diameter (Fig. 6.21a). Before fracturing the specimen, the lower tier (**1**) attached to a small polypropylene (or metal) container (**4**) and the two other parts (**2, 3**) of the assembly are all immersed and cooled down in liquid nitrogen in an expanded polystyrene container (**5**). To obtain single fracture planes, the frozen specimen with its holder is mounted onto the base plate and is fractured using a scalpel or single-edged razor blade (Fig. 6.21b). For complementary replicas, the specimen is mounted within two hollow specimen holders (Fig. 6.2; type I) positioned together in an end-to-end fashion. The frozen specimen is fractured under liquid nitrogen by pulling the holders apart, and the holders are then positioned side-by-side (Fig. 6.21c) in the base plate. Specimen holders of type G (Fig. 6.2) can be fixed by two spring clips attached to the lower block and fracturing under liquid nitrogen is performed by pulling the upper sheet (Brown 1981). After fracturing, the middle cylinder with the two (2.5 mm) shadowing tunnels is locked into position on a pin in the lower cylinder and, finally, the upper tier, which functions as a protection lid, is placed on top of the middle cylinder. The whole assembly, together with some of the liquid nitrogen, is now lifted out of the expanded polystyrene container and transferred to a coating unit (Fig. 6.21d) where it is fitted onto the base plate with the shadowing tunnels in proper alignment with the evaporation sources. The top lid is connected with a metal wire to a small electrically operated crane in the evaporator (Fig. 6.21e), the bell-jar is placed in position and the unit

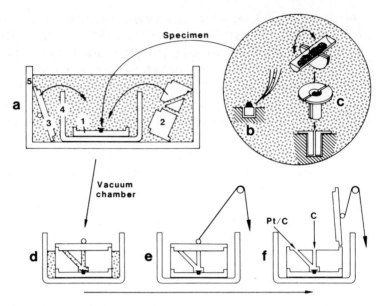

Fig. 6.21.   Procedure for freeze-fracturing under liquid nitrogen using the Bullivant-Ames device (Bullivant and Ames 1966). (a) The three-tiered apparatus comprises a base-block (**1**) which is fixed to a small container (**4**); a middle piece with tunnels (**2**); and a lid (**3**). The whole assembly is cooled in liquid nitrogen before loading. (b,c) Specimens are loaded into holders under liquid nitrogen and then fractured either by using a knife (b) or by tensile stress (c) and are placed in the central cylindrical hole in the base-block (**1**). (d) The complete and assembled apparatus, with the specimen in position, is filled with liquid nitrogen and transferred into a high-vacuum coating unit. (e) The lid of the apparatus (**3**) is attached to a lifting mechanism (e.g. a simple electric motor or a rotary vacuum lead-through) and the work chamber is evacuated. (f) When the nitrogen has sublimed away (it usually solidifies during evacuation) and the chamber has attained high vacuum, the lid is lifted away and the frozen-fractured surface is shadowed with platinum/carbon (**Pt/C**) and then replicated with carbon (**C**). (Redrawn after Bullivant and Ames 1966.)

is rough-pumped. Under reduced pressure, the nitrogen first solidifies and then sublimes away. At a pressure of 1.33×10^{-1} Pa (1×10^{-3} Torr) the diffusion pump is opened to the vacuum and, once a vacuum of better than 1.33×10^{-3} Pa (1×10^{-5} Torr) is reached, the lid is lifted (Fig. 6.21f) and the exposed specimen fracture plane is shadowed and replicated. Measurements have shown that the specimen temperature on lifting the lid is around 133 K ($-140°$C) and thus no etching takes place. Results obtained with this simple device have demonstrated that the narrow cold-shrouding tunnels provide excellent protection against condensation contamination, even under these moderate vacuum conditions (Bullivant 1980b).

Rather different from the Bullivant-Ames technique, which can be adapted to almost any evaporation unit, is the Geymayer (1966) method, which was especially developed for the Leybold-Heraeus EPA 100 EM preparation plant (Geymayer 1967). This is now manufactured by Paar K.G. (Austria) as the EPA 101 (Appendix 2). The system uses the facilities provided by a revolving specimen stage and a cylindrical cold trap. Specimens are mounted and frozen on specimen supports of type C, H or I (Fig. 6.2) in such a way that the material is slightly protruding (Fig. 6.2; Ca). Frozen specimens are mounted on a carrier plate immersed in liquid nitrogen and fractured with a single-edged razor blade attached to a simple Reichert cryomicrotome (Fig. 6.22a). After fracturing, the specimens are transferred, using a special transport device (Fig. 6.22b), while protected by the liquid nitrogen trapped between the hood and the carrier plate, into the evaporation unit (Fig. 6.22c). This is often done through an air-lock system to avoid hoar frost formation on the cold shields. During pumping down, the specimen is kept very close to a cold trap. After a high vacuum has been obtained, the specimen is either replicated immediately or, if etching is required, firstly allowed to warm up. Platinum/carbon, or other heavy metals, can be evaporated at any angle between 0 and 90° and the carbon backing layer is deposited while the specimen both swivels and rotates. The Geymayer technique was later adapted (Sleytr 1970a,b) to allow the production of complementary fracture faces.

A special device (see Fig. 6.17d in §6.3.3b) has also been developed (Sleytr et al. 1981) for the contamination-free transport of specimens freeze-fractured under liquid nitrogen into the Leybold-Heraeus Bioetch 2005 freeze-fracture replication unit. The lid mechanism, mounted on the Bioetch universal specimen cup, is closed under liquid nitrogen and acts as a cold shroud for the specimens during transfer to the air-lock system. After the specimen cup has been mounted on the specimen stage, the concentric cold shrouds are closed (§6.3.3b(i)). The lid is opened electromagnetically at a preset point in the automatic preparation cycle of the Bioetch 2005.

(ii)   Freeze-fracturing under liquid helium
Attempts to develop methods which allow the evaluation of fracture faces obtained by freeze-fracturing at temperatures close to absolute zero were stimulated by the problem of the temperature-dependence of plastic deformation during the fracturing process (see §6.8.3 and Sleytr and Robards 1977a). Theoretically, as has already been pointed out, freeze-fracturing

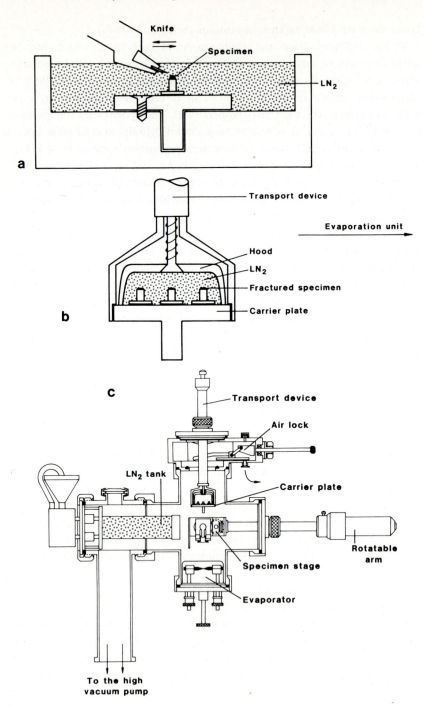

Knife

Specimen

LN₂

a

Transport device

Evaporation unit

Hood

LN₂

Fractured specimen

Carrier plate

b

c

Transport device

Air lock

LN₂ tank

Carrier plate

Rotatable
arm

Specimen stage

Evaporator

To the high
vacuum pump

close to zero Kelvin can be accomplished in one of two different ways: fracturing can either be done under ultra-high vacuum or under the protection of a liquid gas with a very low melting point. For fracturing at 4 K under vacuum, highly complex UHV equipment with an air-lock system (§6.3.3b(ii)) which prevents hoar frosting during specimen transfer, is required. Since there are few such units commercially available, it has been simpler to develop methods for fracturing specimens under liquid helium and then to transport them to liquid nitrogen before loading into the vacuum unit.

A simple method for fracturing specimens under liquid helium and transporting them to a liquid nitrogen container while avoiding contamination of the fracture faces, has been developed by Sleytr and Umrath (1974b). This system also allows the production of complementary fracture faces. Specimens are frozen in type Hb or Ib specimen holders (Fig. 6.2), as used for the production of complementary fracture faces, and inserted under liquid nitrogen into a fracturing device (Fig. 6.23). This device basically consists of two stainless steel tubes which can be twisted and moved relative to each other into certain defined positions (Sleytr 1974). In the position shown in Fig. 6.23a, the specimen can be mounted onto the spring holder on the inner tube. The tubes are then twisted into a position where the specimen is completely surrounded by cold surfaces and the whole device is inserted into a container filled with liquid helium (Fig. 6.23b). Fracturing takes place by pushing the inner rod downwards to knock off the upper specimen holder. The fracturing device is then removed from the liquid helium container, transferred back to a container of liquid nitrogen (Fig. 6.23c) and the specimens are collected for further processing by any of the procedures developed for observing fracture faces obtained under liquid nitrogen (§6.3.3c(i)).

6.3.3d *Freeze-fracturing outside the coating unit at atmospheric pressure*

Dempsey and Bullivant (1976b) made controlled fractures at atmospheric pressure with a cooled glass knife, using the Cryokit attachment of the LKB

Fig. 6.22.   (a) Geymayer device for fracturing under liquid nitrogen. (b) Transport mechanism for transferring the fractured specimens to (c), the Leybold-Heraeus EM preparation plant EPA 100 now manufactured by Paar KG (Austria) in a slightly modified form as EPA 101. (Fig. 6.22a and b redrawn after Geymayer 1966; Fig. 6.22c redrawn from an illustration kindly provided by A. Aldrian.)

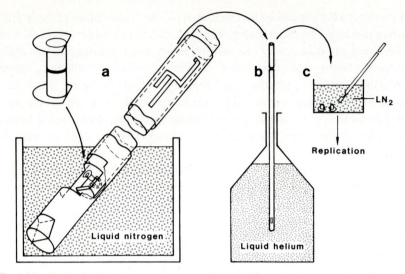

Fig. 6.23.   Device for fracturing specimens under liquid helium and subsequent contamination-free transfer to liquid nitrogen. (a) Specimens are frozen in a pair of opposed hollow rivets and are loaded, under liquid nitrogen, into the spring-holder attached to the inner tube of the fracturing device. After twisting the tubes into a position where the specimen is completely surrounded by cooled surfaces, the device is inserted into the container of liquid helium (b). By pushing the inner rod downwards, the specimen is fractured and the device is then returned to liquid nitrogen so that the specimens can be released (c). (Specimens are then processed by any of the methods used for the preparation of specimens that have been fractured under liquid nitrogen.) (Redrawn and extended after Sleytr and Umrath 1974b.)

ultramicrotome (see Appendix 2) and this procedure can obviously also be used with other cryoultramicrotomes. Specimens were mounted on a specimen holder composed of an LKB silver pin holder inserted into a Teflon-brass plug which was suitable for fitting to the Cryokit attachment (§4.2.2b, Fig. 4.6). Freezing was done according to the method of Van Harreveld and Crowell (1964) by pushing the specimen, mounted on the specimen holder, onto a liquid nitrogen-cooled copper block (§2.4.2). The frozen specimen and holder were then inserted under liquid nitrogen into the LKB Cryokit specimen holder and transferred in a dish of liquid nitrogen to the cooled cryochamber on the ultramicrotome. The chamber was kept filled with dry nitrogen to avoid condensation of water on the specimen during the transfer and fracturing processes. Before fitting the specimen to the microtome arm, the temperature of the arm was held at approximately 103 K ($-170°$C) and the temperature of the knife was usually 113 K ($-160°$C). Controlled fracturing was then performed at a specimen temperature of 93 K ($-180°$C) and a knife temperature of 113 K ($-160°$C) with a 1–3 μm advance per stroke. Immediately after cutting, the specimen holder was removed from the microtome arm and placed in a dish of liquid nitrogen. Dempsey and Bullivant used the Bullivant and Ames block device to carry out shadowing and carbon coating of the fracture faces, but any other method which allows contamination-free replication of specimens fractured under liquid nitrogen (§6.3.3c(i)) would be suitable.

6.4 Shadowing and replication

6.4.1 Evaporation and evaporators

6.4.1a Formation of thin films

The replication process should copy the relief produced by specimen fracturing (§6.3.1), or by fracturing and etching (§6.5), as faithfully as possible. The most usual method for replicating frozen specimen surfaces involves a two-step process. Firstly, an electron-opaque metal is evaporated onto the specimen at an oblique angle (usually around 45°) to the 'average' orientation of the fracture plane to provide contrast. The patches of shadowing material are subsequently bound together and reinforced by deposition of a low electron-scattering supporting layer of carbon or silicon monoxide which is evaporated perpendicular to the fracture plane. Most limitations to spatial resolution and in depicting the accurate surface geometry by replication arise from metal vapour-specimen surface interactions during shadowing. The technical problems, as well as those of interpretation of high resolution shadowing of fracture faces, have been discussed in detail by Abermann et al. (1972) and, with special emphasis on membrane structure, by Zingsheim and Plattner (1976) (see also Willison and Rowe 1980).

Theoretically, if the randomly arriving atoms of heavy metal vapour could be immobilised at the point of their first contact with the specimen, a spatial resolution close to the dimensions of a few atomic layers of heavy metal should be possible. Such resolution has not yet been achieved since atoms migrate laterally while dissipating their thermal kinetic energy until they are eventually trapped, either by strong binding sites on the specimen surface or by previously condensed atoms, crystallites, or clumps of the evaporation material. During continued condensation of atoms, these first nucleation centres grow, so forming stable nuclei with no lateral mobility and eventually merge with neighbours to form a continuous film. Since such films would be too thick for high resolution shadowing, the evaporation process must be stopped at a stage where individual metal aggregates are still distinguishable. Thus, in practice, the main problem of obtaining fine grain shadowing is to decrease the surface mobility of condensing atoms and to inhibit growth of nuclei.

A reduction in grain size and a consequent increase in shadowing resolution, can be obtained by:

(*i*)   lowering the specimen temperature (and thus reducing the lateral diffusion of adsorbed atoms);

(*ii*)   reducing the deposition rate;

(*iii*)   using higher melting point shadowing metals with stronger initial binding capacity; or

(*iv*)   simultaneous evaporation, or combination, of heavy metals with carbon.

In addition to these nucleation parameters, local topological and physicochemical properties of the specimen itself also have a strong influence on the extent of the lateral mobility of condensing atoms. This phenomenon, which is known as *preferential nucleation* or *decoration*, makes it particularly difficult to interpret accurately the geometry of fine structural details of the replica: such random nuclear growth can simulate topological features and lead to misinterpretation. However, local differences in binding strength may help in obtaining information about structural or physicochemical discontinuities on the specimen surface which would not be detectable if geometry alone were to govern the mass distribution of condensing material. Typical examples of this are crystal surfaces or cleavage planes where, due to decoration phenomena, linear accumulations of metal nuclei indicate sites of discontinuities (crystal lattice steps) which are not detectable by high resolution shadowing. It is almost impossible to distinguish between shadowing and decoration phenomena when dealing with fine structural details in replicas of biological specimens (Abermann et al. 1972). In other words, the smaller the structural feature on the replica, the greater becomes the uncertainty that the mass distribution of shadowing material seen in a two-dimensional projection on the microscope screen indicates the actual geometry (topography). These considerations are of particular importance for rotary shadowing at small angles (§6.4.1f).

6.4.1b High resolution shadowing and contamination problems

For high resolution shadowing in freeze-fracture replication, elements with melting points over 2000 K are used. The most common are platinum and iridium, which are evaporated along with carbon and a tantalum-tungsten alloy. A number of relevant physical properties of these elements are given in Table 6.2.

Besides their ability to condense with minimum grain size, the elements listed in Table 6.2 have the necessary advantage that they are, in pure form or as alloys, inert to the chemicals commonly used for replica cleaning (§6.6). Simultaneous evaporation of platinum and carbon, introduced by Bradley (1958, 1959), is still the commonest method for high resolution shadowing. In the last twenty years, various technical modifications have been developed (see Reimer 1967; Glitsch 1969; Abermann et al. 1972) with the main objective of producing films with a consistent thickness and optimum platinum/carbon (Pt/C) composition. Under ideal conditions, the proportion of carbon will be just sufficient to suppress the surface mobility of the platinum atoms, so leading to electron scattering deposits with little crystalline structure. Layers with too much carbon show almost no granularity but lack contrast and resolution. Pt/C can be evaporated either by resistance heating (§6.4.1c) or by electron-beam evaporation (§6.4.1d). Good contrast and low granularity is also observed when a platinum-iridium alloy is evaporated along with carbon (Moor 1973a) using the same methods as for Pt/C evaporation.

Besides Pt/C, one of the most suitable materials for high resolution, fine-grain shadowing is tungsten-tantalum (Ta-W) alloy, although such refractory alloys must be evaporated using the more complex electron-beam heating systems (§6.4.1d) or Penning sputtering (Penning discharge) techniques (Jakopić et al. 1978; Peters 1980; Wildhaber et al. 1984). Since the size of the grain in the shadowing layer decreases with increasing temperature of melting of the material, tungsten (Table 6.2) should give deposits with the finest grain. However, the use of pure tungsten as a shadowing material in freeze-fracture replication is limited due to its chemical instability if replicas need to be cleaned in aqueous solutions. However, deposits of a Ta-W alloy (evaporation occurs roughly in a 1:2 tungsten:tantalum ratio) have a fairly high chemical stability and also have a small granularity (Abermann et al. 1972).

Summarising, high melting point metals can be evaporated using electron-gun heating with good reproducibility and the deposited films show

TABLE 6.2
Physical properties of evaporants

Element	Atomic number	Atomic weight	Density ($\times 10^{-3}$ kg m^{-3})	Thermal conductivity at 300 K (J m^{-1} s^{-1} K^{-1})	Resistivity at 300 K ($\mu\Omega$ cm^{-1})	Melting point (K)	Boiling point (K)	Vapour pressure at melting point		1 Torr Temperature[a] (K)	Latent heat of evaporation (kJ g^{-1})	Specific heat at 300 K (J g^{-1} K^{-1})
								Torr	Pa			
C	6	12	2.26	129	3500	$\approx$4073	4473	–		2940	–	0.712
Ir	47	192.2	22.4	147	6.10	2727	4773	1.1×10^{-2}	1.4	2990	3.31	0.133
Pt	78	195.1	21.5	72	10.0	2043	4573	2.4×10^{-4}	3.2×10^{-2}	2590	2.62	0.133
Ta	73	181.0	16.6	58	13.1	3270	4373	6×10^{-3}	8×10^{-1}	3630	4.17	0.140
W	74	183.9	19.3	174	5.50	3669	6173	3.3×10^{-2}	4.4	3830	4.35	0.133

[a]Temperature at which the vapour pressure is 10^{-1} Torr (13.3 Pa): for high evaporation rates from small sources the vapour pressure should be in this range. (Data from Abermann et al. 1972 and Goldstein et al. 1981.)

high density, fine grain size and no recrystallisation during exposure to the electron beam in the microscope. Ta-W shadowing is superior to Pt/C if precise surface geometry is to be recorded, but with Pt/C, periodicities or discontinuities of the fracture face become visible through decoration effects (Abermann et al. 1972). The decoration effects with Pt/C and Ta-W shadowing can easily be demonstrated by comparing the structural information revealed from specimens onto which metal has been deposited at 90° (perpendicular shadow) or ⩽45° respectively (Gross et al. 1982; Bachmann et al. 1984).

In addition to complex specimen-metal vapour interactions, other factors determine the degree of similarity between the geometry of the fracture plane and the final mass distribution of heavy metal on the replica. Vacuum conditions, particularly the partial pressure of condensable gases in the specimen area during the period between fracturing and the end of replication, are of particular importance (§6.5.2). Condensation contamination can mask, enhance or specifically decorate structural details (Gross et al. 1978b; Walzthöny et al. 1981, 1982). It is frequently overlooked that the evaporators themselves can cause a deterioration of the vacuum since, during evaporation, not only heavy metal vapour but also condensable gases may be expelled. To outgas the evaporators, it is best to pre-heat them under vacuum close to their working temperature and then to cool them down before use. Generally, carbon evaporation should start as soon as the shadowing process is finished. Residual gases in the vacuum system can condense on the heavy metal deposit and cause a separation of the two evaporated layers during thawing and cleaning. This particular artefact is only observed in poor vacuum systems and is easy to recognise since parts of the thin shadowing layer will either be lost during replica cleaning or be seen curled up on the carbon backing layer.

Structural alterations in the fracture face itself may arise from thermal load during evaporation. As can be seen from Table 6.2, high temperatures are required during vacuum evaporation of heavy metals. An energy flow to the specimen is caused by:

(*i*)   dissipation of heat to the specimen surface during condensation of the vapourised metal (heat of condensation);

(*ii*)   thermal radiation from the evaporation source (heat of radiation); and

(*iii*)   ion or electron bombardment when poorly designed electron-beam evaporators are used.

TABLE 6.3
Shadow-casting methods and heat damage

Material	Rate of deposition[a] (0.1 nm s⁻¹)	Type of evaporator	Evaporation area (mm²)	Distance (cm)	Temperature (K)[e]	Radiation (mW cm⁻²)	Condensation (mW cm⁻²)	Total heat flux (mW cm⁻²)	Experiment[f] flux (mW cm⁻²)
Pt	1.0	Electron bombardment[b]	12	15	2560	10.5	0.6	11.1	17
				30	2725	3.2		3.8	4.5
Pt/C	0.8/1.1	Resistance heating[c]	5	15	≈3000	24	1.0	25	40
Ta-W	1.2	Electron bombardment	7	20	3700	15	1.0	16	23
C	10	Resistance heating[c]	5	12	≈3000	45	2.5	47.5	100
C	2.5	Resistance heating[d]	5	20	≈3000	14	0.6	14.6	30
				15		–		–	17

[a] Determined by a quartz-crystal monitor.
[b] Pt wire wound around tip of W anode.
[c] As supplied by manufacturers of freeze-etching apparatus.
[d] Indirect evaporation of carbon, small disc approximately 10 mm², 1 cm in front of the hottest parts.
[e] As determined from vapour pressure data.
[f] By a calibrated thermistor in the specimen position.

The amount of energy flow to specimens under different evaporation conditions has been calculated by Zingsheim et al. (1970a,b) and relevant data are given in Table 6.3.

The heat of condensation varies with the shadowing material and is about the same for deposition of equal thicknesses of tantalum-tungsten and Pt/C; for all practical purposes this amounts to less than 5% of the total thermal load (Bachmann et al. 1969; Zingsheim et al. 1970b; Abermann et al. 1972). The intensity of radiant heat for a given element is determined from the Stefan-Boltzmann law and depends on the temperature of the evaporation source and the ratio of the source radius to the source-specimen distance. As the equation below shows, the radiant heat is proportional to the fourth power of the source temperature.

$$I = \left(\frac{R}{D}\right)^2 E\varphi T^4 \qquad (6.1)$$

where

I = intensity of the radiation reaching the specimen
T = source temperature
R = source radius
D = source-specimen distance
E = total emissivity
φ = Stefan-Boltzmann constant

As can be seen from Fig. 6.24, the vapour pressure of the shadowing material increases much more steeply with temperature than the released energy. Consequently, the energy flow as radiant heat can be considerably reduced by using a smaller source or by increasing the specimen-source distance (Fig. 6.25). In other words, it is recommended to use the highest possible source temperature, together with the smallest possible source diameter and the greatest possible source-specimen distance (Zingsheim et al. 1970b). However, there are practical limits to the reduction of radiant heat, since deposition rates are affected to the same extent as the radiation intensity. Short shadowing periods (5–10 s) are often necessary in order to guarantee rapid stabilisation of the specimen surface. This is of particular importance if fracture faces are replicated at etching temperatures (§6.5.1), or if a freeze-fracture replication method is used in which it is difficult to maintain contamination-free conditions (§6.5.2).

Different designs of electron gun are commercially available, but most work with a source diameter of about 2 mm and a source-sample distance

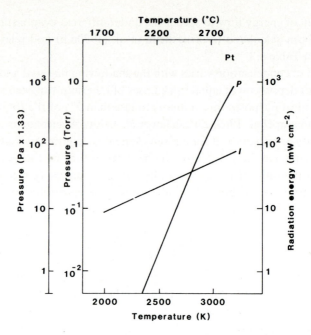

Fig. 6.24.   Heat radiation is proportional to the fourth power of the source temperature (Stefan-Boltzmann law). The vapour pressure (*P*) depends much more strongly on the temperature than the radiant energy (*I*). *I* was calculated for platinum and for a source-to-specimen distance of 10 cm. Between 2000 and 3000 K, the vapour pressure and hence the evaporation rate rises by a factor of 1000, whereas the radiation energy rises only by a factor of 5. (Redrawn from Zingsheim et al. 1970b.)

of about 14–15 cm (§6.4.1d). Although total elimination of radiant energy is possible by introducing rotating apertures between the evaporation source and the specimens (Horn 1962; Aldrian and Horn 1974), only prototypes of this apparatus have been produced. In contrast to thermal radiation, charged particles (ions and electrons), created by electron bombardment evaporation, are adsorbed within a very thin surface layer and pose a much greater threat of heat damage to the fracture face. Disadvantages of early designs of electron-beam evaporator have been overcome by introducing effective charged-particle deflection systems which totally eliminate this source of thermal load (Zingsheim et al. 1970a). As shown by Zingsheim et al. (1970b), with such a well-designed electron-beam evaporator, Ta-W shadowing can be performed within 5–10 s with only 50% of the heat load produced by a commercial resistance-heated Pt/C evaporator (Table 6.3).

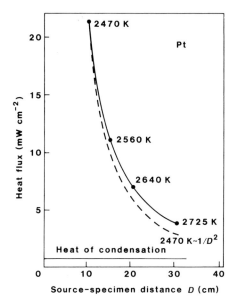

Fig. 6.25. The total heat flux at the specimen as a function of source-specimen distance (*D*). Calculated for a constant deposition of 0.1 nm s⁻¹ of platinum (solid line). The deviation from the $1/D^2$ dependence (dashed line), which would hold for constant temperature, is very small. If, for instance, the source is placed at a distance of 30 cm instead of 20 cm, heat flux and deposition rate are approximately halved. The source temperature must be increased from 2640 K to 2725 K to restore the original deposition rate. This rise in temperature increases the heat flux at the specimen again, but only by 10%. At large distances the total heat flux diminishes towards the value for the heat of condensation. (Redrawn from Zingsheim et al. 1970b.)

6.4.1c   Methods of resistance heating

Resistance heating can be used for both Pt/C shadowing and carbon replication (see Willison and Rowe 1980). It involves passing a current between two electrodes the resistance of which causes heating, so leading to evaporation of the desired material which may either be the electrode itself (e.g. carbon) or a substance attached to it (e.g. platinum). Common electrode assemblies for Pt/C evaporation are illustrated in Fig. 6.26. In most freeze-fracture replication units the evaporators are positioned above, or at the same level as, the specimen. This limits the choice of resistance evaporation methods to types (a) and (b) (Fig. 6.26). In contrast, the very simple and reliable resistance platinum/carbon evaporation method, developed by Glitsch (1969) (Fig. 6.26c), is only usable if the evaporators can be installed below the specimen stage. Electrodes are usually located at least 15 cm from the specimen plane. An exception to this is the evaporation procedure developed by Bank and Robertson (1976) (Fig. 6.26b) which, in combination with a resistance monitor (Steere et al. 1977), allows high resolution Pt/C layers to be deposited at a source-specimen distance of only 7 cm. Results obtained with delicate specimens clearly demonstrated that this short distance had no deleterious effects on the apparent resolution of specimen topography (Steere 1982).

When platinum is evaporated from opposed spring-loaded carbon rods, different ways of shaping the tips to accommodate the platinum wire have been suggested. One tip may be reduced to a cylinder (using a small lathe or special 'sharpener') of about 1 mm diameter, while the other is formed to a steep cone. Alternatively, both tips may be cone-shaped with flat, 1.0 mm diameter tips (Moor 1959) (Fig. 6.26a). An asymmetric assembly of carbon rods (Fig. 6.26b), designed for automatic regulation of shadowing via a resistance monitor, was suggested by Steere et al. (1977) (see also Bank and Robertson 1976). The platinum used for evaporation (obtainable as pure wire from suppliers – see Appendix 2) is usually in a tight coil which can easily be made by winding the wire around a cylindrical, 1 mm diameter

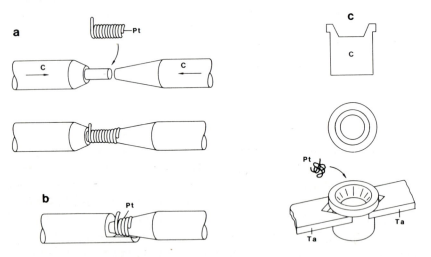

Fig. 6.26.   Common electrode assemblies for platinum/carbon resistance evaporation. (a,b) Two different methods of shaping the tips of carbon (**C**) rods to accommodate a helical coil of platinum (**Pt**) wire. (c) 'Glitsch'-type evaporator consisting of a carbon crucible (**C**) fixed between two spring-loaded tantalum (**Ta**) cheek pieces. The crucible is loaded with a length of tightly coiled platinum wire.

former. With the type (a) (Fig. 6.26) carbon rod assembly, 5–7 cm of 0.1 mm diameter platinum wire gives a shadow of sufficient contrast. With the type (b) (Fig. 6.26) assembly, 8 mm of 0.127 mm diameter platinum wire is coiled around a cone-shaped (0.5 mm diameter tip) carbon rod. This rod is forced by spring tension against the opposite rod which has a flat end and an extended edge. Both rods in this system are made from 3.0 mm diameter graphite rods. To obtain good results with both type (a) and type (b) (Fig. 6.26) assemblies, it is essential to maintain firm contact, using spring pressure and correct alignment of the rods, during heating. Also, special emphasis must be given to the correct spring tension. Even so, with the possible exception of the design suggested by Steere et al. (1977) (Fig. 6.26b), considerable variation in the quality of shadowing is often observed. Part of this problem can be overcome by pre-melting the platinum under vacuum. After venting the vacuum chamber, the carbon rods can then be realigned in such a way that the platinum, which has melted to a blob, points towards the specimen.

The Glitsch-type evaporator (Fig. 6.26c), used in the Leybold-Heraeus freeze-fracture replication unit, consists of a 3.0 mm diameter carbon crucible fixed between two spring-loaded tantalum cheeks. The crucible is loaded with 4–6 mm of 0.1 mm diameter platinum wire, coiled into a small ball, and can be used for at least three evaporations when reloaded for each run with the same amount of platinum. This type of evaporator is unsurpassed in ease of operation, reproducibility and fine grain of the deposited film. However, it has the limitation that it cannot be adapted for use with most other freeze-fracture replication units since it can only evaporate in an upward direction as the melted platinum would flow out if the evaporator were to be mounted upside-down.

If commercial units are used, then evaporation is best done according to the manufacturer's specifications. However, a good starting point for choice of conditions is to use pointed carbon electrodes (1.0 mm diameter tips), a current of 55–60 A with a voltage of about 7.0 V, and a 5.0 s evaporation (the time can be selected either simply by turning the appropriate switch on and off or, on some machines, by using an automatic timer). For the type (b) (Fig. 6.26) assembly, a resistance monitor with sensors and automatic power cut-off have been developed to control the thickness of the deposited film (Steere et al. 1977).

The second stage in the replication process involves the formation of a continuous carbon backing layer on top of the heavy metal shadowing layer. Carbon evaporation using resistance heating is generally less of a problem than is the shadowing procedure. The electrodes can be trimmed to the same shape as for platinum evaporation but untrimmed thin carbon rods with flat ends are also often used satisfactorily. As for platinum evaporation, good spring-controlled physical contact of the rods is essential for successful evaporation which is normally accomplished within 6–12 s using 60 A and 7.0 V. Both Pt/C and carbon evaporations are best done through shields with apertures to reduce contamination of the work chamber. For checking the correct alignment of the apertures in relation to the specimen, a small current, just sufficient to produce light from the carbon tips, is passed through the electrodes so that the point of evaporation can be seen. Care should be taken during all alignment checks and evaporation procedures that the bright source is only viewed through dark filters (e.g. neutral density filters).

6.4.1d   *Methods of electron-beam evaporation*

Electron-beam bombardment heating can be used successfully for both heavy metal shadowing and carbon replication. Since no materials are available from which crucibles can be formed to allow resistance heating of tungsten and tantalum to give adequate evaporation rates, the development of electron guns was particularly important in allowing improvements in shadowing methods. Electron-beam heating was used in electron microscopy for shadowing as long ago as the 1960s (Bachmann et al. 1960; Westmeyer and Lorenz 1960; Bachmann 1962) but it was about ten years before guns were used successfully for high resolution shadowing of frozen specimens. Deflection of charged particles with an electrostatic field was shown to be particularly important in avoiding serious heat damage to the fracture face (Zingsheim et al. 1970a,b). The basic design of an electron-beam evaporator is shown in Fig. 6.27. It consists of a heated tungsten cathode filament (**1**) which emits electrons. The electrons are accelerated and focused toward the tip of a rod-shaped anode (**2**) which is made from the material to be evaporated. The transferred energy causes the tip to melt and eventually to evaporate. A focusing aperture (**3**) prevents electrons from reaching the upper part of the anode and its holder. The front aperture (**4**) has an opening which is small enough to protect the specimen from radiation emitted from the heated filament. Finally, ion-deflection plates (**5**) deflect charged particles generated by electrons on their way to the anode. For convenience, one plate is kept at ground potential and the other at the anode potential. It is possible to decrease the amount of energy needed for evaporation by improving the focusing of the electrons. This increases the temperature within a smaller area of the tip of the anode. In addition, the evaporation material can be mounted on a low thermal conductivity support, such as a stainless steel tube, to increase the temperature gradient (Tesche 1975). For Pt/C evaporation a small platinum cylinder is fitted into a hole at the tip of a carbon anode rod (Fig. 6.28a). When the guns are positioned *above* the specimen the anode must first be heated in the upright position (Fig. 6.28a) to pre-melt the platinum (otherwise the platinum would drop out). The electron gun can usually be fired several times, giving shadows of consistently good quality and with a Pt/C ratio of approximately 95:5, as shown by microprobe analysis (Moor 1973a). Although close examination reveals that the grain size is about the same as with good resistance evaporation, it is both the reproducibility and reliability that make electron-beam evaporation more attractive for Pt/C shadowing. For Ta-W evaporation, the anode is prepared by winding a tantalum wire around the reduced tip of a cylindrical tungsten rod (Fig. 6.28b). On electron bombardment, the tip melts and a Ta-W alloy is formed. This evaporation source can be used several times. The tantalum content of the deposited film, which is about 70% at the first evaporation, decreases with subsequent evaporations without obvious changes in grain size or chemical stability (Zingsheim et al. 1970a).

To avoid condensation contamination on the fracture faces, electron guns

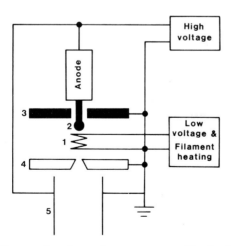

Fig 6.27. Schematic diagram of an electron-beam evaporator. (For details see text.) (Redrawn after Zingsheim et al. 1970a.)

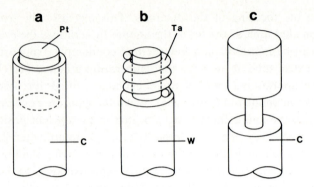

Fig. 6.28.   Anode assemblies for the electron-beam evaporation of various metals. (a) Plati-
num(**Pt**)/carbon(**C**) source. (b) Tungsten(**W**)/tantalum(**Ta**) source. (c) Shaped carbon rod to
improve temperature gradient.

should be pre-heated under vacuum. During this cycle all parts of the assembly should reach a temperature close to that reached during the actual evaporation. With a carbon rod as the anode, electron guns can also be used
to evaporate the carbon backing layer. To improve the temperature gradient, carbon rods with a narrow 'neck' a few millimetres below the tip can
be used (Fig. 6.28c). However, for reasons of economy, a combination of
an electron gun for shadowing and a resistance source for carbon evaporation is often used.

Electron-beam evaporators for freeze-fracture replication are commercially available from Balzers Union and Cressington (Appendix 2). These
guns have basically the same design as that described by Zingsheim et al.
(1970a) and Moor (1971). In addition to the freeze-fracture apparatus for
which they were designed, they have also been fitted to many other high
vacuum units. Generally, guns are fired using 1600–1900 V and 60–80 A for
6–13 s to give suitable shadowing or coating deposition rates. A simple electron-beam gun was designed by Hagler et al. (1977) for use with the Denton
freeze-fracture replication unit. This evaporator is easily built in a workshop
but only gives deposition rates adequate for Pt/C evaporation since it works
at only 1000 V.

6.4.1e   Monitoring film thickness

When resistance evaporation (§6.4.1c) is used with standardised conditions
of voltage and current, the thickness of the shadowing layer is determined

by the amount of platinum wire wrapped around the carbon electrodes (Fig. 6.26a,b) or placed in a crucible (Fig. 6.26c). With the exception of Steere's technique (Steere et al. 1977), platinum evaporation from pin-pointed carbon electrodes is less reproducible than from a crucible (Glitsch 1969). The advantage of the Glitsch method is that the direction of evaporation is determined by the shape of the crucible. Again with the exception of the method suggested by Steere et al. (1977), settings for resistance Pt/C evaporation from a crucible lead to total platinum evaporation.

It is possible to deposit a film of the desired thickness by comparing the contrast in the final electron microscope image with the density of shadowing material seen as metallic grey deposit on a piece of white paper or porcelain-fragment placed beside, or around, the specimen during evaporation. After one or two initial trials, it is then quite easy to maintain reproducible shadow thicknesses in subsequent runs. A drop of high vacuum oil (e.g. diffusion pump oil) is placed on the surface of a white porcelain fragment. The covered area appears to remain 'clean' (white) during the evaporation and thus allows a comparative estimate of the thickness of the evaporated film to be made (Reimer 1967). Although this only provides a rough estimation, it is still the most frequently used method for monitoring resistive Pt/C shadowing. Instead of comparing the 'greyness' of Pt/C deposits on an indicator paper or white enamel, the thickness of the deposits can also be judged by the interference colours seen on gold-coated metal or glass sheets (Reiter 1961). A mid-purple colour is close to an ideal Pt/C evaporation. Carbon coating is generally more reproducible than platinum shadowing if the predetermined voltage is applied for a given time (usually between about 8 and 15 s). A carbon layer of 10 nm thickness is *just* noticeable as a difference in density between the oil-covered area and the rest of the white porcelain surface and a carbon deposit with sufficient stability for freeze-fracture replicas appears light grey-brown. With gold-coated metal sheets or glass the following interference colours relate to the thickness of the carbon layer (Reiter 1961):

red	=	20 ± 4.0 nm
blue	=	24 ± 4.0 nm
blue-green	=	35 ± 3.5 nm
green-silver	=	42 ± 3.5 nm

Thus, a carbon deposit for normal freeze-fracture replicas will give a red (tending to blue) interference colour. Due to the low electron-scattering power of carbon, differences in the thickness of the carbon layer are much less obvious in the microscope than the heavy metal shadow.

For accurate measurement of both film thickness and deposition rate of heavy metal shadowing and carbon coating, quartz crystal film-thickness monitors are used (see Appendix 2 for suppliers). Quartz crystal monitors measure the weight per unit surface area (mass thickness) of deposits and are mainly used in combination with electron-beam evaporators. The increasing thickness of film deposited onto the crystal changes its frequency of oscillation which can be calibrated to give a direct indication of film thickness. If accurate film thicknesses are to be determined, it must be remembered that, with Pt/C evaporation, a mixture of approximately 5% (w) carbon and 95% (w) platinum is evaporated simultaneously. With Ta-W, a mixture of about 20% (w) tungsten and 80% (w) tantalum evaporates from the anode tip (Fig. 6.28b in §6.4.1d) but, with each successive evaporation, the proportion of tungsten increases. For accurate measurements of film thickness, it is essential that the quartz crystal and the specimen are both positioned in a precisely defined and similar alignment with respect to the evaporation source. If this is not done, incorrect mass distributions will be measured due to differences in the directional distribution of the metal vapour (Simpson 1979). It is best to evaporate using a shutter placed between the specimen and the electron-beam evaporator. This shutter is opened once the gun is providing a constant evaporation rate and is closed when the desired film thickness (=mass density) is reached. The use of a shutter has the additional advantage that the specimen is protected from heat load and discharged gases during the heating-up phase of the evaporator. If a quartz crystal film-thickness monitor is not available, it is still possible to produce reasonably consistent film thicknesses with electron-beam evaporators if constant evaporation times are used and if special care is given to positioning the anode with respect to the coils of the heated filament. In the absence of a film-thickness monitor, an indicator paper, porcelain fragment or gold-coated metal (aluminium) foil is used to estimate film thickness as described above.

For routine freeze-fracture replication, a heavy metal film of 1.5–2 nm is used. The optimal thickness of the carbon backing layer is strongly dependent on the type of specimen and the required stability during thawing and cleaning, but is generally in the range of 10–35 nm. However, it is also necessary to note that, for highest resolution shadowing (Ta-W) a metal film thickness of 0.7–1 nm will give sufficient contrast. An increase in film thickness increases the contrast but, simultaneously, resolution is sacrificed as a result of a decrease in *relative* contrast variations and self-shadowing effects.

Self-shadowing arises when the shadowing material piles up on a structure and eventually casts its own shadow and consequently nothing is generally gained in resolution if fracture faces are shadowed at angles much lower than 45° (Zingsheim and Plattner 1976). Since fracture faces of most specimens have great variations in surface geometry (topography), the optimum shadowing angle cannot be obtained over the whole surface.

6.4.1f Rotary shadowing

It is not possible to provide contrast in all directions using uni-directional shadowing. Furthermore, the shadow cast by one structure may obscure smaller neighbouring structures. To overcome these problems, which are inherent in uni-directional shadowing, Heinmets (1949) introduced the technique of rotary shadowing which involves rotating the specimens during evaporation. Generally, specimens are rotated at 60–150 revolutions per minute with the evaporation source at an angle of 6–45° to the specimen surface. The method is most frequently used in electron microscopy to demonstrate isolated filamentous molecules, such as DNA (Kleinschmidt 1960) but it is also applicable to the shadowing of fracture faces (Elgsaeter 1974; Margaritis et al. 1977; Heuser and Salpeter 1979; Heuser 1980, 1981, 1983a,b; Hirokawa et al. 1982; Heuser and Cooke 1983; Espevik and Elgsaeter 1984). A special rotary replication unit for the Balzers freeze-fracture apparatus has been described by Margaritis et al. (1977) and is now commercially available (Appendix 2) and a completely new design of freeze-etch unit with a rotating specimen stage has been described by Elgsaeter (1978). Most freeze-etching units now have at least the potential to be modified to accept a stage for rotary shadowing (see §6.3.3.).

Using T4 polyheads, erythrocyte ghosts and chloroplast membranes as model systems, it was shown by Margaritis et al. (1977) that resolution *per se* is not improved by rotary shadowing. When Pt/C was evaporated with an electron-beam gun, 2.5 nm periodicities were resolved in both uni-directional and rotary shadowed specimens. Nevertheless, due to the radially symmetrical contrast and enhanced decoration phenomena on rotary shadowed specimens, the subunit structure of particles seemed to be demonstrated more clearly in some areas.

As pointed out by Heinmets in 1949, the optimal angle for rotary shadowing depends on the geometry of the object. With ferritin as a model system and using Ta-W as the shadowing material, Neugebauer and Zingsheim (1979) showed that the appearance of globular molecules can vary dramatically depending on the shadowing angle and the thickness of the metal film. As illustrated in Fig. 6.29, ferritin appears more and more like a doughnut as the shadowing angle decreases and the metal deposit increases, while the background granularity concomitantly increases. By varying the different shadowing parameters, the *apparent* diameter of ferritin can be changed from 12 nm to more than 26 nm. These model experiments clearly show the potential danger of misinterpretation of results obtained by rotary shadowing (see also Willison and Rowe 1980). Such considerations are of particular importance if attempts are made to interpret the subunit structure of intra-membraneous particles (Branton and Kirchanski 1977).

In summary, it appears that, under very well-defined experimental conditions, rotary shadowing may provide useful images of fracture faces but that the technique cannot be seen as a substitute for routine uni-directional shadowing. Rotary evaporation with Ta-W at an angle of $\simeq 45°$ has shown that the genuine structural details of a fracture face are sometimes depicted more easily by symmetrical metal deposition. The images obtained by low angle

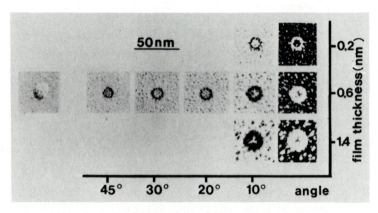

Fig. 6.29.   Appearance of ferritin molecules after rotary shadowing under different conditions. Ferritin (30 μg ml^{-1}) in distilled water was dried onto freshly cleaved mica and shadowed with varying amounts of tantalum-tungsten at various angles in a vacuum unit at 133 μPa (Zingsheim et al. 1970a). The amount of metal deposited (0.2 nm, 0.6 nm, 1.4 nm) refers to the plane of the support surface. During rotary shadowing the specimen table revolved at about 240 revs min^{-1}. A carbon layer, approximately 3 nm thick, was evaporated on top of the shadowing layer. Replicas were cleaned in 70% sulphuric acid (2 h). A molecule shadowed uni-laterally at 45° is shown for comparison on the left. The series of molecules shadowed at 10° is also shown with reversed contrast. Micrographs were taken at $\times 36,000$ magnification close to Gaussian focus with a Philips EM 301 at 80 kV. (Reproduced, with permission, from Neuge-bauer and Zingsheim 1979.)

(6–12°) Pt/C rotary shadowing, seem to be mostly the result of specific decoration of small protruding structures and self-shadowing phenomena. Finally, in interpreting images of rotary shadowed specimens, it is necessary to be aware that very few areas of freeze-fractured (and etched) specimens are either flat or parallel to the top of the rotating specimen stage and also that, when more than one specimen is used, the specimens are not all rotated about their own central axis.

In addition to its use in the study of fracture faces, rotary shadowing is also particularly suitable for demonstrating the fine structure of filamentous (e.g. cytoskeletal) and other very sculptured, structures (e.g. coated vesicles) exposed by deep etching (§6.5) (Heuser and Kirschner 1980; Heuser 1980, 1983a,b; Espevik and Elgsaeter 1984).

A method of 'double-axis rotary replication' has been described by Shibata et al. (1984). In this method, as well as the specimen rotating, the shadowing angle is changed continuously during replication. The authors consider that this leads to improved replicas from deep etched material.

6.4.1g Bi-directional shadowing

As an alternative to either uni-directional or rotary shadowing, Willison and Moir (1983) have suggested that bi-directional shadowing can be used to demonstrate structural features of fracture faces. They showed the value of both portrait shadow-casting using a 90° inter-shadowing rotation of the specimen (Williams and Smith 1958) as well as the classical technique for particle size analysis using a 180° inter-shadowing rotation (Williams and Wyckoff 1944; Kahler and Lloyd 1950).

6.4.2 Replica strengthening methods

Replicas of heterogeneous tissues frequently disintegrate into small pieces during thawing and cleaning (§6.6). The explanation for this is that there are differential dimensional changes in the specimen and replica arising from swelling and shrinkage (Robards 1978a). More stable replicas are produced with a thicker carbon backing layers but this limits the resolution of the replica. Consequently, methods have been suggested which involve the application of an additional strengthening layer to provide support for the heavy metal/carbon replica during thawing and cleaning but which can be dissolved away again afterwards.

6.4.2a   Silver technique

One of the simplest and most efficient means of strengthening the replica is to use an additional layer of silver or gold (Robards and Umrath 1978) which is evaporated under high vacuum at the end of the normal freeze-fracture replication procedure (§6.4.1). Although this technique was developed for the Leybold-Heraeus Bioetch 2005 unit (see §6.3.3b(i)) it is adaptable to other freeze-fracture units which either provide enough space for the installation of a third evaporator which evaporates upwards, or possess an airlock system for exchanging evaporators or specimens.

Silver is used as an evaporant because it is easy to evaporate, provides sufficient strength, is relatively inert to the commonest cleaning solutions and can easily be removed later from Pt/C replicas. The evaporation parameters can be preset in the unit. For a range of test specimens (including plant tissue) prepared in the Leybold-Heraeus unit, 18 mg of silver evaporated from a molybdenum boat provides a supporting layer of sufficient stability. After removing the silver-backed replica from the apparatus and dissolving the organic material, the silver backing-layer is removed using a cyanide solution (§6.6(iii)) and the replica is then mounted on a grid (§6.7).

6.4.2b   Other reinforcement (and replica-stabilising) methods

Both to keep replicas floating during thawing and cleaning and to provide additional strength, Bullivant (1973) suggested that the replica side of replicated specimens should be dipped in a 1% solution of Collodion (cellulose nitrate) in amyl acetate. An alternative method is to apply the Collodion solution to the replica using a loop. After cleaning and mounting the replica (§6.6 and §6.7), the Collodion backing-layer is gently dissolved by applying successive loops of amyl acetate (Bordi 1979). The Collodion method has the disadvantage that, to apply the strengthening layer, the specimen must be removed from the vacuum and condensed water may therefore prevent firm attachment of the backing-layer to the replica.

Stolinski et al. (1983) found that polystyrene is superior to Collodion as a strengthening layer because is is considerably tougher when dried and is more resistant to cleaning solutions, such as acids, alkalis and dimethylformamide. A 2% (w/v) solution of polystyrene (e.g. from plastic Petri dishes) in chloroform is used. A flat layer of this solution is picked up on a 4.0 mm diameter copper wire loop and is frozen on a metal block cooled with liquid nitrogen (Fig. 6.30a). When the freeze-fracture run is finished, the chamber is vented with dry nitrogen gas and the frozen solution is lowered onto the still frozen sample which is maintained at about 153 K ($-120°$C) (Fig. 6.30b). The chamber is then closed again and rough-pumped for a short time while the temperature of the stage is raised to 218 K ($-55°$C). After melting at about 213 K ($-60°$C), the polystyrene solution spreads over the top of the specimen while the chloroform evaporates. After about 15 min, the specimen is removed from the vacuum chamber and transferred to a cooled container in which is a frozen solution of the original liquid with which the specimen was treated (e.g. a glycerol solution). By placing this container under a lamp (Fig. 6.30c), it is possible to allow the tissue to thaw slowly while, at the same time, the polystyrene layer dries on top of the replica. Stolinski et al. (1983) subsequently cleaned the specimen by floating it, reinforcing film uppermost, on a sequence of cleaning solutions such as 5–10% sodium hypochlorite, 25% dimethylformamide and 50% nitric acid (see also §6.6). After mounting on a Formvar-coated grid with the polystyrene layer uppermost, the replica and film are dried and then gently immersed in amyl acetate to dissolve the reinforcing layer. After drying, the specimen is ready for examination.

Since fragmentation of replicas often occurs during thawing (§6.6), a two-step carbon evaporation has been found to be useful for obtaining large intact replicas of delicate specimens. After heavy metal shadowing, approxi-

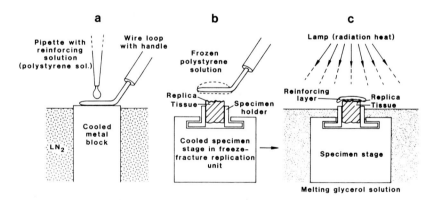

Fig. 6.30. Application of a strengthening layer of polystyrene to a replica before digestion of the specimen. (a) Freezing a thin layer of polystyrene solution on a cooled metal block. (b) Deposition of the frozen reinforcing solution onto the factured and replicated specimen maintained at 153 K. After warming to 213 K under vacuum to melt the polystyrene solution, the specimen is transferred to a melting glycerol solution (c) and the reinforcing layer is dried by radiant heat. (Redrawn after Stolinski et al. 1983.)

mately 20% of the total carbon layer is first evaporated. The specimen is then slowly warmed up to 263 K ($-10°$C) or room temperature to allow freeze-drying to occur and then, finally, an additional carbon layer is evaporated to build up a replica of greater stability and to connect together cracks generated in the first replica during drying (or thawing) (Griepp et al. 1978).

Before experimenting with any of these rather more laborious replica stabilisation methods, it should be remembered that shrinkage or swelling of delicate specimens during thawing and/or cleaning can often be minimised by placing the frozen specimen, together with the replica, *directly* into acid, alkali or freeze-substitution fluid (see also §6.6).

6.5   Etching

The specimen surface *can* be replicated directly after fracturing. Often, however, the surface topography is further enhanced by allowing a limited degree of etching. This involves the controlled sublimation of volatile materials from the specimen surface under vacuum. This superficial freeze-drying process lowers the surface of the ice crystals exposed during fracturing. Consequently, the effect of etching as seen in the final image of the replica is determined by the distribution of non-volatile components in the specimen and, especially, the eutectic or hypoeutectic network produced during freezing (§2.1.2). For biological specimens, etching is a particularly valuable process since additional faces and structures, such as membranes, cell surfaces and cytoplasmic structures, may be revealed.

6.5.1   Physical conditions for sublimation and condensation

The basic physical conditions for the sublimation and condensation of water have already been outlined in the Chapter concerned with freeze-drying (§5.2). Here we consider the particular aspects that must be taken into account when controlled freeze-drying of a superficial layer of a frozen specimen is required and when condensation contamination of exposed fracture faces must be avoided.

The process of accurate etching requires that the physical conditions of the specimen are precisely monitored and controlled (Dunlop et al. 1972; Robards 1974; Umrath, 1977). If the saturation vapour pressure of ice in the specimen is higher than the partial pressure of water vapour in the immediately ambient vacuum, then sublimation will take place. In contrast, if the partial pressure of water vapour in the vacuum is higher than the saturation vapour pressure of ice in the specimen, then condensation of water vapour from the vacuum onto the specimen surface will result. Consequently, the two parameters of prime importance are specimen temperature, which determines the vapour pressure of the ice (water), and the partial pressure of water in the vacuum around the specimen. As can be seen from Fig. 6.31, as well as in Table 5.2 in §5.2.1a, the saturation vapour pressure of ice varies markedly with temperature.

For pure ice, the net rate of sublimation is a function of the ice temperature and the partial pressure of water above the ice. In the presence of its own vapour at a pressure (P), vapour molecules will impinge on the ice surface with a flux expressed by the equation:

$$J_i = P(2\pi m \varphi T)^{-0.5} \qquad (6.2)$$

where

J_i	= impinging flux (dimensions: molecules per unit area per unit time)
m	= mass of vapour molecules ($H_2O = 18.016$)
φ	= gas constant (Stefan-Boltzmann constant)
T	= temperature (K)
P	= water vapour pressure (Pa)

When the water vapour pressure (P) in the system is equal to the saturation vapour pressure (P_s) for a given temperature, the total departing flux (J_d) will be equal to the impinging flux (J_i). In other words, the constantly evaporating and recondensing water molecules will maintain a state of equilibrium. If the partial pressure of water vapour in the vacuum system is reduced, this state of equilibrium is disturbed and a net rate of sublimation equivalent to $J_d - J_i$ occurs (Southworth et al. 1975).

Many modern freeze-fracture replication and etching units have liquid nitrogen-cooled cold traps around the specimen to reduce the partial pressure of water vapour in the specimen area (§6.3.3b(i)). At 77 K, the saturation vapour pressure of ice is of the order of 1.33×10^{-22} Pa (1×10^{-24} Torr). Thus molecules leaving the specimen are readily trapped providing that the vacuum is sufficiently good and the condensation (trapping) coefficient is 1 (Umrath 1983). At 1.33×10^{-3} Pa (1×10^{-5} Torr), the mean free path of gas molecules is greater than 600 cm and most subliming molecules will travel unimpeded to the trapping surface. Since the water flux from liquid

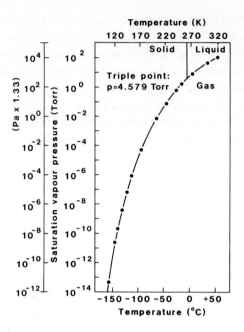

Fig. 6.31.   Variation of the saturation vapour pressure of ice with temperature. (Redrawn from Umrath 1977.)

nitrogen-cooled cold traps can be practically neglected, as seen from the equation:

$$J_{net} \simeq J_d = P_s(2\pi m \varphi T)^{-0.5} \tag{6.3}$$

(terms as defined in Eq. (6.2)), the net water vapour flux (J_{net}) is only determined by the specimen temperature. The maximum rate of ice evaporation in relation to temperature as derived by Davy (1969) is shown in Fig. 6.32.

As already discussed in §5.2, determination of the maximum evaporation (etching) rate by the method described above can only be applied to pure ice and where there is a large temperature difference between the specimen and the surrounding cold traps. The theoretical values for sublimation rates between 213 and 113 K (-60 and $-160°C$) are given in Table 5.2 (§5.2.1a). A variety of factors, such as gas collisions, surface cooling, surface roughness, impurities and special molecular properties associated with the vapourisation of ice all affect the sublimation rate (Davy and Branton 1970; Southworth et al. 1975; Mellor 1978) which may be reduced to negligible proportions when cryoprotective agents are present. Nevertheless, as can be seen from Fig. 6.32 and Table 5.2, under ideal conditions a layer approximately 100 nm thick will sublime from the surface of pure water (ice), maintained at 173 K ($-100°C$), in 1 min. This is a useful rate as it is controllable and provides adequate etching for most specimens. However, as can also be seen from Fig. 6.32 and Table 5.2, minor fluctuations in specimen temperature cause considerable changes in the saturation vapour pressure of water and, therefore, in the etching rate. Even a change of as little as 1.0 K alters the rate by about 25%; at 183 K etching is excessively fast (approximately 800 nm min^{-1}), while at 163 K less than 15 nm sublimes within a minute. It is thus obvious that, to obtain specified rates of etching, the specimen stage temperature must be controlled very accurately and calibrated regularly (Dunlop and Robards 1972). Depending on the actual *amount* of etching, it is usual to distinguish between normal etching ($\leqslant 100$ nm) and deep etching (up to a few μm).

The effects of normal and deep etching on fractured specimens composed of both etchable (volatile) structures, such as ice crystals, in a non-etchable

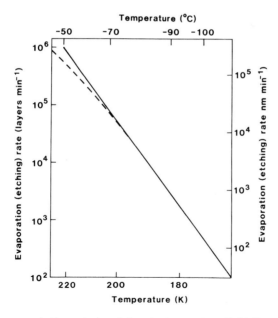

Fig. 6.32. Ice vapourisation rate in relation to temperature. Solid line: theoretical rate. Dashed line: measured rate derived from microbalance techniques. (Redrawn after Davy 1969 and Davy and Branton 1970.)

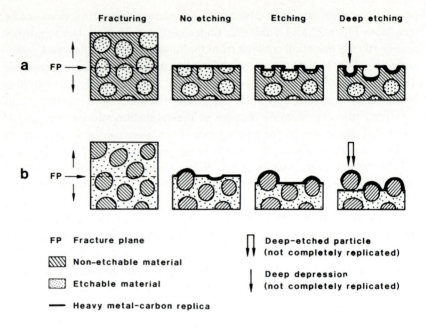

Fig. 6.33.   The effects of conventional and deep etching on fractured specimens composed of: (a) etchable structures in a non-etchable matrix; and, (b) non-etchable structures in an etchable matrix.

matrix (e.g. a pseudoeutectic network), and non-etchable structures surrounded by an etchable matrix (e.g. water) are illustrated in Fig. 6.33.

When using deep etching, it is important to understand that structures may become exposed which, for geometric reasons, cannot be completely replicated (Fig. 6.33). The incomplete replicas of such structures either collapse or get lost during the replica cleaning procedure. During deep etching it is also possible to 'lose' small liberated structures, before there is time for them to be replicated, through the effects of the water vapour wind (Southworth et al. 1975).

6.5.2   Etching methods and contamination phenomena

To understand the phenomena of etching and contamination, it is important to note that the nature and amount of contaminant material (§6.8.3) arriving on the specimen during exposure of the fracture plane depends on:

(*i*)   the general condition of the vacuum system;
(*ii*)   the specimen temperature;
(*iii*) the position of the fracture plane in relation to outgassing warmer surfaces which act as water vapour sources;
(*iv*)   the effectiveness of any cold traps which are kept at a lower temperature than the specimen itself; and
(*v*)   the nature of the fracturing method (simple fracturing or using a knife).

If efficient cold traps are used in the vacuum system then, even in a relatively poor general vacuum, the quantity of condensable gases can be very low (Dunlop and Robards 1972; Robards 1974; Umrath, 1977). As shown by mass spectrometer measurements, water vapour is the main source of specimen contamination in a vacuum system pumped by routine methods (rotary pump and oil diffusion pump, or turbo-molecular pump) without cold traps. Under ideal conditions, this water comes only from atmospheric moisture adsorbed to surfaces during venting the unit. In practice, however, other sources of water need to be considered, such as:

(*i*)   degassing surfaces which have become covered with hoar frost during exposure to moist air;
(*ii*)   leaks in the system; and
(*iii*) localised outbursts of small amounts of water vapour arising from heat generation in water-containing systems (e.g. specimen fracturing or cutting, hoar-frosted moveable parts of the apparatus, specimen chips generated during fracturing and in poor thermal contact with cold surfaces).

Other potential sources of specimen contamination are hydrocarbons liberated from gaskets, finger prints (always use gloves) and back-streaming from oil diffusion pumps into the vacuum chamber (although this should not be a problem in a well-designed system using liquid nitrogen-cooled baffles above the diffusion pump).

As discussed earlier (§6.3.3b(i)), liquid nitrogen- or helium-cooled shrouds provide ideal protection for a specimen during etching. Such cold traps, kept at a temperature below that of the specimen, efficiently prevent contamination of the fracture plane both from gases produced by desorption from warmer surfaces and from gases in the system that condense at temperatures lower than that of the specimen itself. Such cold shrouds, which are only opened to the outer vacuum during replication, are, for example, installed in the Denton and Leybold freeze-fracture replication units and, in a different form with a permanent opening to the evaporators, in the Balzers unit (Figs. 6.15–6.17 in §6.3.3). With such designs, the drastic reduction in condensable gases in the specimen area has been demonstrated by long-term exposure experiments. No detectable contamination from condensation onto exposed fracture faces was seen when specimens were kept for up to 15 min at temperatures close to 100 K (Sleytr and Messner 1978; Sleytr et al. 1981). In the Bullivant freeze-fracture device (§6.3.3c(i)), the specimens seem to be well protected by the cold block, although it is constantly open to the outer vacuum through small evaporation tunnels. Since only those gas molecules moving along a path parallel to the tunnel axis reach the exposed fracture plane, the contamination rate is very low even when the apparatus is in a relatively poor general vacuum. Attempts to perform controlled etching (§6.5.2) with the Bullivant type of apparatus have been generally unsuccessful since there is no simple way of maintaining a constant specimen temperature. Deep etching *can* be obtained with this apparatus by using one of two methods. With the first, the specimen is fractured as normal under liquid nitrogen and its temperature is monitored with a thermocouple positioned in the lower block. On average, it takes about 45 min for the lower block and specimen to reach 173 K ($-100°C$). The time necessary to give a good depth of etching must be determined by experiment (Bullivant 1973; Dempsey and Beever 1979). Alternatively, the specimen can be fractured with a spring-loaded blade when the correct etching temperature has been reached within the chamber. This gives more reproducible results because the start of etching is determined by the actual fracturing process (Bullivant 1980b).

In some freeze-fracture replication units, such as the Balzers apparatus (§6.3.3a(iii)), a cooled protection surface is provided by the microtome knife and, in the latest unit (§6.3.3a(iii)), by an additional cold shield. For etching, it is important to stop the blade and holder, immediately after the last cut, in a position over the specimen so that the blade lies about 2.0 mm directly above the specimen with the edge of the knife about 4.0 mm forward. As shown by Staehelin and Bertaud (1971), the knife should not be too far forward or backward during etching or else water subliming from specimen chips which are in poor thermal contact with the knife edge may severely contaminate the fracture plane (Fig. 6.34). Also, a large gap between the knife and the specimen decreases the effectiveness of protection since molecules desorbing from warmer surrounding surfaces will be able to reach the fracture plane. If the microtome assembly is used both for fracturing and for contamination shielding, then it is important to realise that the specimen may be deeply fractured during an initial passage of the knife while the following pass(es) may fail to expose new faces in these deep regions (Fig. 6.34). Condensation contamination of exposed parts of the specimen is greatly reduced by the liquid nitrogen-cooled hood installed in the Balzers unit illustrated in Fig. 6.15c and d in §6.3.3a. Nevertheless, the possibility

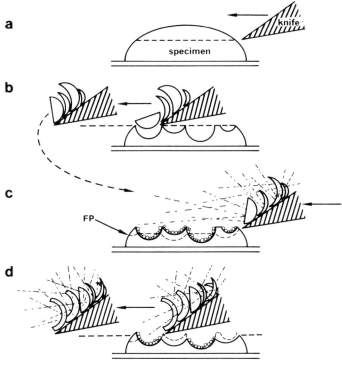

Fig. 6.34. Diagram illustrating the fracturing of a frozen specimen and how sublimation of water vapour from the specimen chips remaining on the edge of the microtome knife may contaminate the exposed surface. The chipping of the specimen shown in stages (a) and (b) leads to the formation of both deep and shallow fractures. During the following pass of the microtome knife (stages (c) and (d)), which produces the fracture plane (**FP**), the shallow fractures are cleaved away while the deep ones remain unaffected. At the same time the sublimation of water vapour from the chips on the knife edge leads to the contamination of the specimen surface in front of the advancing knife (stage (c)). Surfaces of deep fractures that are not cleaved away during the last pass of the microtome knife before replication can therefore be contaminated (stage (d)). Similar events can lead to etching in supposedly unetched specimens. (Redrawn after Staehelin and Bertaud 1971.)

of contamination from water vapour generated by frictional heating as the knife edge moves over the specimen must still be considered (Walzthöny et al. 1982).

In common with cold-shroud devices, ultra-high vacuum units allow contamination-free etching of fracture faces, since the amount of adsorbed gases at all surfaces in the system is greatly reduced by baking at 500–700 K and by using an air-lock system for specimen transfer (to reduce the formation of hoar frost on the specimen to a minimum (§6.3.3b(ii)).

It has been assumed that fracture faces produced under, or exposed to, liquid gases (He, N_2) become masked with a 10 nm layer of adsorbed material which cannot be removed even by etching under the best vacuum conditions (Moor 1973b). However, in practice, no differences in structural resolution or in the effects of etching have been observed when such specimens are compared with those freeze-fractured under improved vacuum conditions (Gross et al. 1978a; Bullivant 1980a; Sleytr et al. 1981; Steere 1982). As shown by Gross et al. (1978b) and Walzthöny et al. (1981), the condensation of water, the main contaminant, is usually determined by the physicochemical properties of the exposed fracture face. Such condensation phenomena have been observed in different model systems (Deamer et al. 1970; Staehelin and Bertaud 1971; Moor 1971; Walzthöny et al. 1981) and often lead to specific decoration effects (§6.9.7) or to the enhancement of defined, obviously hydrophilic, areas which may mimic membrane-intercalated particles (Gross et al. 1978b). Similarly, although not yet demonstrated, it might be expected that there is specific decoration (contamination) of hydrophobic areas by hydrocarbons in the vacuum. Finally, warts or whiskers, of a similar size to membrane-associated particles (§6.8.2), can be observed in replicas of etched ice (Davy and Branton 1970). These structures cannot be ascribed to impurities of the water, but may reflect crystallographic imperfections. Such warts are observed less frequently in spray-frozen specimens (Zingsheim and Plattner 1976).

6.5.3   Etching of non-aqueous systems

It is also possible to etch specimens if they are suspended in, or infiltrated with, suitable non-aqueous media, such as ethanol or dioxane, which sublime under similar conditions to those applicable to ice. (*Freeze-drying* from non-aqueous frozen solvents is described in §5.4.)

Due to the low vapour pressure of glycerol, the most commonly used cryoprotectant, deep etching only reveals the extended, unetchable eutectic network that forms during freezing. To avoid this masking effect, attempts have been made to use volatile cryoprotectants (Buchheim 1976, 1982; Schiller and Taugner 1980; Sawada and Yamada 1981). For example, infiltration of glutaraldehyde-fixed animal tissue with 30–40% ethanol provides good cryoprotection and also has the advantage that, on etching at 173 K to 178 K (-100 to $-95°C$), in addition to the normal membrane faces (see Fig. 6.38 in §6.8.2), extended areas of cell surface are revealed (Schiller and Taugner 1980). Frozen ethanol has a higher vapour pressure than ice (see Table 5.1a in §5.1 and Fig. 5.14 in §5.4.1) and this necessitates very accurate control of the specimen temperature to avoid excessive surface drying of the specimen. Using muscle as a model system, Sawada and Yamada (1981) found that the concentration of the ethanol has a significant influence on the fracturing behaviour of the specimen. The fracture plane through tissue infiltrated with 100% ethanol (which obviously causes considerable lipid extraction) is flat, like a mirror. The lower the concentration of ethanol, the greater the likelihood of the tissue splitting preferentially along membranes, as in specimens cryoprotected with glycerol. However, by lowering the concentration of ethanol, the chance of ice-crystal formation is increased. Sawada and Yamada found that infiltration with 65% ethanol provided a reasonable compromise between these two factors.

Buchheim (1976, 1982) used pure *p*-dioxane or *p*-dioxane/water mixtures as the suspending medium to study the surface structure of casein micelles. In common with the results from ethanol-treated specimens, changes in the fracturing pattern associated with differences in dioxane concentration were noted. To sublime the dioxane, etching is done at 173 to 163 K (-100 to $-110°C$). The specimens are then cooled back down to 143 K ($-130°C$) for replication. Despite the potential value of these techniques, the formation of specific artefacts (especially lipid-extraction artefacts) such as with non-volatile cryoprotectants, must constantly be borne in mind.

6.6 Thawing and replica cleaning

After the freeze-fracture replication process has been completed, the whole vacuum chamber, or the air-lock chamber, is vented and the specimen holders, or whole fracturing device (e.g. complementary replica device), are removed. The subsequent thawing and replica-cleaning processes are among the most delicate steps in the whole procedure. Heavy metal/carbon replicas of the required thickness are extremely fragile and brittle and are strongly attached to the surface that has been replicated. Consequently, dimensional changes in the specimen during thawing and cleaning can easily cause fragmentation of the adhering replica. This is generally less of a problem with suspensions than with tissues and the effects can be reduced by using one of the replica strengthening methods described in §6.4.2. It is best to thaw the specimen in a solution with the same composition as that in which the specimen was originally frozen, although, in some instances, it is advantageous to thaw the specimen in the first of the cleaning solutions. To inhibit disintegration of replicas during thawing it is possible to transfer the still frozen, replicated specimen to a plastic scintillation vial containing frozen methanol. The methanol is then allowed to thaw so that the specimen is simultaneously freeze-substituted. The specimens are then rinsed once in water and transferred through sodium hypochlorite and any other cleaning solutions until the replicas float off.

Thawing, cleaning and final washing are best done in small Petri dishes or in depressions of porcelain 'spotting' plates. Spotting plates have the advantage that the white background makes it easier to see replica fragments. Whereas replicas of suspensions often float when they are detached from the support by immersion at an angle of about 45° to the fluid surface, pieces of tissue have a tendency to sink. Spray-frozen samples must first be immersed in pure acetone to dissolve the butyl benzene (Bachmann and Schmitt-Fumian 1973). The replicas, which float in the acetone, are subsequently transferred, using a pipette, into a 10–20% acetone/water mixture where they spread out on the surface. The rest of the cleaning procedure for the replica can then follow any suitable schedule. Replicas which continue to float during thawing are best transferred from thawing to cleaning fluids using a 3.0–4.0 mm diameter platinum loop made of 0.2 mm diameter wire. If the replicas sink, they can be carefully transferred with a pipette.

Unless completely water-soluble specimens are replicated, adhering material must be cleaned away from the replica before examination. This is done by oxidation and hydrolysis in acids and alkalis, which do not corrode the replica, or by enzymatic digestion. A variety of replica-cleaning schedules has been proposed and it is often necessary to employ an empirical approach to determine the best cleaning agent and the correct length of time for replica cleaning for a particular biological specimen. When more than one cleaning reagent is used, care must be taken to minimise replica fragmentation during transfer of the replica. Particularly if H_2SO_4 is employed, transfer through a gently graded dilution series can be crucial. With both floating and sunken replicas, the concentration of the cleaning solution in a dish can be almost continuously altered by using two pipettes simultaneously, one for adding and another for removing liquids. If different cleaning reagents are used, the first solution is diluted all the way down to pure water before the other one is added in an increasing concentration. Washing chambers to provide a continuous concentration gradient through cleaning solutions have been designed and constructed by Jensen et al. (1973) and by Hohenberg and Mannweiler (1980).

(i)   Cleaning schedule for cell fragments and soft biological specimens
With most cell organelles and macromolecules, clean replicas are obtained after 1–2 h treatment at room temperature with 6% sodium hypochlorite or a 30% aqueous solution of chromic acid. Red blood cells, for example, are readily dissolved in 6% sodium hypochlorite within 20 min (Southworth et al. 1975). To clean replicas of adhering viruses Nermut (1973) recommends that the replica is floated on alternating solutions of 6% sodium hypochlorite and 30% sulphuric acid. Using this procedure, replicas of unfixed viruses are usually clean after 2 h. For fixed viruses, which may require longer cleaning, combined treatment with 70% sulphuric acid or 40% chromic acid (for 2 h) followed by hypochlorite (for 1–2 h) is necessary.

(ii)   Cleaning schedule for microorganisms
Replicas of most bacteria, fungi and algae can be cleaned successfully with a 30–40% solution of chromic acid within 1–3 days at room temperature. Faster cleaning with chromic acid solutions is achieved by incubation at 50–60°C. Other methods involve treatment for several hours, firstly with a 10% sodium hypochlorite solution followed by 70% sulphuric acid (Sleytr 1970b) or *vice versa* (Moor 1970; Lickfeld 1977). Specimens which are difficult to dissolve in 70% H_2SO_4 are often more readily dissolved in 40% chromic acid. Chromic acid treatment has the general advantage that further cleaning with 6% sodium hypochlorite is usually not necessary.

(iii)   Cleaning schedules for plant tissue
Cells of the mature tissues of higher plants are extremely heterogeneous compared with most other specimens. Fracture planes of compact cell walls become more firmly attached to the replica than, for example, the large watery vacuoles. This variation in composition, and the resulting differential swelling or shrinking, often causes fragmentation of the replica during thawing and cleaning. Therefore, one of the best methods of retaining large intact replicas is to provide an additional supporting layer (Robards and

Umrath 1978; §6.4.2) that can be removed once the organic material has been cleared away. Specimens attached to silver-backed replicas are best cleaned away by immersion in 30–40% chromic acid for 1–2 days. After rinsing in distilled water (three changes), the strengthened replica is treated with a mixture containing 9% potassium cyanide and 10% hydrogen peroxide (3% aqueous stock solution) in water for 1–2 h to dissolve the silver. All treatments are carried out at room temperature. The solutions are then either diluted with water or the replica is transferred directly into clean water.

If no technical facilities for evaporating a backing layer are available, the typical schedules for digesting plant material given by Fineran (1978) should be used. To obtain clean replicas within a few hours he suggests that, after thawing, the water under the replica is first gradually replaced over half an hour by concentrated nitric acid and then left for 1 h. The concentrated acid is then replaced by fuming nitric acid, which is used for 1 h either at room temperature or at 75°C. On cooling, the acid is slowly diluted with water.

In an alternative schedule (Fineran 1978), the replica is transferred, after thawing in distilled water, to 75% sulphuric acid for 4 h. The acid is then diluted to <10% and the replica is transferred into a mixture of 5% chromic and 5% nitric acid and left overnight. As for the cleaning procedure for micro-organisms, replicas can firstly be placed in bleach for 2–4 h and then transferred to 75% H_2SO_4 for up to 24 h, or the treatment can be reversed, using the acid first.

Pearce (1983) suggested a cleaning method that ensures the recovery of replicas of fractured leaf cells. This involves a preliminary dehydration step. The replicated specimens are placed in methanol and then transferred to chloroform for 15 min. They are then transferred to sodium hypochlorite for 1 h and finally rinsed with water. A loop is used to transfer the material to a carbon-coated gold grid and this grid is then floated, with the specimen on the upper side, on 50% chromic acid for three days. The gold grid usually separates from the carbon layer and can then be used to transfer the carbon layer and replica to a rinsing bath of distilled water. The cleaned replica and adherent carbon layer are then collected on the original gold grid or on another support. Without the initial wash in chloroform, the cuticle is carbonised by the chromic acid and forms an extensive layer that obscures much of the rest of the replica.

Procedures for dissolving and handling replicas of cutinised leaves and other plant tissue which yield comparatively large replicas have been described by Platt-Aloia and Thomson (1982). Dissolution of the tissue involves the sequential use of alcoholic KOH (12% KOH in 95% ethanol), sulphuric acid (70%) and chromic acid (half-strength glassware cleaning solution).

(iv)   Cleaning schedules for animal tissue

Replicas of most soft and unfixed tissue can easily be cleaned with 30–40% chromic acid or 6% sodium hypochlorite in 5–24 h. As with other specimens, shorter cleaning times are required when a CrO_3 solution is used at 60°C. For tissues that have a tendency to shrink, it is recommended that the cleaning procedure is started in a mixture of 10% nitric acid and 10% sulphuric acid, or a 5–10% solution of caustic soda (Stolinski 1977). Depending on the type of tissue, this initial treatment lasts for a few hours or overnight. Subsequent cleaning steps involve strong acids (40% chromic acid or 70% H_2SO_4) or alkaline solutions at room temperature or 60°C. The optimal time for the final cleaning step must be checked experimentally for different specimens. The more complex and prolonged cleaning procedures, given above, are mainly used for difficult unfixed tissues, such as skin or peripheral nerve, and for all fixed tissues, which are more difficult to digest.

(v)   Cleaning schedule for specimens with a high lipid content

For replicas of these specimens, it is recommended that the cleaning procedure starts with 1–2 h in a 2:1 mixture of chloroform/methanol before oxidising agents are used (Moor 1970; Pearce 1983). When indexed gold grids are used as an aid to the observation of complementary replicas of suspensions, the Vaseline which may be used to hold the two grids in register (see Fig. 6.2; type Bh) should be removed with chloroform before the specimen is digested. This is done, after thawing, by drying the grids, with the attached replicas, and then washing them in chloroform, with frequent agitation, for a few minutes (Hess and Blair 1972).

Lipid or Vaseline contamination can be also removed from mounted replicas by placing the grid over boiling chloroform and allowing the chloroform vapour to condense on the replica surface (Vail and Stollery 1978). Revel et al. (1971) suggest that the replicated specimen is treated with 25% dimethylformamide in distilled water overnight, to extract lipid, before specimen digestion and replica cleaning procedures.

6.7   Mounting of replicas on grids

After cleaning, the replicas are carefully rinsed in several changes of distilled (preferably double-distilled) water. During this final step, it is particularly important to avoid secondary contamination of the replica from dirty water, dishes and handling tools (tweezers, loops, etc.). Floating replicas are manipulated with a platinum wire loop while sunken ones are gently transferred with a pipette (e.g. a shortened pasteur pipette). After the final wash, the replica is picked up on a coated or uncoated grid. This step is omitted when replicas remain attached to the grids which are used in the preparation of complementary fracture faces (§6.2.2d).

Replicas of suspensions of isolated cells or cell organelles have considerable redundancy of information and are picked up on uncoated low transmission grids with wide bars. With more complex specimens, such as animal or plant tissues, much useful information would be lost beneath the grid bars. Replicas from this type of specimen are picked up on high transmission grids. Grids with a single, large central hole are especially useful. These grids are coated with a support film, preferably of Formvar which is more stable in the electron microscope than Collodion. Carbon reinforcement is not necessary, since the heavy metal/carbon replicas provide sufficient stability themselves.

In practice, there are many different ways of transferring the replica from the final wash in distilled water to the centre of a grid. It is often best to carry out such manipulations under a dissecting microscope. If it is possible to keep the replica floating during the cleaning procedure, then it can be picked up from the surface by bringing a coated (or uncoated) grid, held in clean tweezers, down on it (Fig. 6.35a). The grid need not be pushed down into the trough and the fluid surface should only be depressed, not broken. The grid is subsequently lifted to a horizontal position and the surplus water is blotted off with filter paper, either by touching the side to the filter paper or by placing the grid with the replica uppermost flat on the filter paper. It is important to be sure that there is no water left between the tips of the tweezers before releasing the dry grid. Floating replicas can also be picked up with a wire loop (about 4 mm diameter) and thus transferred to coated or uncoated grids. Finally, either floating or sunken replicas can be picked up by placing the grid underneath the replica and then slowly raising it. This is generally easier using uncoated grids. With coated grids the floating replica will often drift away. Working under a dissecting microscope provides the possibility of using an eyelash fixed to a rod as a mount-

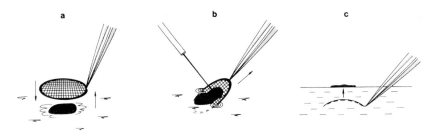

Fig. 6.35. Techniques for transferring a replica from the final wash onto a grid. (a) Bringing a grid down on top of a floating replica: the fluid surface should not be broken and the grid should be lifted off horizontally. (b) Holding the replica against the grid, to prevent it floating away whilst the grid is removed from the fluid, with a mounted eyelash. (c) Raising a grid from underneath the replica, parallel to the fluid surface. The grid has been bent slightly to present a convex surface to the replica in order to help centralise the replica on the grid.

ing aid (Fig. 6.35b). When an uncoated grid is raised from beneath a floating or sunken replica, the chances of picking up the replica in a central position are much better if the grid is bent slightly convex in relation to the replica and then raised parallel to the water surface (Fig. 6.35c). Excess liquid is drained away by placing the grid on filter paper.

Sometimes replicas have a tendency to curl up and sink during the cleaning process. To unroll a replica on the water surface it is possible to transfer the replica (with a pipette) first to 100% ethanol or acetone, and then back, within a drop of the organic solvent, to the distilled water. Due to differences in surface tension, the replica uncurls on touching the water but it also often breaks into smaller pieces. However, these pieces usually contain enough information to be useful, particularly if suspensions are examined. In other words, it is better to save a small piece of something useful than a large piece that cannot be studied because it is curled up!

6.8 Interpretation of freeze-fracture replicas

6.8.1 General considerations

Heavy metal/carbon replicas obtained by freeze-fracture replication are studied by transmission electron microscopy. The final image seen on the microscope screen is determined by the topography of the fracture face and represents a two-dimensional projection of the mass distribution in the replica. Thus, to interpret the latent three-dimensional information encoded in the electron micrographs, it is important to know the direction and angle of shadowing. The direction of shadowing is generally indicated on published micrographs by an arrow.

The correct appearance of the replica relief (surface sculpturing) is easily obtained by orientating electron micrographs with the shadows extending *away from* the observer or with the shadows pointing from the bottom to the top of the picture. Other orientations may give an unpractised observer the optical illusion of reversed relief (i.e. depressions will appear as protruberances and *vice versa*). Since no heavy metal is deposited in the lee of prominent structures, shadows appear as white areas on prints when electron microscope negatives are directly used for photographic printing. Some authors publish reversed images in order to obtain dark shadows. This gives pictures which provide the illusion that the freeze-fracture relief is illuminated by light. A procedure has also been described which allows printing of freeze-fracture pictures with black shadows without the necessity of preparing an intermediate negative (Schulz and Reynolds 1977). In fact, little is gained by this procedure and the danger of misinterpretation of fine structural detail is obvious. As pointed out earlier (§6.4.1a), the faithful reproduction of surface structure information by a shadowing method is limited by the condensation process of heavy metal atoms. Reversed contrast of areas in which the accumulation of heavy metal does not accurately reflect the surface geometry (decoration phenomena) is particularly misleading.

The fracturing process frequently splits biological specimens within membranes, so leading to views of extended membrane faces. With the possible exceptions of isolated organelles and procaryotic cells, the fracture faces produced may have little similarity with images from thin sections. Particularly for the interpretation of fracture planes through tissue or isolated eucaryotic cells, it is important to have previous knowledge of the structure of the specimen obtained from other electron (or light) microscopical methods. Unless replica reinforcement methods are used (§6.4.2), it is difficult to maintain coherent replicas of fracture faces through a heterogeneous tissue. It is commonly necessary for several pieces of replica to be collected to obtain a complete view of structures exposed by a single fracture. The best aid in the understanding of replica fragments is the examination of thin sections of the same material, since the gross morphological features of thin sections and freeze-etching images are very similar. However, considerable differences in the preservation of fine structural details may be expected since chemical fixation and freeze-fixation act very differently (§2.1 and §2.3.1).

Provided that the fracturing process has exposed areas of interest, a high quality freeze-fracture replica can be characterised by the following criteria:

(*i*)   No major structural changes have been introduced during sampling and pretreatment (see §2.3.2).

(*ii*) The freezing procedure has limited ice crystal growth to a size which does not interfere with morphological analysis (§2.3.1). It must be remembered that disruptive ice crystals may also develop if the specimen temperature has risen above the recrystallisation point during preparation (§2.1).

(*iii*) Shadowing must have reproduced the surface geometry of the fracture face as faithfully as possible. Structural details must show sharp shadows with good contrast. Replicas with too great a contrast reveal less structural detail although they may provide a better three-dimensional illusion of the topography at low magnification (§6.4.1a). With resistance evaporation techniques (§6.4.1c), lack of contrast is frequently due to insufficient evaporation of platinum in relation to carbon. Such replicas have a soft appearance and show poor resolution of structural details.

(*iv*) The replica must be free of incompletely dissolved specimen material which is seen in the microscope as irregular dark patches with diffuse outlines.

These criteria are generally easy to establish but do not take into account the formation of specific artefacts which will be discussed in §6.8.3.

Not all areas of a replica necessarily show high quality fracture faces. Parts of the replica may show the structure of the unfractured edges of the specimen or the surface of the specimen support. Furthermore, most replicas of specimens smoothed down with a microtome exhibit areas with knife marks or flat areas generated by local friction between the knife and the specimen surface. These areas are devoid of any information and are interpreted as melted regions (§6.3.1b). Similarly, even under optimal conditions, the quality of shadowing may vary in different parts of the replica. This is especially the case if fracturing has created a coarse topography with some regions orientated at an extremely low angle to the evaporation (shadowing) source. Images of fine structures that have been shadowed at such an unfavourably low angle may be considerably improved by tilting the replica in a microscope equipped with a goniometer stage.

Comparison of etched and unetched fracture faces of the same specimen

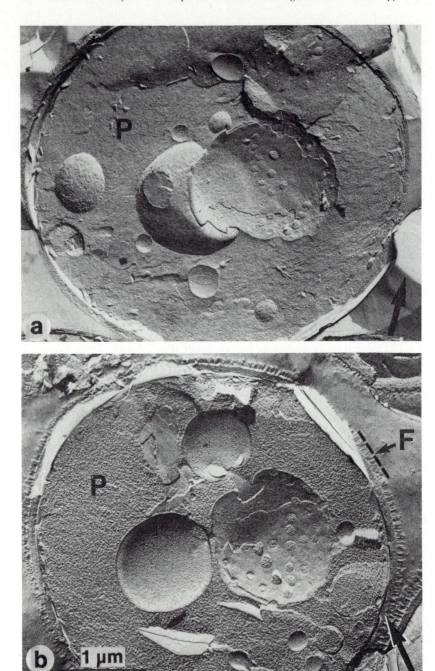

can be of great value in the interpretation of the structure of different cell components. Steere and Erbe (1979) have even developed a method that allows the evaluation of etched and unetched images of complementary fracture faces. Under condensation-free conditions, the structural differences between etched and unetched images are entirely due to the loss of volatile material which, for biological specimens, is predominantly water. Since moderate segregation phenomena (§2.1.2) occur with most freezing techniques, etching roughens hydrated regions by exposing the pseudo-eutectic network (see Fig. 6.33a in §6.5.1 and Fig. 6.36). Thus spatial differences in ice crystal size can reflect differences in water content (degree of hydration). An obvious example of this is the contents of vacuoles which frequently show a much rougher etched structure than the surrounding cytoplasm. Similarly, in procaryotic cells which have been chemically fixed before freezing, the nucleoplasm is easily distinguished from the surrounding matrix by its rougher structure after etching (Lickfeld 1968; Nanninga 1973). By comparing both etched and unetched complementary areas of freeze-fractured virus crystals, Steere and Erbe (1979) were able to show that etching may reveal useful information even in specimens which have been treated with high concentrations of cryoprotectant (25% glycerol plus 25% sucrose).

Cell structures are not always enhanced by etching, however. Fine structural detail in regions which fracture but are not in the main fracture plane may become obscured by lowering the surface of exposed ice crystals. For example, structural components of the cytoplasm or nucleoplasm which have been pulled up or deformed by the fracturing process are only clearly distinguished if the background is smooth, as in an unetched fracture face (Bullivant et al. 1972). However, deep etching in combination with rotary shadowing may help to reveal, for example, the fine structural detail of the cytoskeletal network of a cell (Heuser 1981).

Membrane *face* views are generally not affected by etching but become easier to distinguish from the hydrated surroundings which change from smooth to rough as etching progresses (Fig. 6.36). Nevertheless, it has been

---

Fig. 6.36. Frozen-fractured yeast cells before and after etching. Yeast cells (*Saccharomyces cerevisiae*) were rapidly frozen (as a 30 μm thick sandwich) by dipping in liquid Freon 22 and were then fractured and replicated without etching (a) or with 20 s etching at 173 K (b). The membrane fracture faces of the etched and unetched specimens reveal the same structural detail. In well-frozen specimens, etching changes the appearance of the cross-fractured cytoplasm (**P**) from smooth to rough. Etching has also changed the dense coat of radially arranged fibrous material (**F**) on the cell surface. The arrows indicate the direction of shadowing.

shown that prolonged etching can result in the formation of small holes in membranes (Meyer and Winkelmann 1970) which are possibly the sites of concentrations of transmembrane proteins (Tranum-Jensen and Bhakdi 1983). Cross-fractured membranes, which appear in unetched specimens as single lines, show a double-line structure after lowering the ice matrix. It must always be remembered that a two-dimensional image of a replicated three-dimensional structure is seen on the microscope screen. Consequently, as with other structures, the image of cross-fractured single or multi-layered

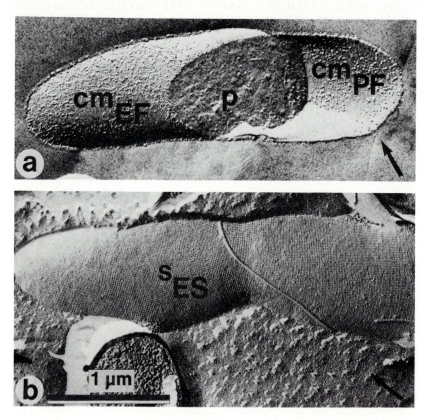

Fig. 6.37.   Fracture faces of a bacterial cell suspension (*Desulfotomaculum nigrificans* suspended in growth medium) (a) without and (b) with etching. In non-etched cell suspensions only the fracture faces of the cytoplasmic membrane **cm**$_{EF}$ and **cm**$_{PF}$ can be observed. (For nomenclature see Fig. 6.41 in §6.8.2.) **cm**, cytoplasmic membrane; **p**, cross-fractured protoplast; **F**, fracture faces; **E**, surfaces facing extracellular space; **P**, surfaces facing cytoplasm. The paracrystalline cell wall surface structure (**S**) is only exposed by deep etching (e.g. 90 s at 173 K). Arrows indicate the direction of shadowing.

membranes strongly depends on the thickness of the evaporated layer, angle of shadowing and orientation of the membrane in relation to the evaporation source (Lickfeld et al. 1972).

Surfaces of single cells or isolated cell components (e.g. membranes), which are not favoured by the fracture plane, can be exposed by deep etching (§6.5.2) provided the specimen can be suspended in distilled water. For many bacteria it has been shown that extended areas of regular arrays of macromolecules on the cell surface are clearly visible after deep etching (Fig. 6.37), even when cells have been frozen while suspended in the growth medium (Sleytr 1978; Sleytr and Glauert 1975, 1982).

The study of complementary areas on both fracture faces created by a single fracture (§6.3.3a(ii)) has been shown to be particularly valuable in the interpretation of freeze-fracture images and has provided considerable insight into the generation of artefacts (§6.8.3). As an aid to localising complementary areas of fractured suspensions, replicas are mounted on gold finder grids. For finding complementary areas on replicas which were not originally attached to finder grids, the pattern of distribution of marker cells (such as yeast), or easily identifiable features of tissue fracture faces, may be examined in a conventional or, preferably, a projection light microscope (Sleytr and Umrath 1974a). The replicas on grids are placed on a microscope slide and drawings or micrographs of the replica are made. If a projection microscope is available, the drawings can be made by tracing on transparent paper. As another aid to finding identical areas more easily, it has been suggested that one of the complementary replicas should be mounted upside down so that the two images in the microscope are comparable and not mirror images of each other (Bullivant 1973).

6.8.2 Interpretation of membrane structures and nomenclature

The dominating feature of freeze-fracture replicas of biological specimens is often membrane face views (Fig. 6.38). It is now well established, with the exception of the cytoplasmic membranes of those Archaebacteria which have abnormal lipid composition (Langworthy et al. 1982), that membranes split along an interior plane and that true membrane surfaces are only exposed by etching (Fig. 6.38) (for reviews see Branton 1971; Verkleij and Ververgaert 1975, 1978; Zingsheim and Plattner 1976; Bullivant 1977). With few exceptions, both membrane fracture faces are studded with particles and depression (pits) of various sizes (Figs. 6.38 and 6.39). Particles seen on membranes are described as membrane-intercalated or membrane-associated particles, or just membrane particles. Convincing evidence has accumulated that they reflect discontinuities in the bimolecular leaflet of the membrane lipids, generated by protein, glycoprotein or lipoprotein structures, or even by inverted lipid micelles (Zingsheim and Plattner 1976; Verkleij and Ververgaert 1975, 1978; Ververgaert and Verkleij 1978; Verkleij et al. 1980). By comparing different membrane types it has been shown that the number and distribution of particles seen on both membrane fracture faces

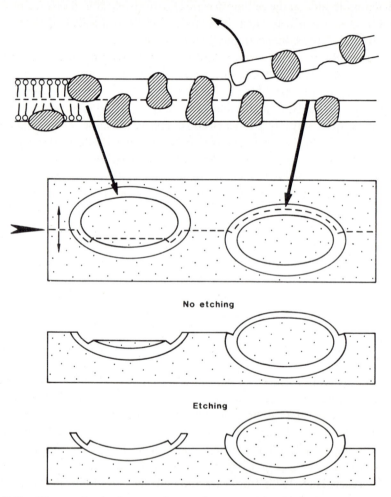

Fig. 6.38.   Diagram showing how two fracture faces are produced by splitting a biological membrane along a unique internal plane. Non-etched specimens only reveal membrane fracture faces whereas, on the sublimation of ice, membrane surface views become exposed.

is a fixed characteristic of a given membrane type (Branton 1971). The composition of most particles is not known but their distribution between the two complementary fracture faces reflects an asymmetric organization of biological membranes. Besides the two faces of the interior of the membrane, both true surfaces of a membrane can be exposed by etching provided that they are in intimate contact with ice (Fig. 6.38). Thus, extended areas of membrane surface are only revealed by deep etching (§6.5.2) when isolated membranes have been suspended in water, etchable organic solvents (§6.5.3), or a volatile buffer. Otherwise, as a result of segregation phenomena, the true membrane surfaces may become masked by non-etchable material which is pushed against the membrane during ice crystal growth (§2.3.1).

Several systems of nomenclature have been used in the literature to designate the two internal membrane faces exposed by fracturing and the two surfaces only revealed by etching. The most common, among others, were A, B, C and D faces (Bullivant 1973), ($+$) and ($-$) faces (Meyer and Winkelmann 1969) and convex and concave faces (Pinto da Silva and Branton 1970). As a logical consequence of this confusion of descriptions, a unifying system was suggested for plasma membranes by a group of scientists active

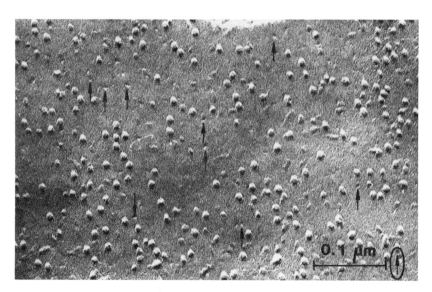

Fig. 6.39. Protoplasmic fracture face (PF; see Fig. 6.40) of a yeast vacuolar membrane. Membrane intercalated particles, and holes (pits; arrowed), from which similar particles have been removed, can be seen.

in the field (Branton et al. 1975). It was proposed that, for any membrane that can be split (Fig. 6.40), the half closest to the cytoplasm, nucleoplasm, chloroplast stroma or mitochondrial matrix be designated the protoplasmic half, abbreviated P, and that the complementary half, closest to the extracellular space, or endoplasmic space, be designated accordingly the extracellular, exoplasmic or endoplasmic half, and abbreviated E. The exoplasmic and endoplasmic space would include the interior of endocytic vacuoles, phagosomes, primary and secondary lysosomes, food vacuoles, tonoplasts, cisternae and vesicles of dictyosomes and the endoplasmic reticulum and the cisternae formed between the inner and outer nuclear membranes. The abbreviation E is also used to designate the half-membranes lying closest to the space between the inner and outer membranes of mitochondria and chloroplasts, as well as the intrathylakoid space. Views of the fracture face of that half of the membrane associated with the protoplasm are referred to as the protoplasmic face (PF) and that associated with the extracellular, exoplasmic or endoplasmic space as the extracellular face (EF). The true membrane surfaces, which may be revealed by etching, are then referred to as the P surface (PS) and E surface (ES), the former representing the membrane surface originally in contact with the protoplasm, the latter with the exoplasm.

Since the designation E or P is not applicable to procaryotic cells because of their intermembraneous spaces, it was suggested by Schmid et al. (1980) that the terms extracellular space (E) and protoplasm (P) should only be used at the reference level. In this way, the fracture faces (F) and surfaces (S) of the membranes and non-membraneous layers of Gram-positive and Gram-negative bacterial envelopes can be clearly designated (Fig. 6.41).

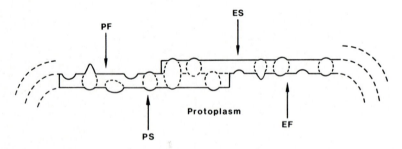

Fig. 6.40.   Nomenclature for the freeze-fracture faces of membranes exposed by fracturing and for the membrane surfaces exposed by etching. **PF**, Fracture face of that half of the membrane associated with the protoplasm. **EF**, Fracture face of that half of the membrane associated with the extra-cellular space. **PS** and **ES**, Surfaces of the membrane halves associated with the protoplasmic or extra-cellular sides, respectively. (After Branton et al. 1975.)

This method for the designation of frozen-fractured bacterial envelopes is independent of the number of membranes or non-membraneous layers and spaces and helps to overcome the diversity of incoherent topographical designations previously used in the literature. As can be seen from examples of typical Gram-positive (Fig. 6.41a) and Gram-negative (Fig. 6.41b) envelopes, membranes or non-membraneous layers are firstly labelled according to their description from thin sections, using specific data on the organism studied. In general, the following abbreviations are used: outer membrane (om), peptidoglycan (pg), cell wall (cw), cytoplasmic membrane (cm) and, if present, surface layer (s). Secondly, an index is added which clearly indicates whether a face is directed towards the cell cytoplasm (P) or towards the extracellular space (E). Thirdly, a supplementary index is added which clearly indicates whether an unfractured surface (S) or a fractured face (F)

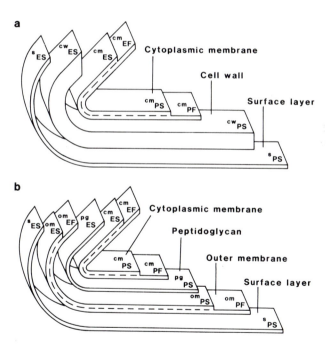

Fig. 6.41. Nomenclature system for the fracture faces and surfaces of procaryotic cell envelopes: (a) shows a typical Gram-positive bacterial envelope and (b) shows a typical Gram-negative envelope. **E**, Surfaces facing the extracellular space; **P**, surfaces facing the cytoplasm; **F**, fracture faces; **S**, etched surfaces; **cm**, cytoplasmic membrane; **cw**, cell wall; **pg**, peptidoglycan; **om**, outer membrane; **s**, surface layer. (Redrawn after Schmid et al. 1980.)

is shown. The illustrations in this volume have been labelled according to the nomenclature suggested by Branton et al. (1975) and Schmid et al. (1980).

6.8.3   Artefacts

As pointed out at the beginning of this Chapter, the early hope that freeze-fracture replication methods might give artefact-free images of biological structures was over-optimistic. It was soon realised that the individual preparation steps can each lead to the production of artefacts and that, for correct interpretation of the information encoded in the final two-dimensional image of a freeze-fracture replica as seen on the microscope screen, it is essential to understand fully the factors involved in determining structural changes (Branton 1973; Zingsheim and Plattner 1976; Bullivant 1977; Sleytr and Robards 1977a, 1982; Kellenberger and Kistler 1979; Lepault and Dubochet 1980; Plattner and Zingsheim 1983; Verkleij 1984). Since cryofixation methods which minimise segregation phenomena are now available for most biological specimens and these consequently retain the distribution of molecules close to that pertaining in the native state, artefacts are mainly caused by subsequent preparation steps, particularly fracturing and replication. The ability to produce and evaluate replicas from both halves of a frozen-fractured specimen (§6.3.3a(ii)), has proved a valuable asset in helping to interpret the nature and mechanism of artefact formation. While gross features of complementary fracture planes match, the finer details may not exhibit the anticipated mirror-image relationship. The most likely explanations for the lack of complementarity between originally opposite areas are:

(*i*)   condensation of material on the freshly fractured surface in the period between fracturing and replication (§6.5.2);

(*ii*)   deformation and dislocation of specimen components during freeze-fracturing (§6.3.1);

(*iii*) collapse and aggregation phenomena resulting from dehydration during etching; and

(*iv*)   thermal load-induced alterations during replica formation (§6.4.1b).

In freeze-fracture replication, the greatest variations in specimen preparation are in the fracturing process (see Fig. 6.1 in §6.1) but, independently of which freeze-fracture method is used, replication of the specimen fracture face must be carried out in a vacuum unit (§6.3.2). Thus condensation of material onto the fracture face of the freshly cleaved specimen in the period between fracturing and replication is one of the main sources of artefact. Condensation contamination is frequently not easy to detect unless it has led to obvious structural alterations by masking or enhancing fine structural details or has created prominent new structures which have no complementary partners on the opposite fracture face. Warts or flat plaques seen in areas which are normally devoid of such structures (e.g. smooth lipid or membrane fracture faces) are strong indications of heavy contamination of the fracture face with water or hydrocarbon vapour (Deamer et al. 1970; Staehelin 1973; Bullivant 1973; Walzthöny et al. 1981). It has been suggested that high quality replicas can be judged, for instance, from the number of small holes that correspond to the pattern of membrane particles on opposite fracture faces (Staehelin 1973). Although the existence of these membrane pits is an excellent indication of a good (high resolution) preparation, their absence is not necessarily due to specific condensation contamination. Membrane pits may also be destroyed by deformation or dislocation during freeze-fracturing or during shadowing (Sleytr and Robards 1982). In contrast, specific decoration phenomena (§6.9.7) may easily mimic membrane particles (Bullivant 1973; Zingsheim and Plattner 1976).

Considering that condensation-free conditions can be maintained in the specimen area by using either cold shrouds or ultra-high vacuum technology, the possibility of artefact formation should be restricted to the fracturing and replication processes. Nevertheless, it is also necessary to be aware that dehydration during etching can cause the aggregation of particles and collapse phenomena (Kistler and Kellenberger 1977; Lepault and Dubochet 1980; Sleytr and Robards 1982).

It is important to remember that, using the right technical methods, specimens freeze-fractured under liquid nitrogen (§6.3.3c(i)) or liquid helium (§6.3.3c(ii)) show the same structural details on fracture faces as specimens fractured at similar temperatures under high vacuum (Bullivant 1973, 1980a; Sleytr and Messner 1978; Bullivant et al. 1979; Sleytr et al. 1981). In other words, the formation of a disturbing, or resolution-limiting, non-etchable adsorption layer as predicted by Moor (1973b) is not detectable even when specimens have been freeze-fractured under liquid helium and transferred via liquid nitrogen into evaporation units for replication (Sleytr et al. 1981). Steere and Erbe (1979) reported that even a layer of contaminating ice can be sublimed from fracture faces under etching conditions using a cold shroud device (§6.3.3b(i)) without detectable loss in resolution compared with non-contaminated specimens. This correlates with the observation that, after partial freeze-drying (deep etching) of two-dimensional arrays of macromolecules (proteins), involving the sublimation of a layer of ice several micrometres thick before exposing the frozen specimens, the shadowed preparations reveal the same structural details as vacuum-fractured and etched or negatively stained specimens (Crowther and Sleytr 1977; Sleytr 1978; Kistler et al. 1978; Studer et al. 1980; Gross et al. 1982; Wildhaber et al. 1982).

As discussed earlier (§6.3.1), experiments have shown that ductile and brittle regions may alternate along the fracture plane and polymers may de-

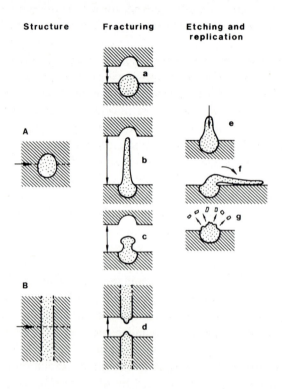

Fig. 6.42.   Schematic illustration of possible events during fracturing and subsequent preparation steps of a two-phase system (a polymer in an ice matrix). (A) The globular polymer may either separate 'cleanly' from the surrounding matrix (a) or may undergo complete (b) or partial (c) deformation. (B) Fibrillar polymer structures generally deform in a symmetrical fashion (d). Exposed deformed structures can be affected during etching and/or replication. Structural changes, such as elastic recontraction (healing) (e), collapse (f) or total disintegration (g) may occur. (Redrawn from Sleytr and Robards 1982.)

form during freeze-fracturing even at the temperature of liquid helium (for general reviews see Sleytr and Robards 1977a, 1982 and, with reference to membranes, Zingsheim and Plattner 1976). Plastically deformed components frequently exhibit a fibrillar structure and show a clear lack of complementarity on opposite fracture faces. Fig. 6.42 is a schematic illustration of the possible events during the fracturing and subsequent preparation steps of a two-phase system (polymers in an ice matrix) (Sleytr and Robards 1982). The deformation products may either stick out from the fracture plane or collapse onto the nearest surface. Strong evidence exists that, due to thermal load during evaporation, the survival chances of delicate fibrillar deformation products sticking out from the fracture plane are very small. This means that, generally, only fibrils which have collapsed onto a large dominant structure and are in close thermal contact with it will be seen on the final replica. During specimen fracturing, deformed structures may have the ability to recontract if freeze-fracturing is done at higher temperatures, but may re-freeze in the extended position at lower fracturing temperatures. This may give an apparent increase of plastically deformed structures when freeze-fracturing is done at low temperatures (Steere et al. 1980; Sleytr and Robards 1982). Structural distortion and collapse phenomena may also occur during etching. For instance, in replicas of fractured and deep-etched bacterial suspensions, most flagella are seen in close contact with the cell surface (Sleytr and Robards 1977a; Sleytr 1978; Kellenberger and Kistler 1979). Furthermore, the heat dissipation during local deformation processes or replication may be sufficient to facilitate elastic recontraction of the deformed structure.

In conclusion, with regard to artefacts it can be stated that:

(*i*)   Freeze-fracture replication methods are available which have the potential, even when freeze-fracturing is carried out at 4 K, to reduce the possibility of artefact formation to the fracturing and replication stages.

(*ii*) Structural deformation during fracturing may be expected at temperatures down to 4 K. This clearly shows that the freeze-fracture replication technique has a significant methodological limitation.

(*iii*) With some structures the degree of deformation is temperature-dependent and may be reduced by lowering the fracturing temperature, by chemical fixation procedures (introducing covalent bounds), by infiltration with glycerol, or by a combination of all of these.

(*iv*)   A number of fine structural details seen on freeze-fracture replicas must be interpreted as the result of a multi-event process, involving plastic deformation, reconstruction, collapse phenomena and thermal load-induced alterations during replica formation. All these factors, which are involved in producing the final topography of the fracture face, may enhance or compensate for each other.

(*v*)   Finally, in characterizing artefacts it must be remembered that it is a *two-dimensional* projection of a replica that is examined in the microscope and that, during shadowing, the heavy metal atoms are not necessarily trapped at their first sites of impact (§6.4.1a; also Abermann et al. 1972).

6.9   Cytochemical and quantitative methods

Freeze-fracture replication techniques produce inert replicas. This means that, with the exception of autoradiography, histochemical methods are not applicable for determining the chemical nature and function of structural features once the continuous replica has been deposited on the fracture plane. As summarised in Table 6.4, the cytochemical techniques which can be combined with freeze-fracture and/or replication and etching methods can be classified into two major groups (see also Severs 1984). The pre-fracturing methods involve techniques where labels or specimen constituent-dependent alterations are introduced *before* freezing the specimen. This can be done using radioactive labels which are then detected by freeze-fracture autoradiography after specimen fracturing and replication (§6.9.1). Specific sites on the specimen surface may also be labelled with morphologically discernible markers which later, after fracturing, become exposed during etching (§6.9.2). Similarly, specific sites on the specimen surface may be labelled with electron-opaque markers (e.g. conjugates of antibody with ferritin or colloidal gold) which, after fracturing, replication and embedding the specimen in plastic resin, can be detected in thin sections cut parallel to the replica/tissue interface (§6.9.3). Finally, selective pre-freezing alterations of the organisation of specimen components (particularly membranes) into recognisable patterns by physicochemical means may provide valuable insights into the chemical nature of such structures (§6.9.4).

Post-fracture methods (Table 6.4) include techniques where labelling or structural changes are carried out on the freshly exposed fracture face (§6.9.5). Labelling with morphologically discernable or electron-opaque

TABLE 6.4

Cytochemical methods for specimens examined
by freeze-fracture replication

Pre-freezing methods (labelling or specific alterations before freezing)
1. Freeze-fracture autoradiography
    1.1. Freeze-fracture autoradiography of bulk specimens
    1.2. Freeze-fracture autoradiography of monolayers
    1.3. Freeze-fracture autoradiography in combination with the sectioned replica technique
2. Labelling with morphologically detectable markers for replica techniques
3. Labelling with electron-opaque markers for the sectioned replica technique
4. Specific pre-freezing alterations of the molecular organisation of specimen (membrane) components into recognisable patterns
    4.1. Thermal phase transitions
    4.2. Changes in pH and/or ionic strength
    4.3. Interactions with lipid perturbers

Post-fracture methods (labelling or specific alterations of fracture faces)
1. Thin sectioning of post-fracture labelled specimens
2. Replication of freeze- or critical point-dried post-fracture labelled specimens
3. Thin sectioning of post-fracture, post-shadow labelled and replicated specimens
4. Replication of freeze- or critical point-dried post-fracture, post-shadow labelled specimens
5. Lipid extraction of fractured frozen specimens
6. Specific decoration of fracture faces *in vacuo* with water vapour

markers can be done directly after the fractured specimen has been thawed (§6.9.5a and b) or after shadowing and subsequent thawing (§6.9.5c and d). In either method, the labelled specimens can be studied by replication or thin sectioning. Information about the chemical composition of structures exposed on fracture faces may also be obtained by specific low temperature extraction methods (§6.9.6). Finally, specific decoration of fracture faces with water vapour *in vacuo* (§6.9.7) may provide valuable insights into the physicochemical properties of exposed structures.

6.9.1   *Freeze-fracture autoradiography*

Freeze-fracture autoradiography, introduced by Fisher and Branton (1976) and Rix et al. (1976), involves the application of a monolayer of an autoradiographic emulsion to replicated fracture faces and subsequent processing. The autoradiographic principles are the same as those described fully by Williams (1977). The technique represents a promising approach for the

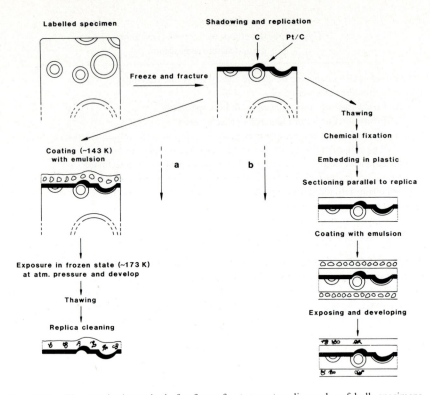

Fig. 6.43.   The two basic methods for freeze-fracture autoradiography of bulk specimens. (a) Fractured and replicated specimens are coated directly with an autoradiographic emulsion. (b) Fractured and replicated specimens are processed for thin sectioning, after which the auto-radiographic emulsion is applied.

analysis of labelled water-soluble and diffusible compounds and transmembrane concentrations of molecules without encountering the problem of diffusion artefacts introduced by chemical fixation, dehydration and embedding. Freeze-fracture autoradiography can be applied to both fractured bulk specimens or monolayers.

6.9.1a   Freeze-fracture autoradiography of bulk specimens

The two basic methods for the study of bulk specimens by freeze-fracture autoradiography are illustrated in Fig. 6.43. The fractured and replicated specimens are coated directly (*in vacuo* or *ex vacuo*) with an autoradiographic emulsion (Fig. 6.43a) or, alternatively, are first processed for thin sectioning before applying the emulsion (Fig. 6.43b). With tissue blocks, the problem of the coarse topography of a fracture face making it difficult to obtain uniform contact between the film and the replica is often encountered. In addition, resolution may be limited by radiation from deeper levels in the block. This is a particular drawback if labelled membrane fracture faces are to be studied but may be partially overcome by using short-range radiation such as tritium or [125]I (Schiller et al. 1978, 1979). Replicated fracture faces of bulk specimens can be coated with dry monolayers of autoradiographic emulsions *in vacuo* or *ex vacuo*. The practicability of *ex vacuo* coating was shown by Fisher and Branton (1976) and Rix et al. (1976).

With the Fisher and Branton technique (Fisher and Branton 1976; Fisher 1978) specimens are fractured and replicated in a Balzers-unit (§6.3.2) and subsequently transferred, without protecting the replica from hoar frost formation, to the surface of a cold pillar and then to a dark room. After defrosting the replicas in flowing dry cold nitrogen gas, the autoradiographic emulsion mounted on a wire loop is immediately applied to the replica and pushed against the replica contours using the pressure of a stream of cooled dry nitrogen. Nuclear emulsion monolayers made from Ilford L-4 (Ilford; see Appendix 2) can be strengthened and stabilized using Collodion films. Samples are stored for exposure at 193 K ($-80°C$) in gas- and light-tight containers and then thawed before developing the emulsion and dissolving the specimen remnants adhering to the replica. Fisher and Branton (1976) reported an attempt to apply emulsions *in vacuo* but their method produced inexplicably high background levels.

Compared with the method described above, the *ex vacuo* coating procedure developed independently by Rix et al. (1976) has the advantage that, using the Leybold EPA 100 freeze-fracture replication unit (§6.3.3c(i)) with a transport device (e.g. Figs. 6.17d and 6.22b in §6.3.3), less condensation contamination of the fracture face occurs during transfer of specimens to a cryostat for coating with a monolayer of emulsion. With this method, the mechanical stability of the emulsion-coated replica is increased by adding, in the cryostat, a layer of gelatin. After exposure in the dark at 253, 223 or 173 K (-20, -50 or $-100°C$) and photographic processing of the emulsion, the specimens are transferred back to the vacuum unit. In a vacuum of 7×10^{-5} Pa (5.25×10^{-7} Torr) and at a specimen temperature of 188 K ($-85°C$) the monolayer is freeze-dried and, finally, strengthened by evaporating a carbon layer on top of it.

A reliable and routine method for *in vacuo* emulsion coating of freeze-fractured specimens has been developed by Schiller et al. (1978, 1979) using a special coating device designed for the Leybold Heraeus EPA 100 freeze-fracture unit (§6.3.3c(i)). With this method (Fig. 6.44) tritium-labelled specimens are frozen in type I holders (Fig. 6.2; §6.2.2) and subsequently fractured under careful control with a Reichert (Austria) OM P microtome under liquid nitrogen (Fig. 6.22a) using the Geymayer (1966) technique (§6.3.3c(i)). Samples with flat fracture faces are then transferred using the transport device (Fig. 6.22b) under the protection of liquid nitrogen to the stage of the vacuum unit. They are warmed up to 173 K ($-100°C$) for etching, shadowing and carbon replication at 1.33×10^{-5} Pa (1×10^{-7} Torr). Subsequently, specimens are cooled to 143 K ($-130°C$) and coated with a monolayer of autoradiographic emulsion in a vacuum of about 1.33×10^{-2} Pa (1×10^{-4} Torr) using a specially designed coating device (Fig. 6.44) carrying 3.0 mm plastic coating tubes. Before use, the coating tubes must be freshly loaded with air-dried monolayers of Ilford L-4 emulsion (10 g Ilford L-4; 15 g distilled water; 1.6 g glycerol) on stainless steel loops. To aid adhesion of the emulsion, 1 μl of 1-butanol or a 1-butanol/2-butanol mixture (1:1) is applied to each of the monolayers shortly before introducing the coating device into the vacuum chamber through the air-lock system by means of the transport device. After applying the emulsion to the replicated specimen fracture face, samples are warmed from 143 K ($-130°C$) to 173 K ($-100°C$) and kept at this temperature until the butanol is completely evaporated and the monolayer of emulsion is in close contact with the replica. Subsequently, specimens are transferred from the freeze-fracture unit into a cold chamber and kept under atmospheric pressure at 163 K ($-110°C$) for exposure. Photographic processing was performed by Schiller et al. (1978) in tissue-isotonic solutions of standard photographic chemicals (Amidol with Hostapon), diluted fixer, and 0.6% sodium chloride for rinsing. Before dissolving and separating the tissue from the emulsion-covered replica in 1 N KOH, a carbon protection layer can be evaporated on top of the specimen, as in the *ex vacuo* coating procedure described above (Rix et al. 1976).

The Schiller et al. (1978, 1979) technique for *in vacuo* coating has demonstrated that the autoradiographic background of the film emulsion is not significantly higher than in *ex vacuo* controls. The authors claim that no histochemographic artefacts are observed and that the method definitely overcomes potential *ex vacuo* coating hazards, such as ice contamination of the replicated fracture face before coating with the autoradiographic emulsion and recrystallisation or thawing of the specimen ice during transfer to the cryostat for coating.

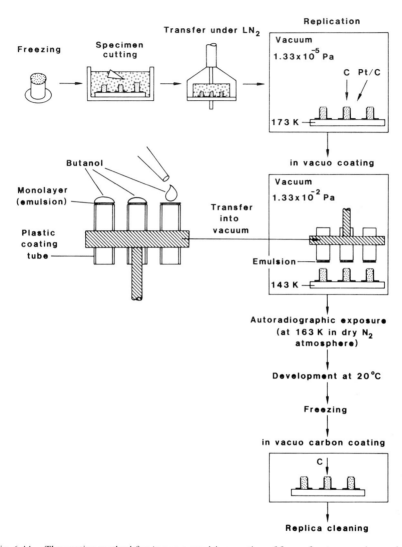

Fig. 6.44. The routine method for *in vacuo* emulsion coating of freeze-fracture specimens developed by Schiller et al. (1978). Tritium-labelled specimens are fractured, etched and replicated before a drop of emulsion containing a small proportion of 1-butanol is applied to the surface. The specimen is warmed to 173 K at which temperature the butanol evaporates leaving the emulsion as a monolayer on the specimen surface. The specimens are exposed at 163 K at atmospheric pressure and the emulsion developed at 293 K. A protecting layer of carbon can be evaporated onto the specimen before the specimen is digested away. (Redrawn from Schiller et al. 1978.)

*6.9.1b   Freeze-fracture autoradiography in combination with the sectioned
replica technique*

This technique, suggested by Rash and co-workers (1980), can help to overcome the limits to resolution known to arise in autoradiography of bulk specimens (§6.9.1a) where isotopes located at sub-membrane levels may still contribute to the exposure of silver grains. As described below (§6.9.3 and §6.9.5a), the sectioned replica technique allows an unambiguous correlation of freeze-fracture and thin section images within a single electron micrograph. As illustrated in Fig. 6.43b, the sectioned replica technique, in combination with autoradiography, involves the following steps: the radioactively labelled tissue is first fractured and replicated by standard methods but, instead of digesting the tissue, the replica-tissue sandwich is post-fixed in a solution of 30% buffered glycerol containing 1% osmium tetroxide; the sandwich is then post-stained, embedded in a plastic and sectioned parallel to the replica/tissue interface. Thin sections from the replica/tissue interface are then coated on both sides with Ilford L-4 emulsion. After exposure and development by standard methods (see Williams 1977), specimens are examined stereoscopically in the transmission electron microscope to determine which of the two emulsion layers has been developed. Because autoradiographic grains are produced and recorded on both sides of the sections, at least one-half of the low energy emission is recorded without having to cross the electron-opaque platinum layer. Rash et al. (1980) demonstrated that 2.5 nm diameter ^{125}I-α-bungarotoxin molecules are separated from the only class of intramembrane particles present in junctional fold membranes and so it can be concluded that half-membrane resolution is attainable by sectioned replica autoradiography. Despite these advantages, this technique is not suitable for the analysis of the distribution of labelled water-soluble and diffusible compounds, since extraction and displacement artefacts occur after thawing the specimen.

6.9.1c   Freeze-fracture autoradiography of monolayers

There are a number of problems involved in freeze-fracture autoradiography of membranes in suspension or in bulk specimens, including those arising from the smallness of exposed membrane areas, the irregular topography of fracture faces, the disturbances by isotopes underlying membrane fracture faces, and the necessity for long exposure of the emulsion at specimen temperatures close to 173 K ($-100°C$). Consequently, Fisher (1978) developed a different approach known as monolayer freeze-fracture autoradiography (MONOFARG). MONOFARG is a combination of monolayer freeze-fracture (Fisher 1975, 1976b; for review see Nermut 1982a,b, and Nermut 1984) and electron microscope autoradiography (Salpeter and Bachmann 1972; Fisher 1976a, 1978, 1980; Salpeter et al. 1977). As originally conceived (Fisher 1976a) this technique can provide both qualitative and quantitative information about the transmembrane or translayer distribution of radioisotopic molecules (for review see Fisher 1982a).

The procedure is particularly suitable for the study of intact cells and protoplasts, membrane vesicles (e.g. erythrocyte ghosts and liposomes) and model bilayers or non-vesicular membrane fragments (Fisher 1982c). A synopsis of the MONOFARG technique is given in Fig. 6.45. The initial step in the procedure involves the specific binding of the labelled specimen to a support. Since most biomembranes possess a net negative surface charge, they can be attached to positively charged solid supports (e.g. glass or mica) by electrostatic interactions (see §6.2.2). Although these techniques are relatively simple, good binding (with the production of dense monolayers) can only be achieved when the cells are free from any contamination with exogeneous protein or cell debris which may compete for positive sites on the support. Nermut (1982b) recommends washing the cells three to five times with isotonic buffer just before use and then resuspending them at an increased concentration. It is useful to centrifuge some specimens down onto a support. For making preparations of intact membranes (Fig. 6.45a) involving cell lysis and the unrolling of membranes it is desirable to bind cells when they are widely separated.

As an alternative to the attachment of cells or isolated membranes to a positively charged surface, covalent binding of cell surface (membrane) proteins to glutaraldehyde-coated supports (see Robinson et al. 1971; Büechi and Bächi 1979; Bächi and Büechi 1981; Aplin and Hughes 1981) or binding to supports coated with a specific ligand such as concanavalin A (see Aplin and Hughes 1981) is also possible.

After binding the radioactively labelled cells or membrane preparations to the support, the MONOFARG method involves two separate processes. For examining intact membrane preparations (Fig. 6.45a), cells are first lysed. Removal of excess unbound membranes can be done using a stream of buffer solution from a syringe fitted with a hypodermic needle (Lang et al. 1981; Nermut 1982b). After freezing, specimens are transferred to a coating unit and freeze-dried. Subsequent steps are identical to those for the preparation of split membranes.

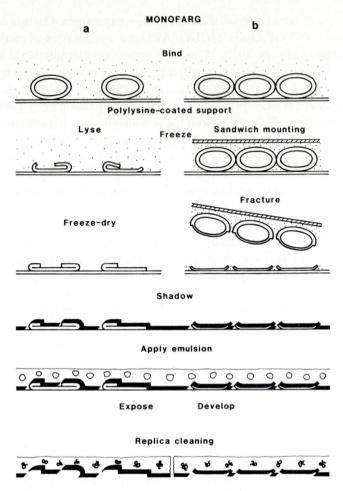

Fig. 6.45.   The monolayer freeze-fracture autoradiography (MONOFARG) technique (Fisher 1978). This technique can be used to examine intact membranes (a) after lysis and freeze-drying; or fractured membranes (b) after freeze-fracture. In both cases, specimens attached to glass supports are shadowed and the emulsion is applied at room temperature. After exposure and development the replica and emulsion are removed from the support by floating onto dilute hydrofluoric acid. (Redrawn from Fisher 1978.)

Half-membrane fracture faces (Fig. 6.45b) are obtained by sandwiching the cell monolayer with a thin (0.2–0.5 mm) copper sheet or an uncoated glass coverslip. Before use, the copper sheet is treated with 35% nitric acid for 10 s, rinsed with distilled water and dried with clean nitrogen gas. Glass coverslips are cleaned as for the poly-L-lysine coating method (§6.2.2d). Freezing of the sandwich is done either in propane or in Freon (§2.4). After fracturing, which can be done using either single or complementary methods, the specimen is freeze-dried. Both intact and split membrane preparations are then paired and processed together. After shadowing and warming up to room temperature, specimens are removed from the vacuum and transferred to a dark room. A Parlodion-supported Ilford L-4 photographic emulsion is applied to the replica at room temperature and the samples are stored together with non-radioactive controls in light-tight containers in a dry atmoshere at $+5°C$. After exposure, photographic processing is carried out using standard methods for electron microscope autoradiography (Salpeter and Bachmann 1972; Williams 1977; Salpeter et al. 1977, 1978). The replica, with the emulsion, is removed from the glass support by floating it onto dilute hydrofluoric acid (HF:water ratio of 1:1). It is then rinsed in distilled water, mounted and examined.

Fisher (1982b) has also studied the quantitative aspects of MONOFARG using, as a model system, human erythrocyte membranes treated with radio-iodinated and fluoresceinated concanavalin A (^{125}I-FITC-Con-A). The background was regularly $< 2.5 \times 10^4$ grains μm^{-2} day^{-1}. The highest overall efficiency was between 25% and 45% and grain density and efficiency were dependent on radiation dose for ^{125}I and D19 emulsion (Eastman Kodak Co.; see Appendix 2) development. The corrected grain densities were linearly proportional to ^{125}I concentration. Analysis of autoradiographs showed that silver grain densities over split extracellular half membranes (Fig. 6.45b) were identical to those over intact unsplit single membranes (Fig. 6.45a). These results demonstrate that ^{125}I-FITC-Con-A remains exclusively attached to the extracellular half of the membrane following freeze-fracturing and can thus be used as a quantitative marker for the proportion of extracellular split membrane halves. As well as this excellent transmembrane resolution of membrane constituents, a unique feature of MONOFARG lies in its potential to determine the position of the radio-isotope in the plane of the split membrane. Preliminary studies indicate that the in-plane resolution of MONOFARG for ^{125}I is about 150 nm (Fisher 1982b). Thus, for membranes, MONOFARG clearly overcomes the resolution limits that apply to bulk specimens where isotopes at levels in the specimen lower than the membranes may contribute to the formation of silver grains.

6.9.2   Labelling with morphologically detectable markers for replica techniques

This cytochemical technique involves labelling specimens with marker molecules before freezing. The markers must have a recognisable structure and must be large enough to be detectable in high resolution replicas of deep-etched (freeze-dried) or critical point-dried specimens.

The most commonly used techniques involve specific antibodies or agglutinins coupled to marker molecules such as ferritin, hemocyanin or colloidal metals (Horisberger and Rosset 1977; Perkins and Koehler 1978; Horisberger 1979; Roth 1982). Among the most commonly used immunochemical markers for demonstrating cell surface antigens in electron microscopy are IgG-ferritin conjugates (Morgan 1972). This method has successfully been used to label antigens on different membrane surfaces exposed by deep etching such as the surface of erythrocytes (Pinto da Silva et al. 1971; Howe and Bächi 1973; Elgsaeter et al. 1976), lymphocytes (Karnovsky et al. 1972; Unanue et al. 1973), or chloroplasts (Miller and Staehelin 1976). The size of the IgG-ferritin conjugates (approximately 20 nm) allows a positive identification of membrane surface components which are more than 25 nm apart, provided that no genuine membrane structures resemble the marker in shape and dimensions. The question of how small a topographical marker can be and still be consistently recognisable on freeze-etched membrane faces has been studied by Carter and Staehelin (1979). Using sheep erythrocyte ghosts as a model system, they compared the labelling patterns of IgG molecules, their antigen-binding fragments, IgG-horseradish peroxidase conjugates and also varying amounts of associated diaminobenzidine precipitates. The authors concluded that an organic surface marker for freeze-etched membranes must have a diameter of more than 15 nm if it is to be seen consistently over extended areas and against the background granularity of red blood cell ghosts. Markers with a distinct shape (e.g. inorganic crystals) coupled to Fab′ fragments of IgG can give even better resolution.

Specific labelling of cell surface glycoproteins may be done using receptor-specific proteins of non-immune origin, such as phytohaemagglutinins, protectins and lectins (Gold and Balding 1975). Lectins, a group of haemagglutinins derived from plants, have specific affinity for certain mono- or oligosaccharide groups and are used for labelling the carbohydrate residues of cell membrane constituents (for references see Cook 1976; Hubbard and Cohn 1976; Sutherland 1977; Roth 1978, 1982). Using ferritin-conjugated phytohaemagglutinin, Tillack et al. (1972) were able to analyse specific erythrocyte membrane antigens and to correlate the pattern of distribution of surface markers with particles seen on the fracture plane.

With periodic structures the topographical resolution of specific surface markers may be considerably improved since optical or computer filtering techniques can be used on the final image. For example, on freeze-dried and tungsten or platinum/carbon shadowed T4 polyheads, individual antigenic sites can be localised by labelling them with Fab′ fragments alone (Kistler et al. 1978). Unlabelled antibodies have also been observed on fast-frozen bacterial flagella (Munn et al. 1980). Finally, charged particles such as polycationic or polyanionic ferritin can be used for labelling charged specimen structures on cell and membrane surfaces. Polycationic (cationised) ferritin is commercially available (Miles Israel; Appendix 2) but may also be produced by attaching aliphatic amines to the carboxylic groups of ferritin using carbodiimides as condensing agents (Danon et al. 1972). Experiments with bacterial cells (Fig. 6.46) have shown that the marker molecules stick to negatively charged sites even during prolonged etching (Sleytr and Friers 1978). Similarly, other suitable marker molecules (e.g. haemocyanin, horseradish peroxidase) can be specifically charged for labelling experiments. By analogy with the use of polycationic macromolecules, polyanionic topographical markers should be usable as probes for positively charged cell constituents (Rennke et al. 1978).

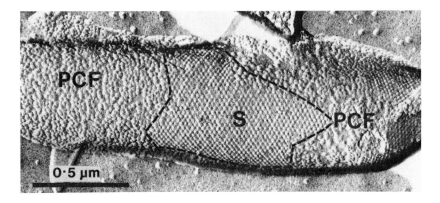

Fig. 6.46. Cell surface of the bacterium *Clostridium thermohydrosulfuricum* (revealed by deep etching) with patches of hexagonal arrays of glycoproteins. The polycationic ferritin molecules (**PCF**) are only bound to areas where the negatively charged peptidoglycan layer is not masked by the glycoprotein molecules. **S**, S-layer. (Reproduced, with permission, from Sleytr and Friers 1978.)

6.9.3   Labelling with electron-opaque markers for the sectioned replica technique

The initial step involves labelling the specimen with an electron-opaque marker. Both immunoferritin and colloidal gold have been used (Rash et al. 1982). Labelled specimens are then frozen, fractured and replicated using standard freeze-fracture methods. The subsequent preparations steps follow the schedule used for the sectioned replica technique described in §6.9.1b. In thin sections cut parallel to the replica tissue interface, the distribution of the electron-opaque markers can be correlated with the superimposed freeze-fracture and thin section images. Rash et al. (1980, 1982) have shown that, particularly with the help of stereoscopic images, it is possible to determine which half of the membrane retains the electron-opaque markers after fracturing. This technique, together with post-fracture labelling methods (§6.9.5), has proved to be a valuable tool for analysing individual intramembrane particles.

6.9.4   Specific pre-freezing alterations of the molecular organisation of membranes

These methods involve a variety of pre-freezing treatments of specimens with the objective of inducing a specific change in the organisation of membrane components into recognisable patterns.

The simplest procedures involve temperature changes (cooling) which, through phase transition phenomena, cause phase separations in the membrane lipids (transition from the liquid crystalline to the gel state) and aggregation of membrane-intercalated particles (Verkleij and Ververgaert 1975, 1978; Maul 1979). Although such segregation is a common undesirable artefact caused by inadequate cryofixation (see §6.8.3), it can also provide valuable information about the physicochemical properties of membrane lipids.

As well as temperature shifts, a variety of other treatments (e.g. gravitational forces, hydrostatic pressure, changes in ionic strength, pH and gas composition, infiltration with cryoprotectants, chemical fixation; see Table 2.12 in §2.3.2) can induce particle aggregation. These changes in the molecular organisation of membrane constituents seen on fracture faces can also be used for studying the topography of membrane surfaces (Pinto da Silva et al. 1971; Shotton et al. 1978; Maurer and Mühlethaler 1981) by correlating the pattern of aggregation of membrane particles seen on fracture faces with the distribution of specific marker molecules on membrane surfaces exposed by deep etching.

Considering the variety of immunological or affinity markers that are available (see §6.9.2), it is obvious that these simple techniques have great potential in studies on membrane architecture and, particularly, for trans-membrane features.

Finally, membrane components, particularly membrane lipids, may be specifically complexed into characteristic structures detectable in surface fractures (protruberances, ripples, pits, etc.). To reduce the redistribution of these complexed membrane structures, probes are often used in combination with aldehyde fixatives.

As a probe to reveal β-hydroxy sterols, the polyene antibiotic filipin (Tillack and Kinsky 1973; Verkleij et al. 1973; De Kruijf and Demel 1974; Kitajima et al. 1976; Elias et al. 1979; Montesano et al. 1979; Robinson and Karnovsky 1980; Garcia-Segura et al. 1982) or saponins (e.g. digitonin and tomatin) (Elias et al. 1978, 1979; Robenek et al. 1982; Severs and Simons 1983) have been used (see Severs 1984 for a review of this area).

The antibiotic, bacitracin (which forms a complex with C_{55} prenol pyro-phosphate) has been shown to induce the formation of rod-like elevations in the plasma membranes of procaryotic cells (Sleytr et al. 1976).

6.9.5 Post-fracture labelling procedures

The cytochemical labelling techniques for membrane or cytoplasmic consti-tuents exposed by freeze-fracturing follow two principles (Fig. 6.47). One approach involves direct labelling of fractured and thawed specimens (§6.9.5a, b) whereas, with the other technique, fracture faces are first stabi-lised by the deposition of heavy metal shadows before the specimens are thawed for labelling (§6.9.5c, d).

6.9.5a Thin sectioning of post-fracture labelled specimens

This method is illustrated schematically in Fig. 6.47a and involves the grind-ing of frozen samples while immersed in liquid nitrogen, followed by thaw-ing, cytochemical labelling of fracture faces and processing for thin section-ing (Pinto da Silva et al. 1981b,c; Pinto da Silva 1982; Torrisi and Pinto da Silva 1984). Tissues or cell suspensions that have been fixed in glutaraldehyde (e.g. 1% glutaraldehyde for 15–20 min at 4°C) before embedding in a protein matrix (e.g. 30% bovine serum albumin) cross-linked with glutaraldehyde (1% for 30 min at 25°C) are often used. The tissues and gels are then sliced into pieces approximately $1 \times 2 \times 2$ mm, gradually impregnated with 30% glycerol and frozen by immersion in partially solidified Freon 22 cooled with liquid nitrogen. The frozen pieces are then transferred to a polished glass container (e.g. the base of a tissue homogeniser) filled with liquid nitrogen and immersed in a slush of liquid nitrogen and solid carbon dioxide. Drops of a 30% glycerol:1% glutaraldehyde solution in 310 mOsm phosphate buffer, pH 7.5, frozen in liquid nitrogen, are added in approximately equal amounts to the gels or tissues. After the pieces have all sedimented to the bottom of the container, they are repeatedly fractured with a glass pestle at 77 K ($-196°C$) until pulverised into fine fragments. The glass container is then removed from the liquid nitrogen, placed at ambient temperature until the volume of the liquid nitrogen is reduced to approximately one tenth, and 2–3 ml of the glycerol/glutaraldehyde buffer is added in liquid form. The glass container is immersed briefly in water at 30°C and, on thawing of the glycerol, is transferred to an ice bucket for 15 min. The fragments of the gels and tissues are then gradually deglycerinated by dropwise addition of 1.0 mM glycyl-glycine in 310 mOsm phosphate buffer at pH 7.5 and are then washed twice in 310 mOsm phosphate buffer, also at pH 7.5. Centrifugation and pelleting of the gel or tissues pieces is not necessary because they settle rapidly.

Conventional cytochemical labelling techniques can then be used to label the membrane faces exposed during freeze-fracturing. Later, the samples are fixed in glutaraldehyde and osmium tetroxide using standard procedures (Glauert 1974), dehydrated and embedded for thin sectioning. Before final trimming of the blocks, thick sections are used to localise suitable areas of the specimen for fine sectioning. Most cytochemical techniques involving electron-opaque markers can be used and good results have been obtained in labelling anionic sites with cationised ferritin and concanavalin A and wheat germ receptor sites using ferritin conjugates or colloidal gold and iron labels (Pinto da Silva et al. 1981b,c; Pinto da Silva 1982).

In summary, although a considerable rearrangement of membrane constituents can occur during thawing and labelling (fractured membranes often reveal an image of unit membrane-like segments), the technique allows a direct, high resolution, chemical and immunological characterisation of plasma and intracellular membranes.

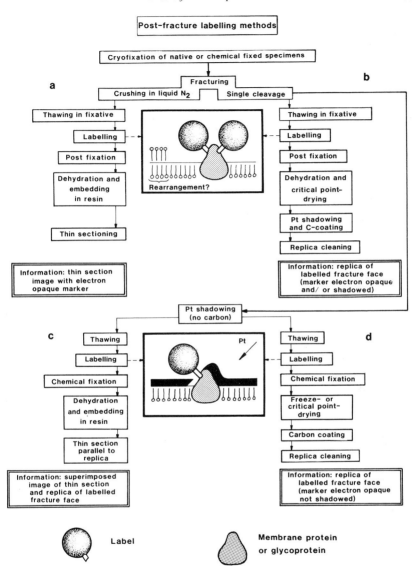

Fig. 6.47. Post-fracture labelling procedures. (a) Thin sectioning of post-fracture labelled specimens. Frozen samples are ground in liquid nitrogen followed by thawing in fixative, cytochemical labelling and processing for thin sectioning. (b) Replication of critical point-dried, post-fracture labelled specimens. Frozen specimens are fractured, thawed and labelled as in (a) and then critical point-dried. The fracture faces are replicated and the specimen digested away. (c) Frozen samples are fractured and then a thin layer of platinum is evaporated onto the fracture surface. The specimen is then thawed, labelled and processed for thin sectioning. (d) The same procedure as in (b) except that a thin layer of platinum is evaporated onto the fractured surface prior to thawing.

6.9.5b   Replicating critical point-dried, post-fracture labelled specimens

This technique (Fig. 6.47b) was developed by Pinto da Silva and Torrisi (1982) and resembles the previous method except that the frozen tissues and cells within gels are not fragmented in a cold tissue homogeniser. Instead, slices of gels and tissues are immersed in a Petri dish filled with liquid nitrogen and fractured with a liquid nitrogen-cooled scalpel. After thawing and labelling as described above (§6.9.5a), the tissue and gel fragments are fixed in buffered glutaraldehyde/osmium, dehydrated in ethanol and critical point-dried. After drying, the specimens are orientated on a specimen carrier with the fracture faces pointing upwards, shadowed with heavy metal (2.0 nm Pt/C) and subsequently replicated with carbon. The replicas are separated and cleaned by digesting the specimen in a 5.0% solution of sodium hypochlorite, rinsed in water and collected on grids. The fine structure of the membrane fracture faces revealed by this technique differs from that of conventional freeze-fracture replicas. Nevertheless, the electron-opaque markers (e.g. colloidal gold) can clearly be distinguished from the rough membrane background structure. By analogy with the interpretation of post-fractured sectioned material, it is necessary to take into account the fact that, on thawing in an aqueous environment, significant rearrangement of membrane components can occur. In conclusion, post-fracture labelling techniques allow structural and cytochemical dissection of membrane surface sites and, in particular, the identification of those sites associated with transmembrane proteins (Pinto da Silva et al. 1981a; Pinto da Silva 1982).

6.9.5c   Replicating freeze- or critical point-dried, post-fracture, post-shadow labelled specimens

This post-fracture labelling technique, together with the one that follows (§6.9.5d), was devised by Rash and collaborators (Rash et al. 1978, 1980, 1982). They basically resemble the techniques described in the previous section except that the labelling reagents must penetrate a thin but discontinuous layer of platinum superimposed on the molecules of interest (Fig. 6.47d). The methods follow the standard freeze-fracture replication schedules until heavy metal (Pt/C) has been evaporated. The specimen is then thawed and labelled by one of the standard methods involving ferritin-antibody complexes or colloidal gold-affinity ligands. In a study of P-face intramembrane particles corresponding to the acetyl-choline receptor complex, Rash and colleagues have shown that the unshadowed side of these intramembrane particles, exposed by the fracturing process, retain sufficient biochemical information to permit both immuno-specific and neurotoxin-specific labelling, despite initial formaldehyde fixation. After labelling, samples are fixed in 1% osmium tetroxide (1.0 min), rinsed and post-fixed with 0.5% aqueous uranyl acetate (Silva et al. 1968). The labelled surfaces are now stablised, either by critical point-drying or by freeze-drying. Dried specimens are then rotary carbon coated to protect the labels during subsequent rehydration and specimen digestion. Carbon films of 15–20 nm have been found to be sufficient to stabilise freeze-dried specimens. However, because the critical point-dried specimens have a strong tendency to swell and fragment during rehydration, it is necessary to increase the thickness of the carbon stabilising coat to about 50–80 nm, which reduces resolution. Nevertheless, both techniques allow the recovery of large intact replicas without underlying tissue remnants and with clearly visible electron-opaque labels directly above the individual membrane-intercalated particles.

Comparing all the post-fracture labelling techniques, it appears from the data currently available that the post-shadow labelling techniques give more stability to the fracture face and, thus, inhibit excessive rearrangement of membrane components during thawing and incubation with markers. However, the shadowing step masks a considerable proportion of the potential binding sites for the label. This is a particular problem if the angle of shadowing is not favourable.

6.9.5d   Thin sectioning of post-fracture, post-shadow, labelled and replicated specimens

Instead of critical point- or freeze-drying, post-shadow labelled specimens can be post-fixed, stained and embedded in resin for thin sectioning (Rash 1979; Rash et al. 1982). Thin sections are cut parallel to the replica/tissue interface and are best examined as stereo-pairs. The sectioned, labelled replica method permits the simultaneous imaging of freeze-fracture replicas, their subadjacent tissue thin sections and their attached electron-opaque labels (Fig. 6.47c). It can be demonstrated that, although the shadowed intramembraneous particles can be labelled specifically, the relative amount of ligand binding to these particles exposed by membrane fracturing is substantially reduced from that observed on sub-adjacent non-fractured membranes (Rash et al. 1982). Despite such intrinsic difficulties, this cytochemical method can be expected to have a wide application.

6.9.6   *Lipid extraction from fracture faces*

It has been shown by Richter and Sleytr (1971) that lipids may be extracted specifically from fracture faces at very low temperatures before replication. Yeast species with lipid vacuoles (*Rhodotorula, Candida*) and oil droplets suspended in a glycerol/gelatin/water matrix were used as a model system. For this procedure specimens are best mounted and frozen in holders suitable for preparing complementary replicas (Fig. 6.2). Freeze-fracturing is done under liquid nitrogen with two pairs of cooled tweezers applying tensile stress. The fractured specimens are then transferred in a small container (<2 ml), under the protection of liquid nitrogen, to a receptacle filled with diethyl-ether or other extraction liquids (see §7.2), maintained at 195 K (−78°C) in a dry ice/methanol bath (Fig. 6.48) or at lower temperatures (173 K; −100°C) using a liquid nitrogen-cooled cold box. During extraction, which may involve treatment for a few hours or up to several days, the diethyl-ether vessel is closed with a lid to avoid condensation of water vapour from the atmosphere. After lipid extraction (freeze-substitution is

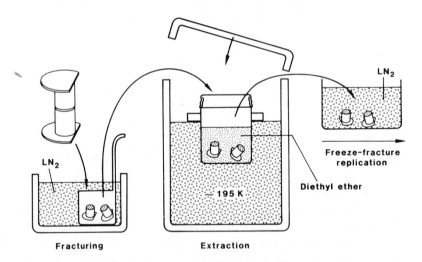

Fig. 6.48.   Method for specific extraction of lipids from fracture faces at low temperatures. Specimens fractured under liquid nitrogen are transferred, under the protection of liquid nitrogen, to a small container filled with a pre-cooled extraction liquid (e.g. diethyl ether). After the extraction, specimens are transferred back to liquid nitrogen and processed following schedules suitable for the preparation and examination of specimens fractured under liquid nitrogen (see §6.3.3c). (Derived from Richter and Sleytr 1971.)

generally not a problem since diethyl-ether is a very poor solvent for ice at this temperature; see §7.2), the specimens are transferred with cooled forceps under the protection of dry nitrogen gas back to liquid nitrogen. Any remnants of the extraction liquid solidify on the specimen holder (the freezing point of diethyl-ether is 157 K; $-116°C$) and are removed from the holder base with a cold dissection needle before inserting the holders into the transfer device. The subsequent replication is done by any method which allows controlled etching and replication of fracture faces obtained under liquid nitrogen (§6.3.3c(i)). Etching at approximately 163 K ($-100°C$) is important to remove, by sublimation, the solidified diethyl-ether from the fracture face. With this method, controlled extraction of lipids below the recrystallisation point of the specimen and a combination of morphological examination of fracture faces and chemical analysis of the extracted lipids by gas chromatography is possible.

6.9.7 *Specific decoration of fracture faces* in vacuo *with water vapour*

From the analysis of condensation artefacts on lipid/water and glycerol/water mixtures, it was realised that residual gases in the vacuum chamber can condense on fracture faces in a non-random distribution (Deamer et al. 1970; Staehelin and Bertaud 1971). As in the condensation of shadowing material (§6.4.1a), impinging atoms or gas molecules can migrate on a fracture face before being immobilised. By this lateral surface diffusion, stable nuclei at sites with high binding energy will be formed and eventually lead to discrete structures. Experiments by Gross and Moor (1978) and Gross et al. (1978b) have shown that the paracrystalline regions on E faces of yeast plasmalemma exposed by fracturing at 77 K ($-196°C$) in a vacuum of 1.33×10^{-7} Pa (1×10^{-9} Torr) (§6.3.3b(ii)) can be decorated specifically with water vapour. In their experiments, pure water vapour generated by heating copper sulphate-pentahydrate was stored in a vessel and released through a gas admittance valve into the ultra-high vacuum system. The pattern of water condensation, seen as small cubic ice crystals, depended on the incident flux of water molecules let into the system and the introduction of other simultaneously impinging (nitrogen) molecules. The fact that water vapour condensation on freshly exposed specimen fracture faces is highly substrate-specific was later confirmed by condensation experiments on chemically well-defined specimens. Using multi-lamellar cardiolipin liposomes and stearic acid crystals which, after fracturing, reveal hydrophobic faces and hydrophilic cross-fractured steps, as well as multi-lamellar octatriacontane crystals which, after fracturing, expose hydrophobic faces and geometrically similar but hydrophobic steps, it was demonstrated that hydrophilic surface features in hydrophobic surroundings can be specifically decorated by particle-like ice crystals (Walzthöny et al. 1981). This clearly shows that, under ultra-high vacuum conditions, the controlled deposition of suitable condensing gases (specific decoration) may be a valuable tool for labelling regions on fracture faces according to their physicochemical properties. It also became obvious from these experiments that condensing water vapour may lead to the formation of small ice crystals which resemble particles seen on membrane fracture faces.

6.9.8   Molecular and particle weight determination

A method for determining weights of macromolecules or colloid particles by particle counting on replicas of spray-frozen, fractured and etched samples was developed by Bachmann et al. (1974). The technique was tested by the authors for particle molecular weights ranging from 10^4 to 10^8 daltons. Theoretically, the only requirements for accurate measurement are:

(*i*)   that the random distribution of particles or macromolecules in the sample is preserved during freezing;

(*ii*)   that the particles are suspended in a volatile solvent;

(*iii*) that the particles do not sublime during controlled etching of the fracture face and remain on the frozen surface;

(*iv*)   that no lateral movement of the particles occurs during etching; and

(*v*)   that the particles have a detectable size.

The particle distribution seen on the replica will then represent a two-dimensional projection of their original three-dimensional distribution in the solvent layer which has been removed by controlled etching (§ 6.5). The observed number of particles per area (particle density) is then representative for the sample and the molecular weight can be derived from the equation:

$$M = \frac{S N_A}{n_v} \tag{6.4}$$

where

M = molecular weight

S = dry weight concentration of solution

n_v = number of particles per unit volume
N_A = number of molecules per mole (Avogadro's constant or Losch-midt's number = 6.02×10^{23})

Since relatively large errors can be expected as a result of insufficiently accurate sublimation rates (§6.5.1), two solutions, one containing the sample and the other containing a protein of known molecular weight (standard) are spray-frozen (§2.4.1c), one after the other in the same propane vessel. As standards, molecules such as ferritin or bovine serum albumin are suitable (Junger and Bachmann 1977). After freeze-fracturing and etching the frozen mixture, the replicas show side-by-side spray frozen droplets of the sample and the standard, both etched under the same conditions. The unknown molecular or particle weight of the sample is then:

$$M_x = \frac{n_{st} \times S_x}{n_x \times S_{st}} \times M_{st} \qquad (6.5)$$

where
M_x = molecular or particle weight of the sample
M_{st} = molecular weight of standard
n_x = number of particles per unit area of sample
n_{st} = number of particles per unit area of standard
S_x = concentration of the sample
S_{st} = concentration of the standard

The optimal concentrations increase with the particle size and the following figures have been recommended by Bachmann et al. (1974): cytochrome c (MW = 12,500) 0.001 mg/ml, ferritin (MW = 750,000) 0.2 mg/ml, Latex (100 nm) 2 mg/ml. Under the conditions described above, the accuracy of the method depends mainly on the accuracy of the dry weight protein assay (Junger and Bachmann 1977).

6.9.9 *Freeze-fracturing and freeze-fracture replication in combination with negative staining*

Negative staining has been used in combination with freeze-fracture techniques primarily to facilitate the classification of intramembraneous particles according to their size and shape (§6.8.2). The method of Fujikawa (1979) consists of initial replication of fractured membrane faces with a uniform,

20 nm, carbon layer and treatment of the cleaned replica with a negative-staining solution. Before negative staining, the replicas, mounted on uncoated grids with the specimen side to the surface, are made wettable by glow discharge. The staining solution (2% phosphotungstic acid adjusted to pH 6.9 with KOH) is applied to the surface of the replica and the excess liquid is removed by touching the edge with a piece of filter paper. After drying, the specimen is ready for examination in the transmission electron microscope. The method shows a clear distribution of intramembraneous particles without interference from neighbouring structures and makes it easy to discriminate between contiguous elements by means of density differences between the intramembraneous particles and the flat lipid monolayer. It has even been claimed that density differences between individual particles can be seen.

A method for negative staining of freeze-fractured membranes has been reported by Nermut and Williams (1976), Nermut (1977) and Nermut et al. (1978). Specimens are frozen as a sandwich between two sheets of mica, or mica in combination with other materials (e.g. filter paper; see also §6.2.2) and freeze-fractured by hand under liquid nitrogen by pulling the sandwich apart. For negative staining, the mica with the adhering fractured membrane halves is removed from liquid nitrogen and a drop of negative stain is added to the melting surface. The released material is adsorbed onto carbon-coated grids. Alternatively, to avoid possible precipitation of negative stain, a droplet of distilled water may be added upon thawing and, subsequently, covered with a coated grid which is transferred after a few minutes onto a drop of negative stain. The value of the method has been demonstrated with *Escherichia coli* envelopes and the purple membrane of *Halobacterium halobium* (Nermut 1982a). Both pits, with a diameter of 4–6 nm and particles of 6–8 nm diameter were demonstrated.

6.9.10   Quantification of structural details and interpretation aids

Information about membrane architecture is obtainable by analysing the size, geometry and density distribution of membrane-intercalated particles (§6.8.2). As discussed earlier (§6.4.1b), the accuracy with which the size and geometry of fine structural details are depicted on fracture planes depends on the shadowing procedure. Furthermore, any correlation between morphologically distinguishable membrane constituents (e.g. membrane-intercalated particles) and the *in vivo* or *in situ* structure has to consider the possibility of preparation artefacts (Pricam et al. 1977; §6.8.3). Furthermore, for a shadowed object, the shadow itself causes an increase in size (§6.4.1a). To minimise such deviations, the diameters of uni-directionally shadowed particles are measured at right angles to the shadowing direction (Gulik et al. 1982; Schmidt and Buchheim 1982). Willison and Moir (1983) suggested the use of the classical technique for particle size analysis involving a 180° inter-shadow rotation of the specimen (bi-directional shadowing according to Williams and Wyckoff 1944; Kahler and Lloyd 1950). A method of reconstructing the surface topography of isolated heavy metal-shadowed particles by computer techniques has been described by Smith and Kistler (1977) and Smith and Ivanov (1980) (see also Guckenberger 1984). Finally, for an automated system for the quantitative analysis of membrane-intercalated particles on membrane fracture faces, attempts have been made to create a reference library of computer-generated synthetic electron-optical images of morphologically defined particles prepared using a variety of different shadowing conditions (Weinstein et al. 1979).

For the accurate determination of particle size and the dimensions of membrane fracture face edges and cross-fractured bilayer membrane edges, Ruben and Telford (1980) recommend the use of colloidal gold as a calibration standard. A method for calculating the diameters of rotary-shadowed particles which, under ideal conditions, cast annular shadows, has been described by Simpson (1979).

Characterisation of particle populations is best done by means of particle size histograms (Staehelin 1973). So far, counts of particle densities per unit area have been done without corrections for local tilting angles, although freeze-fracture replicas are generally copies of non-planar specimens. Methods are available (see below) for determining the local angle of the replica in relation to to the electron beam.

Critical testing of topographical distribution by statistical methods (Gershon et al. 1979; de Laat et al. 1981; Van Winkle and Entman 1981; Duniec et al. 1982; Niedermayer and Wilke 1982) is often required in order to identify significant changes in topography which are unconvincing on casual inspection of electron micrographs. A method based on a statistic, the so-called 'coefficient of dispersion', has been used by Weinstein et al. (1976) and Robards et al. (1981) to determine if the two-dimensional distribution of membrane-intercalated particles approximates to a statistically random distribution. In instances of non-randomness, the method indicates whether the particles tend to be in a regular array or, alternatively, if the individual units are statistically aggregated (clumped).

A stereological method, allowing an estimation of the relative surface area of a certain membrane type in a freeze-fracture preparation of subcellular fractions has been used by Weibel and collaborators (Weibel et al. 1976). The method exploits the fact that each membrane type has a characteristic range of numerical density of membrane-intercalated particles. It is necessary that the membrane vesicles are spherical and that the size distribution of the various types of vesicle is known.

Numerous methods are available to facilitate the interpretation of the topographical undulations recorded in the replica. Using a goniometer stage, many replica areas that are not at first well resolved can be set to give best resolution by tilting in the appropriate direction (Steere et al. 1974; Hörandner 1976; Steere and Rash 1979; Chalcroft 1984). This is of particular value if complementary replicas (§6.3.3a(ii)) are examined. One replica may look excellent while the complementary part, due to mechanical distortion or displacement during attachment to the grid, may deviate considerably from an ideal orientation. The best definition of a specimen is generally obtained when the areas to be examined are normal to the electron beam (Steere 1971). With coarse undulations in the replica, well-shadowed areas can even be hidden by protruding parts of a replica. By tilting replicas, a coincidence of the viewing angle with the shadowing angle is recognised by the minimisation of the contrast (and by concentric shadows of membrane-intercalated particles). From this position the optimum viewing angle is more easily obtained by tilting the replica in the appropriate direction (Hörandner 1976).

A method which allows a topographical map to be produced from digitised micrographs of a heavy metal-shadowed periodic specimen was reported by Smith and Kistler (1977). The calculations are based on the theoretical model that the thickness of metal deposited along the line of the incident metal deposits itself uniformly onto the specimen surface without diffusing or aggregating. Another method, which allows the three-dimensional reconstruction of surface profiles from microdensitometric data of electron micrographs, has been described by Krbecek et al. (1979).

The evaluation of stereo-pairs represents a simple procedure for better interpretation of the spatial relationships of regular structures (Waldhäusel 1978). Steere (1971, 1973) reports that good stereo-pairs can be obtained with 15° tilt between the two micrographs but, generally, a total of 7.5° tilt is sufficient. For best results he recommends tilting the specimen at right angles to the direction of the shadow. To avoid the reversed appearance of depressions and elevations, special care must be taken to ensure that the prints are correctly orientated (shadows pointing towards the bottom) and that the left and right hand micrographs are not transposed (Allen et al. 1978).

The examination of stereo-pairs of complementary areas (Steere 1973) can be a valuable approach for analysing artefacts, such as plastic deformation (§6.8.3) or condensation contamination (§6.5.1), which often contribute to an apparent lack of complementarity. A more accurate and objective comparison of complementary periodic structures is possible by digital image processing methods, as shown for the regular arrays on yeast plasmalemma fracture faces by Kübler et al. (1978).

References

Abermann, R., M.M. Salpeter and L. Bachmann (1972), High resolution shadowing, in: Principles and techniques of electron microscopy, M.A. Hayat, ed. (Van Nostrand Reinhold Company, New York), p. 197.

Akahori, H., S. Okamura, M. Nishiura and K. Uehira (1972), Proc. 30th Ann. Meeting EMSA, p. 294.

Akahori, H. and M. Nishiura (1972), Rev. Microsc. elect. *1*, 124.

Aldrian, A.F. and H.R.F. Horn (1974), Trennung von Dampf- und Wärmestrahlung während des Aufdampfens beim thermischen Verdampfen empfindlicher Objekte, Proc. 8th Int. Congr. Electron Microscopy, Canberra *1*, 406.

Allain, M. and J. Ardonceau (1983), Modification of the anticontamination device of a Reichert-Jung freeze-fracture apparatus, J. Microscopy *134*, 217.

Allen, J.V., J.S. Gardner and W.M. Hess (1978), Pseudoscopic illusions in stereo views of freeze-etch replicas, Protoplasma *93*, 473.

Aplin, J.D. and R.C. Hughes (1981), Protein derivatised glass coverslips for the study of cell to substratum adhesion, Analyt. Biochem. *113*, 144.

Bächi, T. and M. Büechi (1981), Internal structural differentiation of the plasma membrane in the Sendai virus maturation, in: The replication of negative strand viruses, D.H.L. Bishop and R.W. Compans, eds. (Elsevier/North-Holland, Amsterdam), p. 553.

Bachmann, L. (1962), Shadow casting using very high melting metals, Proc. 5th Int. Congr. Electron Microscopy, Philadelphia *1*, FF3.

Bachmann L. and W.W. Schmitt-Fumian (1973), Spray freezing and freeze-etching, in: Freeze-etching, techniques and applications, E.L. Benedetti and P. Favard, eds. (Soc. Française de Microscopie Électronique, Paris), p. 73.

Bachmann, L., R. Abermann and H.P. Zingsheim (1969), Hochauflösende gefrierätzung, Histochemie *20*, 133.

Bachmann, L., H. Fritzmann and W.W. Schmitt-Fumian (1974), Molecular and particle weight determination by counting, Proc. 8th Int. Congr. Electron Microscopy, Canberra *2*, 18.

Bachmann, L., G. Leupold, M. Barth and W. Baumeister (1984), Decoration of freeze-etched catalase crystals, Proc. 8th Eur. Reg. Conf. Electron Microscopy, Budapest *2*, 1289.

Bachmann, L., W.H. Orr, T.N. Rhodin and B.M. Siegel (1960), Determination of surface structure using ultrahigh vacuum replication, J. appl. Phys. *31*, 1458.

Bank, H. and J.D. Robertson (1976), A simple electrode for metallic replication, J. Microscopy *106*, 343.

Barnett, J.R. (1969), Physical studies of cellulose biosynthesis, Ph.D. Thesis, University of Leeds.

Bendicht, U.P., K.E. Kuettner and R.S. Weinstein (1979), Intercellular junctions in FANFT induced carcinomas of rat urinary bladder in tissue culture: *in situ* thin-section, freeze-fracture and scanning electron microscopy studies, J. Microscopy *115*, 271.

Benedetti, E.L. and P. Favard, eds. (1973), Freeze etching, techniques and applications (Soc. Française de Microscopie Électronique, Paris).

Bordi, C. (1979), Parlodion coating of highly fragile freeze-fracture replicas, Micron *10*, 139.

Bradley, D.E. (1958), Simultaneous evaporation of platinum and carbon for possible use in high-resolution shadow-casting for the electron microscope, Nature, Lond. *181*, 875.

Bradley, D.E. (1959), High resolution shadow casting technique for the electron microscope using the simultaneous evaporation of platinum and carbon, Br. J. appl. Phys. *10*, 198.

Branton, D. (1966), Fracture faces of frozen membranes, Proc. natl. Acad. Sci. *55*, 1048.

Branton, D. (1971), Freeze-etching studies of membrane structure, Phil. Trans. R. Soc. Ser. B *261*, 133.

Branton, D. (1973), The fracture process of freeze-etching, in: Freeze-etching, techniques and applications, E.L. Benedetti and P. Favard, eds. (Soc. Française de Microscopie Électronique, Paris), p. 107.

Branton, D. and S. Kirchanski (1977), Interpreting the results of freeze-etching, J. Microscopy *111*, 117.

Branton, D., S. Bullivant, N.B. Gilula, M.J. Karnovsky, H. Moor, K. Mühlethaler, D.H. Northcote, L. Packer, B. Satir, P. Satir, V. Speth, L.A. Staehelin, R.L. Steere and R.S. Weinstein (1975), Freeze-etching nomenclature, Science *190*, 54.

Brown, J.N. (1981), Replication of cell monolayers with the Bullivant-Ames freeze-fracture apparatus, Stain Technol. *56*, 267.

Bucheim, W. (1976), Freeze-etching of dehydrated biological material, Proc. 6th Eur. Reg. Congr. Electron Microscopy, Jerusalem *2*, 122.

Bucheim, W. (1982), Aspects of sample preparation for freeze-fracture/freeze-etch studies of proteins and lipids in food systems. A review, Food Microstruct. *1*, 189.

Büechi, M. and T. Bächi (1979), Immunofluorescence and electron microscopy of the cytoplasmic surface of the human erythrocyte membrane and its interaction with Sendai virus, J. Cell Biol. *83*, 338.

Bullivant, S. (1973), Freeze-etching and freeze-fracturing, in: Advanced techniques in biological electron microscopy, J. Koehler, ed. (Springer, Berlin), p. 67.

Bullivant, S. (1977), Evaluation of membrane structure facts and artefacts produced during freeze-fracturing, J. Microscopy *111*, 101.

Bullivant, S. (1980a), Interpretation of yeast plasma membrane structure, Micron *11*, 359.

Bullivant, S. (1980b), Advantages and limitations of a simple cold block freeze-fracture technique, Proc. 38th Ann. Meeting EMSA, p. 748.

Bullivant, S. and A. Ames (1966), A simple freeze-fracture replication method for electron microscopy, J. Cell Biol. *29*, 435.

Bullivant, S., R.S. Weinstein and K. Someda (1968), The type II simple freeze-cleave device, J. Cell Biol. *39*, 19a.

Bullivant, S., D.G. Rayns, W.S. Bertaud, J.P. Chalcroft and G.F. Grayston (1972), Freeze-fractured myosin filaments, J. Cell Biol. *55*, 520.

Bullivant, S., P. Metcalf and K.P. Warne (1979), Fine structure of yeast plasma membrane after freeze-fracturing in a simple shielded device, in: Freeze-fracture: methods, artefacts and interpretations, J.E. Rash and C.S. Hudson, eds. (Raven Press, New York), p. 141.

Carter, D.P. and L.A. Staehelin (1979), Evaluation of IgG molecules, Fab fragments and IgG-horseradish peroxidase conjugates as surface labels for freeze-etched membranes, J. Microscopy *117*, 363.

Chalcroft, J.P. (1984), Towards rational terminology and improved techniques in the quantitative analysis of surface replicas, Proc. 8th Eur. Reg. Conf. Electron Microscopy, Budapest *2*, 1297.

Chalcroft, J.P. and S. Bullivant (1970), An interpretation of liver cell membrane and junction structure based on observation of freeze-fracture replicas of both sides of the fracture, J. Cell Biol. *47*, 49.

Collins, T.R., J.C. Bartholemew and M. Calvin (1975), A simple method for freeze-fracture of monolayer cultures, J. Cell Biol. *67*, 904.

Cook, G.M.W. (1976), Techniques for the analysis of membrane carbohydrates, in: Biochemical analysis of membranes, A.H. Maddy, ed. (Chapman and Hall, London), p. 283.

Costello, M.J. and J.M. Corless (1978), The direct measurement of temperature changes within freeze-fracture specimens during rapid quenching in liquid coolants, J. Microscopy *112*, 17.

Crowther, R.A. and U.B. Sleytr (1977), An analysis of the fine structure of the surface layers from two strains of *Clostridia*, including correction of distorted images, J. Ultrastruct. Res. *58*, 41.

Danon, D., L. Goldstein, Y. Marikovsky and E. Skutelsky (1972), Use of cationized ferritin as a label of negative charges on cell surfaces, J. Ultrastruct. Res. *38*, 500.

Davy, J.G. (1970), Vaporization mechanism of single ice crystals, Ph.D. Thesis, University of California, Berkeley.

Davy, J.G. and D. Branton (1970), Subliming ice surfaces: freeze-etch electron microscopy, Science *168*, 1216.

Deamer, D.W., R. Leonard, A. Tardieu and D. Branton (1970), Lamellar and hexagonal lipid faces visualized by freeze-etching, Biochim. Biophys. Acta *219*, 47.

De Kruijff, B. and R.A. Demel (1974), Polyene antibiotic-sterol interactions in membranes of *Acholeplasma laidlawii* cells and lecithin liposomes. III. Molecular structure of the polyene antibiotic-cholesterol complexes, Biochim. Biophys. Acta *339*, 57.

De Laat, S.W., L.G.J. Tertoolen and J.G. Bluemink (1981), Quantitative analysis of the numerical and lateral distribution of intramembrane particles in freeze-fractured biological membranes, Eur. J. Cell Biol. *23*, 273.

Dempsey, G.P. and R.E. Beever (1979), Electron microscopy of the rodlet layer of *Neurospora crassa* conidia, J. Bacteriol. *140*, 1050.

Dempsey, G.P. and S. Bullivant (1976a), A copper block method for freezing non-cryoprotected tissue to produce ice-crystal-free regions for electron microscopy, J. Microscopy *106*, 251.

Dempsey, G.P. and S. Bullivant (1976b), A copper block method for freezing non-cryoprotected tissue to produce ice-crystal-free regions for electron microscopy. II. Evaluation using freeze-fracturing with a cryoultramicrotome, J. Microscopy *106*, 261.

Demsey, A., H. Frank, H. Schwartz and E. Henne (1974), Complementary freeze-etch replicas by adopting the Denton apposed specimen tooling to the Balzers unit, J. Microscopy *100*, 169.

Dubochet, J., M. Ducommun, M. Zollinger and E. Kellenberger (1971), A new preparation method of dark-field electron microscopy of biomacromolecules, J. Ultrastruct. Res. *34*, 147.

Duniec, J.T., D.J. Goodchild and S.W. Thorne (1982), A method of estimating the radial distribution function of protein particles in membranes from freeze-fracture micrographs, Comput. Biol. Med. *12*, 319.

Dunlop, W. and A.W. Robards (1972), Some artefacts of the freeze-etching technique, J. Ultrastruct. Res. *40*, 391.

Dunlop, W., G.R. Parish and A.W. Robards (1972), Temperature, vacuum and fracturing in the freeze-etch technique, Proc. 5th Eur. Reg. Conf. Electron Microscopy, Manchester, p. 248.

Elgsaeter, A. (1974), The role of spectrin in determining the lateral distribution of the particles exposed by freeze-fracure of isolated human erythrocyte plasma membranes, Ph.D. Thesis, University of California, Berkeley.

Elgsaeter, A. (1978), A new freeze-etch unit for freeze-etch rotary shadowing, low temperature freeze-fracturing and conventional freeze-etching, J. Microscopy *113*, 83.

Elgsaeter, A., D. Shotton and D. Branton (1976), Intramembrane particle aggregation in erythrocyte ghosts, Biochim. Biophys. Acta *426*, 101.

Elias, P.M., J. Goerke and D.S. Friend (1978), Freeze-fracture identification of sterol-digitonin complexes in cell and liposome membranes, J. Cell Biol. *78*, 577.

Elias, P.M., D.S. Friend and J. Goerke (1979), Membrane sterol heterogeneity. Freeze-fracture detection with saponins and filipin, J. Histochem. Cytochem. *27*, 1247.

Escaig, J., D. Gache and G. Nicolas (1980), Rapid freeze-fracturing at down to 12 K, and freeze-drying under ultrahigh vacuum, Proc. 7th Eur. Reg. Conf. Electron Microscopy, The Hague *2*, 654.

Escaig, J. and G. Nicolas (1976), Cryo-fractures de matériel biologique réalisées à trés basses températures en ultra-vide, C. r. Acad. Sci., Paris *283*, 1245.

Espevik, T. and A. Elgsaeter (1981), In situ propane jet-freezing and freeze-etching of monolayer cell cultures, J. Microscopy *123*, 105.

Espevik, T. and A. Elgsaeter (1984), Liquid propane jet-freezing, freeze-drying and rotary replication of the cytoskeleton and plasma membrane associated structures in human monocytes, J. Microscopy *134*, 203.

Fineran, B.A. (1978), Freeze-etching, in: Electron microscopy and cytochemistry of plant cells, J.L. Hall, ed. (Elsevier/North-Holland, Amsterdam), p. 279.

Fisher, K.A. (1975), "Half" membrane enrichment: verification by electron microscopy, Science *190*, 983.

Fisher, K.A. (1976a), Autoradiography of membrane "halves": ^{3}H-cholesterol labelled erythrocytes, J. Cell Biol. *70*, 218a.

Fisher, K.A. (1976b), Analysis of membrane halves: cholesterol, Proc. natl. Acad. Sci. U.S.A. *73*, 173.

Fisher, K.A. (1978), Split membrane lipids and polypeptides, 9th Int. Congr. Electron Microscopy, Toronto *3*, 521.

Fisher, K.A. (1980), Split membrane analysis, Ann. Rev. Physiol. *42*, 261.

Fisher, K.A. (1982a), Monolayer freeze-fracture autoradiography: origins and directions, J. Microscopy *126*, 1.

Fisher, K.A. (1982b), Monolayer freeze-fracture autoradiography: quantitative analysis of transmembrane distribution of radioiodinated Concanavalin A, J. Cell Biol. *93*, 155.

Fisher, K.A. (1982c), Preparation of planar membrane monolayers for spectroscopy and electron microscopy, Meth. Enzymol. *88*, 230.

Fisher, K.A. and D. Branton (1976), Freeze-fracture autoradiography: feasibility, J. Cell Biol. *70*, 453.

Fisher, K.A., K. Yanagimoto and W. Stoeckenius (1978), Oriented adsorption of purple membrane to cationic surfaces, J. Cell Biol. *77*, 611.

Fujikawa, S. (1979), Negative staining-carbon replica methods, a new procedure to reveal ultrastructure of freeze-fractured membrane, J. Microscopy *117*, 261.

Garcia-Segura, L.M., D. Baetens and L. Orci (1982), Freeze-fracture cytochemistry of neuronal membranes: inhomogenous distribution of filipin-sterol complexes in perikarya, dendrites and axons, Brain Res. *234*, 494.

Gershon, N.D., A. Dempsey and C.W. Stackpole (1979), Analysis of local order in the spatial distribution of cell surface molecular assemblies, Exp. Cell Res. *122*, 115.

Geymayer, W. (1966), Die gleichzeitige elektronenmikroskopische Erfassung von Oberfläche und Querschnitt gefriergesschokter kolloidaler Systeme, Proc. 6th Int. Congr. Electron Microscopy, Tokyo *1*, 577.

Geymayer, W. (1967), Die elektronenmikroskopische Untersuchung temperaturempfindlicher Kolloide, Staub-Reinhaltung Luft *27*, 237.

Glauert, A.M. (1974), Fixation, dehydration and embedding of biological specimens, in: Practical methods in electron microscopy, Vol. 3, A.M. Glauert, ed. (North-Holland, Amsterdam).

Glitch, S. (1969), Verdampfung von Platin aus Kohletiegeln zur elektronenmikroskopischen Schrägbedampfung, Naturwissenschaften *56*, 559.

Gold, E.R. and P. Balding (1975), Receptor-specific proteins – Plant and animal lectins, (Excerpta Media, Amsterdam).

Goldstein, J.I., D.E. Newbury, P. Echlin, D.C. Joy, Ch. Fiori and E. Lifshin (1981), Scanning electron microscopy and X-ray microanalysis (Plenum Press, New York, London), p. 646.

Griepp, E.B. and M.R. Bernfield (1978), Acquisition of synchronous beating between embryonic heart cell aggregates and layers, Exp. Cell Res. *113*, 263.

Griepp, E.B., J.H. Peacock, M.R. Bernfield and J.P. Revel (1978), Morphological and functional correlates of synchronous beating between embryonic heart cell aggregates and layers, Exp. Cell Res. *113*, 273.

Gross, H. (1977), Gefrierätzung im Ultrahochvakuum, UHV, bei −196°C, PhD. Dissertation, Eidgenössiche Technische Hochschule, Zürich.

Gross, H. and H. Moor (1978), Decoration of specific sites on freeze-fractured membranes, Proc. 9th Int. Congr. Electron Microscopy, Toronto *2*, 140.

Gross, H., E. Bas and H. Moor (1978a), Freeze-fracturing in ultra high vacuum (UHV), at −196°C, J. Cell Biol. *76*, 712.

Gross, H., O. Kuebler, E. Bas and H. Moor (1978b), Decoration of specific sites on freeze-fractured membranes, J. Cell Biol. *79*, 646.

Gross, H., D. Studer and I. Wildhaber (1982), Information contained in metal replicas, Proc. 10th Int. Congr. Electron Microscopy, Hamburg *3*, 17.

Guckenberger, R. (1984), Reconstruction of surface profiles from shadowing data, Proc. 8th Eur. Reg. Conf. Electron Microscopy, Budapest *2*, 1281.

Gulik, A., L.P. Aggerbeck, J.C. Dedieu and T. Gulik-Krzwicki (1982), Freeze-fracture electron microscopic analysis of solutions of biological molecules, J. Microscopy *125*, 207.

Gulik-Krzwicki, T. and M.J. Costello (1978), Use of low temperature X-ray diffraction to evaluate freezing methods in freeze-fracture electron microscopy, J. Microscopy *112*, 103.

Haefer, R.A. (1981), Kryo-Vakuumtechnik (Grundlagen und Anwendungen), (Springer-Verlag, Berlin, Heidelberg, New York).

Hagler, H.K., W.W. Schulz and R.C. Reynolds (1977), A simple electron beam gun for platinum evaporation, J. Microscopy *110*, 149.

Hall, C.E. (1950), A low temperature replica method for electron microscopy, J. appl. Physics *21*, 61.

Heinmets, F. (1949), Modification of silica replica technique for study of biological membranes and application of rotary condensation in electron microscopy, J. appl. Physics *20*, 384.

Heinzmann, U. (1974), Verfeinerte Präparation biologischer Gewebeproben für die Raster-Elektronenmikroskopie, Microscopica Acta *76*, 145.

Hess, W.M. and R.L. Blair (1972), Production and cleaning of freeze-etch replicas which show complementary surfaces of fractured fungus spores and hyphae, Stain Technol. *47*, 249.

Heuser, J.E. (1980), Three-dimensional visualisation of coated vesicle formation in fibroblasts, J. Cell Biol. *84*, 560.

Heuser, J.E. (1981), Preparing biological samples for stereomicroscopy by the quick-freeze, deep-etch, rotary replication technique, in: Methods in cell biology, Vol. 22, J.N. Turner, ed. (Academic Press, New York), p. 97.

Heuser, J.E. (1983a), Structure of the myosin cross bridge in insect flight muscle, J. mol. Biol. *169*, 123.

Heuser, J.E. (1983b), Procedure for freeze-drying molecules adsorbed to mica flakes, J. mol. Biol. *169*, 155.

Heuser, J.E. and R. Cooke (1983), Actin-myosin interactions visualised by the quick-freeze, deep-etch replica technique, J. mol. Biol. *169*, 97.

Heuser, J.E. and M.W. Kirschner (1980), Filament organisation revealed in platinum replicas of freeze-dried cytoskeletons, J. Cell Biol. *86*, 212.

Heuser, J.E. and S.R. Salpeter (1979), Organisation of acetylcholine receptors in quick-frozen, deep-etched, and rotary replicated *Torpedo* postsynaptic membrane, J. Cell Biol. *82*, 150.

Heuser, J.E., T.S. Reese and D.M.D. Landis (1976), Preservation of synaptic structure by rapid freezing, Cold Spring Harbour Symp. Quant. Biol. *40*, 17.

Hirokawa, N., L.G. Tilney, K. Fujiwara and J.E. Heuser (1982), Organisation of actin, myosin and intermediate filaments in the brush border of intestinal epithelial cells, J. Cell Biol. *94*, 425.

Höhenberg, H. and K.C. Mannweiler (1980), Semi-automatic washing device for simultaneous cleaning of surface replicas under identical conditions, Mikroskopie *36*, 145.

Hörandner, H. (1976), Evaluation of corresponding freeze-etch replicas with the goniometer stage and by densitometry, Proc. 6th Eur. Reg. Conf. Electron Microscopy, Jerusalem *2*, 125.

Horisberger, M. (1979), Evaluation of colloidal gold as a cytochemical marker for transmission and scanning electron microscopy, Biol. Colloids *36*, 253.

Horisberger, M. and J. Rosset (1977), Colloidal gold, a useful marker for transmission and scanning electron microscopy, J. Histochem. Cytochem. *25*, 295.

Horn, H.R.F. (1962), Thermische Bedampfung ohne Licht- und Wärmestrahlungsbelastung des Objektes, Proc. 5th Int. Congr. Electron Microscopy, Philadelphia *1*, A9.

Howe S.C. and T. Bächi (1973), Localization of erythrocyte membrane antigens by immune electron microscopy, Exp. Cell Res. *76*, 321.

Hubbard, A.L. and Z.A. Cohn (1976), Specific labels for cell surfaces, in: Biochemical analysis of membranes, A.H. Maddy, ed. (Chapman and Hall, London), p. 427.

Jakopić, E., A. Brunegger, R. Essl and G. Windisch (1978), A sputter source for electron microscopic preparation, Proc. 9th Int. Congr. Electron Microscopy, Toronto *1*, 150.

Jensen, T.E., M.A. Smith and W.M. Hess (1973), Washing apparatus for freeze-etch replica fragments, Stain Technol. *48*, 43.

Johnson, R.P.C. (1968), Microfilaments in pores between frozen-etched sieve elements, Planta *81*, 314.

Johnson, R.P.C., A. Freundlich and G.F. Barclay (1976), Transcellular strands in sieve tubes; what are they? J. exp. Bot. *27*, 1117.

Junger, S. and L. Bachmann (1977), Structure of pyruvate dehydrogenase complex: comparison between freeze-etching and negative staining, Biochim. Biophys. Acta *481*, 364.

Kahler, H., and B.J. Lloyd, Jr. (1950), Metallic evaporation and the diameter of tobacco mosaic virus with the electron microscope, J. appl. Phys. *21*, 699.

Karnovsky, M.J., E.R. Unanue and M. Leventhal (1972), Ligand-induced movements of lymphocyte membrane macromolecules, J. exp. Med. *136*, 907.

Kellenberger, E. and J. Kistler (1979), The physics of specimen preparation, in: Advances in structure research, W. Hoppe and R. Mason, eds. (Friedrich Vieweg and Sohn Verlagsgesellsch., Wiesbaden) *7*, 49.

Kistler, J. and E. Kellenberger (1977), Collapse phenomena in freeze-drying, J. Ultrastruct. Res. *59*, 70.

Kistler, J., U. Aebi, L. Onorato, B. ten Heggeler and M.K. Showe (1978), Structural changes during the transformation of bacteriophage T4 polyheads: charaterization of the initial and final states by freeze-drying and shadowing Fab-labelled preparations, J. mol. Biol. *126*, 571.

Kitajima, Y., T. Sekiya and Y. Nozawa (1976), Freeze-fracture and ultrastructural alterations induced by filipin, pimaracin, nystatin and amphotericin B, Biochem. Biophys. Acta *445*, 452.

Kleinschmidt, A., D. Lang and R.K. Zahn (1960), Darstellung molekularer von DNS, Naturwissenschaften *47*, 16.

Knoll, G., G. Oebel and H. Plattner (1982), A simple sandwich-cryogen-jet procedure with high cooling rates for cryofixation of biological materials in the native state, Protoplasma *111*, 161.

Koehler, J.K. (1966), Fine structure observations in frozen-etched bovine spermatozoa, J. Ultrastruct. Res. *16*, 359.

Koehler, J.K. (1972a), The freeze-etching technique, in: Principles and techniques of electron microscopy, Vol. 2, M. A. Hayat, ed. (Van Nostrand Reinhold Company, New York), p. 53.

Koehler, J.K. (1972b), Human sperm head ultrastructure: a freeze-etching study, J. Ultrastruct. Res. *39*, 520.

Krah, S., L.A. Staehelin and E. Nettersheim (1973), A new type of storage container for freeze-etch specimens, J. Microscopy *99*, 349.

Krbecek, R., C. Gebhardt, H. Gruler and E. Sackmann (1979), Three dimensional microscopic surface profiles of membranes reconstructed from freeze-etching electron micrographs, Biochim. Biophys. Acta *554*, 1.

Kreutziger, G.O. (1970), The cryopump freeze-etch device, J. Cell Biol. *47*, 111A.

Kübler, O., H. Gross and H. Moor, (1978), Complementary structures of membrane fracture faces obtained by ultra high vacuum freeze-fracturing at −196°C and digital image processing, Ultramicroscopy *3*, 161.

Lang, R.D.A., M.V. Nermut and L.D. Williams (1981), Ultrastructure of sheep erythrocyte plasma membranes bound to solid supports – effects of selective removal of membrane components, J. Cell Sci. *49*, 383.

Langworthy, T.A., T.G. Tornabene and G. Holzer (1982), Lipids of archaebacteria, Zbl. Bakt. Hyg., I. Abt. Orig. C. *3*, 228

Lepault, J. and J. Dubochet (1980), Freezing, fracturing and etching artefacts in particulate suspensions, J. Ultrastruct. Res. *72*, 223.

Lickfield, K.G. (1968), Der frostgeätzte Bakterienkern. Ein Beitrag zur Klärung seiner Tertiärstruktur, Z. Zellforsch. *88*, 560.

Lickfield, K.G. (1977), Transmission electron microscopy of bacteria, in: Methods in microbiology, Vol. 9, J.R. Norris, ed. (Academic Press, London), p. 160.

Lickfield, K.G., M. Achterrath and F. Hentrich (1972), The interpretation of images of cross-fractured frozen-etched and shadowed membranes, J. Ultrastruct. Res. *38*, 279.

Margaritis, L.H., A. Elgsaeter and D. Branton (1977), Rotary replication for freeze-etching, J. Cell Biol. *72*, 47.

Maul, G.G. (1979), Temperature dependent changes in intramembrane particle distribution, in: Freeze-fracture: methods, artefacts and interpretation, J.E. Rash and C.S. Hudson, eds. (Raven Press, New York), p. 37.

Maurer, A. and K. Mühlethaler (1981), Specific labelling of glycoproteins in yeast plasma membrane with Concanavalin A, Eur. J. Cell Biol. *25*, 58.

Mazia, D., W.S. Sale and G. Schatten (1974), Polylysine as an adhesive for electron microscopy, J. Cell Biol. *63*, 212a.

McAlear, J.H. and G.O. Kreutziger (1967), Freeze-etching with radiant energy in a simple cold block device, Proc. 25th Ann. Meeting EMSA, p. 116.

Mellor, J.D. (1978), Fundamentals of freeze-drying (Academic Press, London).

Menold, R., B. Lüttge, W. Kaiser and A. Schmidt (1972), Gefrierbruchtechnik zur Untersuchung von Suspensionen und Emulsionen, Chem. Ing. Techn. *44*, 1226.

Menold, R., B. Lüttge and W. Kaiser (1976), Freeze-fracturing, a new method for the investigation of dispersions by electron microscopy, Adv. Colloid Interface Sci. *5*, 281.

Meryman, H.T. (1950), Replication of frozen liquids by vacuum evaporation, J. appl. Phys. *21*, 68.

Meryman, H.T. and E. Kafig (1955), The study of frozen specimens, ice crystals and ice crystal growth by electron microscopy, Naval Med. Res. Inst. Rept. NM 000 018.01.09 *13*, 529.

Meyer, H.W. and H. Winkelmann (1969), Die Gefrierätzung und die Struktur biologischer Membranen, Protoplasma *68*, 253.

Meyer, H.W. and H. Winkelmann (1970), Die Darstellung von Lipiden bei der Gefrieratzpraparation und ihre Beziehung zur Strukturanalyse biologischer Membranen, Exp. Pathol. *4*, 47.

Miller, K.R. and L.A. Staehelin (1976), Analysis of the thylakoid outer surface. Coupling factor is limited to unstacked membrane regions, J. Cell Biol. *68*, 30.

Montesano, R., A. Perrelet, P. Vassalli and L. Orci (1979), Absence of filipin-sterol complexes from large coated pits on the surface of culture cells, Proc. natl. Acad. Sci. U.S.A. *76*, 6391.

Moor, H. (1959), Platin-Kohle-Abdruck-Technik angewandt auf den Feinban der Milchröhren, J. Ultrastruct. Res. *2*, 393.

Moor, H. (1970), Die Gefrierätztechnik, in: Methodensammlung der Elektronenmikroskopie, G. Schimmel and W. Vogell, eds. (Wissenchaftliche Verlagsanstallt, Stuttgart), p. 1.

Moor, H. (1971), Recent progress in the freeze-etching technique, Phil. Trans. R. Soc. Ser. B. *261*, 121.

Moor, H. (1973a), Evaporation and electron guns, in: Freeze-etching, techniques and applications, E.L. Benedetti and P. Favard, eds. (Soc. Française de Microscopie Électronique, Paris), p. 27.

Moor, H. (1973b), Etching and related problems, in: Freeze-etching, techniques and applications, E.L. Benedetti and P. Favard, eds. (Soc. Française de Microscopie Électronique, Paris), p. 21.

Moor, H., K. Mühlethaler, H. Waldner and A. Frey-Wyssling (1961), A new freezing ultramicrotome, J. biophys. biochem. Cytol. *10*, 1.

Morgan, C. (1972), The use of ferritin-conjugated antibodies in electron microscopy, Int. Rev. Cytol. *32*, 291.

Mühlethaler, K. (1973), History of freeze-etching, in: Freeze-etching, techniques and applications, E.L. Benedetti and P. Favard, eds. (Soc. Française de Microscopie Électronique, Paris), p. 1.

Mühlethaler , K., E. Wehrli and H. Moor (1970), Double fracturing methods for freeze-etching, Proc. 7th Int. Congr. Electron Microscopy, Grenoble *1*, 449.

Mühlethaler, K., W. Hauenstein and H. Moor (1973), Double fracturing method for freeze-etching, in: Freeze-etching, techniques and applications, E.L. Benedetti and P. Favard, eds. (Soc. Française de Microscopie Électronique, Paris), p. 101.

Müller, T., H. Gross and H. Moor (1984), A double replica device for freeze-fracturing and freeze-etching at defined specimen temperatures, Proc. 8th Eur. Reg. Conf. Electron Microscopy, Budapest *3*, 1735.

Müller, M., T. Marti and S. Kriz (1980a), Improved structural preservation by freeze-substitution, Proc. 7th Eur. Reg. Conf. Electron Microscopy, The Hague *2*, 720.

Müller, M., N. Meister and H. Moor (1980b), Freezing in a propane jet and its application in freeze-fracturing, Mikroskopie *36*, 129.

Munn, E.A., L. Bachmann and A. Feinstein (1980), Structure of hydrated immunoglobulins and antigen-antibody complexes. Electron microscopy of spray-freeze-etched specimens, Biochim. Biophys. Acta *625*, 1.

Nanninga, N. (1973), Freeze-fracturing of microorganisms. Physical and chemical fixation of *Bacillus subtilis*, in: Freeze-etching, techniques and applications, E.L. Benedetti and P. Favard, eds. (Soc. Française de Microscopie Électronique, Paris), p. 151.

Nermut, M.V. (1973), Freeze-drying and freeze-etching of viruses, in: Freeze-etching, techniques and applications, E.L. Benedetti and P. Favard, eds. (Soc. Française de Microscopie Électronique, Paris), p. 135.

Nermut, M.V. (1977), Negative staining in freeze-drying and freeze-fracturing, Micron *8*, 211.

Nermut, M.V. (1982a), The "cell monolayer technique" in membrane research, Eur. J. Cell Biol. *28*, 160.

Nermut, M.V. (1982b), Freeze-fracture of red blood cells attached to solid supports: structural and chemical analysis of the outer membrane leaflet, Proc. 10th Int. Congr. Electron Microscopy, Hamburg *3*, 203.

Nermut, M.V. (1984), Ultrastructural and chemical analysis of intramembranous particles, Proc. 8th Eur. Reg. Conf. Electron Microscopy, Budapest *3*, 1737.

Nermut, M.V. and L.D. Williams (1976), Freeze-fracturing of monolayers of cells, membranes and viruses, Proc. 6th Eur. Reg. Conf. Electron Microscopy, Jerusalem *2*, 131.

Nermut, M.V. and L.D. Williams (1977), Freeze-fracturing of monolayers (capillary layers) of cells, membranes and viruses: some technical considerations, J. Microscopy *110*, 121.

Nermut, M.V., I.D.J. Burdett and L.D. Williams (1978), Negative staining of freeze-fractured envelopes of *Escherichia coli* K12, J. Microscopy *114*, 229.

Neugebauer, D.Ch. and H.P. Zingsheim (1979), Apparent holes in rotary shadowed proteins: dependence on angle of shadowing and replica thickness, J. Microscopy *117*, 313.

Nickel, E., G. Oebel and P. Pscheid (1978), Freeze-fracturing of biological material with a new specimen table, J. Microscopy *113*, 101.

Niedermeyer, W. (1982a), Freeze-fracturing at low temperatures. I. A device for fracturing biological specimens at 77–10 K under high vacuum, J. Microscopy *125*, 307.

Niedermeyer, W. (1982b), A simple airlock system for hoar frost free specimen exchange in freeze-etch vacuum plants, J. Microscopy *125*, 299.

Niedermeyer, W. and H. Wilke (1982), Quantitative analysis of intramembrane particle (IMP) distribution on biomembranes after freeze-fracture preparation by a computer-based technique, J. Microscopy *126*, 259.

Parish, G.R. (1973), The application of the freeze-etch technique to the study of plant vascular cells, D. Phil. Thesis, University of York.

Pauli, B.U., R.S. Weinstein, L.W. Soble and J. Alroy (1977), Freeze-fracture of monolayer cultures, J. Cell Biol. *72*, 763.

Pearce, R.S. (1983), A cleaning method ensuring recovery of delicate freeze-fracture replicas of cereal leaf cells, J. Microscopy *129*, 229.

Peng, H.B. and L.F. Jaffe (1976), A simple selective method for freeze-fracturing spherical cells, J. Cell Biol. *71*, 674.

Perkins, W.D. and J.K. Koehler (1978), Antibody-labelling techniques, in: Advanced techniques in biological electron microscopy, Vol. 2, J.K. Koehler, ed. (Springer Verlag, Berlin), p. 39.

Peters, K.R. (1980), Penning sputtering of ultrathin metal films for high resolution electron microscopy, Scanning Electron Microscopy 1980, *1*, 143.

Pfenninger, K.H. and E.R. Rinderer (1975), Methods for the freeze-fracturing of nerve tissue cultures and cell monolayers, J. Cell Biol. *65*, 15.

Pinto da Silva, P. (1982), Freeze-fracturing cytochemistry, labelling of plasma and intracellular membranes, Proc. 10th Int. Congr. Electron Microscopy, Hamburg *3*, 195.

Pinto da Silva, P. and D. Branton (1970), Membrane splitting in freeze-etching: covalently bound ferritin as a membrane marker, J. Cell Biol. *45*, 598.

Pinto da Silva, P. and M.R. Torrisi (1982), Freeze-fracture cytochemistry: partition of glycophorin in freeze-fracture erythrocyte membranes, J. Cell Biol. *93*, 463.

Pinto da Silva, P., S.P. Douglas and D. Branton (1971), Localization of antigen sites on human erythrocyte ghosts, Nature, Lond. *232*, 144.

Pinto da Silva, P., B. Kachar, M.R. Torrisi, C. Brown and C. Parkison (1981a), Freeze-fracture cytochemistry: replicas of critical point dried cells and tissues after fracture-label, Science *213*, 230.

Pinto da Silva, P., C. Parkison and N. Dwyer (1981b), Freeze-fracture cytochemistry: thin sections of cells and tissues after labelling of fracture faces, J. Histochem. Cytochem. *29*, 917.

Pinto da Silva, P., C. Parkison and N. Dwyer (1981c), Fracture-label: cytochemistry of freeze-fracture faces in the erythrocyte membrane, Proc. natl. Acad. Sci. U.S.A. *78*, 343.

Platt-Alloia, K.A. and W.W. Thomson (1982), Freeze-fracture of intact plant tissues, Stain Technol. *57*, 327.

Plattner, H. and H.P. Zingsheim (1983), Electron microscopic methods in cellular and molecular biology, in: Subcellular biochemistry, Vol. 9, D.B. Roodyn, ed. (Plenum Press, New York), p. 1.

Porter, K. and J. Blum (1953), A study of microtomy for the electron microscope, Anat. Rec. *117*, 685.

Prescott, L. and M.W. Brightman (1978), A technique for the freeze-fracture of tissue culture, J. Cell Sci. *30*, 37.

Pricam, C., K.A. Fisher and D.S. Friend (1977), Intramembranous particle distribution in human e. ythrocytes: effects of lysis, glutaraldehyde and poly-L-lysine, Anat. Rec. *189*, 595.

Pscheid, P. and H. Plattner (1980), Cryofixation of cell monolayers: Avoidance of unusual substrate r..aterials, detachment of cells, chemical fixation and antifreeze impregnation, Proc. 7th Int. Congr. Electron Microscopy, Grenoble *2*, 716.

Pscheid, P., C. Schudt and H. Plattner (1981), Cryofixation of monolayer cell cultures for freeze-fracturing without chemical pretreatments, J. Microscopy *121*, 149.

Rash, J.E. (1979), The sectioned-replica technique: direct correlation of freeze-fracture replicas and conventional thin section images, in: Freeze-fracture: methods, artefacts and interpretations, J.E. Rash and C.S. Hudson, eds. (Raven Press, New York), p. 153.

Rash, J.E. and C.S. Hudson (eds.) (1979), Freeze-fracture: methods, artefacts and interpretations (Raven Press, New York).

Rash, J.E., C.S. Hudson and W.F. Graham (1978), Direct and indirect labelling of particles and pits in freeze-etch replicas, J. Cell Biol. *79*, 232a.

Rash, J.E., T.J.A. Johnson, C.S. Hudson, D.S. Copio, W.F. Graham, M.E. Eldefraw and F.D. Giddings (1980), Identification of intramembrane particles by pre- and post-fracture labelling techniques: a progress report, Proc. 38th Ann. Meeting EMSA, p. 692.

Rash, J.E., T.J.A. Johnson, C.S. Hudson, F.D. Giddings, W.F. Graham and M.E. Eldefraw (1982), Labelled-replica techniques: post-shadow labelling of intramembrane particles in freeze-fracture replicas, J. Microscopy *128*, 121.

Reimer, L. (1967), Elektronenmikroskopische Untersuchung- und Präparationsmethoden (Springer-Verlag, Berlin), p. 303.

Reiter, O. (1961), Ein einfaches Verfahren zur Dickenbestimmung von Kohlenstoff-Aufdampfschichten, Naturwissenschaft *48*, 519.

Rennke, H.G., Y. Patel and M.A. Venkatachalam (1978), Glomerular filtration of proteins: Clearance of anionic, neutral and cationic horseradish peroxidase in the rat, Kidney Int. *13*, 324.

Revel, J.P., A.G. Lee and A.J. Hudspeth (1971), Gap junctions between electronically coupled cells in tissue culture and brown fat, Proc. natl. Acad. Sci. U.S.A. *68*, 2924.

Richter, H. and U.B. Sleytr (1971), Fettextraktion bei $-78°C$: Nachweis im Gefrierätzbild, Z. Naturf. *26*, 470.

Rix, E., A. Schiller and R. Taugner (1976), Freeze-fracture autoradiography, Histochemistry *50*, 91.

Robards, A.W. (1974), Ultrastructural methods for looking at frozen cells, Sci. Prog. (Oxford) *61*, 1.

Robards, A.W. (1978a), Freeze-etching of plant cells, Mikroskopie *34*, 12.

Robards, A.W. (1978b), An introduction to techniques for scanning electron microscopy of plant cells, in: Electron microscopy and cytochemistry of plant cells, J.L. Hall, ed. (Elsevier/ North-Holland, Amsterdam), p. 343.

Robards, A.W. and M.H. Cooper (1971), An apparatus for preparing frozen-etched specimens for electron microscopy, Vacuum *21*, 465.

Robards, A.W. and G.R. Parish (1971), Preparation and mechanical requirements for freeze-etching thick-walled plant tissue, J. Microscopy *93*, 61.

Robards, A.W. and W. Umrath (1978), Improved freeze-etching of difficult specimens, Proc. 9th Int. Congr. Electron Microscopy, Toronto *2*, 138.

Robards, A.W., W.R. Austin and G.R. Parish (1970), The role of the microtome assembly in the freeze-etching technique, Proc. 7th Int. Congr. Electron Microscopy, Grenoble *1*, 447.

Robards, A.W., U.B. Sleytr, G.R. Bullock, P. Sibbons and H. Quine (1980), A thin plate cleavage device for the Leybold-Heraeus Bioetch 2005 freeze-etching unit, Proc. 7th Eur. Congr. Electron Microscopy, The Hague *2*, 740.

Robards, A.W., G.R. Bullock, M.A. Goodall and P.D. Sibbons (1981), Computer assisted analysis of freeze-fractured membranes following exposure to different temperatures, in: Effects of low temperatures on biological membranes, G.J. Morris and A. Clarke, eds. (Academic Press, London)p. 219.

Robenek, H., R. Gebhardt and W. Jung (1982), Freeze-fracture identification of complexes between cholesterol and filipin, digitonin or tomatin in hepatocyte plasma membranes during formation of the biliary pole, Proc. 10th Int. Congr. Electron Microscopy, Hamburg *3*, 457.

Robinson, J.M. and M.J. Karnovsky (1980), Evaluation of the polyene antibiotic filipin as a cytochemical probe for membrane cholesterol, J. Histochem. Cytochem. *28*, 161.

Robinson, P.J., P. Dunnill and M.D. Lilly (1971), Porous glass as a solid support for immobilisation of affinity chromatography of enzymes, Biochim. Biophys. Acta *242*, 659.

Roth, J. (1978), The lectins, molecular probes in cell biology and membrane research (VEB Gustav Fischer, Verlag Jena).

Roth, J. (1982), New approaches for *in situ* localisation of antigens and glycoconjugates on thin sections: the protein A-gold technique and the lectin colloidal gold marker system, Proc. 10th Int. Congr. Electron Microscopy, Hamburg *3*, 245.

Ruben, G.C. and J.N. Telford (1980), Dimensions of active cytochrome C oxidase in reconstituted liposomes using a gold ball shadow width standard: a freeze-etch electron microscopy study, J. Microscopy *118*, 191.

Salpeter, M.M. and L. Bachmann (1972), Autoradiography, in: Principles and techniques of electron microscopy, Vol. 2, M.A. Hayat, ed. (Van Nostrand Reinhold Company, New York), p. 219.

Salpeter, M.M., H.C. Fertuck and E.E. Salpeter (1977), Resolution in electron microscope autoradiography. III. Iodine-125, the effect of heavy metal staining, and a reassessment of critical parameters, J. Cell Biol. *72*, 161.

Salpeter, M.M., F.A. McHenry and E.E. Salpeter (1978), Resolution in electron microscope autoradiography. IV. Application to analysis of autoradiographs, J. Cell Biol. *76*, 127.

Sawada, H. and E. Yamada (1981), A freeze-fracture deep-etching replica method with volatile cryoprotectant, J. Electron Microscopy *30*, 341.

Schiller, A. and R. Taugner (1980), Freeze-fracturing and deep-etching with the volatile cryoprotectant ethanol reveals true membrane surfaces of kidney structures, Cell Tissue Res. *210*, 57.

Schiller, A., E. Rix and R. Taugner (1978), Freeze-fracture autoradiography: The *in vacuo* coating technique, Histochemistry *59*, 9.

Schiller, A., R. Taugner and E. Rix (1979), Freeze-fracture autoradiography: Progress towards a routine technique, J. Histochem. Cytochem. *27*, 1514.

Schmid, E.N., U.B. Sleytr and K.G. Lickfield (1980), Nomenclature of frozen-etched bacterial envelopes, J. Ultrastruct. Res. *70*, 86.

Schmidt, D.G. and W. Bucheim (1982), On the size of small protein particles determined by electron microscopy of unidirectionally shadowed freeze-etched preparations, J. Microscopy *126*, 347.

Schulz, W.W. and R.C. Reynolds (1977), Enhancement of three dimensional appearance of freeze-fracture images by reversal processing of electron microscopy sheet film, J. Microscopy *112*, 249.

Severs, N.J. (1984), Freeze-fracture cytochemistry, Proc. 8th Eur. Reg. Conf. Electron Microscopy, Budapest *3*, 1747.

Severs, N.J. and H.L. Simons (1983), Failure of filipin to detect cholesterol rich domains in smooth muscle plasma membrane, Nature, Lond. *303*, 637.

Shibata, Y., T. Arima and T. Yamamoto (1984), Double-axis rotary replication for deep-etching, J. Microscopy *136*, 121.

Shotton, D., Thompson, K., Wofsy, L. and D. Branton (1978), Appearance and distribution of surface proteins of the human erythrocyte membrane, J. Cell Biol. *76*, 512.

Silva, M.T., F.C. Guerra and M.M. Magalhaes (1968), The fixative action of uranyl acetate in electron microscopy, Experientia *24*, 1024.

Simpson, D.J. (1979), Freeze-fracture studies on barley plastid membranes. III. Location of the light-harvesting chlorophyll-protein, Carlsberg Res. Commun. *44*, 305.

Sleytr, U.B. (1970a), Die Gefrierätzung korrespondierender Bruchhälften ein neuer Weg zur Aufklärung von Membranenstrukturen, Protoplasma *70*, 101.

Sleytr U.B. (1970b), Fracture faces in intact cells and protoplasts of *Bacillus stearothermophillus*. A study by conventional freeze-etching and freeze-etching of corresponding fracture moieties, Protoplasma *71*, 295.

Sleytr, U.B. (1974), Freeze-fracturing at liquid helium temperature for freeze-etching, Proc. 8th Int. Congr. Electron Microscopy, Canberra *2*, 30.

Sleytr, U.B. (1975), A method for adapting the Leybold device for obtaining complementary replicas to the Balzers unit, J. Microscopy *103*, 377.

Sleytr, U.B. (1978), Regular arrays of macromolecules on bacterial cell walls: structure, chemistry, assembly and function, Int. Rev. Cytol. *53*, 1.

Sleytr, U.B. and G.P. Friers (1978), Surface charge and morphogenesis of regular arrays and proteins on bacterial cell walls, Proc. 9th Int. Congr. Electron Microscopy, Toronto *2*, 346.

Sleytr, U.B. and A.M. Glauert (1975), Analysis of regular arrays of subunits on bacterial surfaces: evidence for a dynamic process of assembly, J. Ultrastruct. Res. *50*, 103.

Sleytr, U.B. and A.M. Glauert (1982), Bacterial cell walls and membranes, in: Electron microscopy of proteins, Vol. 3, J.R. Harris, ed. (Academic Press, New York), p. 41.

Sleytr, U.B. and P. Messner (1978), Freeze-fracturing in normal vacuum reveals ring-like yeast plasmalemma structures, J. Cell Biol. *79*, 276.

Sleytr, U.B. and A.W. Robards (1977a), Plastic deformation during freeze-cleaving: a review, J. Microscopy *110*, 1.

Sleytr, U.B. and A.W. Robards (1977b), Freeze-fracturing: a review of methods and results, J. Microscopy *111*, 77.

Sleytr, U.B. and A.W. Robards (1982), Understanding the artefact problem in freeze-fracture replication: a review, J. Microscopy *126*, 101.

Sleytr, U.B. and W. Umrath (1974a), A simple fracturing device for obtaining complementary replicas of freeze-fractured and freeze-etched suspensions and tissue fragments, J. Microscopy *101*, 177.

Sleytr, U.B. and W. Umrath (1974b), A simple device for obtaining complementary fracture planes at liquid helium temperature in the freeze-etching technique, J. Microscopy *101*, 187.

Sleytr, U.B. and W. Umrath (1976), Freeze-etching: technical developments and general interpretation problems, Proc. 6th Eur. Reg. Conf. Electron Microscopy, Jerusalem *2*, 50.

Sleytr, U.B., D. Kerjaschki and L. Stockinger (1972), Gefrierätzpräparation von Gewebestücken in der EPA 100 am Beispiel des Nasenseptums der Maus, Mikroskopie *28*, 257.

Sleytr, U.B., R.C. Oliver and K.J.I. Thorne (1976), Bacitracin-induced changes in bacterial plasma membrane structure, Biochim. Biophys. Acta *419*, 570.

Sleytr, U.B., H. Groesz and W. Umrath (1981), Interpretation of morphological data obtained by freeze-fracturing at very low temperatures, Acta Histochemica, Suppl. *23*, 29.

Smith, P.R. and I.E. Ivanov (1980), Surface reliefs computed from micrographs of isolated heavy metal shadowed particles, J. Ultrastruct. Res. *71*, 25.

Smith, P.R. and J. Kistler (1977), Surface reliefs computed from micrographs of heavy metal-shadowed specimens, J. Ultrastruct. Res. *61*, 124.

Sommer, J.R. (1977), To cationize glass, J. Cell Biol. *75*, 245a.

Southworth, D., K.A. Fisher and D. Branton (1975), Principles of freeze-fracturing and etching, in: Techniques of biochemical and biophysical morphology, Vol. 2, D. Glick and R.M. Rosenbaum, eds. (Wiley Interscience, New York, London), p. 247.

Staehelin, L.A. (1973), Analysis and critical evaluation on the information contained in freeze-etch micrographs, Freeze-etching, techniques and applications, E.L. Benedetti and P. Favard, eds. (Soc. Française de Microscopie Électronique, Paris), p. 113.

Staehelin, L.A. and W.S. Bertaud (1971), Temperature and contamination dependent freeze-etch images of frozen water and glycerol solutions, J. Ultrastruct. Res. *37*, 146.

Steere, R.L. (1957), Electron microscopy of structural detail in frozen biological specimens, J. biophys. biochem. Cytol. *3*, 45.

Steere, R.L. (1969), Freeze-etching simplified, Cryobiology *5*, 306.

Steere, R.L. (1971), Retention of 3-dimensional contours by replicas of freeze-fractured specimens, Proc. 29th Ann. Meeting EMSA, p. 242.

Steere, R.L. (1973), Preparation of high resolution freeze-etch, freeze-fracture, frozen surface, and freeze-dried replicas in a single freeze-etch module, and the use of stereo electron microscopy to obtain maximum information from them, in: Freeze-etching, techniques and applications, E.L. Benedetti and P. Favard, eds. (Soc. Française de Microscopie Électronique, Paris), p. 223.

Steere, R.L. (1982), Reliable use of resistance evaporation of Pt and C for high resolution freeze-fracturing and a crystal surface image complementary to the E-face of yeast plasma membranes, J. Microscopy *128*, 157.

Steere, R.L. (1983), Preparation of freeze-fracture, freeze-etch, freeze-dry and frozen surface replicas for electron microscopy in the Denton DFE-2 and DFE-3 freeze-etch units, in: Current trends in morphological techniques, Vol. 2, J.E. Johnson Jr., ed. (CRC Press, Boca Raton FL).

Steere, R.L. and E.F. Erbe (1979), Complementary freeze-fracture, freeze-etch specimens, J. Microscopy *117*, 211.

Steere, R.L. and J.M. Moseley (1969), New dimensions in freeze-etching, Proc. 27th Ann. Meeting EMSA, p. 202.

Steere, R.L. and J.M. Moseley (1970), Modified freeze-etch equipment permits simultaneous preparation of 2–10 double replicas, 7th Int. Congr. Electron Microscopy, Grenoble *1*, 451.

Steere, R.L. and J.E. Rash (1979), Use of double-tilt device (goniometer) to obtain optimum contrast in freeze-fracture replicas, in: Freeze-fracture: methods, artefacts and interpretation, J.E. Rash and C.S. Hudson, eds. (Raven Press, New York), p. 161.

Steere, R.L., E.F. Erbe and J.M. Moseley (1974), Importance of stereoscopic methods in the study of biological specimens prepared by cryotechniques, Proc. 8th Int. Congr. Electron Microscopy, Canberra *2*, 36.

Steere, R.L., E.F. Erbe and J.M. Moseley (1977), A resistance monitor with power cut-off for automatic regulation of shadow and support film thickness in freeze-etching and related techniques, J. Microscopy *111*, 313.

Steere, R.L., E.F. Erbe and J.M. Moseley (1980), Pre-fracture and cold-fracturing images of yeast plasma membranes, J. Cell Biol. *86*, 113.

Stolinski, C. (1975), A freeze-fracture replication apparatus for biological specimens, J. Microscopy *104*, 235.

Stolinski, C. (1977), Freeze-fracture replication in biological research: Development, current practice and future prospects, Micron *8*, 87.

Stolinski, C., G. Gabriel and B. Martin (1983), Reinforcement and protection with polystyrene of freeze-fracture replicas during thawing and digestion of tissue, J. Microscopy *132*, 149.

Stolinski, C., L. Landmann and H. Abt (1984), Design and application of a freeze-fracture stage with a cleaver device for Balzers 'freeze-etching' apparatus, J. Microscopy *135*, 317.

Studer, D., H. Moor, V. Vogt and H. Gross (1980), High resolution shadowing of freeze-dried purple membrane, Proc. 7th Eur. Reg. Conf. Electron Microscopy, The Hague *2*, 616.

Sutherland, I. (ed.) (1977), Surface carbohydrates of the prokaryotic cell (Academic Press, London).

Tesche, B. (1975), Ein Elektronenstoss-Verdampfer mit geringem Leistungsbedarf für die Herstellung dünner Schichten aus hochschmelzenden Materialien, Vakuum Technik *24*, 104.

Tillack, T.W. and S.C. Kinsky (1973), A freeze-etch study of the effects of filipin on liposomes and human erythrocyte membranes, Biochim. Biophys. Acta *323*, 43.

Tillack, T.W., R.E. Scott and V.T. Marchesi (1972), The structure of erythrocyte membranes studied by freeze-etching. II. Localization of receptors for phytohemagglutinin and influenza virus to the intramembranous particles, J. exp. Med. *135*, 1209.

Tonosaki, A. and T.Y. Yamamoto (1974), Double replicating method for the freeze-fractured retina, J. Ultrastruct. Res. *47*, 86.

Torrisi, M.R. and P. Pinto da Silva (1984), Compartmentalisation of intracellular membrane glycocomponents is revealed by fracture label, J. Cell Biol. *98*, 25.

Tranum-Jensen, J. and S. Bhakdi (1983), Freeze-fracture analysis of the membrane lesion of human complement, J. Cell Biol. *97*, 618.

Umrath, W. (1977), Prinzipien der Kryopräparationsmethoden: Gefriertrocknung und Gefrierätzung, Mikroskopie *33*, 11.

Umrath, W. (1978), Automatische Gefrierätzung, Mikroskopie *34*, 6.

Umrath, W. (1981), An improved technique for the freeze-fracture of cell cultures and monolayers, J. Microskopie *121*, 229.

Umrath, W. (1983), Berechnung von Gefriertrocknungszeiten für die elektronenmikroskopische Präparation, Mikroskopie *40*, 9.

Unanue, E.R., M.J. Karnovsky and H.D. Engers (1973), Ligand-induced movements of lymphocyte membrane macromolecules. III. Relationship between the formation and fate of anti-Ig-surface complexes and cell metabolism, J. exp. Med. *137*, 675.

Vail, W.J. and J.G. Stollery (1978), A convenient method to remove lipid contamination from freeze-fracture replicas, J. Microscopy *113*, 107.

Van Harreveld, A. and J. Crowell (1964), Electron microscopy after rapid freezing on a metal surface and substitution fixation, Anat. Rec. *149*, 381.

Van Winkle, W.B. and M.L. Entman (1981), Accurate quantitation of surface area and particle density in freeze-fracture replicas, Micron *12*, 259.

Verkleij, A.J. (1984), Limitations and perpectives of freeze fracturing and freeze etching, Proc. 8th Eur. Reg. Conf. Electron Microscopy, Budapest *3*, 1729.

Verkleij, A.J. and P.H.J.Th. Ververgaert (1975), The architecture of biological and artificial membranes as visualised by freeze-etching, Ann. Rev. Phys. Chem. *26*, 101.

Verkleij, A.J. and P.H.J.Th. Ververgaert (1978), Freeze-fracture morphology of biological membranes, Biochim. Biophys. Acta *515*, 303.

Verkleij, A.J., B. de Kruijff, W.F. Gerritsen, R.A. Demel, L.L.M. Van Deenen and P.H.J.Th. Ververgaert (1973), Freeze-etch electron microscopy of erythrocytes, *Acholeplasma laidlawii* cells and liposomal membranes after the action of filipin and amphotericin B, Biochim. Biophys. Acta *291*, 577.

Verkleij, A.J., C.J.A. Van Echteld, W.J. Gerritsen, P.R. Cullis and B. de Kruijff (1980), The lipidic particle as an intermediate structure in membrane fusion processes and bilayer to hexagonal H*II* transitions, Biochim. Biophys. Acta *600*, 620.

Ververgaert, P.H.J.Th. and A.J. Verkleij (1978), Lipid intermembranous particles, Proc. 9th Int. Congr. Electron Microscopy, Toronto *2*, 154.

Waldhausl, P. (1978), An approximate solution for the restitution of stereo electron micrographs, Photogrammetr. Eng. Remote Sensing *44*, 1005.

Walzthöny, D., H. Moor and H. Gross (1981), Ice crystals specifically decorate hydrophilic sites on freeze-fractured model membranes, Ultramicroscopy *6*, 259.

Walzthöny, D., H. Gross and H. Moor (1982), Contamination sources in freeze-fracturing, Proc. 10th Int. Congr. Electron Microscopy, Hamburg *3*, 213.

Washburn, E.W. (1924), The vapour pressure of ice and of water below the freezing point, Monthly Weather Rev. *57*, 488.

Wehrli, E., K. Mühlethaler and H. Moor (1970), Membrane structure as seen with a double replica method for freeze-fracturing, Exp. Cell Res. *59*, 336.

Weibel, E.R., G. Losa and R.P. Bolender (1976), Stereological methods for estimating relative membrane surface area in freeze-fracture preparations of subcellular fractions, J. Microscopy *107*, 255.

Weinstein, R.S., W.D. Selles and I.T. Young (1976), Quantitative analysis of membrane topography in normal and malignant epithelium, Proc. 6th Eur. Reg. Conf. Electron Microscopy, Jerusalem *2*, 104.

Weinstein, R.S., D.J. Benefiel and B.U. Pauli (1979), Use of computers in the analysis of intramembrane particles, in: Freeze-fracture: methods, artefacts and interpretation, J.E. Rash and C.S. Hudson, eds. (Raven Press, New York), p. 175.

Westmeyer, H. and E. Lorenz (1960), Die Herstellung von Aufdampfschichten vermittels kleiner elektronenstossgeheitzer Dampfquellen, Optik *17*, 244.

Wildhaber, I., H. Gross, and H. Moor (1982), Surface structure preservation in air- and freeze-dried HPI-layer, Proc. 10th Int. Congr. Electron Microscopy, Hamburg *3*, 19.

Wildhaber, I., H. Gross and H. Moor (1984), High resolution shadowing films produced by electron beam evaporation and ion beam sputtering, Proc. 8th Eur. Reg. Conf. Electron Microscopy, Budapest *2*, 1285.

Williams, M.A. (1977), Autoradiography and immunocytochemistry, in: Practical methods in electron microscopy, Vol. 6, A.M. Glauert, ed. (Elsevier/North-Holland, Amsterdam).

Williams, R.C. and K.M. Smith (1958), The polyhedral form of the *Tipula iridesent* virus, Biochim. Biophys. Acta *28*, 464.

Williams, R.C. and R.W.G. Wyckoff (1944), The thickness of electron microscope objects, J. appl. Phys. *15*, 712.

Willison, J.H.M. and R.M. Brown (1977), An examination of the developing cotton fibre: wall and plasmalemma, Protoplasma *92*, 21.

Willison, J.H.M. and R.D. Moir (1983), Bidirectional shadowing in freeze-etching, J. Microscopy *132*, 171.

Willison, J.H.M. and A.J. Rowe (1980), Replica, shadowing and freeze-etch techniques, in: Practical methods in electron microscopy, Vol. 8, A.M. Glauert, ed. (Elsevier/North-Holland, Amsterdam).

Winkelmann, H. and H.W. Meyer (1968), A routine freeze-etching technique of high effectivity by simple technical means. I. The principle, Exp. Pathol. *2*, 277.

Winkelmann, H. and S. Wammetsberger (1968), Eine mit einfachen Mitteln durchführbare Routinegefrierätztechnik hoher Effektivität. II. Die technische Anordnung, Exp. Pathol. *3*, 113.

Zingsheim, H.P., R. Abermann and L. Bachmann (1970a), An ultra-shadowing unit for the freeze-etching technique, J. Phys. E: scient. Instrum. *31*, 39.

Zingsheim, H.P. and H. Plattner (1976), Electron microscopic methods in membrane biology, in: Methods in membrane biology, Vol. 7, E.D. Korn, ed. (Plenum Publishing Company, New York), p. 1.

Zingsheim, H.P., R. Abermann and L. Bachmann (1970b), Shadow casting and heat damage, Proc. 7th Int. Congr. Electron Microscopy, Grenoble *1*, 411.

Chapter 7

Freeze-substitution and low temperature embedding

7.1 Introduction

Freeze-substitution as an electron microscopical preparation technique is based on rapid freezing of specimens followed by substitution (solution) of the specimen ice by an organic solvent at temperatures which inhibit damaging recrystallisation of the frozen water in the system. Generally, the substituted specimens are subsequently infiltrated with resin monomers and, after polymerisation, sectioned. Alternatively, substituted specimens are dried by critical point-drying or freeze- or evaporation-drying methods (Chapter 5). As with all other low temperature techniques, it is the ultimate goal of freeze-substitution to be able to examine specimen structures and specimen constituents trapped by freezing. The basic technique was first described by Simpson in 1941 for preparing tissue for light microscopical studies and several modifications of the technique for light microscopy followed (see review by Feder and Sidman 1958).

Fernández-Morán (1957, 1959a,b) was the first to test freeze-substitution as a technique for electron microscopical studies. He primarily used helium II at about $-271°C$ (2.0 K) as a quenching fluid for pre-glycerinated tissue samples. The frozen specimens were substituted in alcohol, or in alcohol/acetone/ethyl chloride mixtures, at temperatures varying from $-130°C$ to $-80°C$ (143 to 193 K). Stains and/or fixative substances (e.g. platinum chloride, osmium tetroxide, gold chloride) were sometimes added to these substitution media. Substituted specimens were then infiltrated with mixtures of methacrylate at temperatures between -100 and $-75°C$ (173 and 198 K) and, finally, the polymerisation was initiated by UV light using benzoin as a catalyst at -80 to $-20°C$ (193–253 K) (Fernández-Morán 1960, 1961a,b).

461

Subsequently, Bullivant used Fernández-Morán's technique but modified the procedure by introducing methanol and ethanol as substitution media and also by quenching in liquid propane at $-175°C$ (98 K) since this gave much better cooling rates than helium II (§2.2.5). Bullivant also abandoned the low temperature polymerisation of the methacrylate embedding medium (Bullivant 1960, 1965). At the same time, others became involved in developing freeze-substitution procedures for electron microscopical studies, usually using acetone or ethanol/acetone mixtures for substitution (Van Harreveld and Crowell 1964; Van Harreveld et al. 1965; Rebhun 1972). An extensive review of these early developments is given by Pease (1973), who had himself been exploring the potential use of small glycols as substitution media for 'inert' dehydration at room temperature as well as for freeze-substitution (Pease 1966, 1967, 1968). As already stressed by Pease (1973), the reason why these early attempts (and in many respects this remains applicable to the present state of the art) did not culminate in a single methodology lies in the fact that three distinct, but intertwined, decisions are involved in the freeze-substitution technique:

(*i*)   the selection of the best method of specimen freezing, including pretreatment (although there remain difficulties, recent developments have produced good techniques for many systems, see Chapter 2);

(*ii*)   the question of which substitution liquid should be used, for which period of time, and at what temperature and, also, whether fixatives and/or staining components should be added. Considering all the potential applications of freeze-substitution (Fig. 7.1), it is likely that different processing schedules will be required for optimal results with different specimens;

(*iii*) the choice of the best resins and infiltration procedures to provide thin sections for ultrastructural, analytical and cytochemical studies. The recent development of the polar and non-polar methacrylate-based low temperature resins (§7.5) has led to considerable progress in this area.

The different electron microscopical procedures involving a freeze-substitution step are shown in Fig. 7.1. Following specimen pretreatment (§2.3) and freezing (§2.4) the specimen ice is substituted by an organic solvent (Table 7.1). As a basic requirement, the substitution temperature must be maintained below the recrystallisation temperature of the specimen ice. For most

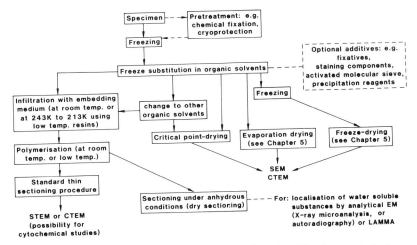

Fig. 7.1.   Sequences of electron microscopical procedures involving freeze-substitution.

biological specimens with an average water content, this critical temperature can be expected to lie within the range -100 to $-80°C$ (173 to 193 K) (§2.1.3) remembering that the recrystallisation rate is itself a temperature-dependent phenomenon. For many biological specimens much higher temperatures might be safe for a short period of time (Steinbrecht 1982). Several substitution liquids and dissolved additives (fixatives and staining and precipitating agents) have been used. As discussed below, the choice strongly depends on the scientific questions posed (e.g. ultrastructural or analytical electron microscopical studies). After completion of freeze-substitution, the substitution liquid may be exchanged against another solvent which is more suitable for the subsequent stage of the preparation procedure.

The procedures for subsequent handling of freeze-substituted specimens are outlined in Fig. 7.1. For straightforward ultrastructural studies, substituted specimens are infiltrated with embedding medium at room temperature or at -30 to $-60°C$ (243 to 213 K), and, after polymerisation of the resin, are thin-sectioned using standard ultramicrotome techniques (see Reid 1974). Alternatively, sectioning under anhydrous conditions (dry sectioning) (§7.4.2) can be used if water-soluble substances are to be localised by means of analytical electron microscopy. Freeze-substituted specimens can also be critical point-, evaporation- or freeze-dried directly for examination in the scanning or transmission electron microscope (Chapter 5).

TABLE 7.1

Liquids used in freeze-substitution[a]

Solvent (+ synonyms)	Formula	Melting Point	
		°C	K
Ethanol (alcohol, ethyl alcohol)	C_2H_5OH	− 117.3	155.7
Methanol (methyl alcohol)	CH_3OH	− 93.9	179.1
Acetone (2-propanone, dimethyl ketone)	CH_3COCH_3	− 95.4	177.6
Diethyl ether (ether, ethyl ether)	$CH_3CH_2OCH_2CH_3$	− 116.2	156.8
1-Propanol (n-propyl alcohol)	$CH_3CH_2CH_2OH$	− 126.5	146.5
1-Butanol	$CH_3CH_2CH_2CH_2OH$	− 89.5	183.5
1-Hexanol (n-hexyl alcohol)	$CH_3(CH_2)_5OH$	− 46.7	226.3
Hexane	$CH_3(CH_2)_4CH_3$	− 95.0	178.0
Propylene oxide (1,2 epoxy propane)	C_3H_6O	− 112.1	160.8
Tetrahydrofuran	C_4H_8O	− 108.5	164.5
Acrolein (Propenal, acrylaldehyde)	$CH_2:CHCHO$	− 87.0	186
Ethylene glycol[b]	$HOCH_2CH_2OH$	− 11.5	261.5

[a] Data taken from 'Handbook of Chemistry and Physics', 60th edition (1980).

[b] 70% Ethylene glycol + 30% water gives a eutectic mixture which freezes below − 50°C (223 K) (Pease 1967a; see also Barlow and Sleigh 1979, p. 88).

7.2 Substitution media

Only those liquids that can dissolve (substitute) the specimen ice at temperatures below the recrystallisation temperature can be used for freeze-substitution. To minimise the extraction or migration of soluble specimen constituents, it is desirable to have the specimen in contact with the substitution liquid for the shortest possible time at the lowest practical temperature. In addition, time becomes particularly important if freeze-substitution is to be a practical alternative to the standard chemical fixation/thin-sectioning procedure. As a more specific requirement, it must be possible to dissolve in the substitution liquid components which stabilise the overall specimen structure (fixatives) or help to 'stain' specifically, or otherwise localise (e.g. by precipitation – see §7.4.1a), certain specimen constituents.

To our knowledge, very few attempts have so far been made to determine the temperature dependence of the rate at which ice dissolves in different substitution media. Zalokar (1966) used a model system consisting of pieces of Millipore filter soaked in an aqueous solution of methylene blue. These pieces were frozen and then placed in various substitution liquids. The completion of the substitution process was determined from the decolouration of the filter. According to Zalokar's observations, acetone is unable to dissolve ice during several days at the temperature of solid CO_2 (dry ice) ($-78°C$; 195 K). Alcohols are much better substitution liquids, and methanol is preferable to ethanol. A mixture of methanol and acrolein is even more efficient. Based on these results, Zalokar used for his work three parts of absolute methanol containing 20% uranyl acetate and one part of acrolein. Acrolein has to be added to the pre-cooled methanol to avoid the rapid formation of acetal. Thus, Zalokar's results from his model system argue strongly against the use of acetone as a suitable substitution liquid. Nevertheless, excellent results have been obtained from its use (e.g. Steinbrecht 1980, 1982). The reason for this discrepancy and for similar controversial observations in the literature relating to other substitution fluids, can be explained as resulting from variations in the following parameters governing the substitution process:

(*i*)   the water content of the substitution liquid at the beginning, during, and at the end of the substitution process;

(*ii*)   the temperature and water content of the substitution liquid at which an equilibrium with the specimen ice is obtained (no substitution should occur below this temperature);

(*iii*)   the presence of a molecular sieve which constantly removes water from the substitution liquid (agitation of the substitution liquid to compensate for the reduced diffusion velocity of the water molecules in cooled (cold) substitution liquids may be of great importance); and

(*iv*)   the rate at which the specimen, immersed in the substitution liquid, is warmed up (as discussed below, the actual substitution may only, or mainly, have occurred during this final stage).

For methanol, acetone and diethyl ether, the most accurate data on substitution of ice as a function of temperature and water content have been obtained by Müller (personal communication), Humbel et al. (1983) and Humbel and Müller (1984) (Fig. 7.2a–c). The purest chemicals (Merck) were used. Acetone and methanol were dried with activated molecular sieve (Linde 0.3 nm; Merck) and diethyl ether containing a piece of sodium. Pieces of filter paper (approximately 3 mm in diameter) were soaked in 1.0 μl of tritiated water (1.0 mCi ml^{-1}), frozen in liquid nitrogen and substituted at -30, -60 and $-90°C$ (243, 213 and 183 K) in 1 ml of substitution liquid. The substitution curves shown in Fig. 7.2a–c were determined by analysing the release of tritiated water with time. In another experiment using frozen aqueous solutions of methylene blue, Müller confirmed the basic observation of Zalokar (1966) that methanol substitutes better than acrolein, which is better than acetone, and that diethyl ether is the worst (Müller et al. 1980; Müller, personal communication). Thus, the most important information to be derived from Fig. 7.2a is that methanol is the fastest substitution liquid at a temperature safely below any possibility of recrystallisation ($-90°C$; 183 K). Nevertheless, both acetone and diethyl ether also substitute ice (very slowly) at this low temperature provided that they are absolutely dry. Humbel et al. (1983) and Humbel and Müller (1984) also examined the substitution properties of methanol and acetone in relation to their water content for the temperatures -30, -60 and $-90°C$ (243, 213 and 183 K). They found that methanol (Fig. 7.2b) substitutes ice at $-90°C$ (183 K) even if it contains 10% water. Using the tritiated water model system, it was shown that a solution of methanol which contains 10% water is capable of substituting water completely within 3 h at $-90°C$ (183 K) (Fig. 7.2b). In comparison, acetone containing 1% water substitutes the same test sample only to 75% after 8 h at the same temperature (Fig. 7.2c). The reason why acetone is a poor substitution fluid at low temperatures is evident from Fig. 7.3, which shows that the amount of water in acetone in equilibrium with ice at $-85°C$ (188 K) is only about 2%. Thus, if this water is not constantly

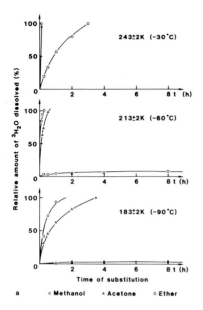

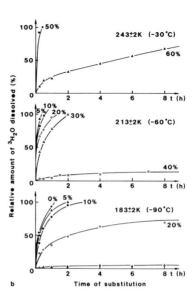

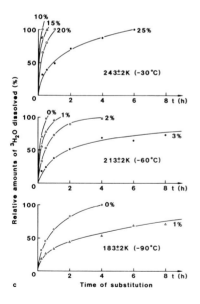

Fig. 7.2. Substitution of ice by methanol, acetone and ether as a function of temperature and water content of the substitution media. The data were obtained by analysing the release of tritiated water from the frozen specimen. (a) Substitution of ice as a function of organic solvent and substitution temperature. Methanol is the fastest substitution liquid at 'crystallisation-safe' temperatures. (b) Substitution of ice by methanol as a function of water content and temperature. Methanol substitutes ice at −90°C even if it contains 10% water. (c) Substitution of ice by acetone as a function of water content and temperature. In comparison to methanol (b), acetone is a much poorer substitution liquid. (Redrawn from Humbel and Müller 1984.)

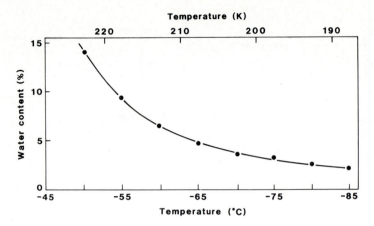

Fig. 7.3.   The water content (volume percent) of acetone in equilibrium with ice as a function
of temperature. (Redrawn from Van Harreveld et al. 1965.)

removed, or if in relation to the amount of specimen ice the volume of substitution liquid has not been properly calculated, the substitution liquid can
soon reach saturation level. It is frequently suggested that molecular sieve
be added to bind the water liberated from the frozen sample (Harvey et al.
1981b; Humbel and Müller 1984). Humbel and Müller (1984) have reported
that a 0.3 nm molecular sieve is very efficient in removing water from diethyl
ether saturated with tritiated water at 213 and 183 K (-60 and $-90°C$).
To our knowledge no systematic studies have yet been done which would
confirm how active sieves are in other substitution liquids at low substitution temperatures, but it can be expected that the addition of molecular sieves supports the substitution capacity of other organic solvents.

According to Ornberg and Reese (1981), the substitution rate also
depends on the ice crystal size within the specimen. These authors determined the substitution rate for acetone at $-83°C$ (190 K) by measuring the
efflux of tritiated water from ferritin samples that had been frozen in Freon
22 or on a helium-cooled metal surface (§2.4.2) (Heuser et al. 1979). As
shown in Fig. 7.4, the ferritin samples frozen on a helium-cooled surface
(which have small ice crystals that extend much deeper into the preparation
than Freon-frozen specimens) substitute much faster than the Freon-frozen
samples. However, the freeze-substitution rates measured by Mackenzie (see
Appendix to Ornberg and Reese 1981) using a light microscope system indicated that the substitution rate is independent of gross ice crystal diameter,

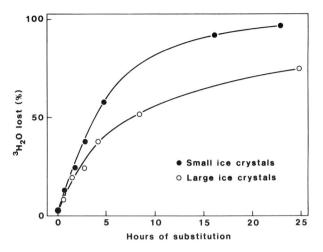

Fig. 7.4. Efflux of tritiated water from quick-frozen (helium-cooled metal block; small ice crystals) and quench-frozen (Freon 22; large ice crystals) ferritin solution at −83°C (190 K) in acetone. (Redrawn from Ornberg and Reese 1981.)

a fact which he explained by assuming that the rate-limiting step is the diffusion of water through the solvent and not the rate of dissolution of ice. To explain the difference shown in Fig. 7.4, he would suggest that the lower freezing rate gives more time during the cooling of any specific volume for recrystallisation, or grain growth, which is a 'sealing-off' process (the branching ice 'necks-off' and produces separate compartments), while faster freezing leaves a continuous skeleton through the spaces of which the substitution proceeds.

In combination with procedures for the localisation of water-soluble substances, other substitution media have also been used (e.g. *n*-hexane, propylene oxide and ethanol; see review by Harvey 1982). However, to our knowledge, no accurate data, such as those available for methanol, acetone and diethyl ether, have been published.

7.3   Freeze-substitution apparatus

Equipment for freeze-substitution must satisfy the following requirements:

(*i*)   the specimen must be kept immersed in the substitution medium at a defined, controlled temperature (between 193 and 178 K; −80 and −95°C) for up to several days;

(*ii*)   during the final stages of substitution it should be possible to control the rate of temperature rise (up to about ambient temperature) and also to maintain certain temperature levels constant over defined periods of time;

(*iii*) the substitution medium must be protected from atmospheric moisture; and

(*iv*)   it should be possible to manipulate both small (e.g. isolated cells) and tissue specimens with relative ease.

Either compressor or liquid nitrogen cooling systems can meet the temperature requirements in (*i*) and (*ii*).

The simplest type of freeze-substitution equipment, which at the same time meets the most important parameters mentioned above, was designed by Sitte (personal communication). As illustrated in Fig. 7.5, a small plastic container fits into a foam-insulated metal container. After the container has been filled about three-quarters full with the substitution medium, the whole assembly is cooled in liquid nitrogen. The frozen specimen is then placed on the surface of the frozen substitution medium. If freezing took place on a cold block (e.g. a metal mirror; §2.4.2), the best frozen part should be placed face-downwards. After closing the container, the metal container is covered with the insulated lid. The insulation characteristics of the foam de-

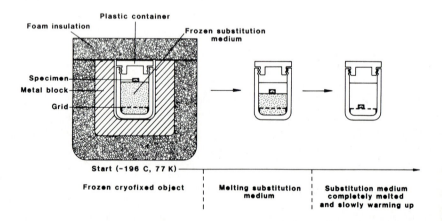

Fig. 7.5.   Simple freeze-substitution method using a foam-insulated pre-cooled metal block (cooled by liquid nitrogen). The frozen specimen is placed on the surface of the frozen substitution medium in the plastic container. The whole assembly gradually warms up allowing the specimen to sink slowly into the molten substitution liquid. (Redrawn after Sitte 1981.)

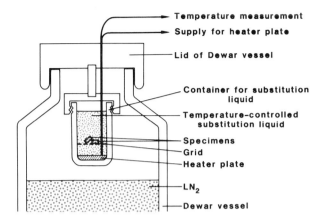

Fig. 7.6. Simple freeze-substitution in a Dewar vessel over liquid nitrogen (cooled by liquid nitrogen vapour). The substitution vessel is connected to the lid of the liquid nitrogen container. A temperature sensor-controlled heating plate ensures that the substitution liquid is maintained at the desired temperature. (Redrawn after Sitte 1981.)

termines the rate at which the whole assembly gradually warms up, thus allowing the specimen to slowly sink into the molten substitution liquid. With this assembly, it is not possible to maintain the specimen at defined temperatures during the warm-up period unless the container is placed in a refrigerator kept at the desired temperature. It can be anticipated that, for many systems, the well-frozen border zones are substituted in a very gentle way. As the substitution temperature increases, additives, such as fixatives and/or staining components, interact with the specimen.

Another simple freeze-substitution apparatus proposed by Edelmann (see Sitte 1981), illustrated in Fig. 7.6, consists of a substitution vessel which can be connected to the lid of a liquid nitrogen container. The substitution vessel is not in contact with the liquid nitrogen surface but is constantly cooled by the liberated nitrogen vapour. A temperature sensor ensures that the heating cartridge at the bottom of the vessel maintains the substitution liquid at the desired temperature. Sitte has stated (personal communication) that the liquid nitrogen consumption of this freeze-substitution unit is not much higher than the evaporation loss of the standard vessel. Both of the freeze-substitution units shown in Fig. 7.5 and 7.6 can easily be made using normal workshop facilities.

For some freeze-substitution procedures, a dry ice/acetone bath with a temperature of about $-78°C$ (195 K) may be adequate, particularly if the

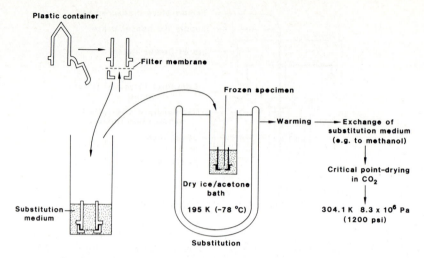

Fig. 7.7.   Simple freeze-substitution procedure in a dry ice/acetone bath for both single cells and small pieces of tissue. Frozen specimens are placed in modified small centrifuge tubes which are placed in small glass tubes containing the pre-cooled substitution medium. (Redrawn after Barlow and Sleigh 1979.)

substituted specimen is to be further processed (e.g. by critical point-drying) for examination in a scanning electron microscope. The flow diagram of a preparative procedure recommended by Barlow and Sleigh (1979) for both single cells and small pieces of tissue is illustrated in Fig. 7.7. The substitution is carried out in a 9 mm (outer diameter) plastic tube (e.g. an embedding capsule, as suggested by Newell and Roath (1975), or a plastic centrifuge tube) with the conical end cut off and the other opening covered with a 20 μm nylon mesh filter. The modified containers are placed in narrow, glass specimen tubes containing 3.0 ml substitution medium cooled by immersion into a dry ice/acetone bath. The frozen specimens are then transferred to the substitution liquid. Freeze-substitution and any subsequent fluid exchange is carried out by draining and filling the tube outside the container. Since the tube containing the substitution liquid is not closed, CO_2 will dissolve in the substitution fluid. If, in the course of preparation, the substitution liquid is warmed too rapidly to room temperature, the specimen may be damaged by liberation of CO_2 gas bubbles. Although not suggested by the authors, this problem could easily be overcome by using a closed container instead of a glass specimen tube. The technique was basically designed for freeze-substitution followed by critical point-drying for analysing delicate structures (ciliated surfaces) in the scanning electron microscope (§7.4.1b).

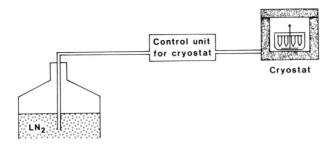

a

Fig. 7.8. Freeze-substitution apparatus developed by Müller (Müller et al. 1980). (a) Overall view of the apparatus. (b) Rack used to support the substitution pots at different temperatures. (Illustration compiled from photographs kindly provided by M. Müller and redrawn after Müller et al. 1980.)

If, during the course of freeze-substitution, defined temperatures are to be maintained according to specific time schedules, an automatically controlled apparatus is most desirable. One such apparatus, which automatically controls substitution time and temperature, has been developed by Müller (Müller et al. 1980). As illustrated in the overall view shown in Fig. 7.8a, the complete device consists of a liquid nitrogen Dewar vessel, the cryostat unit and a control unit. The solid state control unit is designed to keep constant three presettable temperatures (between 0 and $-200°C$; 273 and 73 K) for three presettable times (between 1.0 min and 100 days). The temperature is measured using a 100 ohm platinum resistance element and is regulated within ± 1.0 K by balancing the supply of liquid nitrogen from a Dewar vessel with heat from an electric element. The actual temperature

of the substitution liquid, the particular temperature level (i.e. within a consecutive series of different temperatures) and the time elapsed since the start
of this temperature level are all displayed on the control unit. The cryostat
itself consists of an insulated container into which a rack (Fig. 7.8b) supporting the substitution pots fits. The plastic centrifuge pots will hold 3.0
ml substitution medium and are covered with a Teflon lid during the substitution process. For freeze-substitution, the frozen samples are transferred
from liquid nitrogen (storage after freezing) into the Teflon pots which each
contain 1.5 ml pre-cooled (e.g. −90°C; 183 K) substitution liquid.

The substitution procedure for morphological studies developed by Müller and his co-workers (Müller et al. 1980; Humbel et al. 1983) using this
apparatus is described in detail below. It is anticipated that this type of
apparatus will ensure that freeze-substitution becomes a reproducible, and
consequently routine, procedure for morphological studies of cryofixed biological material. At the moment there is little commercial apparatus available for freeze-substitution, a fact which has probably contributed to the
relatively infrequent use of this technique. Now that the criteria for optimal
freeze-substitution are better understood, we believe that more commercial
equipment will be introduced and that it will, with the possible exception
of using closed circuit cooling systems instead of liquid nitrogen, be similar
in design to the apparatus described by Müller et al. (1980).

Freeze-substitution can also be done directly in a low temperature cabinet
(deep freeze) with the capacity for reaching temperatures down to −90°C
(183 K). Such equipment has the additional advantage that the substituted
specimens can subsequently be very easily processed for low temperature
embedding (see §7.5) including polymerisation at low temperatures, with
UV light.

7.4   Freeze-substitution procedures

7.4.1   Ultrastructural studies

For morphological studies, different substitution media and additives may
have to be used to obtain the optimum preservation of fine structure and
cytological details for either transmission or scanning electron microscopy.
For many systems, results obtained from freeze-fracture replication of
unfixed, non-cryoprotected specimens (see Chapter 6) can serve as valuable
references and comparisons (e.g. Steinbrecht 1976; Browning and Gunning
1979; Hunziker et al. 1984).

7.4.1a   Freeze-substitution for thin sectioning

A variety of schedules has been suggested for freeze-substitution of biological specimens prior to plastic embedding and thin sectioning (see, for example, reviews of Rebhun 1972; Pease 1973; Harvey 1982). The most common criteria used to assess the quality of freeze-substitution results are:

(*i*)   a dense appearance of the cytoplasm, indicating that little extraction of specimen components has occurred;

(*ii*)   that the volume of different cytoplasmic compartments remains unchanged;

(*iii*)   that cell membranes are sharply distinguished and continuous; and

(*iv*)   that there is no 'healing' of the ice crystal damage caused by specimen freezing (Dempsey and Bullivant 1976; Steinbrecht 1980).

Fixation

One of the major problems in ultrastructural work is to stabilise (fix) the cellular (specimen) components after the ice has been removed, i.e. to reduce the extraction of low and high molecular weight cell constituents to a minimum. While gross structures may frequently remain well preserved, organic solvents remove considerable amounts of lipid and affect protein conformation (Glauert 1974; Hayat 1981). Thus methanol is a preferable substitution fluid to acetone because it is a less effective lipid solvent (Glauert 1974). Consequently, for morphological studies it becomes obligatory to add both lipid and protein fixatives to the substitution medium. Aldehydes have been shown to be the most suitable protein fixatives for standard chemical fixation (Glauert 1974) and so acrolein or glutaraldehyde are often added to the substitution media, the former in quite high concentrations (e.g. 30% v/v; Zalokar 1966). In addition to aldehydes, heavy metal salts (such as uranyl acetate or osmium tetroxide) can be added to stabilise and stain lipids and other specimen components.

In order to optimise freeze-substitution schedules it is important to know the temperature-dependent rate of interaction of fixatives with specimen constituents. The dissolved fixatives and staining media penetrate the specimen along with the substitution front. Nevertheless, they cannot be expected to be chemically active at the commonly used ('recrystallisation-safe') substitution temperatures ($\leqslant 80°C$; 193 K). For this reason, most substitution schedules include a stepwise or gradual temperature increase after the initial, low temperature, substitution period. So far, few systematic studies have been carried out to determine the 'activity' of fixatives at different temperatures. Müller and his colleagues (personal communication and Humbel et al. 1983) have investigated the properties of 3% glutaraldehyde in methanol. The fixating capacity of this common cross-linking reagent was determined by injecting 10.0 μl of a 2% bovine serum albumin solution (BSA; Merck) in a 3% solution of glutaraldehyde in methanol at $-90°C$ (183 K). The fixative was then warmed up to the temperatures indicated in Fig. 7.9 and maintained there. The reaction of the glutaraldehyde was stopped by removing the substitution medium and subsequently washing in pure methanol at the temperature of the experiment. The cross-linked protein was pelleted in a Beckman Microfuge B (which can be operated at temperatures down to $-90°C$; 183 K). The specimen was then warmed to room temperature and washed with double-distilled water to ensure that all soluble protein was removed. The cross-linked protein was dissolved in 1 N NaOH and the protein content determined by the Lowry technique. Irres-

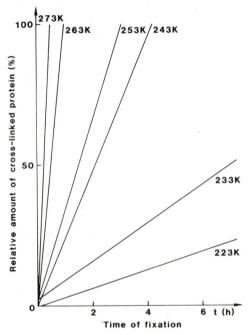

Fig. 7.9.   Fixation properties of glutaraldehyde in methanol at different substitution temperatures. The fixation capacity was determined by injecting a 2% bovine serum albumin solution into a 3% solution of glutaraldehyde in methanol maintained at $-90°C$ (183 K). The fixation solution was then warmed up to the indicated temperatures. (Redrawn from Humbel et al. 1983.)

pective of the fact that the number of cross-linking reactions to prevent structurally observable extraction of protein from an 'average' biological specimen during freeze-substitution remains to be established, Fig. 7.9 clearly shows that there are characteristic temperature-dependent fixation properties for glutaraldehyde in methanol and that, even at $-50°C$ (223 K), about 50% of the protein can be expected to be cross-linked within 24 h. No similar systematic studies have been carried out on OsO_4 as a fixative, but blackening of tissue does not occur below $-25°C$ (248 K) (Van Harreveld et al. 1965) and significant reactions with unsaturated lipids can only be expected above $-30°C$ (243 K) (Müller et al. 1980 and Müller personal communication). However, both uranyl ions and OsO_4 can be expected to react with lipids to some extent even at 183 K ($-90°C$) (Humbel et al. 1983).

Based on his systematic studies both on different substitution media (Fig. 7.2) and on different additives (Fig. 7.9), Müller and his co-workers (Müller et al. 1980, Humbel et al. 1983) concluded that 1% OsO_4, 0.5% uranyl acetate and 3% glutaraldehyde in methanol (containing 3% water) provide the best freeze-substitution 'cocktail' for routine ultrastructural studies on a range of biological specimens. It must be remembered that, as opposed to acetone, methanol substitution occurs at 183 K ($-90°C$) even in the presence of 10% water (Fig. 7.2b). Thus, glutaraldehyde (available as 50–70% aqueous solutions) can be added to the substitution medium. (**Note**: high concentrations of glutaraldehyde have an increased tendency to polymerise during storage.)

Müller's freeze-substitution schedule

(1)   The frozen sample is transferred from liquid nitrogen (storage) into pots containing the substitution liquid (methanol containing fixatives) pre-cooled to 183 K ($-90°C$). If the specimens have been frozen by the sandwich (propane jet) technique (§2.4.1b), the two holders are separated in the substitution liquid or in liquid nitrogen using pre-cooled forceps and the pots are subsequently slightly shaken. 1.5 ml substitution fluid and a substitution time of about 8.0 h is sufficient for a few tens of micrometres thickness of sandwich-frozen specimen. Variations in the substitution time depend on the dimensions of the well-frozen specimen area. Unless high pressure freezing has been used (§2.4.1d), there is no sense in substituting deeper, but poorly frozen, specimen areas at a 'recrystallisation-safe' temperature.

(2)   The temperature is subsequently raised to 213 K ($-60°C$) and maintained here for 8.0 h.

(3)   After a further temperature rise to 243 K ($-30°C$) for a period of 8.0 h, the specimens are removed from the methanol and washed twice in cold (243 K; $-30°C$) acetone. Suspensions are pelleted using a small centrifuge (e.g. Beckman Microfuge B; Appendix 2) operating in a deep freeze at 243 K ($-30°C$). (The Beckman Microfuge B can be operated at temperatures down to 183 K; $-90°C$).

(4)   The substituted pellets or tissue samples are further processed for thin sectioning using standard, or low temperature (§7.5), embedding techniques.

Some typical results obtained with the Müller freeze-substitution technique are illustrated in Figs. 7.10 and 7.11. The *Euglena gracilis* cells (Fig. 7.10) were propane jet-frozen and the frozen suspension was then fractured under liquid nitrogen and subsequently freeze-substituted and embedded in Araldite/Epon at 4°C (277 K) for thin sectioning. Fig. 7.11 shows part of a high pressure-frozen mouse growth cartilage substituted and embedded in the same manner as the *Euglena* cell. Müller (personal communication and Müller 1981) has also found that pure phospholipid structures remain well preserved even if freeze-substituted in methanol without any dissolved fixative (stabilizing additives). Propane jet-frozen lecithin liposomes (Fig. 7.12) were freeze-substituted in methanol for 8 h at temperatures of -90, -60 and $-30°C$ (183, 213 and 243 K) and embedded in Lowicryl HM20 (§7.5). The image contrast arises from post-staining the cut sections with lead citrate and uranyl acetate.

It has also been reported that freeze-substitution in pure methanol, followed by low temperature embedding (§7.5), can give good structural preservation of muscle tissue (Humbel et al. 1983). These authors found that, if freeze-substitution is followed by conventional embedding and heat polymerisation (e.g. in Epon/Araldite), a fixative must be added to the substitution fluid, while, if followed by low temperature embedding techniques, no additional fixative seems to be necessary as long as the temperature is kept below the collapse temperatures of the specimen structures. These temperatures, which are different for different solvents and for different cellular components (MacKenzie 1972), are within the range 213 to 243 K (-60 to $-30°C$). Because methanol is more polar than acetone the collapse temperature in methanol is lower than in acetone. The collapse temperature in polar Lowicryl K4M (§7.5) is probably between that of acetone and methanol.

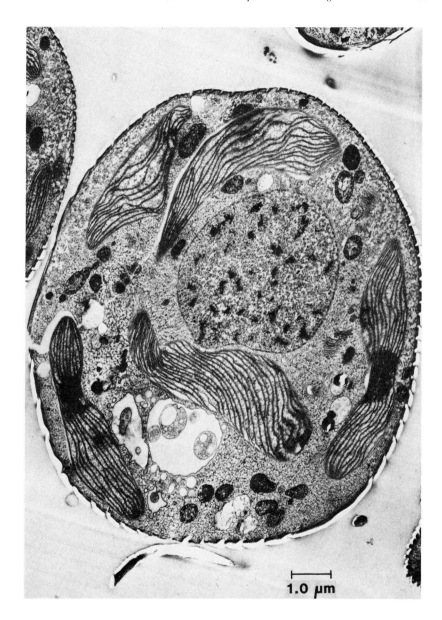

Fig. 7.10 Freeze-substituted *Euglena gracilis* cell. A suspension was propane jet-frozen, fractured under liquid nitrogen and then freeze-substituted and embedded in Araldite/Epon at 277 K (4°C). (Unpublished micrograph kindly provided by M. Müller.)

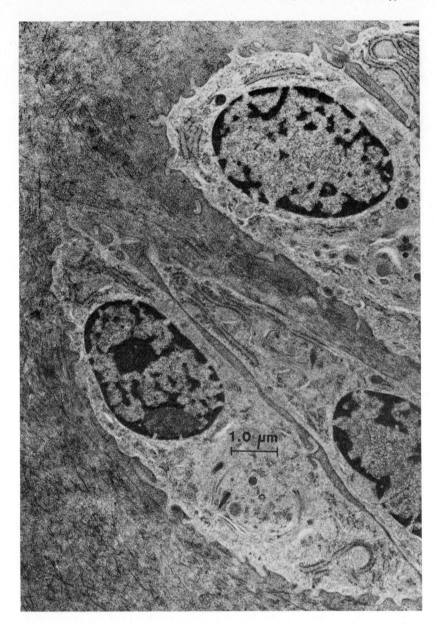

Fig. 7.11.   Freeze-substituted mouse growth cartilage. The tissue was high-pressure frozen and then freeze-substituted and embedded in Araldite/Epon at 277 K (4°C). (Unpublished micrograph kindly provided by M. Müller. For details, see Hunziker et al. 1984.)

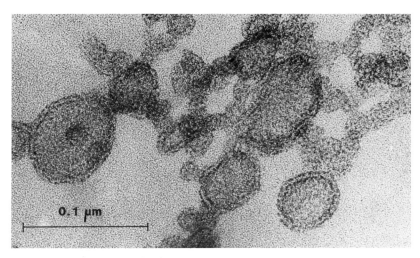

Fig. 7.12.   Propane jet-frozen lecithin liposomes freeze-substituted in methanol and embedded in Lowicryl HM20 at 243 K ($-30\,^\circ$C). (Unpublished micrograph kindly provided by M. Müller.)

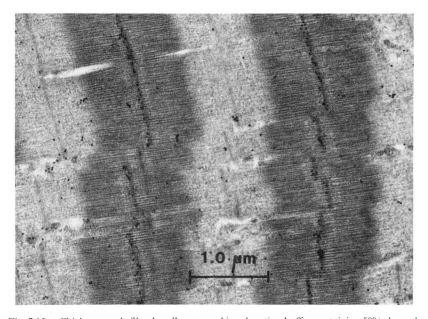

Fig. 7.13.   Chicken muscle fibre bundle prepared in relaxation buffer containing 50% glycerol, frozen in liquid propane and freeze-substituted in methanol before embedding at 243 K ($-30\,^\circ$C) in Lowicryl K4M. The M-line protein was demonstrated by immunolabelling using the protein A-gold complex method. (Reproduced, with permission, from Humbel et al. 1983.)

It was found experimentally that biological material freeze-substituted in the absence of any fixatives is more strongly influenced by the polar K4M than by the apolar HM20 (Humbel et al. 1983). Obtaining thin sections of completely unfixed biological specimens with good preservation of structure and biological activity is of particular importance for immunolabelling techniques. An example (from Humbel et al. 1983) is illustrated in Fig. 7.13 which shows the fine structure of a chicken muscle fibre bundle prepared in relaxation buffer containing 50% glycerol. The specimen was frozen in liquid propane and freeze-substituted in pure methanol before being embedded in K4M at 243 K ($-30°C$). The M-line protein was demonstrated by immunolabelling techniques (using a protein A-gold complex method; Roth et al. 1978).

Van Harreveld's freeze-substitution method

Using silk moth antennae as a model system and freeze-fracture replication to assess quality, Steinbrecht (1980) obtained good structural preservation with freeze-substitution by the method of van Harreveld et al. (1965), which is as follows:

(1)   The frozen sample is transferred from liquid nitrogen (storage) to acetone containing 2% (w/v) OsO_4 maintained at 193 K ($-80°C$) in the presence of molecular sieve (Linde, Type 4A; Appendix 2) in closed vials for three to ten days. With silk moth antennae (diameter < 100 μm) as a test specimen, no significant structural differences were detectable between the longest and shortest durations of substitution.

(2)   The specimens are warmed up to ambient temperature over a period of 4 h (during which the OsO_4 acts as a fixative).

(3)   After two changes of acetone and propylene oxide, specimens are embedded in Epon 812 or another resin at 293 K (20°C).

By comparing freeze-substitution with standard chemical fixation or freeze-etching results, it was shown that, if antennae were fixed in OsO_4 (Zetterqvist buffer, pH 7.4; 300 mOsm) prior to being rapidly frozen and freeze-substituted, the images no longer resembled cryofixed specimens but were almost identical to specimens that were chemically fixed and then conventionally dehydrated.

Using the same model system and substitution medium, Steinbrecht

(1982) also investigated primary freezing damage caused during cooling the specimen and secondary freezing damage caused by recrystallisation when the specimen was warmed up in the substitution medium. The relevant results can be summarised as follows:

(*i*)   The duration of substitution at 194 K ($-79°$C) could be reduced from seven days to 5 min (the shortest time tested) without generating any recrystallisation artefacts provided that the rate of warming up was low (half-time $= 3.0$ h).

(*ii*) Severe recrystallisation occurred when the specimen was substituted for 5.0 min at 194 K ($-79°$C) and subsequently warmed up quickly (half-time $= 30$ s).

(*iii*) No detectable recrystallisation artifacts were seen when the specimen was substituted at 194 K ($-79°$C) for seven days and subsequently warmed up rapidly (half-time $= 30$ s).

These data confirm that acetone dissolves water (ice) slowly at 194 K ($-79°$C) (Fig. 7.2). Unless long substitution times are used at such a temperature, most of the ice will dissolve during the final warming up period. Thus recrystallisation appears to be a minor danger during freeze-substitution provided that the specimens are warmed slowly from low temperatures (allowing the ice to dissolve before any recrystallisation can occur). Consequently, the cold metal block freeze-substitution device (Fig. 7.5; §7.3), which warms up slowly after loading at low temperature, can be expected to give good results. Steinbrecht (1982) also showed that, if water was added to the substitution acetone (5% v/v, without molecular sieve) while otherwise following the standard procedures described above, significant freezing damage occurred in all parts of the tissue. As shown in Fig. 7.3 (§7.2), at 194 K ($-79°$C), acetone containing 5% water is in equilibrium with ice, and so the actual substitution of specimen ice must have occurred at much higher temperatures during the slow warming up period, allowing recrystallisation before substitution. As discussed in §2.1.1 and §2.1.2, great differences in the recrystallisation temperature of different specimens, or even of different compartments within an individual cell, can be expected and a generalisation from Steinbrecht's findings is not recommended. Nevertheless, as Steinbrecht points out, his results are in agreement with the data of MacKenzie (1977) who measured glass transitions and other second order phase behaviour in model systems and a variety of animal tissues and observed no significant recrystallisation below the 'ante-melting' transition at about 223 K ($-50°$C). Again, Steinbrecht (1982) emphasises that, with tissue in an organic solvent, the ante-melting transition may occur at slightly lower temperatures (MacKenzie, personal communication to Steinbrecht, see Steinbrecht 1982).

Summarising the state of the art of freeze-substitution for thin sectioning, it appears that, for most specimens, freeze-substitution in combination with low temperature embedding (§7.5) will be the method of choice. Further studies to optimise methods have the ultimate goal of finding conditions where the only water removed is that which is not necessary for the maintenance of structural and functional integrity. Thus, it is particularly important to learn more about the behaviour of different substitution media with different water contents at different temperatures (Humbel et al. 1983).

7.4.1b   Freeze-substitution for scanning electron microscopy

As illustrated in Fig. 7.1 (§7.1), freeze-substitution can be used for scanning electron microscopy (SEM) as an initial preparation step in combination with critical point-, evaporation- or freeze-drying. As discussed in Chapter 5, once the water has been replaced by an organic solvent, air-drying in a stream of dry air or inert gas (e.g. N_2) is the *simplest* way of removing the solvent. However, severe distortion (e.g. flattening) can be expected with this technique even if the water has been replaced with solvents with a lower surface tension (Revel and Wolken 1973; Albrecht et al. 1976). Alternatively, the substituted specimen can be frozen and subsequently freeze-dried. Nevertheless, the specimen surface structure is usually the main interest in SEM studies and, with properly frozen, fully hydrated specimens, there is little point in not performing the freeze-drying procedure directly. For the SEM, the most common procedure for removing the substitution liquid is critical point-drying (Hayat and Zirkin 1973). If suitable substitution liquids are used it is possible to transfer the specimen directly into the critical point-drying apparatus. This can be done at low temperature and most of the CO_2/intermediate fluid exchanges can be made at low temperature before warming the CO_2 to make the transition through the critical point (Robards 1978).

It has been shown that the substitution mixtures used for TEM, which contain low concentrations of fixatives in pure solvents, are not necessarily suitable for SEM. Particularly for the preservation of delicate surface structures, special substitution mixtures may have to be developed to avoid shrinking and distortion during subsequent critical point-drying. After testing a variety of procedures, Barlow and Sleigh (1979) suggested the following medium for preservation of the metachronal wave pattern of ciliated surfaces of protozoa:

Methanol for the substitution medium is prepared by mixing, at room temperature, one volume of methanol saturated with $CaCl_2$ with one volume of methanol saturated with $HgCl_2$. To 50 parts of this mixture are added 40 parts of pure ethylene glycol and 10 parts of acrolein. The resultant mixture has a freezing point below 193 K ($-80°C$) and, although viscous, is suitable for substitution temperatures as low as 197 K ($-76°C$). The content of the two chlorides in the final mixture is approximately 8% $HgCl_2$ and 3% $CaCl_2$; an increase in concentration causes precipitation at 197 K ($-76°C$). Ethylene glycol is added to the methanol to reduce the extraction of phospholipids from the membrane since, with pure solvents, it was observed that there was insufficient 'stiffness' of structures to resist distortion during critical point-drying. The inclusion of acrolein, $HgCl_2$ and $CaCl_2$ as 'fixatives' proved to be essential to stabilise the delicate ciliated surfaces. In comparison, osmium-fixed material always gave unsatisfactory results because distortion due to shrinkage always occurred during critical point-drying. The end of the substitution process at dry ice temperature can be judged by observing the disappearance of the ice drop. The specimens are then slowly warmed up to room temperature. With soft tissue, a warming up period between 203 and 293 K (-70 and $20°C$) of at least 6 h is recommended and warming up overnight is convenient. After reaching ambient temperature, the substitution medium is gradually exchanged with pure acetone or alcohol. Critical point-drying from the pure acetone or alcohol is then carried out, using CO_2 as the transition fluid.

7.4.2   Localisation of water-soluble substances

In addition to its uses for structural studies, freeze-substitution can also serve as a method for preparing biological specimens for the localisation of water-soluble substances. Thin sections of freeze-substituted, embedded specimens can be evaluated using analytical electron microscopy, autoradiography or LAMMA (Fig. 7.1) (§7.1). As with freeze-dried cryosections, ideal standards for judging the results of freeze-substitution are frozen-hydrated sections (§3.4.1) from the same material. A basic requirement for the success of analytical studies is to minimise the movement and dissolution of specimen components during substitution and the subsequent embedding process. Substitution in the presence of precipitating reagents may provide a means for detecting water-soluble substances by conventional electron microscopy. Unfortunately substitution 'cocktails' which minimise the movement of soluble specimen components often give poor structural preservation. An important difference in sectioning material to be used for ultrastructural studies and for localisation of water-soluble components is that, while the former may be sectioned onto a liquid bath, the latter must be dry-sectioned (even excluding non-polar flotation media) to avoid the loss of soluble materials.

Typical examples of the application of freeze-substitution to the localisation of water-soluble substances in multicellular plants are: Fisher and Housley (1972), Van Steveninck and Van Steveninck (1978), Harvey (1980), Hall et al. (1978), Browning and Gunning (1979) and Harvey and Kent (1981).

In comparison with studies on plant material, relatively little freeze-substitution work has been carried out on animal tissues (e.g. Chandler 1980; Marshall 1980; Ornberg and Reese 1980, 1981).

A comprehensive review of freeze-substitution methods for the localisation of water-soluble substances is provided by Harvey (1982). For relevant discussions of related problems, see also the book on *Microprobe Analysis of Biological Systems* edited by Hutchinson and Somlyo (1981).

7.5   Low temperature embedding techniques

Implicit in almost everything that has been said earlier in this book is the belief that maintenance of low temperatures during the processing of well-frozen specimens is a valuable aid to the preservation of ultrastructure and retention of soluble components close to their *in vivo* positions. Because it is often necessary to cut thin sections of biological specimens for observation in an electron microscope the material must be infiltrated with some form of matrix that can be hardened. This is usually a resin (most commonly an epoxy resin but sometimes an acrylic or polyester). Such resins can themselves be effective lipid solvents, the more so as they are often polymerised at relatively high temperatures. The problems are compounded because the resin viscosity is usually quite high and becomes worse as the temperature falls, so that infiltration can be difficult. The production of a resin that can be infiltrated into biological specimens at sub-zero temperatures and then polymerised, also at low temperature, is clearly an advance of some importance. The Lowicryl resins were developed in Switzerland by

Carlemalm and colleagues (Carlemalm et al. 1982; Armbruster et al. 1982; Carlemalm 1984) for this purpose and are commercially available (see Appendix 2). (At the time of writing, an intermittent newsletter, *Lowicryl letters*, is produced by Chemische Werke Lowi [see Appendix 2] which provides up-to-date information on progress with the use of these resins.) Other low temperature resins, such as London L.R. Gold, are also becoming commercially available (Appendix 2).

7.5.1 Low temperature resins

Two resins are available: Lowicryl K4M and Lowicryl HM20 (subsequently referred to as simply K4M and HM20). These are highly cross-linked acrylate- and methacrylate-based embedding media, specially formulated to provide low viscosity at low temperatures. K4M is a polar (hydrophilic) resin usable down to $-35°C$ (238 K), while HM20 is non-polar (hydrophobic) and can be used as low as $-70°C$ (203 K). Both resins can be polymerised by long wavelength (360 nm) UV light at low temperatures or can be chemically polymerised at $60°C$ (333 K).

The hydrophilic properties of K4M provide some advantages; in particular the specimen can be kept in a partially hydrated state during dehydration and infiltration since K4M can be polymerised with up to 5% (w/w) water in the block. Furthermore, K4M is particularly useful for labelling sections using specific antisera or lectins, since it gives better structural preservation, improved retention of antigenicity and a significantly lower background labelling than HM20.

There are essentially two different processing routes for embedding in these resins: via normal (ambient temperature) fixation with subsequent, gradual, lowering of the temperature during the dehydration stages, and using freeze-substitution so that the frozen specimen is never warmed until it has been fully infiltrated with resin and polymerised at low temperature. The first pathway is the most frequently used although, in principle, the second has much to recommend it.

7.5.2 Fixation and dehydration

Any normal fixation schedule can be used although fixatives which also contribute to contrast (e.g. osmium tetroxide) should be avoided because they interfere with photopolymerisation (by blocking penetration of the UV light) and may also (e.g. OsO_4) react with unsaturated double bonds in the

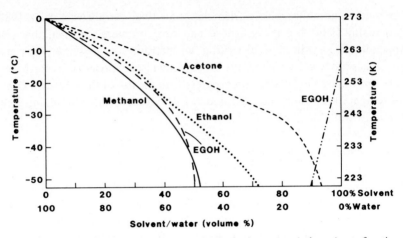

Fig. 7.14. Freezing points of commonly used dehydrating agents (solvents) as a function of concentration. Note the *rise* in the freezing point of ethylene glycol (**EGOH**) at higher temperatures. (Redrawn from Carlemalm, unpublished. Available in details supplied with Lowicryl resins; see Appendix 2.)

resin. Naturally occurring pigments, present in reasonable amounts, do not normally interfere with polymerisation although, if specimens are heavily pigmented with strong absorbers at 360 nm, polymerisation may be less than optimal. It will be clear that photopolymerisation necessitates access of the UV light throughout the specimen block to ensure even polymerisation. Blocks should thus be small ($\leqslant 0.5$ mm³).

If it is decided to pass the specimen from a fixative at ambient (or above 0°C; 273 K) temperature into the low temperature resin, then this involves a stepwise reduction in temperature concomitant with increasing concentration of dehydrating agent. A temperature is selected at each step which is just above the freezing point for the concentration used in the previous stage. This corresponds to the concentration of the dehydrating agent actually in the tissue block as it is introduced into the next higher concentration in the series. Freezing points in relation to concentrations of various dehydrating agents are shown in Fig. 7.14. It is important that the specimens are periodically agitated during dehydration and infiltration, either by gently stirring or by swirling the sample vials. Most polar or non-polar dehydrating agents can be used with either resin although, due to its hydrophobic nature, HM20 is immiscible with ethylene glycol or dimethylformamide. A typical dehydrating schedule, suitable for either K4M or HM20, is given in Table 7.2.

TABLE 7.2
Dehydration at low temperature

Ethanol (vol. % in H_2O)	Temperature (°C)	Time (min)
30	0	30
50	− 20	60
70	− 35 (− 50)[a]	60
95	− 35 (− 50 to − 70)	60
100	− 35 (− 50 to − 70)	60
100	− 35 (− 50 to − 70)	60

[a] Suggested steps for work at lower temperatures — only possible with HM20.

7.5.2a Maintaining low temperatures

When working at a number of different low temperatures, the problem of achieving and maintaining the relevant temperature(s) has often deterred potential users from applying low temperature methods (something that is not only a problem in low temperature embedding). In the present context the following methods for achieving low temperatures during dehydration, infiltration and embedding can be used:

(*i*)   Propriety units are available, such as the Balzers Low Temperature Embedding Apparatus (Appendix 2), which provides four sample-holding blocks, each of which can be independently controlled at temperatures between 0 and − 50°C (273 and 223 K). This unit also allows continuous agitation of the samples.

(*ii*) A temperature of − 20°C (253 K) is obtained by using a 3:1 (w/w) ice/ NaCl mixture. The temperature should be carefully monitored and the mixture periodically replenished to ensure that the temperature is maintained. As with all such low temperature baths, they will maintain their low temperature longer if they are placed inside a glass-lined Dewar vessel which is itself kept in a refrigerator or cold box.

(*iii*) For temperatures in the range down to about − 40°C (233 K) ordinary domestic deep freezers are suitable.

(*iv*) For temperatures below − 35 to − 40°C (238 to 233 K) either special low temperature cabinets (Appendix 2) or dry ice/*o*- and *m*-xylene baths can be used. Obviously, the low temperature cabinets provide greater convenience but are expensive. If dry ice/xylene baths are used, then a range of temperatures between about -26 and $-63°C$ (247 and 210 K) can be obtained depending on the proportions (v/v) of *o*- and *m*-xylene (Fig. 7.15). The xylene mixture appropriate to the temperature required is made up and then crushed dry ice is added to form a thick slurry. This bath will provide a constant temperature for 8–10 h if used in a Dewar vessel. (**Caution:** xylene fumes are toxic – only use in an effective fume hood).

As the samples used in low temperature embedding are almost inevitably small, they are likely to show sharp temperature changes during transfer unless they are suitably protected. For this reason, it is advisable to accommodate the sample vials (or other containers) in holes drilled in an aluminium block (aluminium is a convenient metal to use and has good thermal capacity; see Table 2.7 in §2.2.4). The block can be allowed to equilibrate at a particular temperature before the sample vials are placed in it and, during transfer, its high thermal capacity will prevent any steep temperature gradients within the specimen.

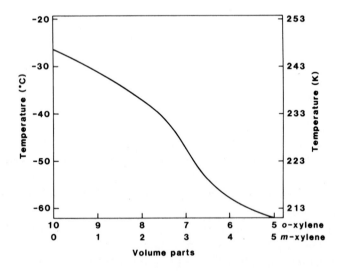

Fig. 7.15.   Temperatures of crushed dry ice/xylene slurries, as a function of the ratio of *o*- to *m*-xylene. (Redrawn from Carlemalm, unpublished. Available in details supplied with Low-icryl resins; see Appendix 2.)

7.5.3   Infiltration and polymerisation

7.5.3a   Preparation of resins and infiltration

Methacrylates, like many other embedding resins, may cause skin reactions in sensitive individuals. Therefore protective rubber gloves should be used. Furthermore, the resins should be mixed in a fume hood and the vapour should not be inhaled. As the resins are of very low viscosity they require little mixing. Indeed, excessive stirring results in the incorporation of too much oxygen into the mixture and this may interfere with polymerisation (especially if thermal polymerisation is to be used).

The monomer and cross-linker should be weighed out into a preweighed vial and then mixed *gently* for 3–5 min by one of the following methods (note that this represents the 'published' formulation; in practice, such accurate weighings are probably unnecessary and, in any case, the mixtures are more easily measured by volume):

(1)   Bubble a stream of dry nitrogen gas through the mixture via a Pasteur pipette (this both mixes the resin and prevents incorporation of oxygen).
(2)   Mix gently with a glass rod.
(3)   If the vial is sealable, rock *slowly* from side to side (avoiding the formation of air bubbles or foaming).

When the monomer and cross-linker have been mixed, the initiator is added and mixing is continued until it has been completely dissolved in the resin.

TABLE 7.3
Mixtures for ultra-violet polymerisation

K4M		HM20	
Component	Amount (g)	Component	Amount (g)
Cross-linker A	2.70	Cross-linker D	3.00
Monomer B	17.30	Monomer E	17.00
Initiator C[a]	0.10	Initiator C[a]	0.10

[a] For polymerisation between $-50°C$ and $0°C$ (223 and 273 K). Above $0°C$, the initiator should be replaced by the same amount of benzoin ethylether.

Mixtures for ultra-violet polymerisation
The mixtures indicated in Table 7.3 produce blocks of average hardness. The hardness can be varied by incorporating more or less cross-linker in the mixtures (more cross-linker gives more hardness). For HM20, the cross-linker concentration can be varied from 5 to 17% (w/w) (1.0 to 3.4 g/20 g resin). For K4M, the cross-linker concentration can be varied from 4 to 18% (w/w) (0.8 to 3.6 g/20 g resin).

Mixtures for thermal (chemical) polymerisation at +60°C
Although HM20 and K4M are primarily designed for use at low temperatures and UV polymerisation, it is also possible to polymerise them by the more conventional thermal method. If this is to be done, then the cross-linker and monomer are first mixed as described in the previous section (preferably using bubbling nitrogen) and then, instead of initiator C, 0.5% w/w (for HM20) or 0.3% w/w (for K4M) dibenzoyl peroxide is used.

Dibenzoyl peroxide is usually supplied as a paste with dibutyl-phthalate or as a powder moistened with water. The extra ingredients should be compensated for so that the resins receive the above amounts of the *peroxide* exclusive of additives.

Infiltration at low temperatures
Infiltration with K4M or HM20 at low temperature is essentially similar to infiltration with other resins at room temperature. The exact procedure will depend on the dehydrating agent used and the temperatures chosen in relation to the viscosities of the dehydrating agent and resin at those temperatures. The schedule in Table 7.4 illustrates a typical route, through ethanol, which is applicable to either K4M or HM20.

TABLE 7.4
Typical infiltration schedule

Resin: Ethanol (v/v)	Temperature (°C)	Time
1:1	−35[a]	60 min
2:1	−35	60 min
Pure resin	−35	60 min
Pure resin	−35	Overnight/4–16 h

[a] Lower temperatures (−50 to −70°C, 223 to 203 K) are possible for infiltration with HM20 and higher temperatures can be used for both resins.

7.5.3b   *Polymerisation*

Samples may be conveniently held for polymerisation in either BEEM-type or gelatin capsules. A suitable holder is required so that capsules receive UV radiation from all sides (e.g. Fig. 7.16a). The polymerisation chamber is made so that it will fit into a chest-type deep freeze or a cold room (Fig. 7.16b). The light source must provide long (360 nm) UV, preferably from two 15 W fluorescent tubes (e.g. Philips TLD 15W/05; see Appendix 2) similar to those used for thin-layer chromatography. To provide diffuse illumination, a right-angle deflector is suspended below the fluorescent tubes. All six inner surfaces, as well as the deflector, should be constructed from UV-reflective material or lined with aluminium foil. The capsules should be positioned about 30–40 cm below the UV source. The whole box should not be made too air-tight: ventilation from top and bottom allows air circulation and help to minimise temperature gradients.

Other UV sources can be used, including some hand-held lamps. However, they must emit at 360 nm and shorter wavelengths should be filtered out. Smaller sources should be positioned correspondingly nearer to the samples (10–15 cm) and the overall dimensions of the polymerisation box should also be reduced. If weak UV sources are used, then radiation from the bottom of the cabinet may help because the light will have less resin to

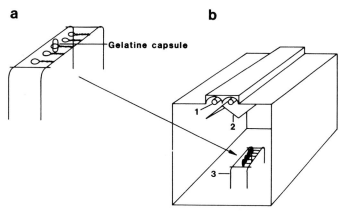

Fig. 7.16.   Apparatus for low temperature polymerisation. (a) A wire capsule holder for UV polymerisation. (b) A polymerisation chamber for indirect UV irradiation. The UV source (**1**) is diffused by a right-angle reflector made, for example, from aluminium foil (**2**). The capsule holder (**3**) is placed 30–40 cm below the UV source. (Redrawn from Carlemalm, unpublished. Available in details supplied with Lowicryl resins; see Appendix 2.)

traverse before polymerising the resin in the sample. Whatever the arrangements for polymerisation, trials without samples should first be attempted. If there is shrinkage and deformation of the blocks, then polymerisation has been too rapid and the distance between the lamps and the capsules should be increased.

Ultra-violet polymerisation at low temperatures

(1)   Fill capsules *to the top* (to minimise dead space filled with air) with fresh, pre-cooled resin.
(2)   Transfer samples to capsules using Pasteur pipettes; close the capsules and allow the temperature to equilibrate for 10–15 min.

(To minimise condensation of water and crystallisation of ice on the sample vials and capsules, all apparatus should be pre-cooled and steps (1) and (2) should be performed *in the cold*.)

(3)   Polymerise for at least 24 h under UV light at -30 to $-50°C$ (the lowest recommended temperature for polymerisation is $-50°C$).
(4)   Remove capsules from the cold and continue 'curing' under UV for 2–3 days at room temperature.

Chemical polymerisation at $+60°C$

(1)   Place fresh resin in gelatin capsules; transfer samples to capsules and fill capsules approximately 3/4 full.
(2)   Close capsules and polymerise at 60°C for 2–3 days.

The use of gelatin capsules is recommended for chemical polymerisation at $+60°C$ since the plasticiser in BEEM capsules may interfere with polymerisation at the edge of the block. The chemical polymerisation of HM20 and K4M with peroxides is an exothermic reaction. To prevent an uncontrolled rise in temperature, the capsules should be in contact with a heat sink such as an aluminium block with predrilled holes into which the capsules fit firmly.

7.5.4   Sectioning and staining

For best results with HM20 or K4M, the blocks should be trimmed to pyramidal faces on a microtome or trimming apparatus. The sides and face should be clean and transparent with the pyramid having an angle of about 28–30° from the cutting face. (See Reid 1974 for details of ultramicrotomy procedures.)

HM20 is a highly cross-linked methacrylate which, when of the correct hardness, sections easily with either glass or diamond knives.

K4M is a hydrophilic resin. Therefore, as with other polar (water-miscible) resins, precautions should be taken to ensure that the block face does not wet during sectioning. This is best accomplished by sectioning with the level of liquid in the trough slightly below normal. In these circumstances the reflection from the trough fluid along the knife edge is slightly darker than the normal bright silver colour. It is important that the fluid level is not reduced so much that the edge becomes dry. This is particularly important with diamond knives because their edges are normally hydrophobic. The best procedure with a diamond knife is to orientate the trimmed block with the knife edge *before* the trough is filled. The specimen arm of the microtome is placed in its lowest position and the trough is overfilled to form a 'reverse meniscus' along the knife edge. The knife is left in this position for 10–15 min. Immediately before sectioning, the level of the trough fluid is lowered to produce a dark silver reflection along the knife edge. The final advance of the knife block should then be made and sectioning commenced. Since K4M is a hydrophilic resin, the sections should be collected as soon as possible after they have been cut. Cutting speeds of 2–5 mm s^{-1} are recommended for both resins.

There are significant differences between the staining properties of the two resins but, as general guidelines, sections may be stained with either aqueous or alcoholic solutions of uranyl acetate, with Reynold's lead citrate, or with Millonig's lead acetate. Completely unstained sections of HM20 have been imaged in STEM instruments using so-called 'Z-contrast' (Carlemalm and Kellenberger 1982; see also Kellenberger 1984). A particularly useful staining combination consists of saturated aqueous uranyl acetate (for 35 min for HM20 or 5–10 min for K4M) followed by Millonig's lead acetate (for 1–3 min for both resins).

Since K4M is hydrophilic, the sections should be incubated on drops of the stains for *short periods of time*. Prolonged staining may cause contamination and distortion of the sections. As with all staining procedures, wash well between the uranyl and lead stains and take precautions against CO_2 during the lead staining and the subsequent rinsing.

K4M has been successfully used in cytochemical and immunocytochemical studies (e.g. Bendayan 1981, 1983; Roth, 1982a, 1983; Roth et al. 1981; Roth and Berger 1982).

References

Albrecht, R.M., D.H. Rasmussen, C.S. Keller and R.D. Hindsill (1976), Preparation of cultured cells for electron microscopy – air drying from organic solvents, J. Microscopy *108*, 21.

Armbruster, B.L., E. Carlemalm, R. Chiovetti, R.M. Garavito, J.A. Hobot, E. Kellenberger and W. Villiger (1982), Specimen preparation for electron microscopy using low temperature embedding resins, J. Microscopy *126*, 77.

Barlow, D.I. and M.A. Sleigh (1979), Freeze-substitution for preservation of ciliated surfaces for scanning electron microscopy, J. Microscopy *115*, 81.

Bendayan, M. (1981), Ultrastructural localization of actin in insulin-containing granules, Biol. Cell *41*, 157.

Bendayan, M. (1983), Ultrastructural localization of actin in muscle, epithelial and secretory cells by applying the protein A-gold immunocytochemical method, Histochem. J. *15*, 39.

Browning, A.J. and B.E.S. Gunning (1979), Structure and function of transfer cells in the sporophyte haustorium of *Funaria hygrometrica* Hedw. II. Kinetics of uptake of labelled sugars and localisation of absorbed products by freeze-substitution and autoradiography, J. exp. Bot. *119*, 1247.

Bullivant, S. (1960), The staining of thin sections of mouse pancreas prepared by the Fernandez-Moran helium II freeze-substitution method, J. biophys. biochem. Cytol. *8*, 639.

Bullivant, S. (1965), Freeze-substitution and supporting techniques, Lab. Invest. *14*, 440.

Carlemalm, E. (1984), Low temperature embedding resins, Proc. 8th Eur. Reg. Conf. Electron Microscopy, Budapest *3*, 1786.

Carlemalm E. and E. Kellenberger (1982), The reproducible observation of unstained embedded cellular material in thin sections: visualization of an integral membrane protein by a new mode of imaging for STEM, EMBO J. *1*, 63.

Carlemalm, E., R.M. Garavito and W. Villiger (1982), Resin development for electron microscopy and an analysis of embedding at low temperature, J. Microscopy *126*, 123.

Chandler, J.A. (1980), Application of X-ray microanalysis in reproductive physiology, Scanning Electron Microscopy 1980, *2*, 475.

Dempsey, G.P. and S. Bullivant (1976), A copper block method for freezing non-cryoprotected tissue to produce ice-crystal-free regions for electron microscopy. I. Evaluation using freeze-substitution, J. Microscopy *106*, 251.

Feder, N. and R.L. Sidman (1958), Methods and principles of fixation by freeze-substitution, J. biophys. biochem. Cytol. *4*, 593.

Fernández-Morán, H. (1957), Electron microscopy of nervous tissue, in: Metabolism of the nervous system, D. Richter, ed. (Pergamon Press, Oxford), p. 1.

Fernández-Morán, H. (1959a), Electron microscopy of retinal rods in relation to localisation of rhodopsin, Science *129*, 1284.

Fernández-Morán, H. (1959b), Cryofixation and supplementary low temperature preparation techniques applied to the study of tissue ultrastructure, J. appl. Physiol. *30*, 2038.

Fernández-Morán, H. (1960), Low temperature preparation technique for electron microscopy of biological specimens based on rapid freezing with liquid helium II, Ann. N.Y. Acad. Sci. *85*, 689.

Fernández-Morán, H. (1961a), Lamellar systems in myelin and photoreceptors as revealed by high resolution electron microscopy, in: Macromolecular complexes, M.V. Edds, Jr., ed. (Ronald Press, New York), p. 113.

Fernández-Morán, H. (1961b), The fine structure of vertebrate and invertebrate photoreceptors as revealed by low temperature electron microscopy, in: The structure of the eye, G.K. Smelser, ed. (Academic Press, New York), p. 521.

Fisher, D.B. and T.L. Housley (1972), The retention of water soluble compounds during freeze-substitution and microautoradiography, Plant Physiol. *49*, 166.

Glauert, A.M. (1974), Fixation, dehydration and embedding of biological specimens, in: Practical methods in electron microscopy, Vol. 3, A.M. Glauert, ed. (North-Holland, Amsterdam).

Hall, J.L., D.M.R. Harvey and T.J. Flowers (1978), Evidence for the cytoplasmic localisation of Betaine in leaf cell of *Suaeda maritima*, Planta *140*, 59.

Harvey, D.M.R. (1980), The preparation of botanical samples for ion localisation studies at subcellular level, Scanning Electron Microscopy, 1980, *2*, 409.

Harvey, D.M.R. (1982), Freeze-substitution, J. Microscopy *127*, 209.

Harvey, D.M.R. and B. Kent (1981), Sodium localisation in *Suaeda maritima* leaf cells using zinc uranyl acetate precipitation, J. Microscopy *121*, 179.

Harvey, D.M.R., J.L. Hall and T.J. Flowers (1981b), The use of freeze-substitution and freeze fracture study of bacterial spore structures, J. Ultrastruct. Res. *76*, 71.

Hayat, M.A. (1981), Fixation for electron microscopy, (Academic Press, New York).

Hayat, M.A. and B.R. Zirkin (1973), Critical point-drying method, in: Principles and techniques of electron microscopy, Vol. 3, M.A. Hayat, ed. (Van Nostrand Reinhold Company, New York), p. 297.

Heuser, J.E., T.S. Reese, M.J. Dennis, Y. Jan, L. Jan and L. Evans (1979), Synaptic vesicle exocytosis captured by quick-freezing and correlated with quantal transmitter release, J. Cell Biol. *81*, 275.

Humbel, B. and M. Müller (1984), Freeze-substitution and low-temperature embedding, Proc. 8th Eur. Reg. Conf. Electron Microscopy, Budapest *3*, 1789.

Humbel, B., T. Marti and M. Müller (1983), Improved structural preservation by combining freeze-substitution and low temperature embedding, Beitr. elektronenmikroskop. Direktabb. Oberfl. *16*, 585.

Hunziker, E.B., W. Herrmann, R.K. Schenk, M. Müller and H. Moor (1984), Cartilage ultrastructure after high pressure freezing, freeze substitution, and low temperature embedding. I. Chondrocyte ultrastructure – implications for the theories of mineralisation and vascular invasion, J. Cell Biol. *98*, 267.

Hutchinson, T.E. and A.P. Somlyo (eds.) (1981), Microprobe analysis of biological systems (Academic Press, London).

Kellenberger, E. (1984), Thin sections with low temperature embedding: successes and remaining limitations, Proc. 8th Eur. Reg. Conf. Electron Microscopy, Budapest *3*, 1781.

Mackenzie, A.P. (1972), Freezing, freeze-drying and freeze-substitution, Scanning Electron Microscopy 1972, *2*, 274.

Mackenzie, A.P. (1977), Non equilibrium freezing behaviour of aqueous systems, Phil. Trans. R. Soc. Ser. B *278*, 167.

Marshall, A.T. (1980), Freeze-substitution as a preparation technique for biological X-ray microanalysis, Scanning Electron Microscopy 1980, *2*, 335.

Müller, M. (1981), Demonstration of liposomes by electron microscopy, in: Membrane proteins, A. Azzi, U. Brodbeck and P. Zahler, eds. (Springer-Verlag, Berlin), p. 252.

Müller, M., T.H. Marti and S. Kriz (1980), Improved structural preservation by freeze-substitution, Proc. 7th Eur. Reg. Conf. Electron Microscopy, The Hague *2*, 270.

Newell, D. and S. Roath (1975), A container for processing of small volumes of cell suspensions for critical point drying, J. Microscopy *104*, 321.

Ornberg, R.L. and T.S. Reese (1980), A freeze-substitution method for localising divalent cations – examples from secretory systems, Fed. Proc. *39*, 2802.

Ornberg, R.L. and T.S. Reese (1981), Quick freezing and freeze-substitution for X-ray microanalysis of calcium, in: Microprobe analysis of biological systems, T.H. Hutchinson and A.P. Somlyo, eds. (Academic Press, London), p. 213.

Pease, D.C. (1966), The preservation of unfixed cytological detail by dehydration with "inert" agents, J. Ultrastruct. Res. *14*, 356.

Pease, D.C. (1967a), Eutectic ethylene glycol and pure propylene glycol as substituting media for the dehydration of frozen tissue, J. Ultrastruct. Res. *21*, 75.

Pease, D.C. (1967b), The preservation of tissue for fine structure during rapid freezing, J. Ultrastruct. Res. *21*, 98.

Pease, D.C. (1968), Unfixed tissue prepared for electron microscopy by glycolic dehydration, Proc. 4th Eur. Reg. Conf. Electron Microscopy, Rome *2*, 11.

Pease, D.C. (1973), Substitution techniques, in: Advanced techniques in biological electron microscopy, J.K. Koehler, ed. (Springer-Verlag, Berlin), p. 35.

Rebhun, L.I. (1972), Freeze-substitution and freeze-drying, in: Principles and techniques of electron microscopy, Vol. 2, M.A. Hayat, ed. (Van Nostrand Reinhold Company, New York), p. 3.

Reid, N. (1974), Ultramicrotomy, in: Practical methods in electron microscopy, Vol. 3, A.M. Glauert, ed. (North-Holland, Amsterdam).

Revel, J.P. and K. Wolken (1973), Electron microscope investigations of the underside of cells in culture, Exp. Cell Res. *78*, 1.

Robards, A.W. (1978), An introduction to techniques for scanning electron microscopy of plant cells, in: Electron microscopy and cytochemistry of plant cells, J.L. Hall, ed. (Elsevier/ North-Holland, Amsterdam), p. 343.

Roth, J. (1982a), New approaches for *in situ* localization of antigens and glycoconjugates on thin sections – the protein A-gold technique and the lectin colloidal gold marker system, Proc. 10th Int. Congr. Electron Microscopy, Hamburg *3*, 245.

Roth, J. (1982b), The protein A-gold (pAg) technique – a qualitative and quantitative approach for antigen localization on thin sections, in: Techniques in immunocytochemistry, Vol. 1, G.R. Bullock and P. Petrusz, eds. (Academic Press, London), p. 107.

Roth, J. (1983), The colloidal gold marker system for light and electron microscopic cytochemistry, in: Techniques in immunocytochemistry, Vol. 2, G.R. Bullock and P. Petrusz, eds. (Academic Press, London, New York), p. 217.

Roth, J. and E.G. Berger (1982), Immunocytochemical localisation of galactaryltranferase in HeLa cells: codistribution with thiamine pyrophosphatase in trans-Golgi cisterna, J. Cell Biol. *83*, 223.

Roth, J., M. Bendayan and L. Orci (1978), Ultrastructure localisation of intracellular antigens by the use of protein A-gold complex, J. Histochem. Cytochem. *26*, 1074.

Roth, J., M. Bendayan, E. Carlemalm, W. Villiger and M. Garavito (1981), Enhancement of structural preservation and immunocytochemical staining in low temperature embedded pancreatic tissue, J. Histochem. Cytochem. *29*, 663.

Simpson, W.L. (1941), An experimental analysis of the Altmann technique of freeze-drying, Anat. Rec. *80*, 173.

Sitte, H. (1981), Ultramikrotomie – Aktuelle Trends und Neuentwicklungen im biologisch-medizinschen Bereich, Git. Labor-medizin *4*, 317.

Steinbrecht, R.A. (1976), Freeze-substitution and freeze-fracturing of insect sensilla without cryoprotectants, Proc. 6th Eur. Reg. Conf. Electron Microscopy, Jerusalem *2*, 111.

Steinbrecht, R.A. (1980), Cryofixation without cryoprotectants – freeze-substitution and freeze-etching of an insect olfactory receptor tissue, Tissue Cell *12*, 73.

Steinbrecht, R.A. (1982), Experiments on freezing damage with freeze-substitution using moth antennae as test objects, J. Microscopy *125*, 187.

Van Harreveld H. and J. Crowell (1964), Electron microscopy after rapid freezing on a metal surface and substitution fixation, Anat. Rec. *149*, 381.

Van Harreveld, H., J. Crowell and S.K. Malhotra (1965), A study of extracellular space in central nervous tissue by freeze-substitution, J. Cell Biol. *25*, 117.

Van Steveninck, R.F.M. and M.E. Van Steveninck (1978), Ion localisation, in: Electron microscopy and cytochemistry of plant cells, J.L. Hall, ed. (Elsevier/North-Holland, Amsterdam), p. 187.

Zalokar, M. (1966), A simple freeze-substitution method for electron microscopy, J. Ultrastruct. Res. *15*, 469.

Chapter 8

Future outlook

Although this is a book dealing with *practical* techniques in low temperature electron microscopy, it would not be serving its full purpose if it did not end by pointing towards some of the more probable future developments. Many low temperature techniques are still relatively new and will continue to develop beyond the level described here. Thus, it is important that the reader should be aware of the probable direction of developments so that he is in a better position to make the right choices in both techniques and equipment.

Much of the apparatus currently available has either been produced as 'one-off' equipment within a single laboratory or is only in the first generation of commercial production. It is usual, in the evolution of techniques, for methodology and apparatus to become simpler for the operator to use and for the equipment to become more fully automated. Such a trend is seen at its best in electron microscopes themselves but can also be discerned in equipment relevant to the topic of this book, such as cryoultramicrotomes, freeze-driers and freeze-etching units. The availability of simple microprocessor control systems will undoubtedly play an important part in making work faster and easier. For example, the availability of a simple microprocessor-based cooling rate meter has resulted in a great improvement over conventional methods for determining rapid cooling rates (§2.2.2). Indeed, it is difficult to see apparatus in the low temperature field where microprocessors could *not* fulfil a useful role! Many methods currently use liquid nitrogen 'by the bucketful'. Not only is this expensive; it also generally exacerbates problems of condensation on cooled surfaces. Excessive use of liquid nitrogen for cooling cold stages and other components is usually a good indication of poor design. Refrigerated cryopumps (§6.3.2; Haefer 1981) have, as yet, made hardly any impact in the area of low temperature electron microscopy and yet, for many purposes, they should be ideal. They can achieve very low temperatures using a closed system and exploration of their merits in applications such as routine freeze-etching is surely long overdue. In general, it can be expected that techniques will become more convenient to apply to different materials and that the choice of the most appropriate technique to solve a particular problem will become easier. In all low temperature work it would be useful to have a suitable test-specimen that could be used to compare techniques from different laboratories: such a specimen would provide great advantages in evaluating the comparative merits of different methods.

8.1   Freezing

Although our current state of knowledge about the mechanisms of freezing biological specimens is very much greater than just a few years ago, there still remains much scope for improvement in techniques to ensure that the best method is used in solving a particular problem. Moreover, more attention will have to be given to sampling from large specimens, such as by the use of improved 'cryoguns' which minimise the otherwise inevitable problems of mechanical damage during sampling. The range of commercially available apparatus for freezing specimens is very limited. It is possible to buy spray-freezing, propane jet-freezing, cold block-freezing and plunge-freezing equipment, although the number of suppliers is restricted and it is to be hoped that additional, competitive versions will become available so that a greater number of research workers can use these techniques and different approaches can be compared. The commercial realisation of a high pressure freezing device (Appendix 2) permits wider investigation of this technique; the advantages of being able to freeze specimens while subjecting them to high pressure are so great that this avenue of research should certainly not be abandoned.

We can thus expect to see the introduction of new commercial apparatus which will make freezing both simpler and more reliable. At the same time, the joint contributions of low temperature physicists, cryogenic engineers and biologists will ensure that there is a greater awareness of the problems involved in freezing and in the selection of appropriate methods for particular problems. We can expect significant improvements in the 'quality' of freezing to result from work such as that of Bald (in press), who has made the first detailed theoretical assessment of the relative merits of different freezing methods. Such work not only suggests ways of optimising present methods but also points to new techniques that should be tried (e.g. freezing in super-critical nitrogen, §2.2.3a).

We will always have to fight against the inflexible limits of the physical properties of water. It may well be that future work will involve further experiments into deep-cooling specimens without allowing them to freeze: if ice formation were to be prevented, no crystal damage could occur.

8.2 Direct viewing methods

While low temperature SEM techniques are now fairly routine and the apparatus is commercially available, it cannot yet be said that low temperature TEM is as widely used. The instrumental problems are more severe, as are the difficulties of preparing specimens. Furthermore, there are a number of demands made of low temperature stages in EMs: in addition to the simple requirement of keeping a specimen cold while it is viewed and/ or analysed, far more specialised (TEM) cold stage will be needed where high resolution and protection from radiation damage are the major objectives. While it is to be expected that many laboratories will acquire cold stages for ordinary TEM and SEM use, the more sophisticated systems will only be available in relatively few laboratories with advanced ancillary facilities. In such circumstances, as in other fields, it is important for the 'normal' user to be aware of the results from the more specialised equipment since these very often have a direct bearing on the use of the simpler systems.

We will also have to learn more about the physical and structural behaviour of ice in the electron microscope. Under the conditions of cryomicroscopy (frozen-hydrated specimens) only 4 of the 12 known types of solid water need to be considered. These are two forms which are crystalline (hexagonal and cubic ice) and two forms of amorphous water (generally called vitreous ice because it retains the order of liquid water). Since temperature- and electron dose-dependent transitions between the four forms of solid water occur (Fig. 8.1), it will be worth studying the relevance of these transitions to changes in the specimen structure in the course of high resolution studies of frozen-hydrated specimens.

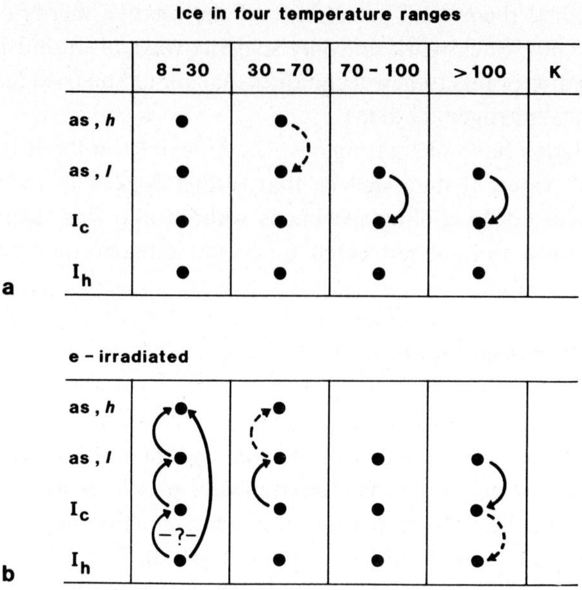

Ice in four temperature ranges

	8 - 30	30 - 70	70 - 100	>100	K
as , h	●	●			
as , l	●	●	●	●	
I_c	●	●	●	●	
I_h	●	●	●	●	

a

e - irradiated

	8 - 30	30 - 70	70 - 100	>100	
as , h	●	●			
as , l	●	●	●	●	
I_c	●	●	●	●	
I_h	●	●	●	●	

b

Fig. 8.1.   (a) Temperature transition ranges for the 4 types of solid water relevant to cryomicroscopy. **as,h**, high density amorphous ice (specific density 1.1 g cm^{-3}); **as,l**, low density amorphous ice (specific density 0.94 g cm^{-3}); I_c, cubic ice; I_h, hexagonal ice. (b) Possible transitions induced by electron bombardment. (Diagram constructed from data kindly provided by H. Heide and E. Zeitler (personal communication) and from Heide and Zeitler (1984).)

8.3   Cryoultramicrotomy

Among the whole gamut of low temperature microscopical techniques, there cannot be one that remains fraught with more technical problems than cryoultramicrotomy. There is now some very fine commercial apparatus available for cutting ultra-thin sections of frozen specimens (§4.2). However, much remains to be done to determine how best to section different tissues with different water contents. The influence of the frozen specimen itself on the results obtained is nowhere seen more clearly than with this technique. The (relatively unsupported) school of thought that believes that sectioning can take place at about 240 K without adverse effects from ice recrystallisation is countered by the protagonists of the belief that the specimen temperature must always be maintained below about 200 K. Furthermore, the nature of the ice crystals within the specimen seem to have an important effect on the sectioning properties of the specimen. Thus, while very good apparatus is now available, it cannot be used to its fullest advantage until we have more fundamental information about the relationship between the nature of a frozen specimen and the sectioning/fracturing process. As this becomes available, new cryoultramicrotomes will doubtless be produced to take advantage of the latest knowledge.

8.4   Freeze-drying

Freeze-drying has been with us for a longer time than most low temperature methods but it still remains far from perfected in most apparatus as far as ultrastructural research is concerned. The primary requirements to ensure reproducible and effective freeze-drying are that the specimen should be held at a *precise*, *known* temperature and that the ambient vacuum should also be well controlled. Once these conditions can be met, then there appears every reason to think that freeze-drying will show an increased application to biological specimens, not least because of criticisms of critical point-drying as a preparation method for SEM (see §5.4).

8.5   Freeze-etching

Freeze-fracture and freeze-etching have now become extremely well established techniques in numerous laboratories and have consequently made a major contribution to our understanding of biological ultrastructure. Nevertheless, they remain methods that require a high level of technical expertise. Furthermore, they are time-consuming and the interpretation of results is a feature requiring the most detailed consideration. The cost of good freeze-etching equipment is high and, partly for this reason, there is little choice of apparatus. This is a pity because further competition would undoubtedly stimulate still further improvements in both the applications of the technique and in the ease of use of the equipment. With modern technology it should be possible to produce freeze-fracture/etching equipment that would be less expensive to run (more economical in use of liquid nitrogen), would process many specimens within a short period of time and would not require 'high-level' operator expertise. Using modern techniques of microprocessor control, the latest information on fracturing devices, refrigerator cryopumps and vacuum/cryotechnology, the 'ultimate' freeze-etching device is waiting to be constructed. By fracturing and replicating at 4 K it should be possible to minimise artefacts.

8.6   *Freeze-substitution*

As with freeze-drying, freeze-substitution has been undergoing something of a revival. This is partly because the latest substitution methods do allow complete solvation of ice at very low temperatures and partly because, when combined with low temperature embedding methods, a highly advantageous processing pathway of sections for TEM is available. Unfortunately, the very superiority of fixation during freeze-substitution means that infiltration with resin can be difficult or even impossible because membranes are so well preserved. Nevertheless, this is a potentially important preparation route and, therefore, one worthy of much further work. Many are frightened away from freeze-substitution because of the lack of appropriate apparatus. Reference to Chapter 7 should dispel any notion that freeze-substitution is particularly difficult to use. Performed carefully, it should provide results from many specimens that challenge the very best that can be produced by conventional techniques – with the added advantage that extraction of soluble components is *very* much reduced.

8.7   *Conclusion*

It is inevitable that low temperature techniques will play an increasingly important role in the elucidation and understanding of biological ultrastructure. Not only this, but they will also open up new possibilities for analytical and cytochemical techniques so that new types of information can be obtained. In turn, the improvement in preparation methods, with the concomitant increase in 'quality' of data, means that our attention *must* also turn towards the greater use of quantitative methods in the analysis of purely structural data. Only in this way is it possible to have any long-term confidence in the results that are obtained from any micoscopical technique.

References

Bald, W.B. (In press), Heat transfer processes in cryofixation (Adam Hilger, Bristol).
Haefer, R.A. (1981), Kryo-Vakuumtechnik (Grundlagen und Anwendungen) (Springer-Verlag, Berlin).
Heide, H.G. and E. Zeitler (1984), Water in Cryomicroscopy, Proc. 8th Eur. Reg. Conf. Electron Microscopy, Budapest *2*, 1388.

Appendix 1

Safety

Low temperatures are inherently dangerous. In addition to direct hazards from low temperatures, workers in laboratories involving cryogenic methods in microscopy will often be concerned with extremes of pressure (high vacuum apparatus such as electron microscopes or freeze-fracture apparatus; high pressure equipment such as gas cylinders and gas/liquid supply lines); with very high voltages; with toxic chemicals; with highly inflammable materials; and with ionising radiation. The EM laboratory is potentially a very unhealthy place! Nevertheless, sensible laboratory procedures ensure that risks are minimised. It is particularly important that, when a scientist embarks on a new field of study (for example, low temperature methods), he or she attempts to become fully acquainted with the particular hazards associated with the work. By *understanding* potential difficulties they are usually easily avoided. This Appendix cannot cover *all* aspects of safety in an EM laboratory although the references at the end will serve as an introduction for those who are concerned to enquire further into this matter.

The specific hazards of low temperature techniques will now be discussed in more detail. Among the useful references specifically dealing with hazards at low temperatures is the *Cryogenics Safety Manual* published by the British Cryogenics Council and a pamphlet produced by Edwards High Vacuum (see references).

Categories of hazard
It is possible to separate the potential hazards that can be encountered while using low temperature methods into different categories:

(*i*)   Direct contact with cold gas, liquid or solid.
(*ii*)   Problems of inhalation and asphyxiation with cryogenic gases.

(*iii*)   Expansion associated with decompression of liquefied gases.
(*iv*)   Inflammable nature of some cryogenic gases.
(*v*)   Toxic nature of some cryogenic gases.

A1.1   *Direct contact with cold gas, liquid or solid*

Extreme cold destroys cells and tissues to a degree related to the duration and temperature of exposure. Moreover, cold burns can be insidious in that frozen tissue does not necessarily feel painful or look very different from normal tissue. It is only on thawing that pain and damage become evident, a fact that will be well known to anyone who has suffered frost-bite. Because heat transfer to a high thermal conductivity solid, such as a piece of metal, is far more efficient than heat transfer to a cold liquid, it is particularly important not to allow unprotected parts of the body to come into contact with such cold solids. Protection against cold 'burns' thus takes the form of using appropriate protective clothing.

When handling cold liquids care should be taken that any spillage will not accumulate in clothing, such as pockets, insides of boots or insides of gauntlet-type gloves. Therefore protective clothing should be close-fitting so that liquids will run straight off it. Similarly, it should be non-porous and be kept dry. Gloves should be relatively loose-fitting (ease of removal) but without large gauntlets into which liquids might fall. Types of glove vary: if objects at low temperature are to be handled, then good insulation is at a premium; if liquefied gases are being used then the hands should not come into direct contact with the liquid and leather or lined rubber gloves are adequate. Overalls should not have open pockets or trouser turn-ups. Trousers should be worn outside boots. The face, and particularly the eyes, should be protected with a face mask. Use tongs or other insulated handling tools whenever dealing with objects under a liquid cryogen.

If a person has suffered cold 'burns', first attempt to establish the severity. For a small, localised injury treat the affected area with cold water and then with a cold compress. Do not use dry heat. If the skin is blistered, if the eyes are affected or if there are any other untoward symptoms, then medical advice should be obtained as soon as possible. For larger areas of damage, ensure that clothing is not restricting the blood supply and remove the casualty to a sick room or hospital as soon as possible. Place the burn in water with a temperature between 42 and 45°C (higher temperatures or dry heat will further 'burn' the already damaged tissue).

Insufficient protection against cold (for example when working in a cold-room) can lead to hypothermia. If the whole body temperature has become depressed (massive cold exposure) then the patient should be immersed in a bath at the temperature given above (between 42 and 45°C). Beware of the possibility of shock as the patient re-establishes thermal equilibrium.

The patient should not be given alcohol or allowed to smoke after receiving a cold burn because these reduce blood flow to the frozen tissue.

While tissues that are still frozen look waxy and pallid and are usually painless, they become extremely painful, swollen and susceptible to infection when thawed. If a casualty occurs away from the possibility for medical attention, then it is best *not* to make any special attempt to thaw the frozen tissue but to transport the patient to the nearest source of medical help. Severe cold burns may require the administration of pain-killers and/or tranquilisers during thawing. If a severe cold burn has thawed before receiving medical attention, then cover with a large sterile dressing to reduce the likelihood of infection. It may be advisable to administer an anti-tetanus booster injection after a severe burn; medical advice should be sought.

A1.2 Problems of inhalation and asphyxiation with cryogenic gases

Cold gas is not likely to be inhaled voluntarily over long periods but long-term inhalation can seriously damage the lung.

Asphyxia can occur under any conditions where atmospheric oxygen content falls below its normal level of 21% (v/v). Thus *all* cryogenic gases should be used in well-ventilated surroundings. Liquid gases release an enormous equivalent volume of gas as they change from the liquid to the gaseous state. Thus, nitrogen expands by a factor of about 680 as it changes from liquid to gas. Precautions must be taken to avoid asphyxiation which may start quite suddenly and, in some situations, can be at an advanced stage before the patient is even aware of it. The remedy is to ensure that high gas concentrations are never allowed to build up. Use small quantities of liquid gas in a large, well-ventilated room or fume cupboard. Handling of large quantities of liquid gas should be undertaken only in specially designated areas where there is ample ventilation. Do not travel in a lift with a large (> 1.0 litre) container of cryogenic fluid; if the lift becomes stuck you will be in a confined space with a constantly increasing atmospheric concentration of the cryogen gas. In the event that asphyxiation *does* occur, the victim should as quickly as possible be transferred to a well-ventilated environment (even outdoors) and have clothing around their neck and chest loosened. Medical advice should be sought as quickly as possible.

A1.3   *Expansion associated with decompression of liquefied gases*

As mentioned above, liquefied gases expand greatly when vapourising, a volume increase of about 800-fold being quite typical. If this expansion is not under controlled conditions then it can be both dramatic and dangerous. For example, if a cryogenic liquid such as liquid nitrogen is placed in a *sealed* container, any heat gain by the liquid will serve to boil off a little more of the nitrogen as gas. Ultimately this will lead to an explosion. Thus *all* liquid cryogen containers must be provided with a means of venting to the atmosphere, whether valved or not. Similarly, storage Dewar vessels must never be completely sealed; this also applies to vacuum containers of the type used for insulating hot drinks (it is very easy to forget that the screw seals on such containers must *not* be fully tightened). Large storage containers, obtainable commercially, are built to stringent specifications and are suitable for their specific purpose. Even here, however, care must be taken to ensure that pressure-release valves do not become iced-up and inoperative. Mistakes can be made as liquids are decanted from one container to another, smaller, one. Unprotected vacuum flasks should not be used for fear of implosion. At the very least they should be wrapped around with thick adhesive tape to prevent sharp pieces of glass from flying about if an implosion occurs. In many situations it is simpler, cheaper and safer to use containers made from expanded polystyrene, polyurethane or similar material; stainless steel containers are also satisfactory. Containers designed for use with CO_2 are not generally suitable for use with N_2.

Always label containers clearly as to what they contain. Ensure that containers are clean and dry before filling with a cryogenic liquid.

A1.4   *Inflammable nature of some cryogenic gases*

Some cryogenic fluids, such as liquid oxygen and liquid propane, are highly inflammable in the gaseous state. With increasing use of propane as a cryogen it is most important that the correct procedures for handling it are fully appreciated.

Liquid oxygen, with a boiling point of $-183°C$ (90 K) will condense into any liquid bath colder than this (for example, into liquid nitrogen at $-196°C$ (77 K) or liquid propane at $-189°C$ (84 K)). Although oxygen itself will not burn, it will fuel the combustion of other substances to burn more fiercely and, under some circumstances, explosively. It will also allow materials to ignite at temperatures lower than their normal ignition temperature. As liquid oxygen is seldom, if ever, used in low temperature microscopy the problem is merely one of avoiding the condensation of large quantities of liquid oxygen from the atmosphere. Therefore, low temperature cryogens $< -183°C$ (90 K) should be covered as far as possible to avoid contact with the air (remembering, however, to allow venting to avoid pressure build-up) and, for the same reason, the container should not be left standing around for prolonged periods. Liquid oxygen is usually easily identifiable as, of all the common low temperature liquids, it has a characteristic light blue colour (the others are colourless).

If there is a fire in which oxygen is involved, then shut off the oxygen supply (if possible); use water to extinguish the fire *unless* electrical equipment is involved, in which case use CO_2, dry chemical extinguishant or vapourising liquid. In attempting to extinguish fires involving cryogenic equipment, do not spray water onto pipes, valves, etc. at low temperatures: the water will freeze and may immobilise valves and cause pressure to build up.

Liquid propane, and similar combustible gases require very careful handling. The major problems arise during *condensation* of the gas (usually from a pressurised cylinder) and during *disposal* of the gas because, at these times, the temperature of the gas can rise above the flash point (for liquid propane $-104°C$; 169 K). It is, therefore, important to duct propane gas through coils of thin copper pipe immersed in liquid nitrogen before condensing it into a container (see §2.4.1a). In this way, the propane is always at a temperature $\ll -100°C$ (173 K) when exposed to the atmosphere. In disposing of the propane (and again remembering the 800-fold volume increase) care must be taken to lead the gas away through a spark-protected fume-hood or, alternatively, to tip the liquid onto the ground in the open air well away from all buildings. It goes without saying that there should be no smoking and no ignition of naked flames while working with combustible gases and clear warning signs should be placed on the doors of rooms in which such work is taking place.

A1.5   Toxic nature of some cryogenic gases

Apart from the asphyxiation hazard (§A1.2), most cryogenic liquids/gases are of relatively low toxicity. Some of the hydrocarbons are mild narcotics and may cause dizziness, narcosis and nausea at low concentrations and unconsciousness on continued exposure. Removal from exposure to vapour will cause the symptoms to disappear. There have been reports of the cardiac toxicity of some of the halocarbons used as the propellants in aerosol cans and as cryogens (Freons, Arctons, etc.; Taylor and Harris 1970) but little further information has been subsequently published. Ethane is an anaesthetic gas and requires particular care in use.

A1.6   Conclusions

By understanding the simple physical principles of what is involved and applying common sense and usual laboratory safety measures, most day-to-day operations with low temperature equipment and chemicals are accomplished in absolute safety. However, if apparatus for cryogenic use is to be 'home-made' then it is absolutely essential that a person who is fully acquainted with cryogenic engineering is consulted about the use of appropriate materials, tolerances and insulation.

References

Alderson, R.H. (1975), Design of the electron microscope laboratory, in: Practical methods in electron microscopy Vol. 4, A.M. Glauert, ed. (North-Holland, Amsterdam), p. 114.

Coward, H.F. and G.W. Jones (1952), Limits of flammability of gases and vapours, Bulletin 503, U.S. Bureau of Mines.

Cryogenics Safety Manual – A Guide to Good Practice, Safety Panel, British Cryogenics Council, c/o Institution of Chemical Engineers, 165–171 Railway Terrace, Rugby CV21 3HQ, England.

Humphreys, W.J. (1970), Health and safety hazards in the SEM laboratory. Scanning Electron Microscopy *1*, 537.

Recommended safety precautions for handling cryogenic liquids (1979), Edwards High Vacuum Publication 59-K100-00-880, B.O.C. Cryoproducts, Manor Royal, Crawley, West Sussex RH10 2LW, England.

Taylor G.J. and W.S. Harris (1970), Cardiac toxicity of aerosol propellants, J. Amer. Med. Ass. *214*, 81.

Appendix 2

List of suppliers

N.B. The following list includes the names of suppliers known to the authors at the present time. For reasons of space it has not been possible to include all international addresses or agencies for large companies; there will obviously also be unintentional omissions and we will be pleased to hear of names and addresses of other suppliers for inclusion in later editions.

The first list gives the names of suppliers of the many routine items required for all methods in electron microscopy. Many of these suppliers provide items relevant to low temperatures. Subsequent lists provide the names and addresses for suppliers of specific items. If you cannot find a particular product, then it is often useful to contact one of the major suppliers (§1), who may well be able to help.

1. General suppliers of materials and equipment for electron microscopy

(a) **Agar Aids for Electron Microscopy** (Agar Aids)
(U.K. agents for Ladd)
66a Cambridge Road
Stansted
Essex CM24 8DA
England

(b) **Balzers Union AG** (Balzers)
P.O. Box 75 FL-9496 Balzers
Fürstentum
Liechtenstein

also
Balzers High Vacuum Ltd.
Northbridge Road
Berkhamsted
Hertfordshire HP4 1EN
England

also
Balzers
8 Sagamore Park Road
Hudson
New Hampshire 03051
U.S.A.

(c) **BEEM**
P.O. Box 132
Jerome Avenue Station
Bronx
New York 10468
U.S.A.

(d) **CW French Inc.** (French)
58 Bittersweet Lane
Weston
Massachusetts 02193
U.S.A.

(e) **Denton Vacuum Inc.** (Denton)
Cherry Hill Industrial Center
Cherry Hill
New Jersey 08003
U.S.A.

(f) **Electron Microscope Aids**
6 Lime Trees
Malford
Chippenham
Wiltshire
England

(g) **Electron Microscope Sciences** (EMS)
P.O. Box 251
Fort Washington
Pennsylvania 19034
U.S.A.

(h) **EMscope Laboratories Ltd.** (EMscope)
Kingsnorth Technology Park
Wotton Road
Ashford
Kent TN23 2LN
England

(i) **Ernest F. Fullam Inc.** (EFFA)
P.O. Box 444
Schenectady
New York 12301
U.S.A.

Graticules Ltd.
(U.K. agents for EFFA)
Morley Road
Botany Trading Estate
Tonbridge
Kent TN9 1RN
England

Touzart & Matignon
(French agents for EFFA)
8 Rue Eugene Henaff
94400 Vitry sur Seine
France

(j) **Ladd Research Industries Inc.** (Ladd)
(U.S. agents for Agar Aids)
P.O. Box 1005
Burlington
Vermont 05402
U.S.A.

(k) **Marivac Ltd.**
1872 Garden Street
Halifax
Nova Scotia B3H 3R6
Canada

(l) **Nisshin EM Co. Ltd.**
Espowarl Ichigaya 40-10
Tomihisa-cho Shinjuku-ku
Tokyo 160
Japan

(m) **Polaron Equipment Ltd.** (Polaron)
(U.K. agents for Polysciences)
53–63 Greenhill Crescent
Watford Business Park
Watford
Hertfordshire WD1 8XG
England

Ted Pella Inc. (Pelco)
(U.S. agents for Polaron)
P.O. Box 510
Tustin
California 92681
U.S.A.

(n) **Polysciences Inc.** (Polysciences)
(U.S. agents for Polaron)
Paul Valley Industrial Park
Warrington
Pennsylvania 18976
U.S.A.

Polysciences Ltd.
24 Low Farm Place
Moulton Park
Northampton NN3 1HY
England

(o) **Soquelec Ltd.**
5757 Cavendish, Suite 407
Montreal
Quebec H4W 2W8
Canada

(p) **SPI Supplies** (SPI)
Division of Structure Probe Inc.
P.O. Box 342
West Chester
Pennsylvania 19380
U.S.A.

(q) **Taab Laboratories Equipment Ltd.** (Taab)
40 Grovelands Road
Reading
Berkshire RG3 1BR
England

Extech International Corp. (Extech)
(U.S. agents for Taab)
177 State Street
Boston
Massachusetts 02109
U.S.A.

(r) **Tousimis Research Corp.** (Tousimis)
6000 Executive
Rockville
Maryland 20852
U.S.A.

(s) **Vaughn Electron Microscope Supplies Inc.**
2176 Dunn Road
Memphis
Tennessee 38114
U.S.A.

(t) **Walter McCrone Associates Inc.**
(McCrone)
2820 South Michigan Avenue
Chicago
Illinois 60616
U.S.A.

2. Electron microscope accessories (special stages, mechanisms, etc.)
(your electron microscope manufacturer should also be contacted)

(a) **Advanced Metal Research** (AMR)
Bedford
Massachusetts 01730
U.S.A.

(b) **Balzers**   (*see* 1b)

(c) **Cambridge Instruments Ltd.**
Rustat Road
Cambridge CB1 3QH
England

(d) **EMscope Laboratories** (EMscope)
(*see* 1h)

(e) **Gatan Inc.**
780 Commomwealth Drive
Warrendale
Pittsburgh
Pennsylvania 15086
U.S.A.

also

6678 Owens Drive
Pleasanton
California 94566
U.S.A.

also

Ingolstädter Str.
40 D-8000 München 40
West Germany

(f) **Hexland Ltd.**
W & G Estate
Faringdon Road
East Challow
Nr Wantage
Oxon OX12 9TF
England

(g) **Oxford Instruments**
Osney Mead
Oxford OX2 0DX
England

(h) **Anton Paar K.G.**
P.O. Box 58
Kaerntnerstrasse 322
A8054 Graz
Austria

(i) **Philips Scientific and Industrial Division**
Electron Optics Department
TQ 1114 Eindhoven
The Netherlands

also

Pye Unicam Ltd.
(U.K. agents for Philips)
York Street
Cambridge CB1 2PX
England

(j) **Carl Zeiss (Oberkochen) Ltd.**
P.O. Box 78
Woodfield Road
Welwyn Garden City
Hertfordshire AL7 1LU
England

also

Carl Zeiss
P.O. Box 1369
D-7082 Oberkochen
West Germany

3. Cryoultramicrotomes

(a) *Cryokit for LKB Ultramicrotome I, III and IV also Cryonova*

LKB Instruments Ltd.
232 Addington Road
South Croydon
Surrey CR2 8YD
England

LKB Instrument GmbH
Lochhamer Schlag 5
P.O. Box 1396
D-8032 Gräfelfing
West Germany

LKB Instruments Inc.
12221 Parklawn Drive
Rockville
Maryland 20852
U.S.A.

LKB-Produkter AB
P.O. Box 305
S-161 26 Bromma
Sweden

(b) *FC 150 freezing attachment for Reichert OMU3 and FC4 and FC4D for Ultracut E*

C. Reichert AG
Hernalser Hauptstrasse 219
A-1170 Wien
Austria

Reichert Jung UK
820 Yeovil Road
Slough
Buckinghamshire SL1 4JB
England

(c) *Frozen thin sectioner attachment FS1000 for MT6000 Sorvall ultramicrotome*

Du Pont Company
Instrument Products Division
Sorvall Operations
Newton
Connecticut 06470
U.S.A.

also
Du Pont
Biomedical Products Division
Wedgwood Way
Stevenage
Hertfordshire SG1 4QN
England

also
Du Pont de Nemours France S.A.
9 Rue de Vienne
Paris 8ᵉ
France

(d) *SLEE TUL cryoultramicrotome*

Slee Medical Equipment Ltd.
Lanier Works
Hither Green Lane
London SE13 6QD
England

4. Freeze-etch apparatus

(a) **Balzers Union** (*see* 1b)

(b) **Cressington Scientific Instruments**
34 Chalk Hill
Oxhey
Watford WD1 4BX
England

(c) **Denton Vacuum Inc.** (*see* 1e)

(d) **Edwards High Vacuum**
Manor Royal
Crawley
West Sussex RH10 2LW
England

(e) **Hitachi Ltd.**
4, 1-chrome
Marunouchi Chiyoda-ku
Tokyo
Japan

(f) **Leybold-Heraeus GmbH and Co. KG**
Bonner Strasse 504
D-5000 Köln 51
West Germany

(g) **Polaron** (*see* 1m)

(h) **C. Reichert Ltd.** (*see* 3b)

(i) **Technics Electron Microscopy Systems Inc.**
7950 Cluny Court
Springfield
Virginia 22153
U.S.A.

5. Freeze-drying apparatus
(Refer also to General Suppliers; *see* 1)

(a) *Freeze-Drying Device FDU 001/FDU 010*

Balzers Union   (*see* 1b)

(b) *Freeze-drying apparatus*

Edwards   (*see* 4d)

EMscope   (*see* 1h)

Meditrionics Inc.
P.O. Box 11209
Dallas
Texas
U.S.A.

Polaron   (*see* 1m)

Chemlab Instruments Ltd.
602 High Road
Ilford
Essex
England

(c) *VirTis freeze-dryers*

Techmation Ltd.
58 Edgware Way
Edgware
Middlesex HA8 8JP
England

6. Low temperature freezers

(a) **Gesellschaft für Labortechnik GmbH**
Schulze-Delitzsch-Str. 4
D-3006 Burgwedel
West Germany

(b) **Scientemp Corp.**
Adrian
Michigan
U.S.A.
(Scien Temp Lo-Cold freezer)

(c) **Borolabs Ltd.**
Paices Hill
Aldermaston
Berkshire RG7 4QU
England
(LABCOLD low temperature cabinets)

7. Storage vessels

(a) **Air Products Ltd.**
Coombe House
St. George's Square
New Malden
Surrey KT3 4HH
England

(b) **Air Reel (Cryoproducts) Ltd.**
Riverside Industrial Estate
Bridge Road
Littlehampton
West Sussex BN17 5DF
England

(c) **Day-Impex Ltd.**
Earls Colne
Colchester CO6 2ER
England

(d) **Edwards**   (*see* 4d)

(e) **Jencons (Scientific) Ltd.**
Cherrycourt Way Industrial Estate
Stanbridge Road
Leighton Buzzard
Bedfordshire LU7 8UA
England

(f) **L'Air Liquide**
Division Matériel Cryogenique
57 Av. Carnot
94500 Champigny-s/Marne
France

(g) **Statebourne (Cryogenics) Ltd.**
18 Parsons Road
Parsons Industrial Estate
Washington
Tyne & Wear
England

(h) **Thor Cryogenics**
Henley Road
Berinsfield
Oxford
England

(i) **Union Carbide**
Union Carbide House
P.O. Box 72
Rickmansworth
Hertfordshire WD3 1AS
England

8. Thermocouples

(a) **Centemp**
62 Curtis Road
Houndslow
Middlesex TW4 5PT
England

(b) **Labfacility Ltd.**
26 Tudor Road
Hampton
Middlesex TW12 2NQ
England

(c) **TC Ltd.**
P.O. Box 130
Uxbridge UB8 2YG
England

9. Temperature monitors and controllers

(a) **Bright Instrument Co. Ltd.**
Clifton Road
Huntingdon PE18 7EU
England

(b) **Eurotherm Ltd.**
Faraday Close
Durrington
Worthing
West Sussex BN13 3PL
England

(c) **Hird-Brown Ltd.** (Divn. of MTE Ltd.)
Hartford Works
Weston Street
Bolton
England

(d) **E. Leitz (Instruments) Ltd.**
48 Park Street
Luton
Bedfordshire LU1 3HP
England

(e) **Roach Associates Ltd.**
1A Sidmouth Road
Orpington
Kent BR5 2EG
England

(f) **Rosemount Engineering Co. Ltd.**
Heath Place
Bognor Regis
West Sussex PO22 9SH
England

(g) **SLEE**   (*see* 3d)   (h) **Stanton Redcroft Ltd.**
                                                  Copper Mill Lane
                                                  London SW17 0BN
                                                  England

10. Joule-Thomson cryotips

(a) **Air Products Ltd.**   (*see* 7a)   (b) **The Hymatic Engineering Co.**
                                                  Glover Street
                                                  Redditch
                                                  Worcestershire B98 8BQ
                                                  England

11. Thermoelectric coolers

(a) **Bailey Instruments Inc.**   (d) **Melcor/Materials Electronic Products**
    515 Victor Street   992 Spruce Street
    Saddle Brook   Trenton
    New Jersey 07662   New Jersey 08648
    U.S.A.   U.S.A.

(b) **Cambion**   (e) **Mullard Ltd.**
    Electronic Products Ltd.   Mullard House
    Castleton   Torrington Place
    Sheffield S30 2WR   London WC1E YHD
    England   England

(c) **Eberbach Inc.**
    P.O. Box 1024
    Ann Arbor
    Michigan 48106
    U.S.A.

12. Freezing devices

(a) **Balzers**   (*see* 1b)   (c) **Reichert**   (*see* 3b)
    (Spray-freezing, jet-freezing,   (KF80 Immersion Cryofixation System,
    high-pressure freezing HPM 010   Escaig cold-block device)

(b) **Polaron**   (*see* 1m)
    (Slammer (cold block))

13. Cooling rate meters

(a) **Agar**   (*see* 1a)   (c) **Hexland**   (*see* 2f)

(b) **EMscope**   (*see* 1h)

14. Liquid N₂ worktable (cryoprep stand)

Polaron   (*see* 1m)

15. Cryogens; quenching media

(a) *Liquid nitrogen*
   Air products   (*see* 7a)

   BOC Ltd.
   Great West Road
   Brentford
   Middlesex TW8 9DQ
   England

   CryoService Ltd.
   Units G/H 1-3
   Blackpole Trading Estate East
   Worcester
   West Midlands WR3 8SJ
   England

   L'Air Liquide   (*see* 7e)

   Matheson
   30 Seaview Dr.
   P.O. Box 1587
   Secaucus
   New Jersey 07094
   U.S.A.

(b) *Halocarbons*
Sold under various trade names by most local refrigeration supply companies. Available from a wide variety of sources and in different containers ranging from small pressurised cans to large cylinders. Contact local suppliers or general EM supply companies (*see also* 1).

Du Pont   (*see* 3c)
(sold under the name 'Freon')

Dean and Wood (London) Ltd.
83/85 Mansell Street
London E1
England
(Sold under the name 'Isceon')

ICI Ltd.
Mond Division
Runcorn
Cheshire
England
(Sold under the name 'Arcton')

J.T. Baker Chemical Co.
Phillipsburg
New Jersey
U.S.A.
(Sold under the name 'Ucon')

(c) *Propane; isopentane*

British Drug Houses Ltd. (BDH)
Broom Road
Poole
Dorset BH12 4NN
England
(*see also* companies under 15a and b)

16. Materials for shadowing and making replicas

Most general suppliers   (*see* 1)

(a) *Carbon rods for evaporation*

Johnson Matthey Chemicals Ltd.
Orchard Road
Royston
Hertfordshire SG8 5HE
England

Morganite Special Carbons Ltd.
Northfields
Wandsworth
London SW18 1NG
England

(b) *Formvar*

 G.T. Gurr
Searle Scientific Services
Coronation Road
Cressex Industrial Estate
High Wycombe
Buckinghamshire
England

Shawinigan Ltd.
118 Southwark Street
London SE1
England

(c) *Metals for evaporation*

Johnson Matthey Metals Ltd.
100 High Street
Southgate
London N14 6ET
England

(d) *Platinum-carbon pellets*

Engelhard Industries Inc.
Delancey Street
Newark
New Jersey
U.S.A.

(e) *Tungsten wire*

Lamp Metals Ltd.
East Lane
Wembley
Middlesex
England

The Tungsten Manufacturing Co. Ltd.
Fishergate Works
Portslade
Brighton
Sussex
England

17. Quartz crystal thin-film monitor

Most larger general suppliers (e.g. Agar Aids, *see* 1a; Polaron, *see* 1m; EMscope, *see* 1h)

18. Nebulizers

General suppliers   (*see* 1)

Ted Pella Inc.   (*see* 1m)
(Pelco nebulizer)

19. Specimen grids

(a) General suppliers   (*see* 1)

(b) **Gilder Grids**
23 Macfarlane Road
London W12 7JA
England

(c) **Graticules Ltd.**   (*see* 1i)

(d) **LKB-Produkter AB**   (*see* 3a)

(e) **Smethurst High-Light Ltd.**   (*see* 1a)

(f) **VECO Zeefplatenfabriek NV**
P.O. Box 10
Karel van Gelreweg 22
Eerbeek (GLD)
The Netherlands

20. Chemicals, resins etc.

Most general suppliers   (*see* 1)

G.T. Gurr   (*see* 16b)

Polysciences Inc. (Polysciences)   (*see* 1n)

Taab Laboratories (Taab)   (*see* 1q)

(a) *Alcian blue*
Alcian blue 8 GX p.a.
C.I. 74240
Serva
Heidelberg
G.D.R.

(b) *CA-15*
Delo GmbH.
Munich
West Germany

(c) *Chromerge*
Fisher Scientific Co.
Fair Lawn
New Jersey
U.S.A.

(d) *Dimethyl sulphoxide (DMSO)*

EMscope   (*see* 1h)

Fisons Scientific Apparatus (Fisons)
Bishop Meadow Road
Loughborough
Leicestershire LE11 0RG
England

Polaron   (*see* 1m)

Polysciences   (*see* 1n)

Taab   (*see* 1q)

(e) *Mounting media for cryosectioning*
Raymond A. Lamb
6 Sunbeam Road
London NW10 6JL
England
(Tissue-Tek II)

Polaron   (*see* 1m)
(Cryoembed)

(f) *Ethylene glycol*

Fisons Scientific Apparatus
Bishop Meadow Road
Loughborough
Leicestershire LE11 ORG
England

Fisher Scientific Co.
633 Greenwich Street
New York
New York 10014
U.S.A.

Kodak Ltd. (U.K. agent)
Chemical Division
Kirby
Liverpool
England

(g) *Hystoacryl blue*

Braun-Dexon GmbH.
Melsungen
West Germany

(h) *London L.R. Gold*

London Resin Company Ltd.
P.O. Box 34
Basingstoke
Hampshire RG22 5AS
England

Also available from general suppliers and
agents   (*see* 1).

(i) *Lowicryl*

Chemische Werke Lowi
Gesellschaft mit Beschrankter Haftung
P.O. Box 1660
D-8264 Waldkraiburg
West Germany

Also available from general suppliers and
agents   (*see* 1).

(j) *P13N (Polyimide resin)*

Ciba

(k) *Polycationic (cationised) ferritin*

Miles Scientific
Division of Miles Laboratories Ltd.
P.O. Box 37
Stoke Poges
Slough SL2 4LY
England

also
Miles Laboratories Inc.
Division of Research Products
P.O. Box 2000
1127 Myrtle Street
Elkahart
Indiana 46515
U.S.A.

(l) *Poly-L-lysine*
Sigma Chemical Co.
P.O. Box 14508
St. Louis
Missouri 63178
U.S.A.

also

Sigma Chemical Co.
Fancy Road
Poole
Dorset BH17 7NH
England

Miles   (*see* 20k)

(m) *Polyvinyl alcohol Vinol 205 S*
Air Products and Chemicals Inc.
Allentown
Pennsylvania
U.S.A.

(n) *Skybond 705 (Polyimide resin)*
Monsanto PLC
Edison Road
Dorcan
Swindon SN3 5HN
England

also

Monsanto Co.
800 N. Lindbergh Boulevard
St. Louis
Missouri 63167
U.S.A.

21. Miscellaneous

(a) *Antistatic pistols*
   Agar   (*see* 1a)

(b) *Abrasive unit SBU 010*
   Balzers   (*see* 1b)

(c) *Airbrushes*
   Simair Graphic Equipment
   16 Woodsley Road
   Leeds LS3 1DT
   England
   (Olympos HP 100 Series)

   The DeVilbiss Company Ltd.
   47 Holborn Viaduct
   London EC1
   England
   (Aerograph 'Super 63')

(d) *Beckman Microfuge Centrifuge B*
   Beckman Instruments Inc.
   250 Harbour Boulevard
   P.O. Box 3100
   Fullerton
   California 92634
   U.S.A.

   also

   Div. Beckman-Riic Ltd.
   Turnpike Road
   Cressex Industrial Estate
   High Wycombe
   Buckinghamshire HP4 3NR
   England

(e) *Cats-Neparofa (heat-sealing plastic laminate)*
   Apparently no longer available; alternatively:

   Camvac Ltd.
   Burrell Way
   Thetford
   Norfolk IT24 3QY
   England
   (Melinex 850)

(f) *Diamond knives*
   Diatome
   P.O. Box 551
   CH-2501
   Switzerland

(g) *Elvanol 51-05*
   Dupont de Nemour   (*see also* 3c)
   Div. Analytical and Biomedical Products
   50–52 Route des Acacias
   CH-1211 Geneva 24
   Switzerland

(h) *Flammable gas sensors*
   Draeger Safety
   Draeger House
   Sunnyside Road
   Chesham
   Buckinghamshire HP5 2AR
   England

Sabre Gas Detection Ltd.
Ash Road
Aldershot
Hantshire GU12 4DD
England

Sieger Gas Detection
Fulwood Close
Fulwood Industrial Estate
Sutton-in-Ashfield
Nottinghamshire NG17 2J2
England

(i) *Gloves for low temperature use*
General suppliers and agents   (*see* 1)

Boro Labs Ltd.
Paicas Hill
Aldermaston
Berkshire RG7 4QU
England

Medalnorth Ltd.
Markyate
Hertfordshire
England

(j) *Hardened stainless steel knives*
R. Jung AG
P.O. Box 1120
D-6901 Nussloch bei Heidelberg
West Germany

(k) *Hedin heaters (cartridges)*
Hedin Ltd.
Raven Road
South Woodford
London E18 1HJ
England

(l) *Ilford L-4 emulsion*
Ilford Ltd.
Ilford
Essex
England

(m) *Indium sheet*
Johnson Matthey   (*see* 16a)

Goodfellow Metals
Cambridge Science Park
Milton Road
Cambridge CB4 4DJ
England

(n) *Melinex Sheet*
ICI Ltd.
Plastics Division
P.O. Box 6
Bessemer road
Welwyn Garden City
Hertfordshire AL7 1HD
England

(*see also* **Camvac** 16e)

(o) *Petriperm*
W.C. Heraeus GmbH.
Hanau
West Germany

(p) *Photographic materials*
Eastman Kodak Co.
343 State St.
Rochester
New York 14650
U.S.A.

Ilford   (*see* 22c)

Also widely available from general suppliers and agents   (*see* 1)

(q) *Polystyrene latex beads*
Particle Information Service
Los Altos
California
U.S.A.

Also widely available from general suppliers and agents (*see* 1)

(r) *SIMCO-Single-Spike*
The SIMCO Company Inc.
2257 North Penn Road
Hatfield
Pennsylvania 19440
U.S.A.

(s) *Thermanox sheet*
Lux Scientific Corp.
Thousand Oaks
California
U.S.A.

(t) *Thin nickel screen*
Perforated Products
Brookline
Massachusetts
U.S.A.

(u) *Ultra Turrax Omnimixer*
Sartorius Ltd.
18 Avenue Road
Belmont SM2 6JD
England

Janke & Kunkle
P.O. Box 447813
Staufen
West Germany

(v) *Zeolite 13X molecular sieve*
Leybold-Heraeus (*see* 4f)

Also widely available from general suppliers and agents (*see* 1)

Index for list of suppliers

Numbers referred to are the sections in Appendix 2.

Subject index